Solutions Manual to accompany

General Chemistry
third edition

and

General Chemistry with Qualitative Analysis
third edition

by Whitten • Gailey • Davis

Yi-Noo Tang
Texas A&M University

Wendy Keeney-Kennicutt
Texas A&M University

Saunders Golden Sunburst Series

SAUNDERS COLLEGE PUBLISHING

Philadelphia New York Chicago
San Francisco Montreal Toronto
London Sydney Tokyo

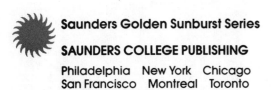

Printed in the United States of America

Solutions Manual to Accompany General Chemistry 3rd edition and General Chemistry with Qualitative Analysis 3 d edition by Whitten, Gailey and Davis

ISBN 0-03-012822-6

Foreword to the Students

This Solutions Manual supplements the textbook, General Chemistry, third edition, by Kenneth W. Whitten, Kenneth D. Gailey and Raymond E. Davis. The solutions of the 1100 even-numbered problems at the end of the chapters have been worked out in a detailed, step-by-step fashion.

Your learning of chemistry serves two purposes: the accumulation of fundamental knowledge in chemistry, and the expansion of your ability to make logical deductions in science by (1) knowing how to reason in a scientific way and, (2) being able to perform the mathematical manipulations that are necessary to solve certain problems. The excellent textbook by Whitten, Gailey, and Davis provides you with a wealth of chemical knowledge, accompanied by good examples of logical scientific deductive reasoning. The problems at the end of the chapters are primarily a review, a practice and, in many cases, a challenge to your scientific problem-solving abilities. It is therefore the fundamental spirit of this Solution Manual to help you to understand the scientific deductive process involved in each problem.

In this manual, we do provide you with a solution and an answer to the numerical problems, but our emphasis always rests on the reasoning behind the mathematical manipulations. In some cases, we present as many as three different approaches to solve the same problem. In stoichiometry as well as many other types of calculations, the "unit factor" method is universally emphasized in general chemistry textbooks. The over-emphasis of this method may train you with a monorail line of logical reasoning, by rushing towards the final answer and clouding the chemical meaning of the intermediate steps. Because the main purpose of a solution manual is to instruct students in understanding how the answer is reached, we have dissected the "unit factor" method for you and introduced chemical meaning into each of the steps.

We gratefully acknowledge the tremendous help provided by Frank Kolar and Gina Pierre in the preparation of this manuscript.

Yi-Noo Tang and
Wendy Keeney-Kennicutt

Department of Chemistry
Texas A&M University

Table of Contents

1 The Foundations of Chemistry

1-2. **Refer to Section 1-1 and the Key Terms for Chapter 1.** . . . .

(a) Matter is anything that has mass and occupies space. An example of matter is your Chemistry professor.

(b) Mass is a measure of the amount of matter in an object and is usually measured in grams or kilograms. A one pound hamburger has a mass of 0.4536 kilograms.

(c) Energy is the capacity for doing work or transferring heat. Solar cells use light energy from the sun to produce electricity.

(d) Potential energy is the energy that matter possesses by virtue of its position, condition, or composition. Your chemistry book on a table has potential energy due to its position. Energy is released if it falls from the table.

(e) Kinetic energy is the energy that matter possesses by virtue of its motion. The kinetic energy belonging to a moving train is easily transferred to a stalled car on the tracks.

(f) Heat is a form of energy that flows between two samples of matter because of their differences in temperature. For example, in winter, heat will flow from a warm house through an uninsulated window to the cold outdoors.

(g) An exothermic process is a process that releases heat energy. The combustion of gasoline is an exothermic process that is used in automobiles for power.

(h) An endothermic process is a process that absorbs heat energy. The boiling of water is a physical process that requires heat and therefore is endothermic.

1-4. **Refer to Section 1-1.** . . . .

(a) $E = mc^2$, where E = energy, m = mass, c = velocity of light.

(b) The Einstein equation, a part of the Theory of Relativity, relates matter to its energy equivalent. It states that the amount of energy that is released when matter is transformed into energy is given by the product of the mass of the matter converted and the speed of light squared. It also applies to the absorption of energy accompanying the conversion of energy into matter.

1-6. **Refer to Section 1-2.** . . . .

<u>Solids</u> - rigid, and have definite shapes; occupy a fixed volume and are thus very difficult to compress; the hardness of a solid is related to the strength of the forces holding the particles of a solid together.

<u>Liquids</u> - occupy essentially constant volume but have variable shape; are difficult to compress; particles can pass freely over each other; boiling points increase with increasing forces of attraction among the particles.

<u>Gases</u> - expand to fill the entire volume of their containers; very compressible; with relatively large separations between particles.

The three states are alike in that they all exhibit definite mass and volume under a given set of conditions. All consist of some combination of atoms, molecules or ions. The differences are stated above. Additional differences occur in their relative densities: gases <<< liquids < solids.

1-8. **Refer to Section 1-3.** . . . .

(a) The melting point of lead is a physical property, since the process of melting does not change its chemical composition. Melting only involves a change in state, from a solid to a liquid.

(b) The hardness of a diamond can be determined without changing its chemical composition and therefore is a physical property of the substance.

(c) The color of a solid depends on the physical and chemical nature of the solid. The color of any substance can be determined without changing its chemical composition. Color is therefore a physical property of the solid.

(d) The color of paint is due to the chemical nature of the pigments in the paint. However, as described in (c), color is a physical property.

(e) The ability of a substance to burn in air can only be determined by testing its flammability. Combustion involves the chemical reaction of the substance with oxygen, producing a change in the chemical composition of the substance. Thus, this is a chemical property.

1-10. **Refer to Sections 1-1 and 1-4, and the Key Terms for Chapter 1.** . . . .

(a) Combustion is an exothermic process in which a chemical reaction releases heat.

(b) The freezing of water is an exothermic process. Heat must be removed from the molecules in the liquid state to cause solidification.

(c) The melting of ice is an endothermic process. The system requires heat to break the attractive forces that hold solid water together.

(d) The boiling of water is an endothermic process. Molecules of liquid water must absorb energy to break away from the attractive forces that hold liquid water together in order to form gaseous molecules.

(e) The condensing of steam is an exothermic process. The heat stored in water vapor must be removed for the vapor to liquify. The condensation process is the opposite of boiling.

1-12. **Refer to Section 1-5.** . . . .

(a) Brass is a homogeneous mixture of metals (copper and zinc) in the solid state, called an alloy.

(b) Tea is a mixture of organic compounds (chlorophyll, tannins, caffeine, etc.) found in the leaves of a tea plant.

(c) Uranium is an element, a pure substance that cannot be decomposed into simpler substances by chemical means.

(d) Iron ore is a mixture of compounds containing iron and other metals.

(e) Methane is a compound, CH_4, consisting of the elements C and H in the fixed atomic ratio, 1:4.

(f) Carbon dioxide is a compound, CO_2, consisting of the elements C and O in the fixed atomic ratio, 1:2.

1-14. **Refer to Section 1-5.** . . . .

(a) Popcorn is a heterogeneous mixture of popped and unpopped kernels, butter and salt.

(b) Milk (raw and unhomogenized) is a heterogeneous mixture of fat, water, vitamins, etc. Homogenized milk is a heterogeneous mixture on the microscopic level.

(c) Gasoline is a homogeneous mixture of various hydrocarbons (compounds containing C and H) and anti-knock additives.

(d) "Milk of Magnesia", an aqueous suspension of magnesium hydroxide, is a heterogeneous mixture.

1-16. **Refer to Sections 1-5 and 1-6.** . . . .

(a) Add water to the mixture of sand and table salt, NaCl, to dissolve the salt. The sand could be separated from the salt solution by filtration. The salt could be recovered by evaporating the water.

(b) A mixture of iron filings and sulfur can be crudely separated by using a magnet to remove the iron filings. A better method would be to use carbon disulfide, CS_2. Sulfur will dissolve in CS_2, while iron is not soluble in this solvent.

1-18. **Refer to Section 1-6.** . . . .

A simple distillation apparatus (Fig. 1-10) is used to separate substances with very different boiling points, i.e. water and salt. The apparatus consists of a distilling flask, a condenser and a receiving flask.

A fractional distillation apparatus (Fig. 1-11) is used to separate two or more volatile liquids with boiling points that are not very different. Instead of a simple condenser, a fractionating column packed with glass beads is used.

1-20. **Refer to Section 1-6.** . . . .

(a) Sugar dissolved in water can be separated using simple distillation to remove the water, leaving the sugar behind.

(b) Alcohol and ether are both volatile liquids. Simple distillation would be ineffective in separating these components. Fractional distillation would accomplish the separation easily.

1-22. **Refer to Section 1-6.** . . . .

In column chromatography, a liquid mixture is passed through a column packed with an adsorbent material. A running stream of solvent is used as a carrier. The components in the mixture will be separated in the order inversely related to their attraction to the absorbent material.

1-24. **Refer to Section 1-9.** . . . .

(a) 7240

(b) 1.23×10^3

(c) 6.98×10^2 (since 8 is an even number)

(d) 5.44×10^{-2} (since 3 in an odd number)

1-26. **Refer to Section 1-9.** . . . .

(a) 0.0278 has 3 significant figures. The leading zeros are not significant.

(b) 1.3 has 2 significant figures.

(c) 1.00 has 3 significant figures. The trailing zeros are significant.

1-28. **Refer to Section 1-9.** . . . .

(a) 1070 = 1.1×10^3 to indicate 2 significant figures.

(b) 43,527 = 4.35×10^4 to indicate 3 significant figures.

(c) 0.000286 = 2.86×10^{-4} to indicate 3 significant figures.

(d) 0.000098765 = 9.876×10^{-5} to indicate 4 significant figures.

1-30. **Refer to Section 1-9.** . . . .

(a) 1.85 + 12.33 = **14.18** (No rounding is necessary.)

(b) 1.234 × 0.247 = 0.304798 = **0.305** (Answer must have only 3 significant figures.)

(c) 8.74/4.3 = 2.0325581 = **2.0** (Answer must be rounded to 2 significant figures.)

1-32. **Refer to Section 1-9.** . . . .

(a) $(1.54 \times 10^2) + (2.11 \times 10^2) = \mathbf{3.65 \times 10^2}$

(b) $(4.56 + 8.7)/(1.23 \times 10^{-3}) = 13.3/1.23 \times 10^{-3} = \mathbf{1.08 \times 10^4}$ (Answer is limited to 3 significant figures.)

1-34. **Refer to Table 1-5.** . . . .

(a) 1×10^3, kilo-

(b) 1×10^{-3}, milli-

(c) 1×10^6, mega-

(d) 1×10^{-1}, deci-

(e) $0.01 = 1 \times 10^{-2}$, centi-

(f) $0.1 = 1 \times 10^{-1}$, deci-

(g) $0.001 = 1 \times 10^{-3}$, milli-

(h) 1×10^{-6}, micro-

1-36. **Refer to Section 1-8.** . . . .

(a) km, length

(b) mL, volume

(c) mm^2 = $length^2$, area

(d) dg, mass

(e) m^3 = $length^3$, volume

(f) cm^2 = $length^2$, area

1-38. **Refer to the conversion factors from Table 1-7.** . . . .

(a) $? \text{ km} = 7.58 \text{ m} \times \frac{1 \text{ km}}{1000 \text{ m}} = \mathbf{7.58 \times 10^{-3} \text{ km}}$

(b) $? \text{ m} = 758 \text{ cm} \times \frac{1 \text{ m}}{100 \text{ cm}} = \mathbf{7.58 \text{ m}}$

(c) $? \text{ g} = 478 \text{ kg} \times \frac{1000 \text{ g}}{1 \text{ kg}} = \mathbf{4.78 \times 10^5 \text{ g}}$

(d) $? \text{ kg} = 9.78 \text{ g} \times \frac{1 \text{ kg}}{1000 \text{ g}} = \mathbf{9.78 \times 10^{-3}\ kg}$

(e) $? \text{ mL} = 1386 \text{ L} \times \frac{1000 \text{ mL}}{1 \text{ L}} = \mathbf{1.386 \times 10^{6}\ mL}$

(f) $? \text{ L} = 3.692 \text{ mL} \times \frac{1 \text{ L}}{1000 \text{ mL}} = \mathbf{3.692 \times 10^{-3}\ L}$

(g) $? \text{ cm}^3 = 1126 \text{ L} \times \frac{1000 \text{ cm}^3}{1 \text{ L}} = \mathbf{1.126 \times 10^{6}\ cm^3}$ (Note: $1 \text{ cm}^3 = 1 \text{ mL}$)

(h) $? \text{ L} = 0.786 \text{ cm}^3 \times \frac{1 \text{ L}}{1000 \text{ cm}^3} = \mathbf{7.86 \times 10^{-4}\ L}$

1-40. **Refer to the conversion factors from Table 1-7.** . . . .

$? \text{ m} = 0.25000 \text{ miles} \times \frac{5280 \text{ ft}}{1 \text{ mile}} \times \frac{12 \text{ in}}{1 \text{ ft}} \times \frac{2.54 \text{ cm}}{1 \text{ in}} \times \frac{1 \text{ m}}{100 \text{ cm}} = \mathbf{4.0234 \times 10^{2}\ m}$

The conversions used here are all exact conversions. Therefore, the number of significant figures is limitless, but only 5 are given.

1-42. **Refer to the conversion factors from Table 1-7.** . . . .

$? \frac{\text{km}}{\text{hr}} = 65 \frac{\text{miles}}{\text{hr}} \times \frac{1.609 \text{ km}}{1 \text{ mile}} = \mathbf{1.0 \times 10^{2}\ \frac{km}{hr}}$

This answer is precise to 2 significant figures since the initial value has only 2 significant figures.

1-44. **Refer to the conversion factors in Table 1-7.** . . . .

All the answers are limited to 3 significant figures since the initial values have only 3 significant figures.

(a) $? \text{ cm}^3 = 1.00 \text{ gal} \times \frac{4 \text{ qt}}{1 \text{ gal}} \times \frac{1 \text{ L}}{1.057 \text{ qt}} \times \frac{1000 \text{ cm}^3}{1 \text{ L}} = \mathbf{3.78 \times 10^{3}\ cm^3}$

(b) $? \text{ mL} = 8.00 \text{ in}^3 \times \frac{(2.54 \text{ cm})^3}{(1 \text{ in})^3} \times \frac{1 \text{ mL}}{1 \text{ cm}^3} = \mathbf{131\ mL}$

(c) $? \text{ mL} = 4.25 \text{ yd}^3 \times \frac{(36 \text{ in})^3}{(1 \text{ yd})^3} \times \frac{(2.54 \text{ cm})^3}{(1 \text{ in})^3} \times \frac{1 \text{ mL}}{1 \text{ cm}^3} = \mathbf{3.25 \times 10^{6}\ mL}$

(d) $? \text{ in}^3 = 1.00 \text{ L} \times \frac{1000 \text{ cm}^3}{1 \text{ L}} \times \frac{(1 \text{ in})^3}{(2.54 \text{ cm})^3} = \mathbf{61.0\ in^3}$

1-46. **Refer to the conversion factors listed in Table 1-7.** . . . .

$? \text{ \$/gal} = \frac{\$0.326}{\text{L}} \times \frac{1 \text{ L}}{1.057 \text{ qt}} \times \frac{4 \text{ qt}}{1 \text{ gal}} = \mathbf{\$1.23/gal}$

1-48. **Refer to Table 1-5 and Table 1-7.** . . . .

(a) $? \text{ g} = 1.00 \times 10^{6} \text{ cg} \times \frac{10^{-2} \text{ g}}{1 \text{ cg}} = \mathbf{1.00 \times 10^{4}\ g \text{ or } 10.0\ kg}$ ($10^3 \text{ g} = 1 \text{ kg}$)

(b) $? \text{ g} = 4.25 \times 10^5 \text{ mg} \times \dfrac{10^{-3} \text{ g}}{1 \text{ mg}} = $ **4.25×10^2 g or 0.425 kg**

(c) $? \text{ g} = 3.0 \text{ ounces} \times \dfrac{1 \text{ g}}{0.03527 \text{ oz}} = $ **85 g or 0.085 kg**

(d) $? \text{ g} = 4.00 \text{ pounds} \times \dfrac{453.6 \text{ g}}{1 \text{ lb}} = $ **1.81×10^3 g or 1.81 kg**

1-50. **Refer to Table 1-5 and Table 1-7 for conversion factors.** . . . .

(a) $? \text{ kg} = 105 \text{ pounds} \times \dfrac{453.6 \text{ g}}{1 \text{ lb}} \times \dfrac{1 \text{ kg}}{1000 \text{ g}} = $ **47.6 kg**

(b) $? \text{ pounds} = 105 \text{ kg} \times \dfrac{1000 \text{ g}}{1 \text{ kg}} \times \dfrac{1 \text{ lb}}{453.6 \text{ g}} = $ **231 lb**

(c) $? \text{ cg} = 8.00 \text{ ounces} \times \dfrac{1 \text{ g}}{0.03527 \text{ oz}} \times \dfrac{1 \text{ cg}}{10^{-2} \text{ g}} = $ **2.27×10^4 cg**

1-52. **Refer to Table 1-7 for conversion factors.** . . . .

(a) The official length of a U.S. basketball court is 94 ft.

$$? \text{ m} = 1 \text{ ft} \times \frac{12 \text{ in}}{1 \text{ ft}} \times \frac{2.54 \text{ cm}}{1 \text{ in}} \times \frac{1 \text{ m}}{100 \text{ cm}} = 0.305 \text{ m}$$

Since 1 ft is approximately 1/3 of a meter, a basketball court is roughly **30 m.**

(b) The interior of a "compact car" is roughly 4 cubic yards.

$$? \text{ m}^3 = 1 \text{ yd}^3 \times \frac{(3 \text{ ft})^3}{(1 \text{ yd})^3} \times \frac{(12 \text{ in})^3}{(1 \text{ ft})^3} \times \frac{(2.54 \text{ cm})^3}{(1 \text{ in})^3} \times \frac{(1 \text{ m})^3}{(100 \text{ cm})^3} = 0.765 \text{ m}^3$$

Since 1 cubic yard is about 3/4 cubic meters, the interior is about **3 cubic meters.**

(c) An average bedroom might be 12 feet square = 144 ft^2.

$$? \text{ m}^2 = 1 \text{ ft}^2 \times \frac{(12 \text{ in})^2}{(1 \text{ ft})^2} \times \frac{(2.54 \text{ cm})^2}{(1 \text{ in})^2} \times \frac{(1 \text{ m})^2}{(100 \text{ cm})^2} = 0.0929 \text{ m}^2$$

Since 1 square foot is about 0.1 square meters, the area of the bedroom is about **14 square meters.**

(d) An average person might weigh 150 lb.

$$? \text{ kg} = 1 \text{ lb} \times \frac{453.6 \text{ g}}{1 \text{ lb}} \times \frac{1 \text{ kg}}{1000 \text{ g}} = 0.454 \text{ kg or about } 0.5 \text{ kg}$$

Therefore, this person has a mass of 150 lb × 0.5 kg/lb or about **75 kg.**

1-54. **Refer to Table 1-7 for conversion factors.** . . . .

Each atom has a diameter = 2 × 1.43 Å = 2.86 Å.

$$? \text{ Al atoms} = 1.00 \text{ inch} \times \frac{2.54 \text{ cm}}{1 \text{ in}} \times \frac{1 \text{ m}}{100 \text{ cm}} \times \frac{1 \text{ Å}}{10^{-10} \text{ m}} \times \frac{1 \text{ atom}}{2.86 \text{ Å}} = \mathbf{8.88 \times 10^7 \text{ atoms}}$$

1-56. **Refer to Section 1-3.** . . . .

An intensive property is one that does not depend upon the amount of material in a sample, e.g., density and temperature. An extensive property is one that does depend upon the amount of material in a sample, e.g., mass, and volume.

1-58. **Refer to Section 1-3.** . . . .

(a) The deep red color of liquid bromine does not depend on the amount present; it is an intensive property.

(b) The mass of coal does depend on the amount of coal; it is an extensive property.

(c) Specific gravity in an intensive property of a substance since it is independent of sample size.

(d) The boiling point of water does not depend on the amount of water; it is an intensive property.

(e) The physical state (gas, liquid or solid) of a substance is determined by a set of conditions and is independent of the amount of substance present. Therefore, physical state is an intensive property.

1-60. **Refer to Section 1-11.** . . . .

$$\text{Density}\left(\text{units} = \frac{g}{cm^3}\right) = \frac{\text{mass (g)}}{\text{volume (cm}^3)} = \frac{97.2\ g}{13.5\ cm^3} = \mathbf{7.20\ \frac{g}{cm^3}}$$

1-62. **Refer to Exercises 1-60 and 1-61.** . . . .

$$\text{Density of chromium} = 7.20\ \frac{g}{cm^3} = \frac{7.20\ g}{1\ cm^3}\text{; density of mercury} = 13.6\ \frac{g}{cm^3} = \frac{13.6\ g}{1\ cm^3}.$$

$$\text{Therefore, } \frac{1\ cm^3 \text{ of mercury}}{1\ cm^3 \text{ of chromium}} = \frac{13.6\ g}{7.20\ g} = \mathbf{1.89}$$

1-64. **Refer to Table 1-8.** . . . .

$$D\left(\frac{g}{cm^3} \text{ or } \frac{g}{mL}\right) = \frac{\text{mass (g)}}{\text{volume (mL)}}\text{; mass (g)} = D \times \text{volume (mL)}$$

Therefore, the most dense substance will have the largest mass, i.e. **gold.** (D of gold = 19.3 g/mL, D of sand = 2.32 g/mL, D of hydrogen = 0.000089 g/mL)

1-66. **Refer to Section 1-11.** . . . .

The specific gravity of a substance is the ratio of its density to the density of water. Since the density of water = 1.00 g/mL in the temperature range of $0^{\circ}C$ to $25^{\circ}C$, the specific gravity is numerically equal to the density.

Method 1: We know that 23 mL of water has 23 g of mass (D of water=1.00 g/mL). To find the volume of alcohol that has a mass of 23 g,

$$D \text{ of alcohol} = \frac{M}{V};\ V\text{ (mL)} = \frac{M\text{ (g)}}{D\text{ (g/mL)}} = \frac{23\ g}{0.79\ g/mL} = \mathbf{29\ mL}$$

Method 2: Dimensional Analysis

$$?\ \text{mL alcohol} = 23\ \text{mL water} \times \frac{1\ \text{g water}}{1\ \text{mL water}} \times \frac{1\ \text{g alcohol}}{1\ \text{g water}} \times \frac{1\ \text{mL alcohol}}{0.79\ \text{g alcohol}} = \mathbf{29\ mL}$$

1-68. **Refer to Section 1-11.** . . . .

Method 1: We realize that the displacement of water, (H_2O) = volume of copper, (Cu).

$$D = \frac{M}{V};\ M = D \times V = 8.92\ \text{g/mL} \times 17.43\ \text{mL} = \mathbf{155\ g}$$

Method 2: Dimensional Analysis

$$?\ \text{g copper} = 17.43\ \text{mL water} \times \frac{1\ \text{mL Cu}}{1\ \text{mL H}_2\text{O}} \times \frac{1\ \text{cm}^3\ \text{Cu}}{1\ \text{mL Cu}} \times \frac{8.92\ \text{g Cu}}{1\ \text{cm}^3\ \text{Cu}} = \mathbf{155\ g}$$

1-70. **Refer to Table 1-7 for conversions.** . . . .

If we convert all units to grams,

0.50 lb: $?\ \text{g} = 0.50\ \text{lb} \times \frac{453.6\ \text{g}}{1\ \text{lb}} = 2.3 \times 10^2\ \text{g}$ (2 significant figures)

0.25 kg: $?\ \text{g} = 0.25\ \text{kg} \times \frac{1000\ \text{g}}{1\ \text{kg}} = 2.5 \times 10^2\ \text{g}$

25 mL: $?\ \text{g} = 25\ \text{mL} \times \frac{19.3\ \text{g}}{1\ \text{mL}} = 4.8 \times 10^2\ \text{g}$ (specific gravity = density)

Therefore, **25 mL of gold** contains the most mass.

1-72. **Refer to Section 1-11 and Table 1-8.** . . . .

(a) volume = length × width × thickness
= 4.72 cm × 3.19 cm × 0.52 cm = $\mathbf{7.8\ cm^3}$ (2 significant figures)

(b) $$D\left(\frac{\text{g}}{\text{mL}}\ \text{or}\ \frac{\text{g}}{\text{cm}^3}\right) = \frac{M\ (\text{g})}{V\ (\text{cm}^3)} = \frac{61.5\ \text{g}}{7.8\ \text{cm}^3} = \mathbf{7.9\ \frac{g}{cm^3}}$$

(c) The difference between this answer (7.9) and the density of iron given in Table 1-8 (7.86) is the number of significant figures in the values. The table gives density to 3 significant figures and this answer is only to 2 significant figures.

1-74. **Refer to Section 1-11.** . . . .

(a) Method 1: $D = \frac{M}{V}$ or $V = \frac{M}{D}$. To find mass (g) of 100 lb of gold,

$$?\ \text{g} = 100\ \text{lb} \times \frac{453.6\ \text{g}}{1\ \text{lb}} = 4.54 \times 10^4\ \text{g}$$

Therefore,

$$V\ (\text{cm}^3) = \frac{M\ (\text{g})}{D\ (\text{g/cm}^3)} = \frac{4.54 \times 10^4\ \text{g}}{19.3\ \text{g/cm}^3} = \mathbf{2.35 \times 10^3\ cm^3}$$

Method 2: Dimensional Analysis

$$?\ \text{cm}^3 = 100\ \text{lb} \times \frac{453.6\ \text{g}}{1\ \text{lb}} \times \frac{1\ \text{cm}^3}{19.3\ \text{g}} = \mathbf{2.35 \times 10^3\ cm^3}$$

(b) For a cube, volume (cm^3) = $(\text{edge})^3$. Therefore

$$\text{edge (cm)} = \text{volume}^{1/3} = (2.35 \times 10^3)^{1/3} = 13.3\ \text{cm}$$

$$\text{edge (in)} = 13.3\ \text{cm} \times \frac{1\ \text{in}}{2.54\ \text{cm}} = \mathbf{5.24\ in}$$

1-76. **Refer to Table 1-8 for densities.** . . . .

(1) mass of mercury (g) $= 57.3 \text{ mL} \times \frac{13.59 \text{ g}}{1 \text{ mL}} = 779 \text{ g}$

(2) mass of water, H_2O (g) $= 772 \text{ mL} \times \frac{1 \text{ g}}{1 \text{ mL}} = 772 \text{ g}$

(3) mass of lead (g) $= 67.1 \text{ cm}^3 \times \frac{11.34 \text{ g}}{1 \text{ cm}^3} = 761 \text{ g}$

(4) mass of gold (g) $= 40.1 \text{ mL} \times \frac{19.3 \text{ g}}{1 \text{ mL}} = 774 \text{ g}$

Therefore, **57.3 mL of mercury** has the greatest mass.

1-78. **Refer to Section 1-11.** . . . .

In Exercise 1-77, the density of the hydrochloric acid solution is quoted as 1.18 g/mL.

Method 1: Plan: (1) Find the mass of solution with $D = 1.23$ g/mL.
(2) Using $D = M/V$ and the known mass, solve for the volume of hydrochloric acid having $D = 1.18$ g/mL.

(1) $D_{\text{solution}} = M/V;\ M = D \times V = 1.23 \text{ g/mL} \times 125 \text{ mL} = 154 \text{ g}$

(2) $V_{\text{hydrochloric acid}} = \frac{M}{D} = \frac{154 \text{ g}}{1.18 \text{ g/mL}} = \mathbf{131 \text{ mL}}$

Method 2: Dimensional Analysis

$$? V = 125 \text{ mL solution} \times \frac{1.23 \text{ g solution}}{1 \text{ mL solution}} \times \frac{1 \text{ g HCl soln}}{1 \text{ g solution}} \times \frac{1 \text{ mL HCl soln}}{1.18 \text{ g HCl soln}}$$

$$= \mathbf{1.30 \times 10^2 \text{ mL HCl solution}}$$

1-80. **Refer to Section 1-12, and Examples 1-17 and 1-18.** . . . .

In determining the correct number of significant figures, note that the following values are exact: 32^oF, $1^oC/1.8^oF$, and $1^oC/1$ K.

(a) $? ^oC = \frac{1^oC}{1.8^oF} \times (105^oF - 32^oF) = \mathbf{41^oC}$

(b) $? ^oC = \frac{1^oC}{1.8^oF} \times (0^oF - 32^oF) = \mathbf{-18^oC}$

(c) $? ^oC = \frac{1^oC}{1 \text{ K}} \times (300.0 \text{ K} - 273.2 \text{ K}) = 26.8^oC$ since $0^oC = 273.15$ K

$? ^oF = \left(26.8^oC \times \frac{1.8^oF}{1^oC}\right) + 32^oF = \mathbf{80.2^oF}$

(d) $? ^oC = \frac{1^oC}{1.8^oF} \times (100.0^oF - 32^oF) = 37.8^oC$

$? \text{ K} = \frac{1 \text{ K}}{1^oC} \times (37.8^oC + 273.2^oC) = \mathbf{311.0 \text{ K}}$

1-82. **Refer to Sections 1-10 and 1-12, and Table 1-7.** . . . .

	Feet	Meters	°F	°C
(a)	1000	<u>305</u>	56°	<u>13°</u>
(b)	<u>4920</u>	1500	<u>41°</u>	5°
(c)	10000	<u>3050</u>	<u>23°</u>	-5°
(d)	<u>14800</u>	4500	5°	<u>-15°</u>
(e)	20000	<u>6100</u>	<u>-15°</u>	-26°
(f)	<u>29500</u>	9000	-47°	<u>-44°</u>
(g)	36087	<u>10999</u>	<u>-69°</u>	-56°

Sample calculations:

(a) $? \text{ m} = 1000 \text{ ft} \times \frac{12 \text{ in}}{1 \text{ ft}} \times \frac{2.54 \text{ cm}}{1 \text{ in}} \times \frac{1 \text{ m}}{100 \text{ cm}} = \mathbf{304.8 \text{ m}}$

$? \,^{o}\text{C} = \frac{1^{o}\text{C}}{1.8^{o}\text{F}} \times (56^{o}\text{F} - 32^{o}\text{F}) = \mathbf{13^{o}C}$

(b) $? \text{ ft} = 1500 \text{ m} \times \frac{100 \text{ cm}}{1 \text{ m}} \times \frac{1 \text{ in}}{2.54 \text{ cm}} \times \frac{1 \text{ ft}}{12 \text{ in}} = \mathbf{4921 \text{ ft}}$

$? \,^{o}\text{F} = \frac{1.8^{o}\text{F}}{1^{o}\text{C}} \times 5^{o}\text{C} + 32^{o}\text{C} = \mathbf{41^{o}F}$

(h) and (i) When elevation (x axis) is plotted against temperature (y axis), the slopes of these lines have units of °F/ft and °C/m. To convert °F/ft to °C/m, only multiplication and addition of constant factors are involved. It is therefore not surprising that the graphs are similar.

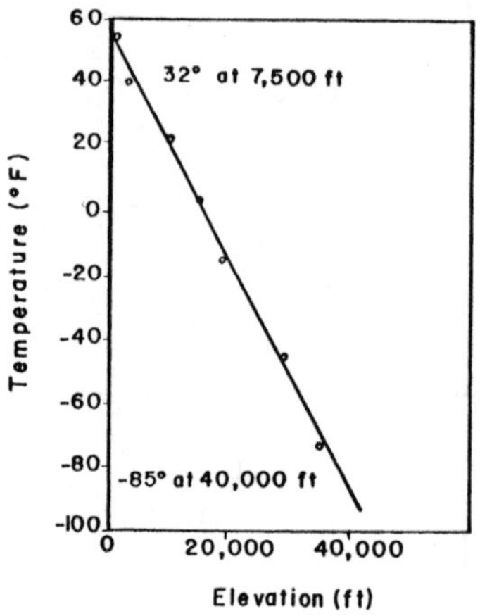

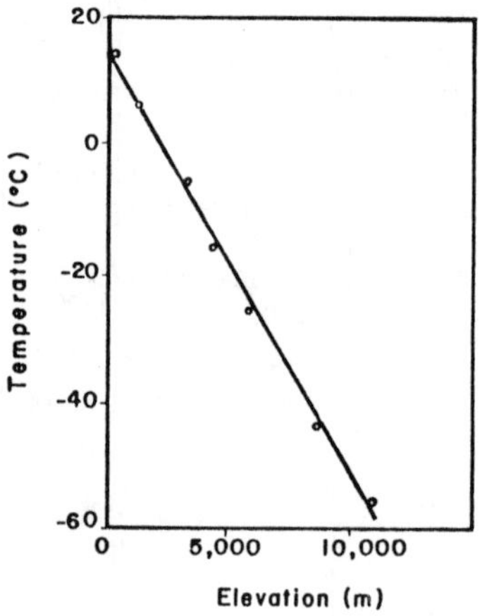

1-84. **Refer to Section 1-13 and Example 1-18.** . . . .

amount of heat <u>removed</u> (J) = (mass of substance)(specific heat)(temp. change)
$= 25.0 \text{ g} \times 4.18 \text{ J/g}^{o}\text{C} \times (60.0^{o}\text{C} - 10.0^{o}\text{C})$
$= \mathbf{5.22 \times 10^3 \text{ J}}$

$? \text{ cal} = 5.22 \times 10^3 \text{ J} \times \frac{1.00 \text{ cal}}{4.18 \text{ J}} = \mathbf{1.25 \times 10^3 \text{ cal}}$

Note that the specific heat of water = 1.00 cal/g°C = 4.18 J/g°C. This gives us a unit factor to convert calories to joules or joules to calories.

1-86. **Refer to Section 1-13.** . . . .

Note that we will follow the convention of representing temperature (°C) as t and temperature (K) as T.

When two substances at different temperatures come into contact, heat is exchanged until both substances come to the same final temperature. Then

the amount of heat lost by Substance 1 = amount of heat gained by Substance 2

As will be discussed in later chapters, "heat lost" is a negative quantity and "heat gained" is a positive quantity. However, the "**amount** of heat lost" and the "**amount** of heat gained" quoted here call for **absolute** quantities without a sign associated with them:

$$|\text{the amount of heat lost by Substance 1}| = |\text{amount of heat gained by Substance 2}|$$

$$|(\text{mass})(\text{Sp. ht.})(\text{temp. change})|_1 = |(\text{mass})(\text{Sp. ht.})(\text{temp. change})|_2$$

Since any "change" is always defined as the final value minus the initial value, we have

$$(\text{temp. change})_1 = (t_{final} - 100^oC) \text{ and } (\text{temp. change})_2 = (t_{final} - 10^oC)$$

In this example, the expected t_{final} would be between 10°C and 100°C.

$$|t_{final} - 10^oC| = |\text{positive value}| = (t_{final} - 10^oC)$$

$$|t_{final} - 100^oC| = |\text{negative value}| = (100^oC - t_{final})$$

The set-up of the equation for this exercise is as follows because the $(\text{Sp. ht.})_1$ and $(\text{Sp. ht.})_2$ are both for liquid water.

$$150\ g \times \cancel{4.18\ J/g^oC} \times (100^oC - t_{final}) = 250\ g \times \cancel{4.18\ J/g^oC} \times (t_{final} - 10^oC)$$

$$15000 - 150 \times t_{final} = 250 \times t_{final} - 2500$$

$$17500 = 400 \times t_{final}$$

$$t_{final} = \mathbf{43.8^oC}$$

1-88. **Refer to Section 1-13.** . . . .

(a) heat gained by mercury = (mass)(Sp. ht.)(temp. change)
= 106 g × 0.138 J/g°C × (75.0°C - 10.0°C)
= **951 J**

(b) heat gained by water = 106 g × 4.18 J/g°C × (75.0°C - 10.0°C)
= **28,800 J**

(c) $$\frac{\text{Sp. ht. of liquid water}}{\text{Sp. ht. of mercury}} = \frac{4.18\ J/g^oC}{0.138\ J/g^oC} = 30.3$$

(d) Answer (c) says that 30.3 times more heat is required to change the temperature of 1.00 g of liquid water by 1.00°C than to change the temperature of 1.00 g of mercury by 1.00°C. This is illustrated by (a) and (b).

1-90. **Refer to Section 1-13 and Exercise 1-86.** . . . .

amount of heat lost by Substance 1 = amount of heat gained by Substance 2

$$|(\text{mass})(\text{Sp. ht.})(\text{temp. change})|_1 = |(\text{mass})(\text{Sp. ht.})(\text{temp. change})|_2$$

$$1000\text{ g} \times \cancel{4.18\text{ J/g}^\circ\text{C}} \times (30.0^\circ\text{C} - 15.0^\circ\text{C}) = \text{mass}_2 \times \cancel{4.18\text{ J/g}^\circ\text{C}} \times (80.0^\circ\text{C} - 30.0^\circ\text{C})$$

$$1000 \times 15.0 = \text{mass}_2 \times 50.0$$

$$\text{mass}_2 = \mathbf{300\text{ g}}$$

1-92. **Refer to Section 1-13 and Exercise 1-86 Solution.** . . . .

amount of heat lost by Substance 1 = amount of heat gained by Substance 2

$$|(\text{mass})(\text{Sp. ht.})(\text{temp. change})|_{\text{metal}} = |(\text{mass})(\text{Sp. ht.})(\text{temp. change})|_{\text{water}}$$

$$75.0\text{ g} \times (\text{Sp. ht.}) \times (75.0^\circ\text{C} - 18.3^\circ\text{C}) = 150\text{ g} \times 4.18\text{ J/g}^\circ\text{C} \times (18.3^\circ\text{C} - 15.0^\circ\text{C})$$

$$(\text{Sp. ht.}) \times 4260 = 2100$$

$$\text{Sp. ht.} = \mathbf{0.49\text{ J/g}^\circ\text{C}}$$

(2 significant figures set by the temp. change of the water)

2 Chemical Formulas and Composition Stoichiometry

2-2. **Refer to the Introduction to Chapter 2.** . . . .

Composition stoichiometry describes the quantitative relationships among elements in compounds, e.g., in water, H_2O, there are 2 hydrogen atoms for every 1 atom of oxygen.

Reaction stoichiometry describes the quantitative relationships among substances as they undergo chemical changes. This will be discussed in Chapter 3.

2-4. **Refer to Section 2-1.** . . . .

Dalton developed the atomic theory, the thesis that all elements are composed of tiny, indivisible particles called atoms that are all alike and have the same atomic weight. Nuclear chemistry now tells us that (1) atoms are divisible, (2) atoms of a given element can have slightly different properties due to having different numbers of neutrons, and (3) atoms of one element can be changed into those of other elements. However, his understanding that elements combine in whole number ratios to form compounds and for a given compound, these ratios are constant, has been proven to be correct. Even with the many shortcomings, Dalton's postulates provided a framework that could be modified and expanded by later scientists.

2-6. **Refer to Sections 2-1 and 2-2.** . . . .

(a) Helium exists as single atoms. It is the monatomic molecule, He.

(b) Elemental oxygen is the diatomic molecule, O_2, consisting of 2 atoms of oxygen bonded together.

(c) A carbon dioxide molecule, CO_2, consists of 2 oxygen atoms bonded to a central carbon atom.

(d) A molecule of ammonia, NH_3, consists of 3 hydrogen atoms, bonded to a central nitrogen atom.

(e) An ethanol molecule consists of 2 carbon atoms, 6 hydrogen atoms and 1 oxygen atom, written as C_2H_5OH. We will learn later about the actual structure.

2-8. **Refer to Section 2-1.** . . . .

Examples of molecules that exist as diatomic:

(1) N_2, O_2, H_2, F_2, Cl_2, Br_2, I_2 in which the atoms are identical (homonuclear).

(2) HF, HCl, NO in which the atoms are different (heteronuclear).

Examples of polyatomic molecules:

(1) P_4, S_8 (homonuclear)

(2) H_2SO_4, H_3PO_4 (heteronuclear)

2-10. **Refer to Section 2-2 and Figure 2-6.** . . . .

(a) Formulas are combinations of symbols that describe the number of each type of atom in the molecules.

(b) Structural formulas show how the atoms in the molecules are bonded together.

(c),(d) Ball-and-stick and space-filling models show not only the bonding, but also the relative sizes and shapes of the molecules.

2-12. **Refer to Section 2-2 and Table 2-1.** . . . .

(a) H_2SO_4 - sulfuric acid

(b) CH_3COOH - acetic acid

(c) C_3H_8 - propane

2-14. **Refer to Section 2-3.** . . . .

Ionic compounds are extended arrays of cations (positively charged ions) and anions (negatively charged ions). Recall that a molecule is the smallest particle of a compound that can have a stable, independent existence. Therefore, the term "molecule" does not apply to ionic substances. Instead, we use "formula unit" to represent the smallest repeating unit of an ionic compound.

2-16. **Refer to Sections 2-1 and 2-3.** . . . .

A polyatomic molecule is a molecule containing more than 2 atoms.

A formula unit of an ionic compound is the smallest repeating unit of an ionic compound. For example, sodium chloride, NaCl, represents a formula unit consisting of one Na^+ and one Cl^- ion.

2-18. **Refer to Table 2-2.** . . . .

(a) Ag^+ - silver ion (b) NH_4^+ - ammonium ion (c) Br^- - bromide ion

(d) SO_4^{2-} - sulfate ion (e) PO_4^{3-} - phosphate ion

2-20. **Refer to Section 2-4.** . . . .

(a) An atomic mass unit (amu) is exactly 1/12 of the mass of an atom of the carbon-12 isotope. It is a unit used for atomic and formula weights.

(b) A ruthenium atom (101.07 amu) is about twice as heavy as a vanadium atom (50.942 amu) since 101.07/50.942 = 1.9840.

2-22. **Refer to Section 2-4 and the inside cover of this textbook.** . . . .

	Element	Atomic Weight	Mass of 1 mole of atoms
(a)	Be	<u>9.012182 amu</u>	<u>9.012182 g</u>
(b)	<u>Cl</u>	35.453 amu	<u>35.453 g</u>
(c)	F	<u>18.9984032 amu</u>	<u>18.9984032 g</u>
(d)	<u>Au</u>	<u>196.9665 amu</u>	196.9665 g

2-24. Refer to Section 2-5 and Example 2-1. . . . .

Method 1: Use the units of atomic weight to derive a formula relating grams, moles and atomic weight:

$$\text{Atomic Weight, AW}\left(\frac{\text{g}}{\text{mol}}\right) = \frac{\text{grams of element}}{\text{moles of element}} \quad \text{and therefore,}$$

$$\text{moles of element} = \frac{\text{grams of element}}{\text{atomic weight (g/mol)}}$$

(a) $? \text{ mol O} = \dfrac{59.4 \text{ g}}{16.0 \text{ g/mol}} = \mathbf{3.71 \text{ mol O}}$

(b) $? \text{ mol Ag} = \dfrac{83.7 \text{ g}}{108 \text{ g/mol}} = \mathbf{0.775 \text{ mol Ag}}$

(c) $? \text{ mol Au} = \dfrac{4.00 \text{ g}}{197 \text{ g/mol}} = \mathbf{0.0203 \text{ mol Au}}$

Method 2: Dimensional Analysis

(a) $? \text{ mol O} = 59.4 \text{ g O} \times \dfrac{1 \text{ mol O}}{16.0 \text{ g O}} = \mathbf{3.71 \text{ mol O}}$

(b) $? \text{ mol Ag} = 83.7 \text{ g Ag} \times \dfrac{1 \text{ mol Ag}}{108 \text{ g Ag}} = \mathbf{0.775 \text{ mol Ag}}$

(c) $? \text{ mol Au} = 4.00 \text{ g Au} \times \dfrac{1 \text{ mol Au}}{197 \text{ g Au}} = \mathbf{0.0203 \text{ mol Au}}$

2-26. Refer to Section 2-6 and Exercise 25. . . . .

(a) From Exercise 25, $\dfrac{\text{number of Cu atoms/1.00 g Cu}}{\text{number of Pt atoms/1.00 g Pt}} = \dfrac{9.48 \times 10^{21}}{3.09 \times 10^{21}} = \mathbf{3.07}$

(b) $\dfrac{\text{mass of one Pt atom}}{\text{mass of one Cu atom}} = \dfrac{195.08 \text{ amu}}{63.546 \text{ amu}} = \mathbf{3.07}$

Answers (a) and (b) are the same since the number of atoms of an element in a given mass is inversely proportional to its atomic weight:

$$? \text{ atoms in 1.00 g} = 1.00 \text{ g} \times \frac{1 \text{ mol}}{\text{atomic weight}} \times \frac{6.022 \times 10^{23} \text{ atoms}}{1 \text{ mol}}$$

2-28. Refer to Section 2-5, Examples 2-1 and 2-2, and Exercise 2-24 Solution. . . .

Method 1:

(a) $? \text{ mol Cl} = \dfrac{\text{g Cl}}{\text{AW Cl (g/mol)}} = \dfrac{1.00 \text{ g}}{35.453 \text{ g/mol}} = \mathbf{2.82 \times 10^{-2} \text{ mol Cl}}$

(b) $? \text{ atoms Cl} = \dfrac{\text{amu Cl}}{\text{AW Cl (amu/atom)}} = \dfrac{35.453 \text{ amu}}{35.453 \text{ amu/atom}} = 1 \text{ atom Cl}$

Recall: Avogadro's Number, $\mathbf{N} = 6.022 \times 10^{23}$ atoms/mol

$$? \text{ mol Cl} = \frac{\text{atoms Cl}}{\mathbf{N} \text{ (atoms/mol)}} = \frac{1 \text{ atom}}{6.022 \times 10^{23} \text{ atoms/mol}} = \mathbf{1.66 \times 10^{-24} \text{ mol Cl}}$$

(c) $? \text{ mol Cl} = \dfrac{\text{atoms Cl}}{N \text{ (atoms/mol)}} = \dfrac{5.66 \times 10^{20} \text{ atoms Cl}}{6.022 \times 10^{23} \text{ atoms/mol}} = \mathbf{9.40 \times 10^{-4} \text{ mol Cl}}$

Method 2: Dimensional Analysis

(a) $? \text{ mol Cl} = 1.00 \text{ g} \times \dfrac{1 \text{ mol Cl}}{35.453 \text{ g Cl}} = \mathbf{2.82 \times 10^{-2} \text{ mol Cl}}$

(b) $? \text{ mol Cl} = 35.453 \text{ amu Cl} \times \dfrac{1 \text{ Cl atom}}{35.453 \text{ amu Cl}} \times \dfrac{1 \text{ mol Cl}}{6.022 \times 10^{23} \text{ Cl atoms}}$

$= \mathbf{1.661 \times 10^{-24} \text{ mol Cl}}$

(c) $? \text{ mol Cl} = 5.66 \times 10^{20} \text{ Cl atoms} \times \dfrac{1 \text{ mol Cl}}{6.022 \times 10^{23} \text{ Cl atoms}}$

$= \mathbf{9.40 \times 10^{-4} \text{ mol Cl}}$

2-30. **Refer to Section 2-6.** . . . .

Formula weight (b) and molecular weight (c) are often used interchangeably when referring to molecular (nonionic) substances, since in this instance, the formula of a substance represents a discrete molecule.

2-32. **Refer to Table 1-7.** . . . .

$? \text{ atoms} = 1.0 \text{ in} \times \dfrac{2.54 \text{ cm}}{1 \text{ in}} \times \dfrac{1 \text{ m}}{100 \text{ cm}} \times \dfrac{1 \text{ Å}}{10^{-10} \text{ m}} \times \dfrac{1 \text{ atom}}{1.48 \text{ Å}} = \mathbf{1.7 \times 10^{8} \text{ atoms H}}$

Note that the diameter of an H atom = 2 × 0.74 Å = 1.48 Å

2-34. **Refer to Section 2-7 and Example 2-9.** . . . .

(a) mass of 1 mol CaO

1 × Ca = 1 × 40.08 g =	40.08 g
1 × O = 1 × 16.00 g =	16.00 g
mass of 1 mol CaO =	56.08 g

percent composition by mass of CaO

% Ca = (40.08/56.08) × 100 =	71.47%
% O = (16.00/56.08) × 100 =	28.53%
total =	100.00%

(b) mass of 1 mol Al_2O_3

2 × Al = 2 × 26.98 g =	53.96 g
3 × O = 3 × 16.00 g =	48.00 g
mass of 1 mol Al_2O_3 =	101.96 g

percent composition by mass of Al_2O_3

% Al = (53.96/101.96) × 100 =	52.92%
% O = (48.00/101.96) × 100 =	47.08%
total =	100.00%

(c) mass of 1 mol $NaHCO_3$

1 × Na = 1 × 22.99 g =	22.99 g
1 × H = 1 × 1.01 g =	1.01 g
1 × C = 1 × 12.01 g =	12.01 g
3 × O = 3 × 16.00 g =	48.00 g
mass of 1 mol $NaHCO_3$ =	84.01 g

percent composition by mass of $NaHCO_3$

% Na = (22.99/84.01) × 100 =	27.37%
% H = (1.01/ 84.01) × 100 =	1.20%
% C = (12.01/84.01) × 100 =	14.30%
% O = (48.00/84.01) × 100 =	57.14%
total =	100.01%

Note: Round off error can produce a total that is not exactly 100.00%.

2-36. **Refer to Section 2-8 and Example 2-10.** . . . .

Plan: (1) If percentage composition instead of sample mass is given, assume a 100 g sample.
(2) Calculate the moles of each element.

(3) Divide each of the mole values by the smallest number obtained as a mole value for the 100 g sample. General Rule: do not round off more than 0.1.

(4) Determine a whole number ratio.

Let us assume we have a 100.0 g sample containing 59.9 g Ti and 40.1 g O.

$? \text{ mol Ti} = \dfrac{\text{g Ti}}{\text{AW Ti}} = \dfrac{59.9 \text{ g}}{47.9 \text{ g/mol}} = 1.25 \text{ mol}$ Ratio $= \dfrac{1.25}{1.25} = 1$

$? \text{ mol O} = \dfrac{\text{g O}}{\text{AW O}} = \dfrac{40.1 \text{ g}}{16.0 \text{ g/mol}} = 2.51 \text{ mol}$ Ratio $= \dfrac{2.51}{1.25} = 2.01 = 2$

Therefore, the simplest formula is **TiO_2**.

2-38. **Refer to Section 2-8, Example 2-11 and Exercise 2-36 Solution.** . . . .

This exercise is similar to 2-36, except that now we do not have to assume a 100 g sample.

$? \text{ mol Fe} = \dfrac{\text{g Fe}}{\text{AW Fe}} = \dfrac{1.116 \text{ g}}{55.85 \text{ g/mol}} = 1.998 \times 10^{-2} \text{ mol}$ Ratio $= \dfrac{1.998 \times 10^{-2}}{1.998 \times 10^{-2}} = 1$

$? \text{ mol O} = \dfrac{\text{g O}}{\text{AW O}} = \dfrac{0.480 \text{ g}}{16.0 \text{ g/mol}} = 3.00 \times 10^{-2} \text{ mol}$ Ratio $= \dfrac{3.00 \times 10^{-2}}{1.998 \times 10^{-2}} = 1.5$

A 1:1.5 ratio converts to a 2:3 ratio by multiplying both numbers by 2. Therefore, the simplest formula is **Fe_2O_3**.

2-40. **Refer to Section 2-9.** . . . .

The answers for Exercises 36 and 37 are TiO_2 and Ti_2O_3, and for Exercises 38 and 39 are Fe_2O_3 and Fe_3O_4.

For example consider Fe_2O_3 and Fe_3O_4. When the number of Fe atoms in both formulas is normalized to 6 while the original Fe:O ratio in each compound is maintained, we obtain:

$Fe_2O_3 \rightarrow (Fe_2O_3)_3 \rightarrow (Fe_6O_9)$

$Fe_3O_4 \rightarrow (Fe_3O_4)_2 \rightarrow (Fe_6O_8)$ and

$$\frac{\text{Mass of O combining with 6 Fe atoms in } Fe_2O_3}{\text{Mass of O combining with 6 Fe atoms in } Fe_3O_4} = \frac{9 \times 16 \text{ amu}}{8 \times 16 \text{ amu}} = \frac{9}{8}$$

It is therefore seen that for the same mass of Fe, the mass of O atoms combining with it in these two different compounds is in the ratio of 9:8.

This result illustrates the **Law of Multiple Proportions** which states that when elements form more than one compound, the ratio of the masses of one element that combine with a given mass of another element in each of the compounds can be expressed by small whole numbers.

The data for TiO_2 and Ti_2O_3 give similar conclusions.

2-42. **Refer to Section 2-8 and Exercise 2-36 Solution.** . . . .

$? \text{ mol Mg} = \dfrac{\text{g Mg}}{\text{AW Mg}} = \dfrac{4.86 \text{ g}}{24.3 \text{ g/mol}} = 0.200 \text{ mol}$ Ratio $= \dfrac{0.200}{0.200} = 1$

$$? \text{ mol S} = \frac{\text{g S}}{\text{AW S}} = \frac{6.42 \text{ g}}{32.1 \text{ g/mol}} = 0.200 \text{ mol} \qquad \text{Ratio} = \frac{0.200}{0.200} = 1$$

$$? \text{ mol O} = \frac{\text{g O}}{\text{AW O}} = \frac{9.60 \text{ g}}{16.0 \text{ g/mol}} = 0.600 \text{ mol} \qquad \text{Ratio} = \frac{0.600}{0.200} = 3$$

The ratio of Mg:S:O is 1:1:3.
Therefore, the simplest formula for this compound is **$MgSO_3$**.

2-44. **Refer to Section 2-8 and Exercise 2-36 Solution.** . . . .

$$? \text{ mol Mg} = \frac{\text{g Mg}}{\text{AW Mg}} = \frac{4.86 \text{ g}}{24.3 \text{ g/mol}} = 0.200 \text{ mol} \qquad \text{Ratio} = \frac{0.200}{0.200} = 1$$

$$? \text{ mol S} = \frac{\text{g S}}{\text{AW S}} = \frac{12.85 \text{ g}}{32.1 \text{ g/mol}} = 0.400 \text{ mol} \qquad \text{Ratio} = \frac{0.400}{0.200} = 2$$

$$? \text{ mol O} = \frac{\text{g O}}{\text{AW O}} = \frac{9.60 \text{ g}}{16.0 \text{ g/mol}} = 0.600 \text{ mol} \qquad \text{Ratio} = \frac{0.600}{0.200} = 3$$

The ratio of Mg:S:O is 1:2:3.
Therefore, the simplest formula for this compound is **MgS_2O_3**.

2-46. **Refer to Section 2-6 and Example 2-6.** . . . .

Method 1: Formula Weight, $\text{FW} \left(\frac{\text{g}}{\text{mol}}\right) = \frac{\text{g substance}}{\text{mol substance}}$

$$? \text{ mol } CH_3OH = \frac{\text{g } CH_3OH}{\text{FW } CH_3OH} = \frac{80 \text{ g}}{32 \text{ g/mol}} = \mathbf{2.5 \text{ mol } CH_3OH}$$

Method 2: Dimensional Analysis

$$? \text{ mol } CH_3OH = 80 \text{ g } CH_3OH \times \frac{1 \text{ mol } CH_3OH}{32 \text{ g } CH_3OH} = \mathbf{2.5 \text{ mol } CH_3OH}$$

2-48. **Refer to Section 2-6 and Example 2-7.** . . . .

Method 1: Recall that Avogadro's number, **N** = 6.02×10^{23} molecules/mol

Plan: g SO_2 $\xrightarrow{(1)}$ moles SO_2 $\xrightarrow{(2)}$ molecules SO_2 $\xrightarrow{(3)}$ atoms O

(1) $? \text{ mol } SO_2 = \frac{\text{g } SO_2}{\text{FW } SO_2} = \frac{160 \text{ g}}{64.1 \text{ g/mol}} = 2.50 \text{ mol } SO_2$

(2) $? \text{ molecules } SO_2 = \text{mol } SO_2 \times \mathbf{N} = 2.50 \text{ mol} \times 6.02 \times 10^{23} \text{ molecules/mol}$
$= 1.51 \times 10^{24} \text{ molecules } SO_2$

(3) $? \text{ atoms O} = 2 \times \text{molecules } SO_2 = \mathbf{3.02 \times 10^{24} \text{ atoms O}}$

Method 2: Dimensional Analysis
Each unit factor corresponds to a step in Method 1.

Step 1 **Step 2** **Step 3**

$$? \text{ atoms O} = 160 \text{ g } SO_2 \times \frac{1 \text{ mol } SO_2}{64.1 \text{ g } SO_2} \times \frac{6.02 \times 10^{23} \text{ molecules } SO_2}{1 \text{ mol } SO_2} \times \frac{2 \text{ atoms O}}{1 \text{ molecule } SO_2}$$

$$= \mathbf{3.01 \times 10^{24} \text{ atoms O}}$$

2-50. **Refer to Section 2-6, Example 2-7 and Exercise 2-48 Solution.** • • • • • • •

FW $Pd(NH_3)_2(OH)_2$ = 106 + (17 × 2) + (17 × 2) = 174 g/mol

Method 1: Plan:

$$g\ Pd(NH_3)_2(OH)_2 \xrightarrow{(1)} mol\ Pd(NH_3)_2(OH)_2 \xrightarrow{(2)} molecules\ Pd(NH_3)_2(OH)_2 \xrightarrow{(3)} atoms\ H$$

(1) $$?\ mol\ Pd(NH_3)_2(OH)_2 = \frac{g\ Pd(NH_3)_2(OH)_2}{FW\ Pd(NH_3)_2(OH)_2} = \frac{69.8\ g}{174\ g/mol} = 0.401\ mol$$

(2) $$?\ molecules\ Pd(NH_3)_2(OH)_2 = 0.401\ mol\ Pd(NH_3)_2(OH)_2 \times 6.02 \times 10^{23}\ molecules/mol$$

$$= 2.41 \times 10^{23}\ molecules\ Pd(NH_3)_2(OH)_2$$

(3) $$?\ atoms\ H = 8 \times molecules\ Pd(NH_3)_2(OH)_2 = \mathbf{1.93 \times 10^{24}\ atoms\ H}$$

Method 2: Dimensional Analysis
Each unit factor corresponds to a step in Method 1.

$$?\ atoms\ H = 69.8\ g\ Pd(NH_3)_2(OH)_2 \times \underbrace{\frac{1\ mol}{174\ g}}_{\textbf{Step 1}} \times \underbrace{\frac{6.02 \times 10^{23}\ molecules}{1\ mol}}_{\textbf{Step 2}} \times \underbrace{\frac{8\ atoms\ H}{1\ molecule}}_{\textbf{Step 3}}$$

$$= \mathbf{1.93 \times 10^{24}\ atoms\ H}$$

2-52. **Refer to Sections 2-8 and 2-9, and Examples 2-12 and 2-13.** • • • • • • • •

(a) in the sample:

$$?\ mg\ C = 627.4\ mg\ CO_2 \times \frac{1\ mmol\ CO_2}{44.01\ mg\ CO_2} \times \frac{1\ mmol\ C}{1\ mmol\ CO_2} \times \frac{12.01\ mg\ C}{1\ mmol\ C} = \mathbf{171.2\ mg\ C}$$

$$?\ mg\ H = 171.2\ mg\ H_2O \times \frac{1\ mmol\ H_2O}{18.02\ mg\ H_2O} \times \frac{2\ mmol\ H}{1\ mmol\ H_2O} \times \frac{1.008\ mg\ H}{1\ mmol\ H} = \mathbf{19.15\ mg\ H}$$

? mg O = mass of sample - (mass of C + mass of H)

= 228.4 mg - (171.2 mg + 19.15 mg) = **38.0 mg O**

(b) $$?\ mol\ C = \frac{171.2\ mg\ C}{12.01\ mg/mmol} = 14.25\ mmol\ C \qquad Ratio = \frac{14.25}{2.38} = 6$$

$$?\ mol\ H = \frac{19.15\ mg\ H}{1.008\ mg/mmol} = 19.00\ mmol\ H \qquad Ratio = \frac{19.00}{2.38} = 8$$

$$?\ mol\ O = \frac{0.0380\ mg}{16.00\ mg/mmol} = 2.38\ mmol\ O \qquad Ratio = \frac{2.38}{2.38} = 1$$

The simplest formula for this compound is $\mathbf{C_6H_8O}$.

2-54. **Refer to Section 2-9.** • • • • • • • • • • • • • • • • • • •

(a) In Exercise 53, the percent composition by mass of calcium chloride was determined:

$$\%\ Ca = \frac{g\ Ca}{(g\ Ca + g\ Cl)} \times 100 = \frac{21.7\ g}{(21.7\ g + 38.3\ g)} \times 100 = 36.2\ \%\ Ca$$

$$\% \text{ Cl} = \frac{\text{g Cl}}{(\text{g Ca} + \text{g Cl})} \times 100 = \frac{38.3 \text{ g}}{(21.7 \text{ g} + 38.3 \text{ g})} \times 100 = 63.8 \% \text{ Cl}$$

Therefore, we know that for any mass of calcium chloride,

$$\frac{\text{g Ca}}{\text{g calcium chloride}} = 0.362$$

Rearranging, we have

$$? \text{ g calcium chloride} = \frac{\text{g Ca}}{0.362} = \frac{30.0 \text{ g}}{0.362} = \mathbf{82.9 \text{ g}}$$

(b) Likewise, in calcium chloride, $\frac{\text{g Cl}}{\text{g Ca}} = \frac{63.8 \text{ g}}{36.2 \text{ g}} = 1.76$

$$? \text{ g Cl} = 1.76 \times \text{g Ca} = 1.76 \times 30.0 \text{ g} = \mathbf{52.8 \text{ g}}$$

Alternative Method: $? \text{ g Cl} = \text{g calcium chloride} - \text{g Ca} = 82.9 \text{ g} - 30.0 \text{ g} = \mathbf{52.9 \text{ g}}$

2-56. Refer to Sections 2-9 and 3-3. . . . .

(a) To produce 100 g of compound, 56.4 g P and 43.6 g O must combine. Let us first determine how much P is needed to react with the entire 5.0 g sample of O. We can set up a ratio:

$$\frac{56.4 \text{ g P}}{43.6 \text{ g O}} = \frac{? \text{ g P}}{5.0 \text{ g O}}$$

Solving, we have $? \text{ g P} = 5.0 \text{ g O} \times \frac{56.4 \text{ g P}}{43.6 \text{ g O}} = 6.5 \text{ g P}$

But we have 10.0 g P. Therefore, in the reaction, all the **oxygen** is consumed completely, and we will have some phosphorus remaining unreacted.

(b) $? \text{ g P unreacted} = \text{g P}_{\text{total}} - \text{g P}_{\text{reacted}} = 10.0 \text{ g} - 6.5 \text{ g} = \mathbf{3.5 \text{ g}}$

(c) $? \text{ g compound produced} = \text{g P}_{\text{reacted}} + \text{g O}_{\text{reacted}} = 6.5 \text{ g} + 5.0 \text{ g} = \mathbf{11.5 \text{ g}}$

2-58. Refer to Section 2-9 and Exercise 2-36 Solution. . . . .

(a) We know there are 44.50 g of Sn in a 100.00 g sample of compound. There is also 1 Sn atom per formula unit, which means there is 1 mol Sn (118.7 g) in a sample with mass equal to the formula weight. We can now set up a ratio:

$$\frac{44.50 \text{ g Sn}}{100.00 \text{ g cmpd}} = \frac{118.7 \text{ g Sn}}{? \text{ FW}}$$

Solving, $? \text{ FW} = 118.7 \text{ g Sn} \times \frac{100.00 \text{ g cmpd}}{44.50 \text{ g Sn}} = \mathbf{266.7 \text{ g per mole Sn}}$

(b) We must determine the simplest formula for hydrated sodium stannate. Assume a 100 g sample.

$$? \text{ mol Sn} = \frac{\text{g Sn}}{\text{AW Sn}} = \frac{44.50 \text{ g}}{118.7 \text{ g/mol}} = 0.3749 \text{ mol Sn} \qquad \text{Ratio} = \frac{0.3749}{0.3749} = 1$$

$$? \text{ mol Na} = \frac{\text{g Na}}{\text{AW Na}} = \frac{17.14 \text{ g}}{22.99 \text{ g/mol}} = 0.7455 \text{ mol Na} \qquad \text{Ratio} = \frac{0.7455}{0.3749} = 2$$

$$? \text{ mol O} = \frac{\text{g O}}{\text{AW O}} = \frac{35.99 \text{ g}}{16.00 \text{ g/mol}} = 2.249 \text{ mol O} \qquad \text{Ratio} = \frac{2.249}{0.3749} = 6$$

$$? \text{ mol H} = \frac{\text{g H}}{\text{AW H}} = \frac{2.27 \text{ g}}{1.01 \text{ g/mol}} = 2.25 \text{ mol H} \qquad \text{Ratio} = \frac{2.25}{0.3749} = 6$$

The simplest formula, $SnNa_2O_6H_6$, contains 1 Sn atom.

Therefore, the simplest formula is the true formula, better written as $\mathbf{Na_2SnO_3 \cdot 3H_2O}$.

2-60. **Refer to Section 2-9 and Exercise 2-36 Solution.** • • • • • • • • • • • •

(a) First, we must calculate the % by mass of N in skatole.
? % N = 100.00 - (% C + % H) = 100.00 - (82.40 + 6.92) = 10.68 % N

To find the simplest formula, assume 100 g of skatole.

$$? \text{ mol C} = \frac{\text{g C}}{\text{AW C}} = \frac{82.40 \text{ g}}{12.01 \text{ g/mol}} = 6.861 \text{ mol C} \qquad \text{Ratio} = \frac{6.861}{0.7623} = 9$$

$$? \text{ mol H} = \frac{\text{g H}}{\text{AW H}} = \frac{6.92 \text{ g}}{1.01 \text{ g/mol}} = 6.85 \text{ mol H} \qquad \text{Ratio} = \frac{6.85}{0.7623} = 9$$

$$? \text{ mol N} = \frac{\text{g N}}{\text{AW N}} = \frac{10.68 \text{ g}}{14.01 \text{ g/mol}} = 0.7623 \text{ mol N} \qquad \text{Ratio} = \frac{0.7623}{0.7623} = 1$$

The simplest formula is the true formula, $\mathbf{C_9H_9N}$.

(b) The molecular weight of skatole:

$9 \times C = 9 \times 12.01$ g = 108.09 g
$9 \times H = 9 \times 1.01$ g = 9.09 g
$1 \times N = 1 \times 14.01$ g = 14.01 g
mass of 1 mol C_9H_9N = **131.19 g**

2-62. **Refer to Section 2-9 and Exercise 2-36 Solution.** • • • • • • • • • • • •

(a) We know there are 11.75 g of oxygen in 100 g of estradiol. Because there are 2 oxygen atoms per molecule, there are 2 moles of O (2 × AW = 32.00 g) in a sample of estradiol with mass equal to the formula weight. We can set up a ratio.

$$\frac{11.75 \text{ g O}}{100.0 \text{ g estradiol}} = \frac{32.00 \text{ g O}}{? \text{ FW}} \qquad \text{Solving, } ? \text{ FW} = 32.00 \text{ g O} \times \frac{100.0 \text{ g cmpd}}{11.75 \text{ g O}}$$

= **272.3 g per mole estradiol**

(b) First, we must find the % H in estradiol.
? % H = 100.0 % - (% C + % O) = 100.0 - (79.3 + 11.75) = 9.0 % H

Now, we determine the simplest formula by assuming a 100 g sample.

$$? \text{ mol C} = \frac{\text{g C}}{\text{AW C}} = \frac{79.3 \text{ g}}{12.01 \text{ g/mol}} = 6.60 \text{ mol C} \qquad \text{Ratio} = \frac{6.60}{0.7344} = 9$$

$$? \text{ mol O} = \frac{\text{g O}}{\text{AW O}} = \frac{11.75 \text{ g}}{16.00 \text{ g/mol}} = 0.7344 \text{ mol O} \qquad \text{Ratio} = \frac{0.7344}{0.7344} = 1$$

$$? \text{ mol H} = \frac{\text{g H}}{\text{AW H}} = \frac{9.0 \text{ g}}{1.01 \text{ g/mol}} = 8.9 \text{ mol H} \qquad \text{Ratio} = \frac{8.9}{0.734} = 12$$

The simplest formula is $C_9H_{12}O$.

However, the true formula contains 2 atoms per molecule. Therefore the formula is $\mathbf{C_{18}H_{24}O_2}$ with formula weight of 272.3 g/mol.

2-64. **Refer to Sections 2-8 and 2-9, and Exercises 2-36 and 2-38 Solutions.** • • • •

(a) $$? \text{ \% C} = \frac{\text{g C}}{\text{g sample}} \times 100 = \frac{0.8077 \text{ g}}{1.6380 \text{ g}} \times 100 = \mathbf{49.31 \text{ \% C}}$$

$$? \text{ \% H} = \frac{\text{g H}}{\text{g sample}} \times 100 = \frac{0.1130 \text{ g}}{1.6380 \text{ g}} \times 100 = \mathbf{6.899 \text{ \% H}}$$

$? \% O = 100.00 \% - (\% C + \% H) = 100.00 - (49.31 + 6.899) = \mathbf{43.79 \% O}$

(b) $? \text{ mol C} = \frac{\text{g C}}{\text{AW C}} = \frac{49.31 \text{ g}}{12.01 \text{ g/mol}} = 4.106 \text{ mol C}$ Ratio $= \frac{4.106}{2.737} = 1.50$

$? \text{ mol H} = \frac{\text{g H}}{\text{AW H}} = \frac{6.899 \text{ g}}{1.008 \text{ g/mol}} = 6.844 \text{ mol H}$ Ratio $= \frac{6.844}{2.737} = 2.50$

$? \text{ mol O} = \frac{\text{g O}}{\text{AW O}} = \frac{43.79 \text{ g}}{16.00 \text{ g/mol}} = 2.737 \text{ mol O}$ Ratio $= \frac{2.737}{2.737} = 1.00$

A ratio of 1.5:2.5:1 converts to a 3:5:2 ratio by multiplying by 2. Therefore, the simplest formula is $\mathbf{C_3H_5O_2}$ with formula weight 73.07 g/mol.

(c) The true formula has a molecular weight of 146.1 g/mol.

$$n = \frac{\text{molecular weight}}{\text{simplest formula weight}} = \frac{146.1}{73.07} = 2$$

The true formula is, therefore, $\mathbf{C_6H_{10}O_4}$.

2-66. **Refer to Sections 2-6 and 2-9.** . . . .

Method 1: Plan: g protein $\xrightarrow{(1)}$ g acid $\xrightarrow{(2)}$ mol acid $\xrightarrow{(3)}$ mol N $\xrightarrow{(4)}$ g N

(1) ? g glutamic acid = 2.50 g protein × 0.164 = 0.410 g glutamic acid

(2) $? \text{ mol glutamic acid} = \frac{\text{g glutamic acid}}{\text{FW glutamic acid}} = \frac{0.410 \text{ g}}{147 \text{ g/mol}} = 0.00279 \text{ mol acid}$

(3) ? mol N = mol glutamic acid = 0.00279 mol N

(4) ? g N = mol N × AW N = 0.00279 mol × 14.0 g/mol = **0.0391 g N**

Method 2: Dimensional Analysis
Each step in Method 1 is a unit factor in dimensional analysis.

$$? \text{ g N} = 2.50 \text{ g protein} \times \underbrace{\frac{16.4 \text{ g acid}}{100 \text{ g protein}}}_{\textbf{Step 1}} \times \underbrace{\frac{1 \text{ mol acid}}{147.0 \text{ g}}}_{\textbf{Step 2}} \times \underbrace{\frac{1 \text{ mol N}}{1 \text{ mol acid}}}_{\textbf{Step 3}} \times \underbrace{\frac{14.0 \text{ g N}}{1 \text{ mol N}}}_{\textbf{Step 4}}$$

$= \mathbf{0.0390 \text{ g N}}$

OR

$$? \text{ g N} = 2.50 \text{ g protein} \times \frac{16.4 \text{ g acid}}{100 \text{ g protein}} \times \frac{14.0 \text{ g N}}{147.0 \text{ g acid}} = \mathbf{0.0390 \text{ g N}}$$

2-68. **Refer to Sections 2-6 and 2-9.** . . . .

Method 1:
Plan: g glycine $\xrightarrow{(1)}$ mol glycine $\xrightarrow{(2)}$ mol C $\xrightarrow{(3)}$ mol CO_2 $\xrightarrow{(4)}$ g CO_2

(1) $? \text{ mol glycine} = \frac{\text{g glycine}}{\text{FW glycine}} = \frac{1.00 \text{ g}}{75.1 \text{ g/mol}} = 0.0133 \text{ mol glycine}$

(2) ? mol C = 2 × mol glycine = 2 × 0.0133 mol = 0.0266 mol C

(3) ? mol CO_2 = mol C = 0.0266 mol CO_2 (since for every mol C in the compound, 1 mol CO_2 can be formed.)

(4) ? g CO_2 = mol CO_2 × FW CO_2 = 0.0266 mol × 44.0 g/mol = **1.17 g CO_2**

Method 2: Dimensional Analysis

Note that every step in Method 1 corresponds to a unit factor.

$$? \text{ g } CO_2 = 1.00 \text{ g glycine} \times \underbrace{\frac{1 \text{ mol glycine}}{75.1 \text{ g glycine}}}_{\textbf{Step 1}} \times \underbrace{\frac{2 \text{ mol C}}{1 \text{ mol glycine}}}_{\textbf{Step 2}} \times \underbrace{\frac{1 \text{ mol } CO_2}{1 \text{ mol C}}}_{\textbf{Step 3}} \times \underbrace{\frac{44.0 \text{ g}}{1 \text{ mol } CO_2}}_{\textbf{Step 4}}$$

$$= \mathbf{1.17 \text{ g } CO_2}$$

2-70. **Refer to Section 2-9 and Example 2-14.** . . . .

(a) in NO: $\frac{? \text{ g O}}{1.00 \text{ g N}} = \frac{16.0 \text{ g O}}{14.0 \text{ g N}}$ Solving, $? \text{ g O} = 1.00 \text{ g N} \times \frac{16.0 \text{ g O}}{14.0 \text{ g N}}$

$= \mathbf{1.14 \text{ g O per } 1.00 \text{ g N in NO}}$

(b) in NO_2: $\frac{? \text{ g O}}{1.00 \text{ g N}} = \frac{32.0 \text{ g O}}{14.0 \text{ g N}}$ Solving, $? \text{ g O} = 1.00 \text{ g N} \times \frac{32.0 \text{ g O}}{14.0 \text{ g N}}$

$= \mathbf{2.29 \text{ g O per } 1.00 \text{ g N in } NO_2}$

2-72. **Refer to Section 2-9 and Example 2-14.** . . . .

(a) in H_2O: $\frac{? \text{ g O}}{1.00 \text{ g H}} = \frac{16.0 \text{ g O}}{2.02 \text{ g H}}$ Solving, $? \text{ g O} = 1.00 \text{ g H} \times \frac{16.0 \text{ g O}}{2.02 \text{ g H}}$

$= \mathbf{7.92 \text{ g O per } 1.00 \text{ g H in } H_2O}$

(b) in H_2O_2: $\frac{? \text{ g O}}{1.00 \text{ g H}} = \frac{32.0 \text{ g O}}{2.02 \text{ g H}}$ Solving, $? \text{ g O} = 1.00 \text{ g H} \times \frac{32.0 \text{ g O}}{2.02 \text{ g H}}$

$= \mathbf{15.8 \text{ g O per } 1.00 \text{ g H in } H_2O_2}$

2-74. **Refer to Sections 2-7 and 2-10.** . . . .

(a) Plan: (1) Calculate the formula weights of $CuSO_4 \cdot 5H_2O$ and $CuSO_4$.
(2) Determine % by mass of $CuSO_4$.

(1) The formula weight of $CuSO_4 \cdot 5H_2O$

$$\begin{aligned} 1 \times Cu &= 1 \times 63.55 \text{ g} = 63.55 \text{ g} \\ 1 \times S &= 1 \times 32.07 \text{ g} = 32.07 \text{ g} \\ 9 \times O &= 9 \times 16.00 \text{ g} = 144.0 \text{ g} \\ 10 \times H &= 10 \times 1.008 \text{ g} = 10.08 \text{ g} \\ \hline \text{mass of 1 mol } CuSO_4 \cdot 5H_2O &= 249.7 \text{ g} \end{aligned}$$

The formula weight of $CuSO_4$

$$\begin{aligned} 1 \times Cu &= 1 \times 63.55 \text{ g} = 63.55 \text{ g} \\ 1 \times S &= 1 \times 32.07 \text{ g} = 32.07 \text{ g} \\ 4 \times O &= 4 \times 16.00 \text{ g} = 64.00 \text{ g} \\ \hline \text{mass of 1 mol } CuSO_4 &= 159.62 \text{ g} \end{aligned}$$

(2) Let us assume that we have 1 mole of $CuSO_4 \cdot 5H_2O$ which contains 1 mole of $CuSO_4$.

$$? \% \ CuSO_4 = \frac{\text{g } CuSO_4 \text{ in 1 mol}}{\text{g } CuSO_4 \cdot 5H_2O \text{ in 1 mol}} \times 100 = \frac{159.6 \text{ g}}{249.7 \text{ g}} \times 100 = \mathbf{63.92 \% \ CuSO_4}$$

(b) We know: $? \% \ CuSO_4 \text{ in ore} = \frac{\text{g } CuSO_4 \text{ in ore}}{\text{g ore}} \times 100$

Let us assume a 100.0 g sample of ore.

$$? \text{ g } CuSO_4 \text{ in ore} = 100.0 \text{ g ore} \times \frac{72.4 \text{ g } CuSO_4 \cdot 5H_2O}{100.0 \text{ g ore}} \times \frac{63.92 \text{ g } CuSO_4}{100 \text{ g } CuSO_4 \cdot 5H_2O}$$

$$= 46.3 \text{ g } CuSO_4$$

Substituting,

$$? \% \ CuSO_4 \text{ in ore} = \frac{46.3 \text{ g } CuSO_4 \text{ in ore}}{100.0 \text{ g ore}} \times 100 = \mathbf{46.3 \% \ CuSO_4 \text{ in ore}}$$

2-76. **Refer to Section 2-10.** . . . .

(a) $? \text{ lb } Al_2O_3 \text{ in ore} = \text{lb ore} \times 0.243 = 775 \text{ lb} \times 0.243 = \mathbf{188 \text{ lb } Al_2O_3 \text{ in ore}}$

(b) $? \text{ lb impurities} = \text{lb ore} - \text{lb } Al_2O_3 = 775 \text{ lb} - 188 \text{ lb} = \mathbf{587 \text{ lb impurities}}$

(c) The fraction by mass of Al in $Al_2O_3 = \frac{2 \times \text{AW Al}}{\text{FW } Al_2O_3} = \frac{2 \times 26.98 \text{ g/mol}}{102.0 \text{ g/mol}} = 0.5290$

$$\begin{aligned} ? \text{ lb Al in ore} &= \text{lb } Al_2O_3 \text{ in ore} \times \text{fraction by mass of Al in } Al_2O_3 \\ &= 188 \text{ lb} \times 0.5290 \\ &= \mathbf{99.5 \text{ lb Al in ore}} \end{aligned}$$

3 Chemical Equations and Reaction Stoichiometry

3-2. **Refer to Section 3-1.** . . . .

The Law of Conservation of Matter states that matter is neither created nor destroyed during an ordinary chemical reaction. Therefore, a balanced chemical equation must always contain the same number of each kind of atom on both sides of the equation.

3-4. **Refer to Section 3-1.** . . . .

Hints for balancing equations:

(1) Use smallest whole number coefficients. However, it may be useful to temporarily use a fractional coefficient, then multiply through by a factor at the end to change the fractions to whole numbers.

(2) Look for special groups of elements that appear unchanged on both sides of the equation, e.g. NO_3, PO_4, SO_4. Treat them as units when balancing.

(3) Begin by balancing the special groups and the elements that appear only once on both sides of the equation. Any element that appears more than once on one side of the equation is normally the last element to be balanced.

(4) When an element has an "odd" number of atoms on one side of the equation and an "even" number on the other side, it is often advisable to multiply the "odd" side by 2.

(a) unbalanced: $Sn + Cl_2 \rightarrow SnCl_4$

Step 1: $Sn + \mathbf{2}Cl_2 \rightarrow SnCl_4$ balance Cl

(b) unbalanced: $Ca(OH)_2 + HNO_3 \rightarrow Ca(NO_3)_2 + H_2O$

Step 1: $Ca(OH)_2 + \mathbf{2}HNO_3 \rightarrow Ca(NO_3)_2 + H_2O$ balance NO_3 units

Step 2: $Ca(OH)_2 + 2HNO_3 \rightarrow Ca(NO_3)_2 + \mathbf{2}H_2O$ balance H,O

(c) unbalanced: $CaF_2 + H_2SO_4 \rightarrow HF + CaSO_4$

Step 1: $CaF_2 + H_2SO_4 \rightarrow \mathbf{2}HF + CaSO_4$ balance H,F

(d) unbalanced: $BF_3 + H_2O \rightarrow B_2O_3 + HF$

Step 1: $\mathbf{2}BF_3 + H_2O \rightarrow B_2O_3 + HF$ balance B

Step 2: $2BF_3 + \mathbf{3}H_2O \rightarrow B_2O_3 + HF$ balance O

Step 3: $2BF_3 + 3H_2O \rightarrow B_2O_3 + \mathbf{6}HF$ balance H,F

(e) unbalanced: $NaOH + SO_2 \rightarrow Na_2SO_3 + H_2O$

Step 1: $\mathbf{2}NaOH + SO_2 \rightarrow Na_2SO_3 + H_2O$ balance Na,O,H

3-6. **Refer to Section 3-1 and Exercise 3-4 Solution.** . . . .

(a) unbalanced: $P_4O_{10} + KOH \rightarrow K_3PO_4 + H_2O$

Step 1: $P_4O_{10} + KOH \rightarrow \mathbf{4}K_3PO_4 + H_2O$ balance P

Step 2: $P_4O_{10} + \mathbf{12}KOH \rightarrow 4K_3PO_4 + H_2O$ balance K

Step 3: $P_4O_{10} + 12KOH \rightarrow 4K_3PO_4 + \mathbf{6}H_2O$ balance H,O

(b) unbalanced: $HBF_4 + H_2O \rightarrow H_3BO_3 + HF$

Step 1: $HBF_4 + H_2O \rightarrow H_3BO_3 + \mathbf{4}HF$ balance F

Step 2: $HBF_4 + \mathbf{3}H_2O \rightarrow H_3BO_3 + 4HF$ balance H,O

(c) unbalanced: $(NH_4)_2Cr_2O_7 \rightarrow N_2 + Cr_2O_3 + H_2O$

Step 1: $(NH_4)_2Cr_2O_7 \rightarrow N_2 + Cr_2O_3 + \mathbf{4}H_2O$ balance H,O

(d) unbalanced: $NH_3 + O_2 \rightarrow NO + H_2O$

Step 1: $\mathbf{2}NH_3 + O_2 \rightarrow NO + \mathbf{3}H_2O$ balance H

Step 2: $2NH_3 + O_2 \rightarrow \mathbf{2}NO + 3H_2O$ balance N

Step 3: $2NH_3 + \frac{5}{2}O_2 \rightarrow 2NO + 3H_2O$ balance O (Hint 1)

Step 4: $\mathbf{4}NH_3 + \mathbf{5}O_2 \rightarrow \mathbf{4}NO + \mathbf{6}H_2O$ need whole number coefficients

(e) unbalanced: $CS_2 + O_2 \rightarrow CO_2 + SO_2$

Step 1: $CS_2 + O_2 \rightarrow CO_2 + \mathbf{2}SO_2$ balance S

Step 2: $CS_2 + \mathbf{3}O_2 \rightarrow CO_2 + 2SO_2$ balance O

3-8. **Refer to Section 3-1 and Exercise 3-4 Solution.** . . . .

(a) unbalanced: $HgCl_2 + SnCl_2 \rightarrow Hg_2Cl_2 + SnCl_4$

Step 1: $\mathbf{2}HgCl_2 + SnCl_2 \rightarrow Hg_2Cl_2 + SnCl_4$ balance Hg,Cl

(b) unbalanced: $N_2O_5 + H_2O \rightarrow HNO_3$

Step 1: $N_2O_5 + H_2O \rightarrow \mathbf{2}HNO_3$ balance H,N,O

(c) unbalanced: $Pb(NO_3)_2 \rightarrow PbO + NO_2 + O_2$

Step 1: $Pb(NO_3)_2 \rightarrow PbO + \mathbf{2}NO_2 + O_2$ balance N

Step 2: $Pb(NO_3)_2 \rightarrow PbO + 2NO_2 + \frac{1}{2}O_2$ balance O

Step 3: $\mathbf{2}Pb(NO_3)_2 \rightarrow \mathbf{2}PbO + \mathbf{4}NO_2 + O_2$ whole number coefficients

(d) unbalanced: $PCl_5 + AsF_3 \rightarrow PF_5 + AsCl_3$

Step 1: $\mathbf{3}PCl_5 + AsF_3 \rightarrow PF_5 + \mathbf{5}AsCl_3$ balance Cl

Step 2: $3PCl_5 + \mathbf{5}AsF_3 \rightarrow \mathbf{3}PF_5 + 5AsCl_3$ balance P,As,F

(e) unbalanced: $P_4O_{10} + Ca(OH)_2 \rightarrow Ca_3(PO_4)_2 + H_2O$

Step 1: $P_4O_{10} + Ca(OH)_2 \rightarrow \mathbf{2}Ca_3(PO_4)_2 + H_2O$ balance P

Step 2: $P_4O_{10} + 6Ca(OH)_2 \rightarrow 2Ca_3(PO_4)_2 + H_2O$ balance Ca

Step 3: $P_4O_{10} + 6Ca(OH)_2 \rightarrow 2Ca_3(PO_4)_2 + 6H_2O$ balance H,O

3-10. **Refer to Section 3-1 and Exercise 3-4 Solution.** . . . .

(a) unbalanced: $C_2H_4 + O_2 \rightarrow CO + H_2O$

Step 1: $C_2H_4 + O_2 \rightarrow 2CO + H_2O$ balance C

Step 2: $C_2H_4 + O_2 \rightarrow 2CO + 2H_2O$ balance H

Step 3: $C_2H_4 + 2O_2 \rightarrow 2CO + 2H_2O$ balance O

(b) unbalanced: $C_7H_6O_2 + O_2 \rightarrow CO_2 + H_2O$

Step 1: $C_7H_6O_2 + O_2 \rightarrow 7CO_2 + H_2O$ balance C

Step 2: $C_7H_6O_2 + O_2 \rightarrow 7CO_2 + 3H_2O$ balance H

Step 3: $C_7H_6O_2 + \frac{15}{2}O_2 \rightarrow 7CO_2 + 3H_2O$ balance O

Step 4: $2C_7H_6O_2 + 15O_2 \rightarrow 14CO_2 + 6H_2O$ whole number coefficients

(c) unbalanced: $FeS_2 + O_2 \rightarrow Fe_2O_3 + SO_2$

Step 1: $2FeS_2 + O_2 \rightarrow Fe_2O_3 + SO_2$ balance Fe

Step 2: $2FeS_2 + O_2 \rightarrow Fe_2O_3 + 4SO_2$ balance S

Step 3: $2FeS_2 + \frac{11}{2}O_2 \rightarrow Fe_2O_3 + 4SO_2$ balance O

Step 4: $4FeS_2 + 11O_2 \rightarrow 2Fe_2O_3 + 8SO_2$ whole number coefficients

(d) unbalanced: $P_4O_{10} + H_2O \rightarrow H_3PO_4$

Step 1: $P_4O_{10} + H_2O \rightarrow 4H_3PO_4$ balance P

Step 2: $P_4O_{10} + 6H_2O \rightarrow 4H_3PO_4$ balance H,O

(e) unbalanced: $XeF_2 + H_2O \rightarrow Xe + O_2 + HF$

Step 1: $XeF_2 + H_2O \rightarrow Xe + O_2 + 2HF$ balance F,H

Step 2: $XeF_2 + H_2O \rightarrow Xe + \frac{1}{2}O_2 + 2HF$ balance O

Step 3: $2XeF_2 + 2H_2O \rightarrow 2Xe + O_2 + 4HF$ whole number coefficients

3-12. **Refer to Section 3-1.** . . . .

(a) $SnCl_2 + Cl_2 \rightarrow SnCl_4$

(b) $2C_6H_{14} + 19O_2 \xrightarrow{\Delta} 12CO_2 + 14H_2O$

(c) $2C_3H_7OH + 9O_2 \xrightarrow{\Delta} 6CO_2 + 8H_2O$

3-14. **Refer to Section 3-1.** . . . .

(a) $2HgO \xrightarrow{\Delta} 2Hg + O_2$

(b) $CaH_2 + 2H_2O \rightarrow Ca(OH)_2 + 2H_2$

(c) $C_4H_{10} \overset{\Delta}{\rightarrow} 2C_2H_2 + 3H_2$

3-16. **Refer to Section 3-1.** . . . .

(a) $3H_2 + N_2 \overset{\Delta}{\rightarrow} 2NH_3$

(b) $2FeCl_3 + SnCl_2 \rightarrow 2FeCl_2 + SnCl_4$

(c) $2SbCl_3 + 3H_2S \rightarrow Sb_2S_3 + 6HCl$

3-18. **Refer to Section 3-1.** . . . .

(a) Two moles of aluminum metal, Al, react with six moles of hydrochloric acid, HCl, to form two moles of aluminum chloride, $AlCl_3$, and three moles of hydrogen gas, H_2.

(b) Four moles of lithium metal, Li, react with one mole of oxygen gas, O_2, to form two moles of lithium oxide, Li_2O.

(c) Two moles of magnesium, Mg, react when heated with one mole of oxygen gas, O_2, to form two moles of magnesium oxide, MgO.

3-20. **Refer to Section 3-2.** . . . .

The balanced equation is: $2KClO_3 \overset{\Delta}{\rightarrow} 2KCl + 3O_2$

Method 1: Use the units of formula weight as an equation.

$$FW\left(\frac{g}{mol}\right) = \frac{g\ substance}{mol\ substance}$$

Plan: mol $KClO_3 \xrightarrow{(1)}$ mol $O_2 \xrightarrow{(2)}$ g O_2

(1) ? mol O_2 = mol $KClO_3 \times 3/2 = 2.50\ mol \times 3/2 = 3.75$ mol O_2

(2) ? g O_2 = mol $O_2 \times$ FW O_2 = 3.75 mol $\times$ 32.0 g/mol = **1.20×10^2 g O_2**

Method 2: Dimensional Analysis
Each unit factor corresponds to a step in Method 1.

$$? \text{ g } O_2 = 2.50 \text{ mol } KClO_3 \times \underset{\textbf{Step 1}}{\frac{3 \text{ mol } O_2}{2 \text{ mol } KClO_3}} \times \underset{\textbf{Step 2}}{\frac{32.0 \text{ g } O_2}{1 \text{ mol } O_2}} = \mathbf{1.20 \times 10^2 \text{ g } O_2}$$

3-22. **Refer to Sections 3-1 and 3-2.** . . . .

The balanced equation is: $2Fe + 3S \overset{\Delta}{\rightarrow} Fe_2S_3$

(a) $? \text{ mol Fe} = 1.00 \text{ mol S} \times \frac{2 \text{ mol Fe}}{3 \text{ mol S}} = \mathbf{0.667 \text{ mol Fe}}$

? g Fe = mol Fe $\times$ AW Fe = 0.667 mol $\times$ 55.85 g/mol = **37.3 g Fe**

(b) $? \text{ mol S} = 1.00 \text{ mol Fe} \times \frac{3 \text{ mol S}}{2 \text{ mol Fe}} = \mathbf{1.50 \text{ mol S}}$

? g S = mol S $\times$ AW S = 1.50 mol $\times$ 32.07 g/mol = **48.1 g S**

(c) $? \text{ mol S} = \dfrac{\text{g S}}{\text{AW S}} = \dfrac{1.00 \text{ g}}{32.07 \text{ g/mol}} = 0.0312 \text{ mol S}$

$? \text{ mol Fe} = \text{mol S} \times \dfrac{2 \text{ mol Fe}}{3 \text{ mol S}} = 0.0312 \text{ mol} \times 2/3 = \mathbf{0.0208 \text{ mol Fe}}$

$? \text{ g Fe} = \text{mol Fe} \times \text{AW Fe} = 0.0208 \text{ mol} \times 55.85 \text{ g/mol} = \mathbf{1.16 \text{ g Fe}}$

(d) $? \text{ mol Fe} = \dfrac{\text{g Fe}}{\text{AW Fe}} = \dfrac{1.00 \text{ g}}{55.85 \text{ g/mol}} = 0.0179 \text{ mol Fe}$

$? \text{ mol S} = \text{mol Fe} \times \dfrac{3 \text{ mol S}}{2 \text{ mol Fe}} = 0.0179 \text{ mol} \times 3/2 = \mathbf{0.0268 \text{ mol S}}$

$? \text{ g S} = \text{mol S} \times \text{AW S} = 0.0268 \text{ mol} \times 32.07 \text{ g/mol} = \mathbf{0.859 \text{ g S}}$

3-24. **Refer to Section 3-2.** . . . .

The balanced equation is: $H_3PO_4 + 3NaOH \rightarrow Na_3PO_4 + 3H_2O$

Method 1: Plan: $\text{g } H_3PO_4 \xrightarrow{(1)} \text{mol } H_3PO_4 \xrightarrow{(2)} \text{mol } Na_3PO_4 \xrightarrow{(3)} \text{g } Na_3PO_4$

(1) $? \text{ mol } H_3PO_4 = \dfrac{\text{g } H_3PO_4}{\text{FW } H_3PO_4} = \dfrac{19.6 \text{ g}}{98.0 \text{ g/mol}} = 0.200 \text{ mol } H_3PO_4$

(2) $? \text{ mol } Na_3PO_4 = \text{mol } H_3PO_4 = 0.200 \text{ mol } Na_3PO_4$

(3) $? \text{ g } Na_3PO_4 = \text{mol } Na_3PO_4 \times \text{FW } Na_3PO_4 = 0.200 \text{ mol} \times 164 \text{ g/mol} = \mathbf{32.8 \text{ g } Na_3PO_4}$

Method 2: Dimensional Analysis

$$? \text{ g } Na_3PO_4 = 19.6 \text{ g } H_3PO_4 \times \underbrace{\frac{1 \text{ mol } H_3PO_4}{98.0 \text{ g } H_3PO_4}}_{\text{Step 1}} \times \underbrace{\frac{1 \text{ mol } Na_3PO_4}{1 \text{ mol } H_3PO_4}}_{\text{Step 2}} \times \underbrace{\frac{164 \text{ g } Na_3PO_4}{1 \text{ mol } Na_3PO_4}}_{\text{Step 3}}$$

$$= \mathbf{32.8 \text{ g } Na_3PO_4}$$

Method 3: Proportion or Ratio Method

$$\frac{? \text{ g } Na_3PO_4}{\text{g } H_3PO_4} = \frac{1 \times \text{FW } Na_3PO_4}{1 \times \text{FW } H_3PO_4}$$

Solving, $? \text{ g } Na_3PO_4 = \text{g } H_3PO_4 \times \dfrac{\text{FW } Na_3PO_4}{\text{FW } H_3PO_4}$

$= 19.6 \text{ g} \times \dfrac{164 \text{ g}}{98.0 \text{ g}}$

$= \mathbf{32.8 \text{ g } Na_3PO_4}$

3-26. **Refer to Section 3-2.** . . . .

The balanced equation is: $CH_3CHOH + 3O_2 \rightarrow 2CO_2 + 3H_2O$

Method 1: Plan: $\text{mol } CH_3CH_2OH \xrightarrow{(1)} \text{mol } O_2 \xrightarrow{(2)} \text{g } O_2$

(1) $? \text{ mol } O_2 = 3 \times \text{mol } CH_3CH_2OH = 3 \times 1.60 \text{ mol} = 4.80 \text{ mol } O_2$ (from coeff.)

(2) $? \text{ g } O_2 = \text{mol } O_2 \times \text{FW } O_2 = 4.80 \text{ mol} \times 32.0 \text{ g/mol} = \mathbf{154 \text{ g } O_2}$

Method 2: Dimensional Analysis

$$? \text{ g } O_2 = 1.60 \text{ mol } CH_3CH_2OH \times \underbrace{\frac{3 \text{ mol } O_2}{1 \text{ mol } CH_3CH_2OH}}_{\text{Step 1}} \times \underbrace{\frac{32.0 \text{ g } O_2}{1 \text{ mol } O_2}}_{\text{Step 2}} = \mathbf{154 \text{ g } O_2}$$

3-28. **Refer to Section 3-2.** . . . .

The balanced equation is: $NH_3 + CH_4 \rightarrow HCN + 3H_2$

Method 1: (a) Plan: $g\ NH_3 \xrightarrow{(1)} mol\ NH_3 \xrightarrow{(2)} mol\ HCN \xrightarrow{(3)} g\ HCN$

(b) Plan: $g\ NH_3 \xrightarrow{(1)} mol\ NH_3 \xrightarrow{(2)} mol\ H_2 \xrightarrow{(3)} g\ H_2$

(a) (1) $? \text{ mol } NH_3 = \frac{g\ NH_3}{FW\ NH_3} = \frac{245\ g}{17.0\ g/mol} = 14.4 \text{ mol } NH_3$

(2) $? \text{ mol HCN} = \text{mol } NH_3 = 14.4 \text{ mol HCN}$

(3) $? \text{ g HCN} = \text{mol HCN} \times \text{FW HCN} = 14.4 \text{ mol} \times 27.0 \text{ g/mol} = \mathbf{389\ g\ HCN}$

(b) (1) $? \text{ mol } NH_3 = \frac{g\ NH_3}{FW\ NH_3} = \frac{245\ g}{17.0\ g/mol} = 14.4 \text{ mol } NH_3$

(2) $? \text{ mol } H_2 = 3 \times \text{mol } NH_3 = 3 \times 14.4 \text{ mol} = 43.2 \text{ mol } H_2$

(3) $? \text{ g } H_2 = \text{mol } H_2 \times FW\ H_2 = 43.2 \text{ mol} \times 2.016 \text{ g/mol} = \mathbf{87.1\ g\ H_2}$

Method 2: Dimensional Analysis
The unit factors correspond to the steps in Method 1.

Step 1 **Step 2** **Step 3**

(a) $? \text{ g HCN} = 245 \text{ g } NH_3 \times \frac{1 \text{ mol } NH_3}{17.0 \text{ g } NH_3} \times \frac{1 \text{ mol HCN}}{1 \text{ mol } NH_3} \times \frac{27.0 \text{ g HCN}}{1 \text{ mol HCN}} = \mathbf{389\ g\ HCN}$

Step 1 **Step 2** **Step 3**

(b) $? \text{ g } H_2 = 245 \text{ g } NH_3 \times \frac{1 \text{ mol } NH_3}{17.0 \text{ g } NH_3} \times \frac{3 \text{ mol } H_2}{1 \text{ mol } NH_3} \times \frac{2.016 \text{ g } H_2}{1 \text{ mol } H_2} = \mathbf{87.2\ g\ H_2}$

Method 3: Proportion or Ratio Method

(a) $\frac{? \text{ g HCN}}{g\ NH_3} = \frac{1 \times FW\ HCN}{1 \times FW\ NH_3}$ Solving, $? \text{ g HCN} = g\ NH_3 \times \frac{1 \times FW\ HCN}{1 \times FW\ NH_3}$

$= 245 \text{ g} \times \frac{27.0 \text{ g}}{17.0 \text{ g}}$

$= \mathbf{389\ g\ HCN}$

(b) $\frac{? \text{ g } H_2}{g\ NH_3} = \frac{3 \times FW\ H_2}{1 \times FW\ NH_3}$ Solving, $? \text{ g } H_2 = g\ NH_3 \times \frac{3 \times FW\ H_2}{1 \times FW\ NH_3}$

$= 245 \text{ g} \times \frac{3 \times 2.016 \text{ g}}{1 \times 17.0 \text{ g}}$

$= \mathbf{87.2\ g\ H_2}$

3-30. **Refer to Section 3-2 and Exercise 3-28 Solution.** . . . .

The balanced equation is: $2NaHCO_3 \xrightarrow{\Delta} Na_2CO_3 + CO_2 + H_2O$

(a) $? \text{ g } NaHCO_3 = 1.00\text{g } Na_2CO_3 \times \frac{1 \text{ mol } Na_2CO_3}{106 \text{ g } Na_2CO_3} \times \frac{2 \text{ mol } NaHCO_3}{1 \text{ mol } Na_2CO_3} \times \frac{84.0 \text{ g } NaHCO_3}{1 \text{ mol } NaHCO_3}$

$= \mathbf{1.58\ g\ NaHCO_3}$

(b) $? \text{ g } Na_2CO_3 = 178 \text{ g } NaHCO_3 \times \frac{1 \text{ mol } NaHCO_3}{84.0 \text{ g } NaHCO_3} \times \frac{1 \text{ mol } Na_2CO_3}{2 \text{ mol } NaHCO_3} \times \frac{106 \text{ g } Na_2CO_3}{1 \text{ mol } Na_2CO_3}$

$= \mathbf{112\ g\ Na_2CO_3}$

(c) $? \text{ g } CO_2 = 178 \text{ g } NaHCO_3 \times \frac{1 \text{ mol } NaHCO_3}{84.0 \text{ g } NaHCO_3} \times \frac{1 \text{ mol } CO_2}{2 \text{ mol } NaHCO_3} \times \frac{44.0 \text{ g } CO_2}{1 \text{ mol } CO_2}$

$= \mathbf{46.6 \text{ g } CO_2}$

3-32. **Refer to Section 3-2 and Exercise 3-28 Solution.** . . . .

Balanced equations: $BaCO_3 \xrightarrow{\Delta} BaO + CO_2$ and $MgCO_3 \xrightarrow{\Delta} MgO + CO_2$

Method 1: The same mass of CO_2 would be produced by the same number of moles of $BaCO_3$ and $MgCO_3$.

Plan: $\text{g } BaCO_3 \xrightarrow{(1)} \text{mol } BaCO_3 \xrightarrow{(2)} \text{mol } MgCO_3 \xrightarrow{(3)} \text{g } MgCO_3$

(1) $? \text{ mol } BaCO_3 = \frac{\text{g } BaCO_3}{\text{FW } BaCO_3} = \frac{88.5 \text{ g}}{197 \text{ g/mol}} = 0.449 \text{ mol } BaCO_3$

(2) $? \text{ mol } MgCO_3 = \text{mol } BaCO_3 = 0.449 \text{ mol } MgCO_3$

(3) $? \text{ g } MgCO_3 = \text{mol } MgCO_3 \times \text{FW } MgCO_3 = 0.449 \text{ mol} \times 84.3 \text{ g/mol} = \mathbf{37.9 \text{ g } MgCO_3}$

Method 2: Proportion or Ratio Method

$\frac{? \text{ g } MgCO_3}{\text{g } BaCO_3} = \frac{1 \times \text{FW } MgCO_3}{1 \times \text{FW } BaCO_3}$ Solving, $? \text{ g } MgCO_3 = \text{g } BaCO_3 \times \frac{1 \times \text{FW } MgCO_3}{1 \times \text{FW } BaCO_3}$

$= 88.5 \text{ g} \times \frac{84.3 \text{ g}}{197 \text{ g}}$

$= \mathbf{37.9 \text{ g } MgCO_3}$

3-34. **Refer to Section 3-2.** . . . .

Balanced equations: $CaCO_3(s) \xrightarrow{\Delta} CaO(s) + CO_2(g)$

$CaO(s) + SO_2(g) \rightarrow CaSO_3(s)$

Method 1: We will work through both equations.

Plan: $\text{g } CaCO_3 \xrightarrow{(1)} \text{mol } CaCO_3 \xrightarrow{(2)} \text{mol CaO} \xrightarrow{(3)} \text{mol } SO_2 \xrightarrow{(4)} \text{g } SO_2$

(1) $? \text{ mol } CaCO_3 = \frac{\text{g } CaCO_3}{\text{FW } CaCO_3} = \frac{1.35 \times 10^6 \text{ g}}{100 \text{ g/mol}} = 1.35 \times 10^4 \text{ mol } CaCO_3$

(2) $? \text{ mol CaO} = \text{mol } CaCO_3 = 1.35 \times 10^4 \text{ mol CaO}$

(3) $? \text{ mol } SO_2 = \text{mol CaO} = 1.35 \times 10^4 \text{ mol } SO_2$

(4) $? \text{ g } SO_2 = \text{mol } SO_2 \times \text{FW } SO_2 = 1.35 \times 10^4 \text{ mol} \times 64.1 \text{ g/mol} = \mathbf{8.65 \times 10^5 \text{ g } SO_2}$

Method 2: Dimensional Analysis

$? \text{ g } SO_2 = 1.35 \times 10^3 \text{ kg } CaCO_3 \times \frac{1000 \text{ g } CaCO_3}{1 \text{ kg } CaCO_3} \times \underbrace{\frac{1 \text{ mol } CaCO_3}{100 \text{ g } CaCO_3}}_{\textbf{Step 1}} \times \underbrace{\frac{1 \text{ mol CaO}}{1 \text{ mol } CaCO_3}}_{\textbf{Step 2}}$

$\times \underbrace{\frac{1 \text{ mol } SO_2}{1 \text{ mol CaO}}}_{\textbf{Step 3}} \times \underbrace{\frac{64.1 \text{ g } SO_2}{1 \text{ mol } SO_2}}_{\textbf{Step 4}} = \mathbf{8.65 \times 10^5 \text{ g } SO_2}$

3-36. **Refer to Section 3-2.** . . . .

Balanced equation: $40H_2S + 16KMnO_4 + \ldots \rightarrow 5S_8 + \ldots$

Method 1: (a) Plan: $g\ S_8 \xrightarrow{(1)} mol\ S_8 \xrightarrow{(2)} mol\ H_2S \xrightarrow{(3)} g\ H_2O$

(b) Plan: $g\ S_8 \xrightarrow{(1)} mol\ S_8 \xrightarrow{(2)} mol\ KMnO_4 \xrightarrow{(3)} g\ KMnO_4$

(a) (1) $?\ mol\ S_8 = \dfrac{g\ S_8}{FW\ S_8} = \dfrac{1.426\ g}{256.5\ g/mol} = 5.560 \times 10^{-3}\ mol\ S_8$

(2) $?\ mol\ H_2S = 40/5 \times 5.560 \times 10^{-3}\ mol\ S_8 = 4.448 \times 10^{-2}\ mol\ H_2S$

(3) $?\ g\ H_2S = mol\ H_2S \times FW\ H_2S = 4.448 \times 10^{-2}\ mol \times 34.08\ g/mol$
$= \mathbf{1.516\ g\ H_2S}$

(b) (1) $?\ mol\ S_8 = \dfrac{g\ S_8}{FW\ S_8} = \dfrac{1.426\ g}{256.5\ g/mol} = 5.560 \times 10^{-3}\ mol\ S_8$

(2) $?\ mol\ KMnO_4 = 16/5 \times 5.560 \times 10^{-3}\ mol\ S_8 = 1.779 \times 10^{-2}\ mol\ KMnO_4$

(3) $?\ g\ KMnO_4 = mol\ KMnO_4 \times FW\ KMnO_4 = 1.779 \times 10^{-2}\ mol \times 158.0\ g/mol$
$= \mathbf{2.811\ g\ KMnO_4}$

Method 2: Dimensional Analysis

(a) $?\ g\ H_2S = 1.426\ g\ S_8 \times \underbrace{\dfrac{1\ mol\ S_8}{256.5\ g\ S_8}}_{\text{Step 1}} \times \underbrace{\dfrac{40\ mol\ H_2S}{5\ mol\ S_8}}_{\text{Step 2}} \times \underbrace{\dfrac{34.08\ g\ H_2S}{1\ mol\ H_2S}}_{\text{Step 3}}$

$= \mathbf{1.516\ g\ H_2S}$

(b) $?\ g\ KMnO_4 = 1.426\ g\ S_8 \times \underbrace{\dfrac{1\ mol\ S_8}{256.5\ g\ S_8}}_{\text{Step 1}} \times \underbrace{\dfrac{16\ mol\ KMnO_4}{5\ mol\ S_8}}_{\text{Step 2}} \times \underbrace{\dfrac{158.0\ g/mol}{1\ mol\ KMnO_4}}_{\text{Step 3}}$

$= \mathbf{2.811\ g\ KMnO_4}$

3-38. **Refer to Section 3-2 and Exercise 3-28 Solution.** . . . .

We see that in every 1 mole of $FeSO_4 \cdot 7H_2O$, there is 1 mole of $FeSO_4$.

(a) Plan: $g\ FeSO_4 \xrightarrow{(1)} mol\ FeSO_4 \overset{(2)}{=} mol\ FeSO_4 \cdot 7H_2O \xrightarrow{(3)} g\ FeSO_4 \cdot 7H_2O$

(b) Plan: $g\ FeSO_4 \cdot 7H_2O \xrightarrow{(1)} mol\ FeSO_4 \cdot 7H_2O \overset{(2)}{=} mol\ FeSO_4 \xrightarrow{(3)} g\ FeSO_4$

Calculations:

(a) (1) $?\ mol\ FeSO_4 = \dfrac{g\ FeSO_4}{FW\ FeSO_4} = \dfrac{91.2\ g}{152\ g/mol} = 0.600\ mol\ FeSO_4$

(2) $?\ mol\ FeSO_4 \cdot 7H_2O = mol\ FeSO_4 = 0.600\ mol\ FeSO_4 \cdot 7H_2O$

(3) $?\ g\ FeSO_4 \cdot 7H_2O = mol\ FeSO_4 \cdot 7H_2O \times FW\ FeSO_4 \cdot 7H_2O$

$= 0.600\ mol \times 278\ g/mol = \mathbf{167\ g\ FeSO_4 \cdot 7H_2O}$

(b) (1) $?\ mol\ FeSO_4 \cdot 7H_2O = \dfrac{g\ FeSO_4 \cdot 7H_2O}{FW\ FeSO_4 \cdot 7H_2O} = \dfrac{112\ g}{278\ g/mol} = 0.403\ mol\ FeSO_4 \cdot 7H_2O$

(2) $?\ mol\ FeSO_4 = mol\ FeSO_4 \cdot 7H_2O = 0.403\ mol\ FeSO_4$

(3) $?\ g\ FeSO_4 = mol\ FeSO_4 \times FW\ FeSO_4 = 0.403\ mol \times 152\ g/mol = \mathbf{61.3\ g\ FeSO_4}$

3-40. **Refer to Section 3-2 and Exercise 3-28 Solution.** . . . .

Balanced equation: $3Ba(OH)_2 + 2H_3PO_4 \rightarrow Ba_3(PO_4)_2 + 6H_2O$

(a) Plan: $g\ H_3PO_4 \xrightarrow{(1)} mol\ H_3PO_4 \xrightarrow{(2)} mol\ Ba_3(PO_4)_2 \xrightarrow{(3)} g\ Ba_3(PO_4)_2$

(b) Plan: $g\ Ba(OH)_2 \xrightarrow{(1)} mol\ Ba(OH)_2 \xrightarrow{(2)} mol\ Ba_3(PO_4)_2$

$\xrightarrow{(3)} g\ Ba_3(PO_4)_2$

Calculations:

(a) (1) $?\ mol\ H_3PO_4 = \dfrac{g\ H_3PO_4}{FW\ H_3PO_4} = \dfrac{39.2\ g}{98.0\ g/mol} = 0.400\ mol\ H_3PO_4$

(2) $?\ mol\ Ba_3(PO_4)_2 = 1/2 \times mol\ H_3PO_4 = 1/2 \times 0.400\ mol = 0.200\ mol\ Ba_3(PO_4)_2$

(3) $?\ g\ Ba_3(PO_4)_2 = mol\ Ba_3(PO_4)_2 \times FW\ Ba_3(PO_4)_2$
$= 0.200\ mol \times 602\ g/mol = \mathbf{1.20 \times 10^2\ g\ Ba_3(PO_4)_2}$

(b) (1) $?\ mol\ Ba(OH)_2 = \dfrac{g\ Ba(OH)_2}{FW\ Ba(OH)_2} = \dfrac{39.2\ g}{171\ g/mol} = 0.229\ mol\ Ba(OH)_2$

(2) $?\ mol\ Ba_3(PO_4)_2 = 1/3 \times mol\ Ba(OH)_2 = 1/3 \times 0.229\ mol = 0.0763\ mol\ Ba_3(PO_4)_2$

(3) $?\ g\ Ba_3(PO_4)_2 = mol\ Ba_3(PO_4)_2 \times FW\ Ba_3(PO_4)_2$
$= 0.0763\ mol \times 602\ g/mol = \mathbf{45.9\ g\ Ba_3(PO_4)_2}$

3-42. **Refer to Section 2-10.** . . . .

Step 1. Find the mass of Cr in 150 g of ore containing chromite, $FeCr_2O_4$.

Plan: $g\ ore \xrightarrow{(1)} g\ chromite \xrightarrow{(2)} mol\ chromite \xrightarrow{(3)} mol\ Cr \xrightarrow{(4)} g\ Cr$

(1) $?\ g\ chromite = g\ ore \times 0.670 = 150\ g \times 0.670 = 100.\ g\ chromite$

(2) $?\ mol\ chromite = \dfrac{g\ chromite}{FW\ chromite} = \dfrac{100.\ g}{224\ g/mol} = 0.446\ mol\ chromite$

(3) $?\ mol\ Cr = 2 \times mol\ chromite = 2 \times 0.446\ mol = 0.892\ mol\ Cr$

(4) $?\ g\ Cr = mol\ Cr \times AW\ Cr = 0.892\ mol \times 52.0\ g/mol = \mathbf{46.4\ g\ Cr}$

Step 2. The mass of Cr in 125 g ore can be obtained by a ratio.

$\dfrac{46.4\ g\ Cr}{150.\ g\ ore} = \dfrac{?\ g\ Cr}{125\ g\ ore}$ Solving, $?\ g\ Cr = 46.4 \times \dfrac{125\ g\ ore}{150.\ g\ ore} = 38.7\ g\ Cr$

Only 87.5% of the Cr can be recovered.
Therefore, ? g Cr recovered = 38.7 g Cr × 0.875 = **33.9 g Cr recovered**

3-44. **Refer to Section 3-4 and Example 3-8.** . . . .

Balanced equation: $3C_3H_3O \rightarrow C_9H_{12} + 3H_2O$

Step 1. Calculate the theoretical yield of C_9H_{12}.

Plan: $g\ C_3H_6O \xrightarrow{(1)} mol\ C_3H_6O \xrightarrow{(2)} mol\ C_9H_{12} \xrightarrow{(3)} g\ C_9H_{12}$

(1) $?\ mol\ C_3H_6O = \dfrac{g\ C_3H_6O}{FW\ C_3H_6O} = \dfrac{143\ g}{58.1\ g/mol} = 2.46\ mol\ C_3H_6O$

(2) $?\ mol\ C_9H_{12} = 1/3 \times mol\ C_3H_6O = 1/3 \times 2.46\ mol = 0.820\ mol\ C_9H_{12}$

(3) $?\ g\ C_9H_{12} = mol\ C_9H_{12} \times FW\ C_9H_{12} = 0.820\ mol \times 120\ g/mol = 98.4\ g\ C_9H_{12}$

Step 2. Solve for the percent yield of C_9H_{12}.

$$? \text{ \% yield} = \frac{\text{actual yield}}{\text{theoretical yield}} \times 100 = \frac{13.4 \text{ g}}{98.4 \text{ g}} \times 100 = \mathbf{13.6\%}$$

3-46. **Refer to Section 3-4.** . . . .

Balanced equation: $C_2H_5OBr + NaOH \rightarrow C_2H_4O + NaBr + H_2O$

Step 1. Solve for the theoretical yield of C_2H_4O.

$$\text{\% yield} = \frac{\text{actual yield}}{\text{theoretical yield}} \times 100 \qquad \text{Substituting, } 89\% = \frac{255 \text{ g}}{? \text{ theor. yield}}$$

$$? \text{ theor. yield} = \frac{255 \text{ g}}{89 \%} \times 100 = 290 \text{ g } C_2H_4O$$

This means that if we start with just enough C_2H_5OBr to theoretically prepare 290 g C_2H_4O, we would actually obtain 255 g C_2H_5O because the percentage yield is only 89%.

Step 2. Solve for the mass of C_2H_5OBr required to prepare 290 g C_2H_4O.

Plan: $\text{g } C_2H_4O \xrightarrow{(1)} \text{mol } C_2H_4O \xrightarrow{(2)} \text{mol } C_2H_5OBr \xrightarrow{(3)} \text{g } C_2H_5OBr$

(1) $? \text{ mol } C_2H_4O = \dfrac{\text{g } C_2H_4O}{\text{FW } C_2H_4O} = \dfrac{290 \text{ g}}{44.1 \text{ g/mol}} = 6.6 \text{ mol } C_2H_4O$

(2) $? \text{ mol } C_2H_5OBr = \text{mol } C_2H_4O = 6.6 \text{ mol } C_2H_5OBr$

(3) $? \text{ g } C_2H_5OBr = \text{mol } C_2H_5OBr \times \text{FW } C_2H_5OBr = 6.6 \text{ mol} \times 125 \text{ g/mol} = \mathbf{820 \text{ g } C_2H_5OBr}$

3-48. **Refer to Section 3-4.** . . . .

Balanced equation: $CaO + 3C \rightarrow CaC_2 + CO$

(a) Plan: (1) Calculate the masses of CaC_2 and CaO in the crude product.
(2) Use stoichiometry and mass balance to determine the initial mass of CaO.

(1) $? \text{ g } CaC_2 \text{ in product} = 0.85 \times \text{g product} = 0.85 \times (2.50 \times 10^5 \text{ g})$
$= 2.1 \times 10^5 \text{ g } CaC_2$

$? \text{ g CaO in product} = \text{g product} - \text{g } CaC_2$
$= 2.50 \times 10^5 \text{ g} - 2.1 \times 10^5 \text{ g} = 4 \times 10^4 \text{ g}$

(2) $? \text{ g CaO}_{\text{reacted}} = 2.1 \times 10^5 \text{ g } CaC_2 \times \dfrac{1 \text{ mol } CaC_2}{64 \text{ g } CaC_2} \times \dfrac{1 \text{ mol CaO}}{1 \text{ mol } CaC_2} \times \dfrac{56 \text{ g CaO}}{1 \text{ mol CaO}}$
$= 1.8 \times 10^5 \text{ g CaO}$

$? \text{ g CaO}_{\text{initial}} = \text{g CaO}_{\text{reacted}} + \text{g CaO}_{\text{unreacted}}$
$= 1.8 \times 10^5 \text{ g} + 4 \times 10^4 \text{ g} = \mathbf{2.2 \times 10^5 \text{ g or } 220 \text{ kg CaO}}$

(b) From (a), the mass of CaC_2 in the crude product is **2.1×10^5 or 210 kg.**

3-50. **Refer to Section 3-4.** . . . .

Method: Dimensional Analysis

$$? \text{ kg TiO}_2 = 1.00 \times 10^6 \text{ g ore} \times \frac{0.25 \text{ g TiO}_2}{100 \text{ g ore}} \times \frac{1 \text{ mol TiO}_2}{79.9 \text{ g TiO}_2} \times \frac{2 \text{ mol TiCl}_4}{2 \text{ mol TiO}_2} \times \frac{65}{100}$$

$$\times \frac{1 \text{ mol TiO}_2}{1 \text{ mol TiCl}_4} \times \frac{92}{100} \times \frac{79.9 \text{ g TiO}_2}{1 \text{ mol TiO}_2} \times \frac{1 \text{ kg}}{1000 \text{ g}}$$

$$= \mathbf{1.50 \text{ kg TiO}_2}$$

3-52. **Refer to Section 3-3 and Examples 3-6 and 3-7.** . . . .

Balanced equation: $N_2 + 3H_2 \rightarrow 2NH_3$

This is a limiting reagent problem.
Plan: (1) Find the limiting reagent.
(2) Do the stoichiometric problem based on amount of limiting reagent.

(1) Convert the mass of reactants to moles and compare the required ratio to the available ratio.

$$? \text{ mol N}_2 = \frac{\text{g N}_2}{\text{FW N}_2} = \frac{77.3 \text{ g}}{28.0 \text{ g/mol}} = 2.76 \text{ mol N}_2$$

$$? \text{ mol H}_2 = \frac{\text{g H}_2}{\text{FW H}_2} = \frac{14.2 \text{ g}}{2.02 \text{ g/mol}} = 7.03 \text{ mol H}_2$$

$$\text{Required ratio} = \frac{3 \text{ mol H}_2}{1 \text{ mol N}_2} = 3 \qquad \text{Available ratio} = \frac{7.03 \text{ mol H}_2}{2.76 \text{ mol N}_2} = 2.55$$

The available ratio < required ratio. Therefore, we do not have enough H_2 to react with all the N_2, and so H_2 is the limiting reagent.

(2) The amount of NH_3 is determined by the amount of limiting reagent, 7.03 moles of H_2.

$$? \text{ g NH}_3 = 7.03 \text{ mol H}_2 \times \frac{2 \text{ mol NH}_3}{3 \text{ mol H}_2} \times \frac{17.0 \text{ g NH}_3}{1 \text{ mol NH}_3} = \mathbf{79.7 \text{ g NH}_3}$$

3-54. **Refer to Section 3-3, Examples 3-6 and 3-7 and Exercise 3-52 Solution.** . . . .

Balanced equation: $3Ca(OH)_2 + 2H_3PO_4 \rightarrow Ca_3(PO_4)_2 + 6H_2O$

This is a limiting reagent problem.

(1) Convert to moles and compare required ratio to available ratio to find the limiting reagent.

$$? \text{ mol Ca(OH)}_2 = \frac{\text{g Ca(OH)}_2}{\text{FW Ca(OH)}_2} = \frac{7.4 \text{ g}}{74 \text{ g/mol}} = 0.10 \text{ mol Ca(OH)}_2$$

$$? \text{ mol H}_3\text{PO}_4 = \frac{\text{g H}_3\text{PO}_4}{\text{FW H}_3\text{PO}_4} = \frac{9.8 \text{ g}}{98 \text{ g/mol}} = 0.10 \text{ mol H}_3\text{PO}_4$$

$$\text{Required ratio} = \frac{3 \text{ mol Ca(OH)}_2}{2 \text{ mol H}_3\text{PO}_4} = 1.5$$

$$\text{Available ratio} = \frac{0.10 \text{ mol Ca(OH)}_2}{0.10 \text{ mol H}_3\text{PO}_4} = 1.0$$

Available ratio < required ratio, and our limiting reagent is $Ca(OH)_2$.

(2) The amount of $Ca_3(PO_4)_2$ formed is determined by 0.10 mol of $Ca(OH)_2$.

$$? \ Ca_3(PO_4)_2 = 0.10 \text{ mol } Ca(OH)_2 \times \frac{1 \text{ mol } Ca_3(PO_4)_2}{3 \text{ mol } Ca(OH)_2} \times \frac{310 \text{ g } Ca_3(PO_4)_2}{1 \text{ mol } Ca_3(PO_4)_2}$$

$$= \mathbf{10 \text{ g } Ca_3(PO_4)_2}$$

3-56. **Refer to Section 3-3, Examples 3-6 and 3-7 and Exercise 3-52 Solution.** . . . .

Balanced equation: $Ca_3(PO_4)_2 + 2H_2SO_4 \rightarrow Ca(H_2PO_4)_2 + 2CaSO_4$

This is a limiting reagent problem.

(1) Convert to moles and compare required ratio to available ratio to find the limiting reagent.

$$? \text{ mol } Ca_3(PO_4)_2 = \frac{\text{g } Ca_3(PO_4)_2}{\text{FW } Ca_3(PO_4)_2} = \frac{250 \text{ g}}{310 \text{ g/mol}} = 0.81 \text{ mol } Ca_3(PO_4)_2$$

$$? \text{ mol } H_2SO_4 = \frac{\text{g } H_2SO_4}{\text{FW } H_2SO_4} = \frac{150 \text{ g}}{98 \text{ g/mol}} = 1.5 \text{ mol } H_2SO_4$$

$$\text{Required ratio} = \frac{2 \text{ mol } H_2SO_4}{1 \text{ mol } Ca_3(PO_4)_2} = 2$$

$$\text{Available ratio} = \frac{1.5 \text{ mol } H_2SO_4}{0.81 \text{ mol } Ca_3(PO_4)_2} = 1.85$$

Available ratio < required ratio; H_2SO_4 is the limiting reagent.

(2) First we will find the mass of $Ca_3(PO_4)_2$ that reacted.

The Law of Conservation of Mass states that the mass of reactants that react equal the mass of products formed. Therefore, we can calculate the mass of superphosphate.

$$? \text{ g } Ca_3(PO_4)_2 = 150 \text{ g } H_2SO_4 \times \frac{1 \text{ mol } H_2SO_4}{98 \text{ g } H_2SO_4} \times \frac{1 \text{ mol } Ca_3(PO_4)_2}{2 \text{ mol } H_2SO_4} \times \frac{310 \text{ g } Ca_3(PO_4)_2}{1 \text{ mol } Ca_3(PO_4)_2} = 240 \text{ g } Ca_3(PO_4)_2$$

$$? \text{ g } [Ca_3(PO_4)_2 + 2CaSO_4] \text{ formed} = \text{g } [H_2SO_4 + Ca_3(PO_4)_2] \text{ reacted}$$

$$= 150 \text{ g } H_2SO_4 + 240 \text{ g } Ca_3(PO_4)_2$$

$$= \mathbf{390 \text{ g superphosphate}}$$

3-58. **Refer to Section 3-2 and Exercise 3-56 Solution.** . . . .

Balanced equation: $CS_2 + 3O_2 \rightarrow CO_2 + 2SO_2$

(1) Law of Conservation of Mass states that the mass of products equals the mass of reactants. If we determine the mass of O_2, then

$$\text{g } CS_2 + \text{g } O_2 = \text{mass of products}$$

$$? \text{ g } O_2 = 33.8 \text{ g } CS_2 \times \frac{1 \text{ mol } CS_2}{76.1 \text{ g } CS_2} \times \frac{3 \text{ mol } O_2}{1 \text{ mol } CS_2} \times \frac{32.0 \text{ g } O_2}{1 \text{ mol } O_2} = 42.6 \text{ g } O_2$$

Therefore, mass of products = g CS_2 + g O_2 = 33.8 g + 42.6 g = **76.4 g products**

(2) Use a mass ratio.

$$\frac{?\ g\ CS_2}{g\ products} = \frac{g\ CS_2\ (from\ Step\ 1)}{g\ products\ (from\ Step\ 1)}$$

$$?\ g\ CS_2 = g\ products \times \frac{g\ CS_2\ (from\ Step\ 1)}{g\ products\ (from\ Step\ 1)} = 54.2\ g \times \frac{33.8\ g}{76.4\ g} = \mathbf{24.0\ g\ CS_2}$$

3-60. **Refer to Sections 3-2 and 2-9.** . . . .

The chemical reaction occurring is $C_xH_yN \rightarrow NH_3 + \ldots$

Using a mass ratio:

$$\frac{g\ C_xH_yN}{g\ NH_3} = \frac{1 \times ?\ FW\ C_xH_yN}{1 \times FW\ NH_3}$$

Solving, $?\ FW\ C_xH_yN = FW\ NH_3 \times \frac{g\ C_xH_yN}{g\ NH_3}$

$$= 17.0\ g/mol \times \frac{150\ mg}{27.4\ mg}$$

$$= 93.1\ g/mol$$

The only compound listed with a formula weight of 93.1 g/mol is **(e) C_6H_7N.**

3-62. **Refer to Section 3-4.** . . . .

Balanced equations: $Ni + 4CO \rightarrow Ni(CO)_4$ and $2Co + 8CO \rightarrow Co_2(CO)_8$

From the data, we realize that 200.0 g of ore reacts with (300.0 - 165.0) g CO = 135.0 g CO. Also, the mass of $Ni(CO)_4$ produced is 15.0 g.

(a) Plan: g $Ni(CO)_4$ $\xrightarrow{(1)}$ mol $Ni(CO)_4$ $\xrightarrow{(2)}$ moles Ni $\xrightarrow{(3)}$ g Ni $\xrightarrow{(4)}$ % Ni

(1) $?\ mol\ Ni(CO)_4 = \frac{g\ Ni(CO)_4}{FW\ Ni(CO)_4} = \frac{15.0\ g}{171\ g/mol} = 0.0877\ mol\ Ni(CO)_4$

(2) $?\ mol\ Ni = mol\ Ni(CO)_4 = 0.0877\ mol\ Ni$

(3) $?\ g\ Ni = mol\ Ni \times AW\ Ni = 0.0877\ mol \times 58.7\ g/mol = 5.15\ g\ Ni$

(4) $?\ \%\ Ni = \frac{g\ Ni}{g\ ore} \times 100 = \frac{5.15\ g\ Ni}{200.0\ g\ ore} \times 100 = \mathbf{2.58\%\ Ni}$

(b) Plan: g $Ni(CO)_4$ $\xrightarrow{(1)}$ mol $Ni(CO)_4$ $\xrightarrow{(2)}$ mol CO $\xrightarrow{(3)}$ g CO react with Ni $\xrightarrow{(4)}$ g CO react with Co $\xrightarrow{(5)}$ mol Co $\xrightarrow{(6)}$ g mol Co $\xrightarrow{(7)}$ g Co $\xrightarrow{(8)}$ % Co

(1) $?\ mol\ Ni(CO)_4 = 0.0877\ mol\ Ni(CO)_4$ (see (1) above)

(2) $?\ mol\ CO = 4 \times mol\ Ni(CO)_4 = 4 \times 0.0877\ mol = 0.351\ mol\ CO$

(3) $?\ g\ CO = mol\ CO \times FW\ CO = 0.351\ mol \times 28.0\ g/mol = 9.83\ g\ CO$ react with Ni

(4) ? g CO react with Co = total CO reacted - g CO react with Ni
= 135.0 g - 9.8 g = 125.2 g CO react with Co

(5) $?\ mol\ CO = \frac{g\ CO}{FW\ CO} = \frac{125.2\ g}{28.0\ g/mol} = 4.47\ mol\ CO$

(6) ? mol Co = 2/8 × moles CO = 1/4 × 4.47 mol CO = 1.12 mol Co

(7) ? g Co = mol Co × AW Co = 1.12 mol × 58.9 g/mol = 66.0 g Co

(8) $? \% \text{ Co} = \frac{\text{g Co}}{\text{g ore}} \times 100 = \frac{66.0 \text{ g Co}}{200.0 \text{ g ore}} \times 100 =$ **33.0% Co**

3-64. **Refer to Section 3-2 and Fundamental Algebra.** . . . .

Balanced equations: $Na_2SO_4 + BaCl_2 \rightarrow 2NaCl + BaSO_4$

$K_2SO_4 + BaCl \rightarrow 2KCl + BaSO_4$

Plan: (1) Set up algebraic expressions for the number of moles of Na_2SO_4 and K_2SO_4.
(2) Determine the algebraic expression for the total mass of BaSO4 and solve.

(1) We know: g Na_2SO_4 + g K_2SO_4 = 1.188 g

Let ? g Na_2SO_4 = x ? g K_2SO_4 = 1.188 - x

$? \text{ mol } Na_2SO_4 = \frac{\text{g } Na_2SO_4}{\text{FW } Na_2SO_4} = \frac{x}{142.0 \text{ g/mol}}$ $? \text{ mol } K_2SO_4 = \frac{\text{g } K_2SO_4}{\text{FW } K_2SO_4} = \frac{1.188 - x}{174.3 \text{ g/mol}}$

(2) We can now calculate

$? \text{ mol } BaSO_4 \text{ from } Na_2SO_4 = \text{mol } Na_2SO_4 = \frac{x}{142.0 \text{ g/mol}}$

$? \text{ g } BaSO_4 \text{ from } Na_2SO_4 = \text{mol } BaSO_4 \times \text{FW } BaSO_4 = \frac{x}{142.0 \text{ g/mol}} \times 233.4 \text{ g/mol}$
$= 1.644\text{ x}$

$? \text{ mol } BaSO_4 \text{ from } K_2SO_4 = \text{mol } K_2SO_4 = \frac{1.188 - x}{174.3 \text{ g/mol}}$

$? \text{ g } BaSO_4 \text{ from } K_2SO_4 = \text{mol } BaSO_4 \times \text{FW } BaSO_4 = \frac{(1.188 - x)}{174.3 \text{ g/mol}} \times 233.4 \text{ g/mol}$
$= 1.591 - 1.339\text{ x}$

The total mass of $BaSO_4$ was 1.739 g, therefore,

g $BaSO_4$ from Na_2SO_4 + g $BaSO_4$ from K_2SO_4 = 1.739 g

1.644x + (1.591 - 1.339x) = 1.739

0.305x = 0.148

? g Na_2SO_4 = x = **0.485 g**

? g K_2SO_4 = 1.188 - x = **0.703 g**

3-66. **Refer to Section 3-2 and Fundamental Algebra.** . . . .

(a) If we examine the equations

(1) $CH_4 + 2O_2 \rightarrow CO_2 + 2H_2O$ and (2) $2CH_4 + 3O_2 \rightarrow 2CO + 4H_2O$

we see that for each equation, 2 moles H_2O are formed for every mole of CH_4. We also know that total mass of H_2O was 18.9 g.
Using ratios,

$$\frac{? \text{ total g } CH_4}{\text{total g } H_2O} = \frac{1 \times \text{FW } CH_4}{2 \times \text{FW } H_2O}$$

Solving,

$$? \text{ total g } CH_4 = \text{total g } H_2O \times \frac{1 \times FW\ CH_4}{2 \times FW\ H_2O} = 18.9 \text{ g} \times \frac{1 \times 16.0 \text{ g}}{2 \times 18.0 \text{ g}} = \mathbf{8.40 \text{ g } CH_4}$$

(b) Plan: (1) Set up the algebraic expressions for masses of CO and CO_2.

(2) Determine the algebraic expression for the mass of CH_4 needed to make CO and CO_2, found in (a), then solve.

(1) We know: $\text{g CO} + \text{g } CO_2 + \text{g } H_2O = 37.2 \text{ g}$ and $\text{g } H_2O = 18.9 \text{ g}$

Therefore, $\text{g CO} + \text{g } CO_2 = 37.2 \text{ g} - 18.9 \text{ g} = 18.3 \text{ g}$

let $? \text{ g CO} = x$ $\qquad$ $? \text{ g } CO_2 = 18.3 - x$

$$? \text{ mol CO} = \frac{\text{g CO}}{FW\ CO} = \frac{x}{28.0 \text{ g/mol}} \qquad ? \text{ mol } CO_2 = \frac{\text{g } CO_2}{FW\ CO_2} = \frac{18.3 - x}{44.0 \text{ g/mol}}$$

(2) Therefore,

$$? \text{ mol } CH_4 \text{ forming CO} = \text{mol CO} = \frac{x}{28.0 \text{ g/mol}}$$

$$? \text{ g } CH_4 \text{ forming CO} = \text{mol } CH_4 \times FW\ CH_4 = \frac{x}{28.0 \text{ g/mol}} \times 16.0 \text{ g/mol} = 0.571x$$

$$? \text{ mol } CH_4 \text{ forming } CO_2 = \text{mol } CO_2 = \frac{18.3 - x}{44.0 \text{ g/mol}}$$

$$? \text{ g } CH_4 \text{ forming } CO_2 = \text{mol } CH_4 \times FW\ CH_4 = \frac{18.3 - x}{44.0 \text{ g/mol}} \times 16.0 \text{ g/mol}$$
$$= 6.65 - 0.364x$$

$$\text{Total mass of } CH_4 = 8.40 \text{ g} = \text{g } CH_4 \text{ forming CO} + \text{g } CH_4 \text{ forming } CO_2$$
$$= 0.571x + (6.65 - 0.364x)$$
$$1.75 = 0.207x$$
$$? \text{ g CO} = x = \mathbf{8.45 \text{ g CO}}$$

(c) $? \text{ g } CO_2 = 18.3 - x = \mathbf{9.8 \text{ g } CO_2}$

(d) Recall that in any reaction, the mass of products equals the mass of reactants due to the Law of Conservation of Mass.

Therefore, $\text{g } CH_4 + ? \text{ g } O_2 = \text{g } CO_2 + \text{g CO} + \text{g } H_2O$

$8.40 \text{ g } CH_4 + ? \text{ g } O_2 = 37.2 \text{ g}$ (given in exercise)

$? \text{ g } O_2 = 37.2 \text{ g} - 8.40 \text{ g} = \mathbf{28.8 \text{ g } O_2}$

3-68. **Refer to Section 3-5 and Example 3-9.** . . . .

Method 1: $\% \text{ by mass} = \frac{\text{g solute}}{\text{g soln}} \times 100$ $\qquad$ g soln = g solute + g solvent

$$? \text{ g } K_2Cr_2O_7 = \frac{\% \text{ by mass}}{100} \times \text{g soln} = \frac{8.65\%}{100} \times 400 \text{ g} = \mathbf{34.6 \text{ g } K_2Cr_2O_7}$$

$$? \text{ g } H_2O = \text{g soln} - \text{g } K_2Cr_2O_4 = 400 \text{ g} - 35 \text{ g} = \mathbf{365 \text{ g } H_2O}$$

Method 2: Dimensional Analysis

$$? \text{ g } K_2Cr_2O_4 = 400 \text{ g soln} \times \frac{8.65 \text{ g } K_2Cr_2O_4}{100 \text{ g soln}} = \mathbf{34.6 \text{ g } K_2Cr_2O_4}$$

$$? \text{ g } H_2O = \text{g soln} - \text{g } K_2Cr_2O_4 = 400 \text{ g} - 35 \text{ g} = \mathbf{365 \text{ g } H_2O}$$

3-70. **Refer to Section 3-5.** . . . .

If an 8.30% NH_4Cl solution contains 8.30 g NH_4Cl per 100.00 g soln then it also contains (100.00 g - 8.30 g) = 91.70 g H_2O per 100.00 g soln.

$$? \text{ g soln} = 100 \text{ g } H_2O \times \frac{100.0 \text{ g soln}}{91.7 \text{ g } H_2O} = \mathbf{109 \text{ g soln}}$$

$$? \text{ g } NH_4Cl = \text{g soln} - \text{g } H_2O = 109 \text{ g} - 100 \text{ g} = \mathbf{9 \text{ g } NH_4Cl}$$

3-72. **Refer to Section 3-5 and Example 3-11.** . . . .

We know $D\left(\frac{g}{mL}\right) = \frac{\text{g soln}}{\text{mL soln}}$ and % by mass $= \frac{\text{g } (NH_4)_2SO_4}{\text{g soln}} \times 100$

$$? \text{ g } (NH_4)_2SO_4 = 350 \text{ mL soln} \times \frac{1.10 \text{ g soln}}{1.00 \text{ mL soln}} \times \frac{18.0 \text{ g } (NH_4)_2SO_4}{100 \text{ g soln}} = \mathbf{69.3 \text{ g } (NH_4)_2SO_4}$$

3-74. **Refer to Section 3-5 amd Exercise 3-72.** . . . .

$$? \text{ mL soln} = 80.0 \text{ g } (NH_4)_2SO_4 \times \frac{100 \text{ g soln}}{18.0 \text{ g } (NH_4)_2SO_4} \times \frac{1.00 \text{ mL soln}}{1.10 \text{ g soln}} = \mathbf{404 \text{ mL soln}}$$

3-76. **Refer to Section 3-5 and Example 3-13.** . . . .

Method 1: Use the units of molarity as an equation: $M\left(\frac{\text{mol}}{\text{L}}\right) = \frac{\text{mol substance}}{\text{L soln}}$

Plan: $\text{g } H_3PO_4 \xrightarrow{(1)} \text{mol } H_3PO_4 \xrightarrow{(2)} M\ H_3PO_4$

$$(1) \quad ? \text{ mol } H_3PO_4 = \frac{\text{g } H_3PO_4}{\text{FW } H_3PO_4} = \frac{490 \text{ g}}{98.0 \text{ g/mol}} = 5.00 \text{ mol } H_3PO_4$$

$$(2) \quad ? \ M\ H_3PO_4 = \frac{\text{mol } H_3PO_4}{\text{L soln}} = \frac{5.00 \text{ mol}}{2.00 \text{ L}} = \mathbf{2.50}\ M\ \mathbf{H_3PO_4}$$

Method 2: Dimensional Analysis

$$? \ M\ H_3PO_4 = \frac{490 \text{ g } H_3PO_4}{2.00 \text{ L soln}} \times \frac{1 \text{ mol } H_3PO_4}{98.0 \text{ g } H_3PO_4} = \mathbf{2.50}\ M\ \mathbf{H_3PO_4}$$

3-78. **Refer to Section 3-5, Example 3-13 and Exercise 3-76 Solution.** . . . . . .

Method 1:

Plan: $\text{g } BaCl_2\cdot 2H_2O \xrightarrow{(1)} \text{mol } BaCl_2\cdot 2H_2O \overset{(2)}{=} \text{mol } BaCl_2 \xrightarrow{(3)} M\ BaCl_2$

$$(1) \quad ? \text{ mol } BaCl_2\cdot 2H_2O = \frac{\text{g } BaCl_2\cdot 2H_2O}{\text{FW } BaCl_2\cdot 2H_2O} = \frac{3.50 \text{ g}}{244 \text{ g/mol}} = 0.0143 \text{ mol } BaCl_2\cdot 2H_2O$$

$$(2) \quad ? \text{ mol } BaCl_2 = \text{mol } BaCl_2\cdot 2H_2O = 0.0143 \text{ mol } BaCl_2$$

$$(3) \quad ? \ M\ BaCl_2 = \frac{\text{mol } BaCl_2}{\text{L soln}} = \frac{0.0143 \text{ mol}}{0.500 \text{ L}} = \mathbf{0.0286}\ M\ \mathbf{BaCl_2}$$

Method 2: Dimensional Analysis

$$? \ M \ BaCl_2 = \frac{3.50 \text{ g } BaCl_2 \cdot 2H_2O}{500 \text{ mL soln}} \times \frac{1000 \text{ mL}}{1 \text{ L}} \times \frac{1 \text{ mol } BaCl_2 \cdot 2H_2O}{244 \text{ g } BaCl_2 \cdot 2H_2O} \times \frac{1 \text{ mol } BaCl_2}{1 \text{ mol } BaCl_2 \cdot 2H_2O}$$

$$= \mathbf{0.0287} \ M \ \mathbf{BaCl_2}$$

3-80. **Refer to Section 3-5 and Example 3-14.**

Method 1: Plan: M, L $ZnCl_2$ soln $\xrightarrow{(1)}$ mol $ZnCl_2$ $\xrightarrow{(2)}$ g $ZnCl_2$

We know $M \ ZnCl_2 = \frac{\text{mol } ZnCl_2}{\text{L soln}}$

(1) ? mol $ZnCl_2$ = M $ZnCl_2$ × L soln = 0.450 M × 5.00 L = 2.25 mol $ZnCl_2$

(2) ? g $ZnCl_2$ = mol $ZnCl_2$ × FW $ZnCl_2$ = 2.25 mol × 136 g/mol = **306 g $ZnCl_2$**

Method 2: Dimensional Analysis

$$? \text{ g } ZnCl_2 = 5.00 \text{ L soln} \times \overbrace{\frac{0.450 \text{ mol } ZnCl_2}{1 \text{ L}}}^{\textbf{Step 1}} \times \overbrace{\frac{136 \text{ g}}{1 \text{ mol } ZnCl_2}}^{\textbf{Step 2}} = \mathbf{306 \text{ g } ZnCl_2}$$

3-82. **Refer to Section 3-5 and Exercise 3-80 Solution.**

Method 1: Plan: g Na_2CO_3 $\xrightarrow{(1)}$ mol Na_2CO_3 $\xrightarrow[\div M]{(2)}$ L soln

(1) $? \text{ mol } Na_2CO_3 = \frac{\text{g } Na_2CO_3}{\text{FW } Na_2CO_3} = \frac{16.4 \text{ g}}{106 \text{ g/mol}} = 0.155 \text{ mol } Na_2CO_3$

(2) $? \text{ L soln} = \frac{\text{mol } Na_2CO_3}{M \ Na_2CO_3} = \frac{0.155 \text{ mol}}{0.320 \text{ mol/L}} = \mathbf{0.484 \text{ L}}$ since $M = \frac{\text{mol } Na_2CO_3}{\text{L soln}}$

Method 2: Dimensional Analysis

$$? \text{ L soln} = 16.4 \text{ g } Na_2CO_3 \times \frac{1 \text{ mol } Na_2CO_3}{106 \text{ g } Na_2CO_3} \times \frac{1 \text{ L soln}}{0.320 \text{ mol } Na_2CO_3} = \mathbf{0.483 \text{ L soln}}$$

3-84. **Refer to Section 3-7 and Example 3-18.**

Balanced equation: $2HNO_3 + Ca(OH)_2 \rightarrow Ca(NO_3)_2 + 2H_2O$

Method 1: Plan: g $Ca(OH)_2$ $\xrightarrow{(1)}$ mol $Ca(OH)_2$ $\xrightarrow{(2)}$ mol HNO_3 $\xrightarrow[\div M]{(3)}$ L HNO_3 soln

(1) $? \text{ mol } Ca(OH)_2 = \frac{\text{g } Ca(OH)_2}{\text{FW } Ca(OH)_2} = \frac{12.61 \text{ g}}{74.09 \text{ g/mol}} = 0.1702 \text{ mol } Ca(OH)_2$

(2) ? mol HNO_3 = 2 × mol $Ca(OH)_2$ = 2 × 0.1702 mol = 0.3404 mol HNO_3

(3) $? \text{ L } HNO_3 \text{ soln} = \frac{\text{mol } HNO_3}{M \ HNO_3} = \frac{0.3404 \text{ mol}}{0.185 \ M} = \mathbf{1.84 \text{ L } HNO_3 \text{ soln}}$ since $M = \frac{\text{mol}}{\text{L}}$

Method 2: Dimensional Analysis

$$? \text{ L } HNO_3 \text{ soln} = 12.61 \text{ g } Ca(OH)_2 \times \overbrace{\frac{1 \text{ mol } Ca(OH)_2}{74.09 \text{ g } Ca(OH)_2}}^{\textbf{Step 1}} \times \overbrace{\frac{2 \text{ mol } HNO_3}{1 \text{ mol } Ca(OH)_2}}^{\textbf{Step 2}}$$

$$\times \overbrace{\frac{1 \text{ L } HNO_3 \text{ soln}}{0.185 \text{ mol } HNO_3}}^{\textbf{Step 3}} = \mathbf{1.84 \text{ L } HNO_3 \text{ soln}}$$

3-86. Refer to Section 3-7 and Example 3-18. . . . .

Balanced equation: $3AgNO_3 + K_3PO_4 \rightarrow Ag_3PO_4 + 3KNO_3$

$$? \text{ L } AgNO_3 \text{ soln} = 18.6 \text{ g } K_3PO_4 \times \frac{1 \text{ mol } K_3PO_4}{212 \text{ g } K_3PO_4} \times \frac{3 \text{ mol } AgNO_3}{1 \text{ mol } K_3PO_4} \times \frac{1 \text{ L } AgNO_3 \text{ soln}}{0.250 \text{ mol } AgNO_3}$$

$$= \mathbf{1.05 \text{ L } AgNO_3 \text{ soln}}$$

3-88. Refer to Section 3-5 and Example 3-15. . . . .

(a) $$\% \text{ by mass} = \frac{\text{g } CaCl_2}{\text{g soln}} \times 100 = \frac{18.0 \text{ g } CaCl_2}{72.0 \text{ g } H_2O + 18.0 \text{ g } CaCl_2} \times 100 = \mathbf{20.0\% \ CaCl_2}$$

(b) Assume we have 1 liter of solution

Plan: 1 L soln $\xrightarrow{(1)}$ g soln in 1 L $\xrightarrow{(2)}$ g $CaCl_2$ in 1 L $\xrightarrow{(3)}$ mol $CaCl_2$ in 1 L

(1) $$? \text{ g soln in 1 L} = 1000 \text{ mL} \times \frac{1.180 \text{ g}}{\text{mL}} = 1180 \text{ g soln}$$

(2) $$? \text{ g } CaCl_2 \text{ in 1 L} = 1180 \text{ g soln} \times 0.200 \% \ CaCl_2 = 236 \text{ g } CaCl_2$$

(3) $$? \text{ mol } CaCl_2 \text{ in 1 L} = \frac{236 \text{ g } CaCl_2 \text{ in 1 L soln}}{111 \text{ g/mol}} = 2.13 \text{ mol } CaCl_2 \text{ in 1 L soln}$$

$$= \mathbf{2.13 \ \mathit{M} \ CaCl_2}$$

3-90. Refer to Section 3-5, Example 3-15 and Exercise 3-88 Solution. . . . . . .

$$? \ M \text{ HF} = \frac{49.0 \text{ g HF}}{100 \text{ g soln}} \times \frac{1 \text{ mol HF}}{20.0 \text{ g HF}} \times \frac{1.17 \text{ g}}{1 \text{ mL}} \times \frac{1000 \text{ mL}}{1 \text{ L}} = \mathbf{28.7 \ \mathit{M} \ HF}$$

3-92. Refer to Section 3-7 and Example 3-19. . . . .

The balanced equation: $10FeSO_4 + 2KMnO_4 + \ldots \rightarrow \ldots$

Method 1:

Plan: M, mL $KMnO_4$ $\xrightarrow{(1)}$ mmol $KMnO_4$ $\xrightarrow{(2)}$ mmol $FeSO_4$ $\xrightarrow[\div M]{(3)}$ mL $FeSO_4$ soln

(1) $? \text{ mmol } KMnO_4 = M \ KMnO_4 \times \text{mL } KMnO_4 \text{ soln} = 0.250 \ M \times 20.0 \text{ mL} = 5.00 \text{ mmol } KMnO_4$

(2) $? \text{ mmol } FeSO_4 = 10/2 \times \text{mmol } KMnO_4 = 5 \times 5.00 \text{ mmol} = 25.0 \text{ mmol } FeSO_4$

(3) $$? \text{ mL } FeSO_4 \text{ soln} = \frac{\text{mmol } FeSO_4}{\text{M } FeSO_4} = \frac{25.0 \text{ mmol}}{0.200 \text{ M}} = \mathbf{125 \text{ mL } FeSO_4 \text{ soln}} \ (M = \frac{\text{mmol}}{\text{mL}})$$

Method 2: Dimensional Analysis

$$? \text{ L } FeSO_4 \text{ soln} = 20.0 \text{ mL } KMnO_4 \text{ soln} \times \frac{0.250 \text{ mmol } KMnO_4}{1 \text{ mL } KMnO_4 \text{ soln}} \times \frac{10 \text{ mmol } FeSO_4}{2 \text{ mmol } KMnO_4}$$

$$\times \frac{1 \text{ mL } FeSO_4 \text{ soln}}{0.200 \text{ mmol } FeSO_4} = \mathbf{125 \text{ mL } FeSO_4 \text{ soln}}$$

3-94. Refer to Section 3-6 and Example 3-16.

For a dilution problem, $M_1 \times V_1 = M_2 \times V_2$

Therefore, $V_1 = \frac{M_2 \times V_2}{M_1} = \frac{2.40\ M \times 3.50\ \text{L}}{12.0\ M} = 0.700\ \text{L conc. HCl soln}$

3-96. Refer to Section 3-6.

We know that $M = \frac{\text{mol Ba(OH)}_2}{\text{L soln}}$ and $\text{mol Ba(OH)}_2 = M \times \text{L soln}$

Therefore, mol $Ba(OH)_2$ in solution 1 = moles $Ba(OH)_2$ in solution 2

$$M_1 \times V_1 = M_2 \times V_2$$

$$V_1 = \frac{M_2 \times V_2}{M_1} = \frac{0.0800\ M \times 120\ \text{mL}}{0.0500\ M} = 192\ \text{mL}$$

3-98. Refer to Section 3-6.

Plan: Find the total number of moles of H_2SO_4 and divide by the total volume in liters to calculate molarity. Let V = volume of solution in liters.

$? \text{ mol } H_2SO_4 \text{ in solution 1} = M_1 \times V_1 = 6.00\ M \times 0.125\ \text{L} = 0.750 \text{ mol } H_2SO_4$

$? \text{ mol } H_2SO_4 \text{ in solution 2} = M_2 \times V_2 = 3.00\ M \times 0.225\ \text{L} = 0.675 \text{ mol } H_2SO_4$

$? \text{ total L } H_2SO_4 \text{ soln} = (125\ \text{mL} + 225\ \text{mL}) \times \frac{1\ \text{L}}{1000\ \text{mL}} = 0.350\ \text{L}$

$? \ M\ H_2SO_4 = \frac{\text{mol } H_2SO_4}{\text{L } H_2SO_4 \text{ soln}} = \frac{(0.750 + 0.675)\ \text{mol}}{0.350\ \text{L}} = 4.07\ M\ H_2SO_4$

4 The Structure of Atoms

4-2. **Refer to Section 4-2 and Figure 4-1.** . . . .

The cathode ray experiment involves the passage of a high voltage electric current through low pressure gases in a cathode ray tube in which two electrodes are sealed. When a high voltage is applied, current flows and rays, found to be negatively charged particles, travel from the cathode to the anode. The area in the tube glows where these rays strike. The rays can be deflected by both magnetic and electric fields.

4-4. **Refer to Section 4-2.** . . . .

The charge-to-mass ratio was determined for cathode rays (electrons) by J.J. Thomson in 1897 by measuring the degree of deflection of cathode rays in different electric and magnetic fields. In 1909, Robert Millikan determined the charge on the electron by performing the "oil drop" experiment, often recreated in undergraduate physics laboratories (see Figure 4-2). Knowing the values for the charge-to-mass ratio and the charge of an electron, the mass was calculated.

4-6. **Refer to Section 4-2 and Figure 4-2.** . . . .

If any oil droplets in Millikan's oil drop experiment had possessed a deficiency of electrons, the droplets would have been positively charged and would have been attracted to, not repelled by, the negatively charged plate. There would have been no voltage setting where the electrical and gravitational forces on the drop would have balanced.

4-8. **Refer to Section 4-3.** . . . .

Canal rays, also produced in the cathode ray tube, move toward the cathode. (The negatiave electrode). Therefore, they must be positively charged. Canal rays are positively charged ions created when cathode rays knock electrons from the gaseous atoms in the tube.

4-10. **Refer to Sections 4-2 and 4-3.** . . . .

(a) We must modify the Millikan oil drop experiment in order to determine the charge-to-mass ratio of the positively charged whizatron by (1) in some way producing an excess of whizatrons on the oil droplets and (2) switching the leads to the plates to make the bottom plate positively charged. The positively charged whizatrons on the oil droplets will be repulsed by the plate.

(b) Since all charges on the droplets will be integral multiples of the charge on the whizatron, we will identify the droplet with the smallest charge and test to see if the other droplets have charges that are multiples of its charge.

Charge on Droplets (coulombs)	Ratio
6.20×10^{-20}	$\frac{6.20 \times 10^{-20}}{6.20 \times 10^{-20}} = 1$
2.17×10^{-19}	$\frac{2.17 \times 10^{-19}}{6.20 \times 10^{-20}} = 3.5$
1.55×10^{-19}	$\frac{1.55 \times 10^{-19}}{6.20 \times 10^{-20}} = 2.5$
1.24×10^{-19}	$\frac{1.24 \times 10^{-19}}{6.20 \times 10^{-20}} = 2.0$
3.10×10^{-19}	$\frac{3.10 \times 10^{-19}}{6.20 \times 10^{-20}} = 5.0$

From these results, we can deduce that the charge on the whizatron is 1/2 of the smallest observed charge: $1/2 \times 6.20 \times 10^{-20} = 3.10 \times 10^{-20}$ coulombs. All the droplets have charges that are integral multiples of **3.10×10^{-20} coulombs**.

4-12. **Refer to Section 4-4.** . . . .

The nuclear atom is composed of mostly empty space with a tiny, positively charged massive center called the atomic nucleus. The nucleus contains an integral number of protons which is equal to the number of electrons in the neutral atom.

4-14. **Refer to Section 4-4.** . . . .

Alpha particles, which are actually the positively charged nuclei of helium atoms, are much more dense than gold. Hence, it was expected that these particles would pass easily through the gold foil with little deflection. The fact that some particles were greatly deflected astounded Rutherford.

4-16. **Refer to Section 4-6.** . . . .

Calculation of the charge-to-mass ratio:

Species	Charge	Mass Number	Charge-to-Mass Ratio
$^{16}O^{+}$	+1	16	1/16 = 0.0625
$^{16}O^{2+}$	+2	16	2/16 = 0.125
$^{17}O^{+}$	+1	17	1/17 = 0.0588
$^{17}O^{2+}$	+2	17	2/17 = 0.118

The order of increasing charge-to-mass ratios is $^{17}O^{+} < {}^{16}O^{+} < {}^{17}O^{2+} < {}^{16}O^{2+}$.

4-18. **Refer to Section 4-6 and Table 4-1.** . . . .

A neutral atom of $^{195}_{78}Pt$ contains 78 electrons, 78 protons and (195-78) = 117 neutrons.

If we assume that the mass of the atom is simply the sum of the masses of its subatomic particles, then

$$\text{mass of } ^{195}\text{Pt} = 78\ e^- \times \text{mass } e^- + 78\ p \times \text{mass } p + 117\ n \times \text{mass } n$$
$$= 78 \times 0.00054858 \text{ amu} + 78 \times 1.0073 \text{ amu} + 117 \times 1.0087 \text{ amu}$$
$$= 196.63 \text{ amu/atom}$$

(a) $\%\text{ by mass } e^- = \dfrac{\text{mass } e^-}{\text{mass Pt}} \times 100 = \dfrac{78\ e^- \times 0.00054858 \text{ amu}/e^-}{196.63 \text{ amu}} \times 100 = \mathbf{0.021761\ \%}$

(b) $\%\text{ by mass } p = \dfrac{\text{mass } p}{\text{mass Pt}} \times 100 = \dfrac{78\ p \times 1.0073 \text{ amu}/p}{196.63 \text{ amu}} \times 100 = \mathbf{39.958\ \%}$

(c) $\%\text{ by mass } n = \dfrac{\text{mass } n}{\text{mass Pt}} \times 100 = \dfrac{117\ n \times 1.0087 \text{ amu}/n}{196.63 \text{ amu}} \times 100 = \mathbf{60.020\ \%}$

4-20. **Refer to Section 4-6, Exercises 4-18 and 4-19.** . . . .

The density of a platinum atom, $D = \dfrac{\text{g Pt atom}}{\text{cm}^3 \text{ Pt atom}}$

To estimate the density of a Pt atom, we must

(1) find the mass of a Pt atom in grams
(2) find the volume of a Pt atom in cubic centimeters
(3) divide the mass by the volume

(1) From Exercise 4-18, 1 atom ^{195}Pt has a mass of 196.63 amu. Then 1 mol ^{195}Pt has a mass of 196.63 g.

$$?\text{ g Pt atom} = \frac{196.63 \text{ g}}{1 \text{ mol Pt}} \times \frac{1 \text{ mol Pt}}{6.022 \times 10^{23} \text{ atoms}} = 3.265 \times 10^{-22} \text{ g Pt}$$

(2) Assuming that a Pt atom is spherical, volume $V = (4/3)\pi r^3$.

$$?\text{ radius of Pt} = 1.38 \text{ Å} \times \frac{10^{-10} \text{ m}}{1 \text{ Å}} \times \frac{100 \text{ cm}}{1 \text{ m}} = 1.38 \times 10^{-8} \text{ cm}$$

$$?\ V = (4/3)\pi r^3 = (4/3)\pi(1.38 \times 10^{-8} \text{ cm})^3 = 1.10 \times 10^{-23} \text{ cm}^3$$

(3) $?\ D \text{ of Pt atom} = \dfrac{\text{g Pt atom}}{\text{cm}^3 \text{ Pt atom}} = \dfrac{3.265 \times 10^{-22} \text{ g}}{1.10 \times 10^{-23} \text{ cm}^3} = \mathbf{29.7\ g/cm^3}$

4-22. **Refer to the Key Terms for Chapter 4.** . . . .

(a) The atomic number of an element is the integral number of protons in the nucleus. It defines the identity of that element. For example, oxygen has an atomic number of 8 and therefore has 8 protons. All oxygen atoms have exactly 8 protons and there is no other element that has 8 protons in its nucleus (**Section 4-4**).

(b) Isotopes are two or more forms of atoms of the same element with different masses. In other words, they are atoms containing the same number of protons but different numbers of neutrons.
^{16}O and ^{17}O are isotopes; both have 8 protons but ^{16}O has (16 - 8) = 8 neutrons while ^{17}O has (17 - 8) = 9 neutrons (**Section 4-6**).

(c) The mass number of an element is the integral sum of the numbers of protons and neutrons in that atom.

The mass number of ^{17}O is 17, the sum of protons and neutrons in the nucleus (**Section 4-6**).

(d) Nuclear charge refers to the number of protons or positive charges in the nucleus. The nuclear charge of an oxygen atom is +8.

4-24. **Refer to Section 4-6.** . . . .

The isotopes of a given element differ in the number of neutrons present in their nuclei, hence the mass numbers of the isotopes differ as well. Consider the major isotopes of uranium:

Uranium Isotopes	number of protons	number of neutrons	mass number
^{238}U	92	146	238
^{235}U	92	143	235
^{234}U	92	142	234

4-26. **Refer to Section 4-6.** . . . .

In order to have similar chemical properties, substances must have similar atomic structures.

(a) H is a single atom; H_2 is a diatomic molecule. These will have dissimilar chemical properties.

(b) H^+ is a hydrogen nucleus since its one electron has been lost. H^- is a hydrogen atom with an extra electron. The charges on the atoms are equal but opposite. They would have very different chemical properties.

(c) ^{12}C and ^{13}C are isotopes of carbon and have different numbers of neutrons in their nuclei. This pair will have similar chemical properties since they are the same element in the same chemical form with the same number of electrons. Their chemical properties will not be exactly identical due to the slight differences in mass; this is called an isotope effect.

4-28. **Refer to Section 4-6, Table 4-2 and the Periodic Table on the inside front cover of the textbook.** . . . .

From the Periodic Table, we see that the atomic number of magnesium is 12. Therefore, each magnesium atom has 12 protons. If it is a neutral atom then it also has 12 electrons. If we assume that these isotopes are neutral, then

Isotope	Number of Protons	Number of Electrons	Number of Neutrons (Mass Number - Atomic Number)
$^{24}_{12}Mg$	12	12	12 (= 24 - 12)
$^{25}_{12}Mg$	12	12	13 (= 25 - 12)
$^{26}_{12}Mg$	12	12	14 (= 26 - 12)

4-30. **Refer to Section 4-6.** . . . .

Remember:

atomic number = number of protons = number of electrons in a neutral atom

mass number = number of protons + number of neutrons

Kind of Atom	Atomic Number	Mass Number	Isotope	Number of Protons	Number of Electrons	Number of Neutrons
manganese	25	55	$^{55}_{25}Mn$	25	25	30
lithium	3	7	$^{7}_{3}Li$	3	3	4
iodine	53	127	$^{127}_{53}I$	53	53	74
rhodium	45	104	$^{104}_{45}Rh$	45	45	59

4-32. **Refer to Section 4-8.** . . . .

The atomic weight of oxygen on the carbon-12 scale is 15.9994 amu compared to exactly 16 on the older scale. Therefore, the atomic weight of cadmium on the older scale should also be slightly higher. We can set up a ratio:

$$\frac{\text{? AW Cd (older scale)}}{\text{AW Cd (C-12 scale)}} = \frac{\text{AW O (older scale)}}{\text{AW O (C-12 scale)}}$$

Solving, $\text{? AW Cd (older scale)} = \text{AW Cd (C-12 scale)} \times \frac{\text{AW O (older scale)}}{\text{AW O (C-12 scale)}}$

$$= 112.411 \text{ amu} \times \frac{16.0000 \text{ amu}}{15.9994 \text{ amu}}$$

= **112.415 amu** (to 6 significant figures)

4-34. **Refer to Section 4-8 and Example 4-1.** . . . .

Plan: (1) For each isotope, convert % abundance to a fraction.
(2) Multiply the fraction of each isotope by its mass and add together to obtain the atomic weight of Fe.

? AW Fe = (mass ^{54}Fe × fraction of ^{54}Fe) + (mass ^{56}Fe × fraction of ^{56}Fe) + (mass ^{57}Fe × fraction of ^{57}Fe) + (mass ^{58}Fe × fraction of ^{58}Fe)

$$\text{? AW Fe} = (53.9396 \text{ amu} \times \frac{5.82}{100}) + (55.9349 \text{ amu} \times \frac{91.66}{100}) + (56.9354 \text{ amu} \times \frac{2.19}{100}) + (57.9333 \text{ amu} \times \frac{0.33}{100})$$

= 3.14 amu + 51.27 amu + 1.25 amu + 0.19 amu

= **55.85 amu**

4-36. **Refer to Section 4-8 and Example 4-2.** . . . .

We know: AW Cu = (mass ^{63}Cu × fraction of ^{63}Cu) + (mass ^{65}Cu × fraction of ^{65}Cu)

let x = fraction of ^{63}Cu and (1 - x) = fraction of ^{65}Cu

Substituting,

$$63.546 \text{ amu} = (62.9298 \text{ amu})x + (64.9278 \text{ amu})(1 - x)$$
$$= 62.9298x + 64.9278 - 64.9278x$$
$$1.998x = 1.382$$

fraction of ^{63}Cu = x = 0.6917 % abundance of ^{63}Cu = 0.6917 × 100 = **69.17 %**

4-38. **Refer to Table 4-3.** . . . .

Boron is approximately 80% ^{11}B with a mass of 11.0 amu and 20% ^{10}B with a mass of 10.0 amu. Therefore the atomic weight of boron using a weighted average should be about **10.8 amu.**

Chlorine is approximately 75% ^{35}Cl with a mass of 35.0 amu and 25% ^{37}Cl with a mass of 37.0 amu. The atomic weight of chlorine can be estimated to be about **35.5 amu.**

4-40. **Refer to Section 4-7.** . . . .

If the mass spectrum were complete for germanium, the calculated atomic weight would be the weighted average of the isotopes:

$$\text{AW (amu)} = \Sigma \frac{\text{(relative abundance)}}{\text{(total relative abundance)}} \times \text{isotope mass (amu)}$$

$$= (5.49/15.90)71.9217 + (1.55/15.90)72.9234 + (7.31/15.90)73.9219 + (1.55/15.90)75.9219$$

$$= \mathbf{73.3}$$

However, the true atomic weight of germanium is 72.59 amu. The observed data gives a value that is too high. Therefore, the spectrum given is incomplete and data must have been lost at the low end of the plot when the recorder jammed.

4-42. **Refer to Section 4-7.** . . . .

(a) $^{20}_{10}Ne^{2+}$ is the ion with the highest charge and lowest mass. It should therefore travel most rapidly under the influence of a particular accelerating voltage.

(b) $^{22}_{10}Ne^{+}$ is the ion with the lowest charge and highest mass. It should travel least rapidly under the influence of a particular voltage.

4-44. **Refer to Section 4-9, Figure 4-11, and the Key Terms for Chapter 4.** . . . . . .

(a) Wavelength is the distance between two corresponding points of a wave during the propagation of an electromagnetic radiation.

(b) Frequency is the number of repeating corresponding points on a wave that pass a given observation point per unit time.

(c) Amplitude describes the intensity or height of a wave and corresponds to the maximum distance of a point on the wave from its center or equilibrium position.

(d) Color is determined by the wavelength or frequency of electromagnetic radiation. The human eye can detect radiation having a wavelength between 400 and 750 nm, known as visible light.

4-46. **Refer to Section 4-9.** . . . .

For electromagnetic radiation: frequency × wavelength = speed of light

$$\nu\ (s^{-1}) \times \lambda\ (m) = c\ (m/s)$$

$$\nu\ (s^{-1}) = \frac{c\ (m/s)}{\lambda\ (m)}$$

(a) $\lambda\ (m) = 9744\ Å \times \frac{10^{-10}\ m}{1\ Å} = 9.744 \times 10^{-7}\ m$

$\nu\ (s^{-1}) = \frac{2.998 \times 10^{8}\ m/s}{9.744 \times 10^{-7}\ m} = \mathbf{3.077 \times 10^{14}\ s^{-1}}$

(b) $\lambda\ (m) = 492\ nm \times \frac{10^{-9}\ m}{1\ nm} = 4.92 \times 10^{-7}\ m$

$\nu\ (s^{-1}) = \frac{3.00 \times 10^{8}\ m/s}{4.92 \times 10^{-7}\ m} = \mathbf{6.10 \times 10^{14}\ s^{-1}}$

(c) $\lambda\ (m) = 4.92\ cm \times \frac{1\ m}{100\ cm} = 0.0492\ m$

$\nu\ (s^{-1}) = \frac{3.00 \times 10^{8}\ m/s}{0.0492\ m} = \mathbf{6.10 \times 10^{9}\ s^{-1}}$

(d) $\lambda\ (m) = 4.92 \times 10^{-9}\ cm \times \frac{1\ m}{100\ cm} = 4.92 \times 10^{-11}\ m$

$\nu\ (s^{-1}) = \frac{3.00 \times 10^{8}\ m/s}{4.92 \times 10^{-11}\ m} = \mathbf{6.10 \times 10^{18}\ s^{-1}}$

4-48. **Refer to Section 4-9, Figure 4-12b and Exercise 4-46 Solution.** . . . . . .

	Wavelength	Region of Electromagnetic Spectrum
(a)	9.744×10^{-7} m	infrared
(b)	4.92×10^{-7} m	visible
(c)	0.0492 m	microwave
(d)	4.92×10^{-11} m	X-ray

4-50. **Refer to Tables 1-5 and 1-7.** . . . .

Plan: (1) Use dimensional analysis to determine how far (in miles) light travels in one year, which is a light year.
(2) Use this as a unit factor to determine the distance (in miles) between Alpha Centauri and our solar system.

(1) $? \text{ miles/light yr} = \frac{3.00 \times 10^{8}\ m}{1\ s} \times \frac{100\ cm}{1\ m} \times \frac{1\ in}{2.54\ cm} \times \frac{1\ ft}{12\ in} \times \frac{1\ mile}{5280\ ft}$

$\times \frac{60\ s}{1\ min} \times \frac{60\ min}{1\ hr} \times \frac{24\ hr}{1\ day} \times \frac{365\ day}{1\ yr}$

$= 5.88 \times 10^{12}$ miles/light yr

Therefore, in 1 year light will travel 5.88×10^{12} miles.

(2) $? \text{ miles} = 4.3 \text{ light years} \times \frac{5.88 \times 10^{12}\ miles}{1\ light\ year} = \mathbf{2.5 \times 10^{13}\ miles}$

4-52. **Refer to Section 4-10.** . . . .

The photoelectric effect is the emission of an electron from a metal surface caused by impinging electromagnetic radiation. This radiation must have a certain minimum energy, i.e. its frequency must be greater than the threshold frequency, which is a characteristic of a particular metal, for current to flow. If the frequency is below the threshold frequency, no current flows. As long as this criterion is met, the current increases with increasing intensity (brightness) of the light.

4-54. **Refer to Sections 4-9 and 4-10.** . . . .

We know: $E = h\nu = \frac{hc}{\lambda}$

Therefore, $\lambda\ (\text{m}) = \frac{hc}{E} = \frac{(6.63 \times 10^{-34}\ \text{J·s})(3.00 \times 10^{8}\ \text{m/s})}{(3.89\ \text{ev})(1.60 \times 10^{-19}\ \text{J/ev})} = \mathbf{3.20 \times 10^{-7}\ m}$

$$\lambda\ (\text{nm}) = 3.20 \times 10^{-7}\ \text{m} \times \frac{1\ \text{nm}}{10^{-9}\ \text{m}} = 320\ \text{nm}$$

4-56. **Refer to Section 4-12 and the Key Terms for Chapter 4.** . . . .

A continuous spectrum contains all wavelengths in a specified region of the electromagnetic spectrum, whereas a line emission or absorption spectrum contains only lines corresponding to certain distinct wavelengths.

4-58. **Refer to Section 4-12.** . . . .

Plan: Use the Rydberg equation, $\frac{1}{\lambda} = R\left(\frac{1}{n_1^2} - \frac{1}{n_2^2}\right)$ where $R = 1.097 \times 10^7\ \text{m}^{-1}$ to evaluate $\frac{1}{\lambda}$, then solve for λ.

For the Balmer series, $n_1 = 2$. The three lowest energy lines correspond to

(1) $n_2 = 3 \rightarrow n_1 = 2$ (2) $n_2 = 4 \rightarrow n_1 = 2$ (3) $n_2 = 5 \rightarrow n_1 = 2$

(1) $\frac{1}{\lambda} = 1.097 \times 10^7\ \text{m}^{-1}\left(\frac{1}{2^2} - \frac{1}{3^2}\right) = (1.097 \times 10^7\ \text{m}^{-1})(0.1389) = 1.524 \times 10^6\ \text{m}^{-1}$

$\lambda = 6.563 \times 10^{-7}\ \text{m} = \mathbf{656.3\ nm}$

(2) $\frac{1}{\lambda} = 1.097 \times 10^7\ \text{m}^{-1}\left(\frac{1}{2^2} - \frac{1}{4^2}\right) = (1.097 \times 10^7\ \text{m}^{-1})(0.1875) = 2.057 \times 10^6\ \text{m}^{-1}$

$\lambda = 4.861 \times 10^{-7}\ \text{m} = \mathbf{486.1\ nm}$

(3) $\frac{1}{\lambda} = 1.097 \times 10^7\ \text{m}^{-1}\left(\frac{1}{2^2} - \frac{1}{5^2}\right) = (1.097 \times 10^7\ \text{m}^{-1})(0.2100) = 2.304 \times 10^6\ \text{m}^{-1}$

$\lambda = 4.340 \times 10^{-7}\ \text{m} = \mathbf{434.0\ nm}$

4-60. **Refer to Section 4-12.** . . . .

Plan: Use the Rydberg equation, $\frac{1}{\lambda} = R\left(\frac{1}{n_1^2} - \frac{1}{n_2^2}\right)$ and solve for n_2.

$\lambda = 95.0\ \text{nm} = 9.50 \times 10^{-8}\ \text{m}$, $R = 1.097 \times 10^7\ \text{m}^{-1}$,

For the Lyman series, $n_1 = 1$.

Substituting, $$\frac{1}{9.50 \times 10^{-8}\ \text{m}} = 1.097 \times 10^7\ \text{m}^{-1}\left(\frac{1}{1^2} - \frac{1}{n_2^2}\right)$$

$$0.960 = \left(1 - \frac{1}{n_2^2}\right)$$

$$\frac{1}{n_2^2} = 4.0 \times 10^{-2}$$ Therefore, $n_2^2 = 25$ and $n_2 = \mathbf{5}$.

4-62. **Refer to Section 4-12.** . . . .

The Bohr model of the hydrogen atom, developed by Neils Bohr in 1913, has three main tenets.

(1) The atom has a number of discrete energy levels.
(2) An electron may move from one energy level to another but in doing so it must emit or absorb a definite (quantized) amount of energy.
(3) The electron moves in a circular path (orbit) around the nucleus. Its motion is restricted by the laws of mechanics and electrostatics.

The main success of the Bohr theory is that it satisfactorily explained the spectra of hydrogen and other charged species containing only one electron. However, it could not explain the observed spectra of more complex species.

4-64. **Refer to Section 4-12 and Example 4-5.** . . . .

The energy loss due to 1 atom emitting a photon is $E = hc/\lambda$.
The energy loss due to 1 mole of atoms each emitting a photon is $E = (hc/\lambda)\mathbf{N}$, where **N** is Avogadro's Number.

Substituting, $$E = \frac{(6.63 \times 10^{-34}\ \text{J}\cdot\text{s})(3.00 \times 10^{8}\ \text{m/s})}{(5.15 \times 10^{3}\ \text{Å})(1 \times 10^{-10}\ \text{m/Å})} \times 6.02 \times 10^{23}\ \text{mol}^{-1}$$

$$= 2.33 \times 10^{5}\ \text{J/mol}$$

$$= \mathbf{233\ kJ/mol}$$

4-66. **Refer to Section 4-12.** . . . .

The energy emitted by 1 photon is $E = hc/\lambda$.
The energy emitted by n photons is $E = (hc/\lambda)\text{n}$.

The energy emitted by this laser in 1 second is

$$E = \text{power} \times \text{time} = 625\ \text{milliwatts} \times \frac{1\ \text{watt}}{1000\ \text{milliwatts}} \times \frac{1\ \text{J/s}}{1\ \text{watt}} \times 1.00\ \text{s} = 0.625\ \text{J}$$

Substituting,

$$0.625\ \text{J} = \frac{(6.63 \times 10^{-34}\ \text{J}\cdot\text{s})(3.00 \times 10^{8}\ \text{m/s})}{(488.0\ \text{nm})(10^{-9}\ \text{m/nm})} \times \text{n}$$

$$\mathbf{n = 1.53 \times 10^{18}\ photons}$$

4-68. **Refer to Section 4-12.** . . . .

The emission spectrum would consist of lines resulting from the following transitions.

		No. Transitions
from $n = 6$:	$n_6 \rightarrow n_5$, $n_6 \rightarrow n_4$, $n_6 \rightarrow n_3$, $n_6 \rightarrow n_2$, $n_6 \rightarrow n_1$	5
from $n = 5$:	$n_5 \rightarrow n_4$, $n_5 \rightarrow n_3$, $n_5 \rightarrow n_2$, $n_5 \rightarrow n_1$	4
from $n = 4$:	$n_4 \rightarrow n_3$, $n_4 \rightarrow n_2$, $n_4 \rightarrow n_1$	3
from $n = 3$:	$n_3 \rightarrow n_2$, $n_3 \rightarrow n_1$	2
from $n = 2$:	$n_2 \rightarrow n_1$	1
	total	15

An emission spectrum of **15 lines** is expected.

4-70. **Refer to Section 4-11.** . . . .

L. de Broglie (1925) proposed that very small particles, e.g. electrons, might display wave-like properties. In 1927, Davisson and Germer demonstrated that electron could be diffracted by a nickel crystal, thereby proving that electrons exhibit wave properties.

4-72. **Refer to Section 4-11.** . . . .

The de Broglie wavelength is given by $\lambda\ (\text{m}) = \dfrac{h(\text{J}\cdot\text{s})}{\text{m(kg)}v(\text{m/s})}$
The units are as stated because

$$1\ \text{J} = 1\ \text{kg}\cdot\text{m}^2/\text{s}^2.$$

The mass of an electron, from Table 4-1, is 0.00054858 amu. Therefore, 1 mole of electrons has a mass of 0.00054858 g.

To find the mass of 1 electron in kg:

$$?\ \text{mass}\ e^-\ (\text{kg}) = \frac{0.00054858\ \text{g}}{1\ \text{mol}\ e^-} \times \frac{1\ \text{mol}\ e^-}{6.02 \times 10^{23}\ e^-} \times \frac{1\ \text{kg}}{1000\ \text{g}} = 9.11 \times 10^{-31}\ \text{kg}$$

The velocity of the electron, $v = 3.00 \times 10^9$ cm/s $= 3.00 \times 10^7$ m/s.

Substituting,

$$\lambda = \frac{6.63 \times 10^{-34}\ \text{J}\cdot\text{s}}{(9.11 \times 10^{-31}\ \text{kg})(3.00 \times 10^7\ \text{m/s})} = \mathbf{2.43 \times 10^{-11}\ m}$$

4-74. **Refer to Section 4-14 and the Key Terms for Chapter 4.** . . . .

Quantum numbers are numbers that describe the energies of electrons in atoms. They are derived from quantum mechanics.

An atomic orbital is a region or volume in space in which the probability of finding electrons is high.

4-76. **Refer to Section 4-14.** . . . .

The subsidiary (or azimuthal) quantum number, ℓ, for a particular energy level as defined by the principle quantum number, n, depends on the value of n. ℓ can take integral values from 0 up to and including (n - 1). For example, when $n = 3$, $\ell = 0$, 1, or 2.

4-78. **Refer to Section 4-14.** . . . .

The values of the magnetic quantum number, m_ℓ, for a particular electron depend upon the value of the subsidiary quantum number, ℓ. Within each sublevel, m_ℓ may take any integral value from $-\ell$ through zero up to and including $+\ell$. For example, when $\ell = 3$, $m_\ell = -3, -2, -1, 0, 1, 2$, or 3.

4-80. **Refer to Section 4-14.** . . . .

The values of the subsidiary quantum number, ℓ, each correspond to a different kind of atomic orbital designated by a letter:

value of ℓ:	0	1	2	3
orbital:	*s*	*p*	*d*	*f*

4-82. **Refer to Sections 4-13 and 4-14, and Table 4-4.** . . . .

There are **9** individual orbitals in the third major energy level, $n = 3$.

n	ℓ	m_ℓ	orbital
3	0	0	3*s*
3	1	-1	3*p*
3	1	0	3*p*
3	1	+1	3*p*

n	ℓ	m_ℓ	orbital
3	2	-2	3*d*
3	2	-1	3*d*
3	2	0	3*d*
3	2	+1	3*d*
3	2	+2	3*d*

4-84. **Refer to Sections 4-14 and 4-15.** . . . .

(a) possible - This set represents an electron in the 1*s* orbital.

(b) impossible - ℓ cannot be equal to n since $\ell = 0, 1, 2 \ldots (n - 1)$.

(c) possible - This set represents an electron in the 3*d* orbital.

4-86. **Refer to Sections 4-15 and 4-16, and the Key Terms for Chapter 4.** . . . . . .

The ground state corresponds to the species in its lowest energy or unexcited state which is the most stable. When an atom is in its ground state electron configuration, its electrons occupy the lowest possible energy levels.

Spherically symmetrical refers, in particular, to the round shape of an *s* orbital. The probability of finding an electron at a certain distance from the nucleus is the same independent of direction.

4-88. **Refer to Section 4-15.** . . . .

The spin of an electron refers to the fact that electrons are spinning about axes through their centers. Hence, they act like tiny magnets. When two electrons have opposite spins and occupy the same orbital, they produce magnetic fields which attract each other and are said to be "spin-paired" or simply "paired".

4-90. **Refer to Section 4-15.** . . . .

The maximum number of electrons per energy level is $2n^2$. Therefore, the number of electrons in the $n = 5$ principle energy level is $2(5)^2 = 50$.

4-92. **Refer to Section 4-15 and Figures 4-20 and 4-21b.** . . . .

All *p* orbitals resemble equal-arm dumbbells. A $2p_x$ and a $2p_y$ orbital are identical in size, shape and energy. They differ only in their orientation: a $2p_x$ orbital lies on the x axis whereas a $2p_y$ lies on the y axis.

4-94. **Refer to Sections 4-15 and 4-16.** . . . .

		1*s*	2*s*	2*p*	3*s*	3*p*	3*d*	4*s*	4*p*
(a)	$_4$Be	↑↓	↑↓						
(b)	$_{15}$P	↑↓	↑↓	↑↓ ↑↓ ↑↓	↑↓	↑ ↑ ↑			
(c)	$_{28}$Ni	↑↓	↑↓	↑↓ ↑↓ ↑↓	↑↓	↑↓ ↑↓ ↑↓	↑↓ ↑↓ ↑↓ ↑ ↑	↑↓	
(d)	$_{42}$Mo	↑↓	↑↓	↑↓ ↑↓ ↑↓	↑↓	↑↓ ↑↓ ↑↓	↑↓ ↑↓ ↑↓ ↑↓ ↑↓	↑↓	↑↓ ↑↓ ↑↓

4*d*	4*f*	5*s*
↑ ↑ ↑ ↑ ↑	__ __ __ __ __ __ __	↑

4-96. **Refer to Sections 4-15 and 4-16.** . . . .

(a) $_{4}Be$ $1s^2 2s^2$

(b) $_{15}P$ $1s^2 2s^2 2p^6 3s^2 3p^3$ or $[Ne]3s^2 3p^3$

(c) $_{28}Ni$ $1s^2 2s^2 2p^6 3s^2 3p^6 3d^8 4s^2$ or $[Ar]3d^8 4s^2$

(d) $_{42}Mo$ $1s^2 2s^2 2p^6 3s^2 3p^6 3d^{10} 4s^2 4p^6 4d^5 5s^1$ or $[Kr]4d^5 5s^1$

4-98. **Refer to Section 4-16.** . . . .

It is known that an electron will occupy the available orbital having the lowest energy. In most atoms, including element 19, K, and element 20, Ca, the 4*s* orbital has slightly lower energy than the 3*d* orbital. Electrons, therefore, will occupy the 4*s* orbital before the 3*d* orbital.

4-100. **Refer to Section 4-16 and Figure 4-28.** . . . .

Paramagnetism is measured by using a balance with its sample arm in a magnetic field to determine the extent to which a sample is attracted to the magnetic field.

4-102. **Refer to Section 4-16 and Exercise 4-101.** . . . .

Given below are the shorthand notations and number of unpaired electrons present for the elements in their ground states.

		Shorthand	**Unpaired Electrons**
(a)	$_{5}B$	$1s^2 2s^2 2p^1$	1
(b)	$_{14}Si$	$[Ne]3s^2 3p^2$	2
(c)	$_{18}Ar$	[Ar]	0
(d)	$_{35}Br$	$[Ar]3d^{10} 4s^2 4p^5$	1
(e)	$_{23}V$	$[Ar]3d^3 4s^2$	3

$_{23}V$ has 3 unpaired electrons and therefore, it exhibits the highest degree of paramagnetism in its ground state among these five elements.

4-104. **Refer to Section 4-16 and Appendix B.** . . . .

More elements as isolated atoms have ground state configurations that are paramagnetic rather than diamagnetic. There are only 16 elements in the first 100 elements that are diamagnetic and contain only paired electrons:

Group IIA elements (6),
Group IIB elements (3), and
the noble gases (6).

All the other elements in the periodic table are paramagnetic and have at least 1 unpaired electron.

4-106. Refer to Sections 4-15 and 4-16, and Examples 4-6 and 4-7.

(a) $_{8}O$ $1s^2 2s^2 2p^4$

Electron	n	ℓ	m_ℓ	m_s
1	1	0	0	+1/2
2	1	0	0	-1/2
3	2	0	0	+1/2
4	2	0	0	-1/2
5	2	1	-1	+1/2
6	2	1	0	+1/2
7	2	1	1	+1/2
8	2	1	-1	-1/2

(b) $_{15}P$ $1s^2 2s^2 2p^6 3s^2 3p^3$

Electron	n	ℓ	m_ℓ	m_s
1	1	0	0	+1/2
2	1	0	0	-1/2
3	2	0	0	+1/2
4	2	0	0	-1/2
5	2	1	-1	+1/2
6	2	1	0	+1/2
7	2	1	+1	+1/2
8	2	1	-1	-1/2
9	2	1	0	-1/2
10	2	1	+1	-1/2
11	3	0	0	+1/2
12	3	0	0	-1/2
13	3	1	-1	+1/2
14	3	1	0	+1/2
15	3	1	+1	+1/2

(c) $_{27}Co$ $1s^2 2s^2 2p^6 3s^2 3p^6 4s^2 3d^7$

Electron	n	ℓ	m_ℓ	m_s
1	1	0	0	+1/2
2	1	0	0	-1/2
3	2	0	0	+1/2
4	2	0	0	-1/2
5	2	1	-1	+1/2
6	2	1	0	+1/2
7	2	1	1	+1/2
8	2	1	-1	-1/2
9	2	1	0	-1/2
10	2	1	+1	-1/2
11	3	0	0	+1/2
12	3	0	0	-1/2
13	3	1	-1	+1/2
14	3	1	0	+1/2
15	3	1	+1	+1/2
16	3	1	-1	-1/2
17	3	1	0	-1/2
18	3	1	+1	-1/2
19	4	0	0	+1/2
20	4	0	0	-1/2
21	3	2	-2	+1/2
22	3	2	-1	+1/2
23	3	2	0	+1/2
24	3	2	+1	+1/2
25	3	2	+2	+1/2
26	3	2	-2	-1/2
27	3	2	-1	-1/2

4-108. Refer to Section 4-16.

IA	ns^1	VA	$ns^2 np_x^{\,1} np_y^{\,1} np_z^{\,1}$
IIA	ns^2	VIA	$ns^2 np_x^{\,2} np_y^{\,1} np_z^{\,1}$
IIIA	$ns^2 np_x^{\,1}$	VIIA	$ns^2 np_x^{\,2} np_y^{\,2} np_z^{\,1}$
IVA	$ns^2 np_x^{\,1} np_y^{\,1}$	0	$ns^2 np_x^{\,2} np_y^{\,2} np_z^{\,2}$

4-110. Refer to Section 4-16 and Appendix B.

(a) $_{32}Ge$ $[Ar]3d^{10}4s^2 4p^2$ The "last" electron entered a $4p$ orbital. Therefore, $n = 4$, $\ell = 1$, and $m_\ell = -1$, 0 or +1.

(b) $_{46}Pd$ $[Kr]4d^{10}$ The "last" electron went into a $4d$ orbital. Therefore, $n = 4$, $\ell = 2$, and $m_\ell = -2$, -1, 0, +1 or +2.

(c) $_{56}Ba$ $[Xe]6s^2$ The "last" electron entered a $6s$ orbital. Therefore, $n = 6$, $\ell = 0$, and $m_\ell = 0$.

(d) $_{92}U$ $[Rn]5f^3 6d^1 7s^2$ The "last" electron entered a $5f$ orbital.

This was determined by comparing the electron configurations of $_{92}U$ with $_{91}Pa$. Therefore, $n = 5$, $\ell = 3$, and $m_\ell = -3$, -2, -1, 0, +1, +2 or +3.

5 Chemical Periodicity

5-2. **Refer to Section 5-1 and Figure 5-1.** . . . .

Mendeleev arranged the known elements in order of increasing atomic weight in sequence so that elements with similar chemical and physical properties fell in the same column or group. To achieve this chemical periodicity, it was necessary for Mendeleev to leave blank spaces for elements undiscovered at that time and to make assumptions concerning atomic weights not known with certainty. The modern periodic table has elements arranged in order of increasing atomic number so that elements with similar chemical properties all fall in the same column.

5-4. **Refer to Section 5-1.** . . . .

The atomic weight of an element is a weighted average of the mass of the naturally occurring isotopes of that element. Therefore, the atoms in a naturally occurring sample of argon must be heavier than the atoms in a naturally occurring sample of potassium. Atoms of argon have 18 protons, whereas atoms of potassium have 19 protons. In order for the atomic weight of argon to be greater than that of potassium, argon atoms must have more neutrons.

Consider the isotopes for these elements:

	Isotope	Percent Composition	Number of Protons	Number of Neutrons
argon (AW: 39.948 amu)	$^{40}_{18}Ar$	99.60%	18	22
potassium (AW: 39.0983 amu)	$^{39}_{19}K$	93.1%	19	20
	$^{41}_{19}K$	6.88%	19	22

From these data, we can see that argon would have a higher atomic weight than potassium.

5-6. **Refer to Section 5-1 and the Periodic Table on the inside front cover of the textbook.** . . . .

Specific heat is an elemental property that is a periodic function of atomic number. The specific heat of antimony can be estimated by averaging the specific heats of its immediate neighbors within a group (vertical column) or a period (horizontal row). Antimony is a Group VA element. Its neighbors in the periodic table and their specific heats (J/g^{o}C) are arranged as shown:

	IVA	**VA**	**VIA**
n = **4**		$_{33}As$ (0.34)	
n = **5**	$_{50}Sn$ (0.23)	$_{51}Sb$ (?)	$_{52}Te$ (0.20)
n = **6**		$_{83}Bi$ (0.14)	

Within Group VA, the specific heat of antimony should be around 0.24 J/g^{o}C. Within the n = 5 period, we can further narrow down the specific heat value to the 0.21 - 0.22 J/g^{o}C range. In fact, the specific heat of antimony is 0.21 J/g^{o}C.

5-8. **Refer to Section 5-1 and the Periodic Table.** • • • • • • • • • • • • •

The periodic trends of the element properties also apply to compound containing the elements. Therefore, the melting points of CF_4, CCl_4, CBr_4 and CI_4 should follow a trend.

If we graph the melting points of these compounds versus molecular weight, we can estimate the melting point of CBr_4.

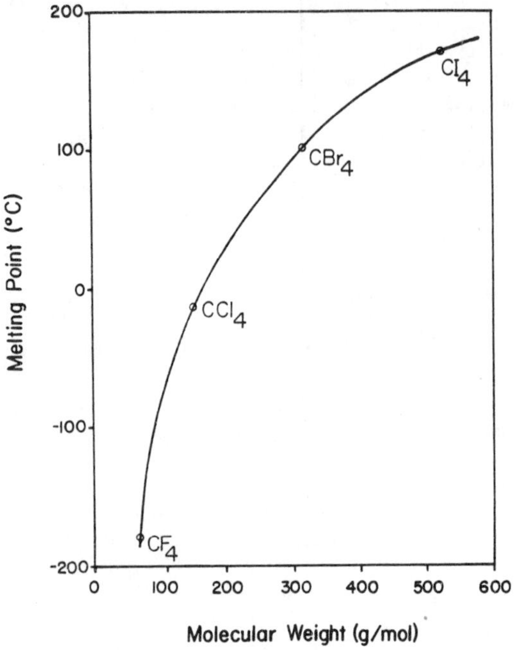

Compound	MW (g/mol)	MP (°C)
CF_4	88.0	-184
CCl_4	153.8	-23
CBr_4	331.6	90.1
CI_4	519.6	171

The estimated melting point of CBr_4 is about 100°C. The actual value is 90.1°C.

5-10. **Refer to Section 5-1.** •

Hydride formulas are related to the group number of the central element, e.g.

IIA	IIIA	IVA	VA	VIA	VIIA
BeH_2	BH_3	CH_4	NH_3	H_2O	HF

Arsine, the hydride of arsenic, has the formula AsH_x. Arsenic, the central element, is a VA element. Therefore, its structure should be similar to NH_3 and is predicted to be AsH_3.

5-12. **Refer to Section 5-1.** •

The deductive method, according to Webster, is a method of reasoning by which concrete applications or consequences are deduced from general principles. Therefore, statement (a) is a better example of deductive logic based on experimental observations.

5-14. **Refer to Section 5-1.** •

(a) alkaline earth metals: beryllium (Be), magnesium (Mg), calcium (Ca), strontium (Sr), barium (Ba) and radium (Ra)

(b) Group IIIA elements: boron (B), aluminum (Al), gallium (Ga), indium (In) and thallium (Tl)

(c) Group VIB elements: chromium (Cr), molybdenum (Mo) and tungsten (W)

5-16. **Refer to Sections 5-3 through 5-7.** . . . .

Properties of elements generally vary in a smooth manner within a group. Ionization energies, and electronegativities generally decrease from top to bottom within a group. Atomic radii, ionic radii, and metallic character generally increase from top to bottom within a group. In addition, electron affinities generally become less negative from top to bottom within a group.

5-18. **Refer to Section 5-8 and the Key Terms for Chapter 5.** . . . .

(a) Metals are the elements below and to the left of the stepwise division (metalloids) in the upper right corner of the periodic table. They possess metallic bonding. Approximately 80% of the known elements are metals, including potassium (K), calcium (Ca), scandium (Sc) and vanadium (V).

(b) Nonmetals are the elements above and to the right of the metalloids in the periodic table, including carbon (C), nitrogen (N), sulfur (S) and chlorine (Cl).

(c) Noble gases are the elements of Group 0, also called the rare gases. They include helium (He), neon (Ne), argon (Ar), krypton (Kr), xenon (Xe) and radon (Rn).

5-20. **Refer to Table 5-2.** . . . .

Period 1 contains two elements since electrons are being added to the $n = 1$ energy level, which can only hold $2n^2 = 2(1)^2 = 2$ electrons.

Period 2 contains eight elements, since electrons are being added to the $n = 2$ energy level. This energy level can hold $2n^2 = 2(2)^2 = 8$ electrons in its $2s$ and $2p$ orbitals.

5-22. **Refer to Table 5-2.** . . . .

The general order in which the energy levels are filled starting with the $n = 3$ major energy level is: $3s$ $3p$ $4s$ $3d$ $4p$ etc. Period 3 includes the elements whose outer electrons are in $3s$ or $3p$ energy sublevels. The maximum number of electrons in these sublevels is a total of 8. Hence Period 3 contains only 8 elements. Period 4 includes the elements whose outermost electrons are in $4s$, $3d$ or $4p$ energy sublevels.

5-24. **Refer to Table 5-2.** . . . .

The atomic number of the yet-to-be discovered alkaline earth element in period 8 is 120. The last portion of its electron configuration after [Rn] (Atomic Number = 86) should be:

$$7s^2 5f^{14} 6d^{10} 7p^6 8s^2$$

5-26. **Refer to Table 5-2.** . . . .

The yet-to-be discovered element 116 is determined to be a Group VIA element by extending the periodic table. Since it lies to the bottom and left of the stepwise division, it is a metal. Its electron configuration should be:

$$[Rn] 7s^2 5f^{14} 6d^{10} 7p^4$$

5-28. **Refer to Section 5-1.** . . . .

Sodium (Na) and potassium (K) are alkali metals with 1 outermost valence electron; sulfur (S) and selenium (Se) are both Group VIA elements. It can be predicted that the formulas for the compounds are similar:

sodium sulfate	Na_2SO_4	sodium selenate	Na_2SeO_4
potassium sulfate	K_2SO_4	potassium selenate	K_2SeO_4

5-30. **Refer to Sections 5-4 and 5-6.** . . . .

Mg^{2+} $1s^22s^22p^6$ Ag^+ $[Kr]4d^{10}5s^0$

Sc^{3+} $1s^22s^22p^63s^23p^6$ In^+ $[Kr]4d^{10}5s^2$

5-32. **Refer to Section 5-2 and Table 5-3.** . . . .

Lewis dot representations for the representative elements show only the electrons in the outermost occupied *s* and *p* orbitals. Paired and unpaired electrons are also indicated.

He: ·C· ·S: :Ar: ·Br: Sr:

5-34. **Refer to Table 5-7.** . . . .

5-36. **Refer to Table 5-8.** . . . .

Metals are located at the left side of the periodic table and therefore have (a) fewer outer shell electrons, (usually 3 or less), (b) less endothermic ionization energies, (c) less exothermic electron affinities, and (d) lower electronegativities than nonmetals of the same period.

5-38. **Refer to Section 5-8.** . . . .

Metallic character increases from top to bottom within a group and from right to left across a period in the periodic table. Therefore, we obtain the following order of increasing metallic character: Br < C < B < Be < Li.

5-40. **Refer to Sections 5-3 and 5-4.** . . . .

Electrons that are in filled sets of orbitals between the nucleus and outer shell electrons shield the outer shell electrons partially from the effect of the protons in the nucleus. As we move from left to right along a period, the outer shell electrons do experience a progressively stronger force of attraction to the nucleus due to the combination of an increase in the number of protons and a constant nuclear shielding by inner electrons. As a result the atomic radii decrease. As we move down a group, the outer electrons are partially shielded from the attractive force of the nucleus by an increasing number of inner electrons. This effect is <u>partially</u> responsible for the observed increase in atomic radii going down a group.

5-42. **Refer to Section 5-3 and Table 5-2.** . . . .

As we move down a group, atomic radii increase <u>primarily</u> because electrons are added to orbitals in higher energy levels which are further and further away from the nucleus.

5-44. **Refer to Section 5-3, and Table 5-2 and Figure 5-2.** . . . .

Atomic radii increase from top to bottom within a group and from right to left within a period. Therefore, in order of increasing size, we have:

(a) Be < Mg < Ca < Sr < Ba < Ra

(b) He < Ne < Ar < Kr < Xe < Rn

(c) Simple prediction: Ne < F < O < N < C < B < Be < Li
Actual Size: F < O < Ne ≃ N < C < B < Be < Li

(d) N < B < Te < Sb < Sr

5-46. **Refer to Section 5-4.** . . . .

(a) The first ionization energy, IE_1, also called the first ionization potential, is the minimum amount of energy required to remove the most loosely bound electron from an isolated gaseous atom to form an ion with a 1+ charge.

$$X\ (g) + IE_1 \rightarrow X^+\ (g) + e^-$$

(b) The second ionization energy, IE_2, is the amount of energy required to remove a second electron from an isolated gaseous atom; i.e. to remove an electron from an ion with a 1+ charge to give an ion with a 2+ charge.

$$X^+\ (g) + IE_2 \rightarrow X^{2+}\ (g) + e^-$$

5-48. **Refer to Sections 5-3 and 5-4.** . . . .

As we move down a given group, the valence electrons are further and further away from the nucleus. The first ionization energies of the elements, which is the energy required to remove an electron from an isolated gaseous atom, decrease while the atomic radii increase.

Likewise, from left to right across a period, the forces of attraction between the outermost electron and the nucleus increase. Therefore the ionization energies increase while the atomic radii decrease.

5-50. **Refer to Sections 5-4 and 4-12.** . . . .

Recall: For 1 atom, $E\ (\text{J/atom}) = h\ (\text{J}\cdot\text{s}) \times \nu\ (\text{s}^{-1})$

For 1 mole of atoms, $E\ (\text{J/mol}) = h\nu\text{N}$ where N is Avogadro's Number

Solving for ν, we have

$$\nu\ (\text{s}^{-1}) = \frac{E}{h\text{N}} = \frac{419\ \text{kJ/mol} \times 1000\ \text{J/kJ}}{(6.63 \times 10^{-34}\ \text{J}\cdot\text{s})(6.02 \times 10^{23}\ \text{mol}^{-1})} = 1.05 \times 10^{15}\ \text{s}^{-1}$$

5-52. **Refer to Section 5-4 and Table 5-4.** . . . .

First ionization energies increase from left to right and bottom to top in the periodic table. However, there are exceptions: elements of Group IIIA generally have lower first ionization energies than elements of Group IIA, and elements of Group VIA generally have lower first ionization energies than elements of Group VA. Therefore, we obtain the following orders of increasing first ionization energies:

(a) Fr < Cs < Rb < K < Na < Li

(b) At < I < Br < Cl < F

(c) Li < B < Be < C < O < N < F < Ne

(d) Cs < Ga < B < Br < H < F

5-54. **Refer to Section 5-4.** . . . .

As atomic radii increase moving down a group, first ionization energies decrease because the valence electrons are further from the attractive force of the nucleus. The force of attraction of the positively charged nucleus for the valence electrons is inversely proportional to the square of the distance between them. This effect overcomes the effect of increasing effective nuclear charges moving down a group.

5-56. **Refer to Section 5-4 and Exercise 5-52 Solution.** . . . .

As we move from left to right across Period 2 of the periodic table, there is an increase in effective nuclear charge and a decrease in atomic radii. Outer valence electrons are held more tightly and first ionization energies generally increase.

5-58. **Refer to Section 5-4.** . . . .

It is difficult to prepare compounds containing Mg^{3+} due to the immense amount of energy that is required to remove a third electron from an atom of magnesium, i.e. there is a very large amount of energy (the third ionization energy) required for this reaction:

$$Mg^{2+}\ (g)\ +\ 7733\ kJ/mol\ \rightarrow\ Mg^{3+}\ (g)\ +\ e^-$$

This energy is not likely to be repaid during compound formation. The reason for such a high third ionization energy for Mg^{3+} is because the electron configuration of Mg^{2+} is $1s^22s^22p^6$ which has a filled set of *p* orbitals. It is the special stability of the filled *p* orbitals which prevents the formation of Mg^{3+} ions.

On the other hand, Al^{2+} has an electron configuration of $1s^22s^22p^63s^1$ where the single e^- in 3*s* could be easily removed to give an Al^{3+} ion and consequently an Al^{3+} compound.

5-60. **Refer to Section 5-4 and Table 5-4.** . . . .

First Ionization Energy for Mg (*k*J/mol) = 738 *k*J/mol
Second Ionization Energy for Mg (*k*J/mol) = 1451 *k*J/mol

Therefore,

$$\begin{array}{lcl} Mg\ (g)\ +\ 738\ kJ/mol & \rightarrow & Mg^{+}\ (g)\ +\ e^- \\ Mg^{+}\ (g)\ +\ 1451\ kJ/mol & \rightarrow & Mg^{2+}\ (g)\ +\ e^- \\ \hline Mg\ (g)\ +\ 2189\ kJ/mol & \rightarrow & Mg^{2+}\ (g)\ +\ 2e^- \end{array}$$

And so, 2189 *k*J/mol of energy is required to produce 1 mole of gaseous Mg^{2+} ions from gaseous Mg atoms. To convert this energy into units of *k*J/g, which is the energy required per 1 gram of gaseous Mg atoms, we apply dimensional analysis:

? energy (*k*J/g) = 2189 *k*J/mol Mg × 1 mol Mg/24.305 g = **90.06 *k*J/g**

5-62. **Refer to Section 5-5.** . . . .

The electron affinity of an element is defined as the amount of energy absorbed when an electron is added to an isolated gaseous atom to form an ion with a 1- charge.

5-64. **Refer to Section 5-5.** . . . .

Elements with very negative electron affinities gain electrons easily to form negative ions. The halogens, with electronic configurations of ns^2np^5, have the most negative electron affinities and easily gain one electron to form stable ions with a filled set of *p* orbitals which are isoelectronic with the noble gases and have noble gas electronic configurations, ns^2np^6. This does not occur when a Group VIA element gains an electron.

5-66. **Refer to Section 5-5 and Table 5-5.** . . . .

		Electronic Configuration
(a) $O (g) + e^- \rightarrow O^- (g) + 142\ kJ/mol$	O	$1s^22s^22p^4$
	O^-	$1s^22s^22p^5$
(b) $Cl (g) + e^- \rightarrow Cl^- (g) + 348\ kJ/mol$	Cl	$[Ne]3s^23p^5$
	Cl^-	[Ar]
(c) $Ca (g) + e^- + 156\ kJ/mol \rightarrow Ca^- (g)$	Ca	$[Ar]4s^2$
	Ca^-	$[Ar]4s^23d^1$

5-68. **Refer to Section 5-6 and Figure 5-2.** . . . .

(a) Within an isoelectronic series, ionic radii increase with decreasing atomic number. Therefore, in order of increasing ionic radii, we have

$$Ga^{3+} < Ca^{2+} < K^+$$

(b) Ionic radii increase down a group. So, $Be^{2+} < Mg^{2+} < Ca^{2+} < Ba^{2+}$

(c) $Al^{3+} < Sr^{2+} < K^+ < Rb^+$ (See Figure 5-2)

5-70. **Refer to Section 5-6 and Figure 5-2.** . . . .

(a) In an isoelectronic series, ionic radii increase with decreasing atomic number. Therefore, in order of increasing ionic radii, we have

$$Cl^- < S^{2-} < P^{3-}$$

(b) Ionic radii increase down a group. So, $O^{2-} < S^{2-} < Se^{2-}$

(c) $N^{3-} < S^{2-} < Br^- < P^{3-}$ (See Figure 5-2)

5-72. **Refer to Section 5-5 and 5-6.** . . . .

The Fe^{2+} ion has 26 protons pulling on 24 electrons, whereas the Fe^{3+} ion has 26 protons pulling on 23 electrons. The electrons in the Fe^{3+} ion are more tightly held and therefore, Fe^{3+} is the smaller ion.

Likewise, the Cu^+ ion has 29 protons pulling on 28 electrons, whereas the Cu^{2+} ion has 29 protons attracting 27 electrons. The electrons in the Cu^{2+} ion are more tightly held and therefore, Cu^{2+} is the smaller ion.

5-74. **Refer to Section 5-7 and Table 5-6.** . . . .

Electronegativities usually increase from left to right across periods and from bottom to top within groups. Exceptions are explained in the textbook.

(a) Al < In < Ga < B

(b) Na < Mg < S < Cl

(c) Bi < Sb < P < N

(d) Ba < Sc < Si < Se < F

5-76. **Refer to Sections 5-6 and 5-7.** . . . .

Both statements, "Chlorine has a high electronegativity," and "It forms chloride ions, Cl^-, readily" are correct. However the quoted sentence is a poor statement. The high electronegativity is part of the fundamental nature of chlorine due to the electron configuration of the element, while the ease of forming chlorides is a resultant property of chlorine due to its high electron pulling force. Therefore, the correct statement should be, "Chlorine forms chloride ions, Cl^-, readily because it has a high electronegativity."

5-78. **Refer to Sections 4-2 and 5-6, and Figure 5-2.** . . . .

Ion	Electronic Charge (coulombs)	Ionic Radii (Å)	$V = (4/3)\pi r^3$ (Å^3)	Charge Density = Charge/V (coulombs/Å^3)
Li^+	$+1(1.60 \times 10^{-19})$ $= 1.60 \times 10^{-19}$	0.60	0.90	1.8×10^{-19}
Mg^{2+}	$+2(1.60 \times 10^{-19})$ $= 3.20 \times 10^{-19}$	0.65	1.2	2.7×10^{-19}
Be^{2+}	$+2(1.60 \times 10^{-19})$ $= 3.20 \times 10^{-19}$	0.31	0.12	2.7×10^{-18}
Al^{3+}	$+3(1.60 \times 10^{-19})$ $= 4.80 \times 10^{-19}$	0.50	0.52	9.2×10^{-19}

The charge densities of Li^+ and Mg^{2+} agree within a factor of 1.5, and the charge densities of Be^{2+} and Al^{3+} agree within a factor of 3.

6 Chemical Bonding and Inorganic Nomenclature

6-2. **Refer to the Key Terms and the Introduction to Chapter 6.** . . . .

Ionic bonding is the chemical bond that results from the transfer of one or more electrons from one atom or group of atoms to another. Covalent bonding is the chemical bond formed by the sharing of one or more electron pairs between two atoms.

In an ionic compound, an ion interacts with all the surrounding ions of both like and opposite charges. It is therefore nondirectional. An atom in a covalent compound is sharing its bonding electrons with surrounding atoms via orbital overlap. Since electron pairs tend to repel each other to a maximum angle, covalent compounds always adopt certain geometries. As a result, covalent bonds stretch out in certain directions and are therefore directional in nature.

6-4. **Refer to *The Handbook of Chemistry and Physics* (67th Ed.) and the Introduction to Chapter 6.** . . . .

In this handbook, you will find the following tabulated data:

Compound	Form	Density	MP (^{o}C)	BP (^{o}C)	Solubility Water	Solubility Alcohol
NaF	colorless cubic or tetragonal crystal	2.558 g/mL (at 41°C)	993	1695	soluble	insoluble
PF_3	colorless gas	3.907 g/L	-151.5	-101.5	decomposes	soluble

NaF is a crystalline solid with relatively high melting and boiling points. It is also soluble in water, a polar solvent and is relatively insoluble in alcohol, a much less polar solvent. Hence it is consistent with the properties of an ionic compound.

PF_3 is a gas with very low melting and boiling points. Unlike NaF, it is soluble in alcohol and decomposes in water. Therefore, it is consistent with the properties of a covalent compound.

6-6. **Refer to Sections 6-1 and 6-3.** . . . .

In general, the bond between a metal and a nonmetal is ionic, whereas the bond between two nonmetals is covalent. In other words, the further apart across the periodic table the two elements are, the more likely they are to form an ionic bond.

(a) Ba (metal) and S (nonmetal) — ionic bond

(b) P (nonmetal) and O (nonmetal) — covalent bond

(c) Cl (nonmetal) and F (nonmetal) — covalent bond

(d) Li (metal) and I (nonmetal) — ionic bond

(e) Si (metalloid) and Br (nonmetal) — covalent bond

(f) Mg (metal) and F (nonmetal) — ionic bond

6-8. **Refer to Sections 6-1 and 6-3.** . . . .

In general, whenever a metal and a nonmetal are together in a compound, it is ionic. If the compound consists only of nonmetals, it is covalent. In other words, the further apart two elements are on the periodic table, the more likely they are to form an ionic compound.

(a)	$CaSO_4$	metal + nonmetals	ionic
(b)	SO_3	nonmetals	covalent
(c)	KNO_3	metal + nonmetals	ionic
(d)	$NiCl_2$	metal + nonmetal	ionic
(e)	H_2CO_3	nonmetals	covalent (H is not a metal)
(f)	NCl_3	nonmetals	covalent
(g)	Li_2O	metal + nonmetal	ionic
(h)	H_3PO_4	nonmetals	covalent
(i)	$SOCl_2$	nonmetals	covalent

6-10. **Refer to Section 6-1 and Appendix B.** . . . .

(a) Co^{3+} $[Ar]3d^6$ (b) Mn^{2+} $[Ar]3d^5$ (c) Zn^{2+} $[Ar]3d^{10}$

(d) Fe^{3+} $[Ar]3d^5$ (e) Cu^{2+} $[Ar]3d^9$ (f) Sc^{3+} $[Ar]$

(g) Ag^+ $[Kr]4d^{10}$

6-12. **Refer to Section 6-1.** . . . .

Stable binary ionic compounds are formed from ions that have noble gas configurations. The following compounds meet this requirement:

BaO ($Ba^{2+} + O^{2-}$), Al_2O_3 ($2Al^{3+} + 3O^{2-}$), RbCl ($Rb^+ + Cl^-$) and Cs_2Se ($2Cs^+ + Se^{2-}$).

The others do not:

MgI ($Mg^+ + I^-$) or ($Mg^{2+} + I^{2-}$), InF_2 ($In^{2+} + 2F^-$) and Be_2O ($2Be^+ + O^{2-}$).

Neither Mg^+, I^{2-}, In^{2+} nor Be^+ have noble gas configurations.

6-14. **Refer to Section 6-1.** . . . .

(a) K· + ·B̤̈r: → K^+[:B̤̈r:]$^-$ (b) 3Na· + ·Ṗ̤· → $3Na^+$,[:P̤̈:]$^{3-}$

(c) Ca: + ·Ṡ̤: → Ca^{2+}[:S̤̈:]$^{2-}$ (d) 2Al̇· + 3·Ö: → $2Al^{3+}$,3[:Ö̤:]$^{2-}$

(e) Mg: + 2·F̤̈: → Mg^{2+},2[:F̤̈:]$^-$ (f) 2Na· + ·S̤̈: → $2Na^+$,[:S̤̈:]$^{2-}$

6-16. **Refer to Section 6-1.** . . . .

All of the following chemical species, O^{2-}, F^-, Ne, Mg^{2+} and Al^{3+}, have 10 electrons and are isoelectronic. The sodium atom, Na, has 11 electrons.

6-18. **Refer to Section 6-1.** •

The following species have 10 electrons and are therefore isoelectronic with neon:

cations: Na^+, Mg^{2+}, Al^{3+} anions: F^-, O^{2-}, N^{3-}

6-20. **Refer to Section 6-1.** •

(a) Cations with ....$3s^23p^6$ electronic configurations are isoelectronic with argon.

Examples: K^+, Ca^{2+}, Sc^{3+}

(b) Cations with ....$6s^26p^6$ electronic configurations are isoelectronic with radon.

Examples: Fr^+, Ra^{2+}

6-22. **Refer to Section 6-1.** •

(a) $2Na + S \rightarrow Na_2S$

$_{11}Na$	[Ne] ↑ (3s)	→	Na^+	[Ne] __ (3s)	1 e^- lost
$_{11}Na$	[Ne] ↑ (3s)	→	Na^+	[Ne] __ (3s)	1 e^- lost
$_{16}S$	[Ne] ↑↓ (3s) ↑↓ ↑ ↑ (3p)	→	S^{2-}	[Ne] ↑↓ (3s) ↑↓ ↑↓ ↑↓ (3p)	2 e^- gained

$2Na([Ne]3s^1) + S([Ne]3s^23p^4) \rightarrow 2Na^+([Ne]) + S^{2-}([Ne]3s^23p^6)$

(b) $Mg + 2Br \rightarrow MgBr_2$

$_{12}Mg$	[Ne] ↑↓ (3s)	→	Mg^{2+}	[Ne] __ (3s)	2 e^- lost
$_{35}Br$	$[Ar]3d^{10}$ ↑↓ (4s) ↑↓ ↑↓ ↑ (4p)	→	Br^-	$[Ar]3d^{10}$ ↑↓ (4s) ↑↓ ↑↓ ↑↓ (4p)	1 e^- gained
$_{35}Br$	$[Ar]3d^{10}$ ↑↓ (4s) ↑↓ ↑↓ ↑ (4p)	→	Br^-	$[Ar]3d^{10}$ ↑↓ (4s) ↑↓ ↑↓ ↑↓ (4p)	1 e^- gained

$Mg([Ne]3s^2) + 2Br([Ar]4s^24p^5) \rightarrow Mg^{2+}([Ne]) + 2Br^-([Ar]4s^24p^6)$

(c) $Al + 3F \rightarrow AlF_3$

$_{13}Al$	[Ne] ↑↓ (3s) ↑ __ __ (3p)	→	Al^{3+}	[Ne] __ (3s) __ __ __ (3p)	3 e^- lost
$_{9}F$	↑↓ (1s) ↑↓ (2s) ↑↓ ↑↓ ↑ (2p)	→	F^-	↑↓ (1s) ↑↓ (2s) ↑↓ ↑↓ ↑↓ (2p)	1 e^- gained
$_{9}F$	↑↓ (1s) ↑↓ (2s) ↑↓ ↑↓ ↑ (2p)	→	F^-	↑↓ (1s) ↑↓ (2s) ↑↓ ↑↓ ↑↓ (2p)	1 e^- gained
$_{9}F$	↑↓ (1s) ↑↓ (2s) ↑↓ ↑↓ ↑ (2p)	→	F^-	↑↓ (1s) ↑↓ (2s) ↑↓ ↑↓ ↑↓ (2p)	1 e^- gained

$Al([Ne]3s^23p^1) + 3F(1s^22s^22p^5) \rightarrow Al^{3+}([Ne]) + 3F^-(1s^22s^22p^6)$

6-24. **Refer to Section 6-1 and Table 6-1.** . . . .

(a) $3M + 2X \rightarrow M_3X_2$ $(3M^{2+}, 2X^{3-})$

(b) $M + X \rightarrow MX$ (M^{2+}, X^{2-})

(c) $M + 3X \rightarrow MX_3$ $(M^{3+}, 3X^{-})$

6-26. **Refer to the Introduction to Covalent Bonding and Figure 6-2.** . . . .

A covalent bond is formed when 2 atoms share one or more pairs of electrons. The electrons, which are negatively charged, preferentially occupy the region between the atoms because they are attracted to the positively charged nuclei of both atoms.

6-28. **Refer to Section 6-3.** . . . .

A heteronuclear diatomic molecule consists of two atoms of different elements covalently bonded together, e.g., HF or NO. A homonuclear diatomic molecule consists of two atoms of the same element covalently bonded together, e.g., H_2 or N_2.

6-30. **Refer To Section 6-5.** . . . .

(a) Lewis dot formulas are representations of molecules, ions or formula units which show the element symbols, the order in which the atoms are connected, the number of valence electrons linking the atoms together, and the number of lone pairs of valence electrons not used for bonding. They do not show the shape of a chemical species.

(b)

```
H2       H:H                  N2      :N:::N:

          ..  ..                        ..
Br2      :Br:Br:              HCl     H:Cl:
          ..  ..                        ..
           ..
HI       H:I:
           ..
```

6-32. **Refer to Section 6-5.** . . . .

```
           ..                        ..
(a) H2O  H:O:             H2S      H:S:
           ..                        ..
           H                         H

           ..
(b) NH3  H:N:H            NH4+       H  +
           ..                        ..
           H                       H:N:H
                                     ..
                                     H

           ..
(c) PH3  H:P:H            PH4+       H  +
           ..                        ..
           H                       H:P:H
                                     ..
                                     H
```

6-34. **Refer to Section 6-6 and Exercise 6-32 Solution.** . . . .

(a) $S = N - A$

(b) S = total number of electrons <u>shared</u> in the molecule or polyatomic ion.

N = number of valence shell electrons <u>needed</u> by all the atoms in the molecule or ion to achieve noble gas configurations.

A = number of electrons <u>available</u> in the valence shells of all the atoms and is the sum of their periodic group numbers.

(c) H_2O $S = N - A$
$= [2 \times 2(\text{for H}) + 1 \times 8(\text{for O})] - [2 \times 1(\text{for H}) + 1 \times 6(\text{for O})]$
$= 12 - 8$
$= 4$ and there are 4 electrons shared in the molecule

H_2S $S = N - A$
$= [2 \times 2(\text{for H}) + 1 \times 8(\text{for S})] - [2 \times 1(\text{for H}) + 1 \times 6(\text{for S})]$
$= 12 - 8$
$= 4$ and there are 4 electrons shared in the molecule

NH_3 $S = N - A$
$= [3 \times 2(\text{for H}) + 1 \times 8(\text{for N})] - [3 \times 1(\text{for H}) + 1 \times 5(\text{for N})]$
$= 14 - 8$
$= 6$ and there are 6 electrons shared in the molecule

NH_4^+ $S = N - A$
$= [4 \times 2(\text{for H}) + 1 \times 8(\text{for N})]$
$- [4 \times 1(\text{for H}) + 1 \times 5(\text{for N}) - 1\ e^-]$
$= 16 - 8$
$= 8$ and there are 8 electrons shared in the polyatomic ion

PH_3 $S = N - A$
$= [3 \times 2(\text{for H}) + 1 \times 8(\text{for P})] - [3 \times 1(\text{for H}) + 1 \times 5(\text{for P})]$
$= 14 - 8$
$= 6$ and there are 6 electrons shared in the molecule

PH_4^+ $S = N - A$
$= [4 \times 2(\text{for H}) + 1 \times 8(\text{for N})]$
$- [4 \times 1(\text{for H}) + 1 \times 5(\text{for N}) - 1\ e^-]$
$= 16 - 8$
$= 8$ and there are 8 electrons shared in the polyatomic ion

6-36. **Refer to Section 6-7 and Example 6-4.** . . . .

(a) HCO_3^- $S = N - A$
$= [1 \times 2(\text{for H}) + 1 \times 8(\text{for C}) + 3 \times 8(\text{for O})]$
$- [1 \times 1(\text{for H}) + 1 \times 4(\text{for C}) + 3 \times 6(\text{for O}) + 1\ e^-]$
$= 34 - 24$
$= 10$ shared electrons

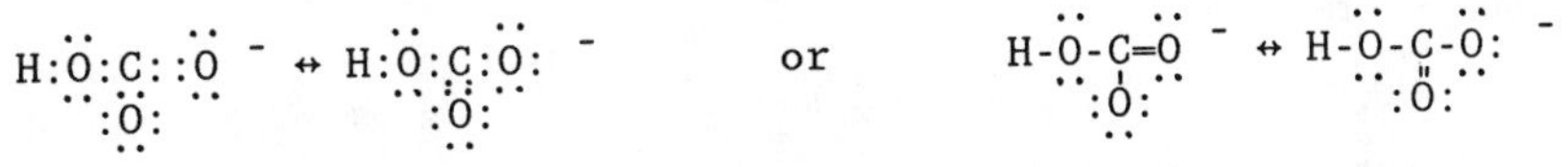

(b) NO_3^- $S = N - A$
$= [1 \times 8(\text{for N}) + 3 \times 8(\text{for O})]$
$- [1 \times 5(\text{for N}) + 3 \times 6(\text{for O}) + 1\ e^-]$
$= 32 - 24$
$= 8$ shared electrons

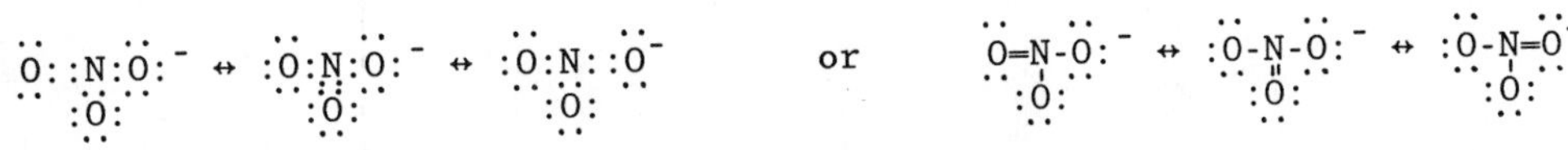

(c) SO_2 $S = N - A$
$= [1 \times 8(\text{for S}) + 2 \times 8(\text{for O})] - [1 \times 6(\text{for S}) + 2 \times 6(\text{for O})]$
$= 24 - 18$
$= 6$ shared electrons

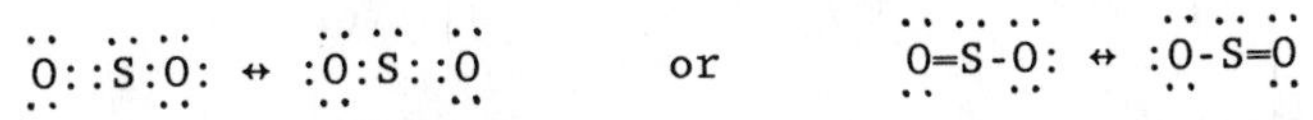

(d) SO_3 $\quad S = N - A$

$= [1 \times 8(\text{for S}) + 3 \times 8(\text{for O})] - [1 \times 6(\text{for S}) + 3 \times 6(\text{for O})]$

$= 32 - 24$

$= 8$ shared electrons

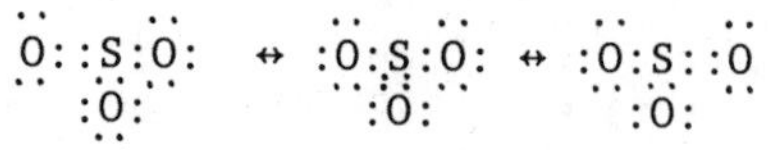

or

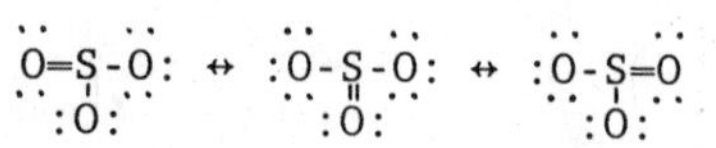

6-38. Refer to Section 6-8 and Examples 6-5, 6-6 and 6-7. . . . .

(a) Only one resonance structure of CO obeys the octet rule.

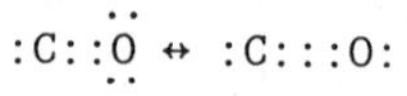

(b) SF_4 violates the octet rule (**Section 6-8**, Limitation 4).

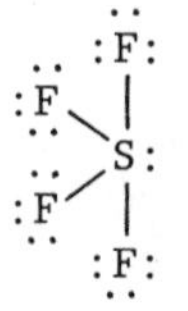

(c) $BeBr_2$ violates the octet rule (**Section 6-8**, Limitation 1).

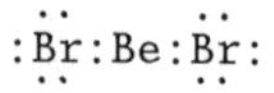

(d) AsH_3 obeys the octet rule.

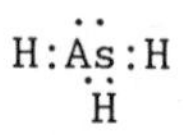

6-40. Refer to Section 6-6. . . . .

S_8 $\quad S = N - A = [8 \times 8(\text{for S})] - [8 \times 6(\text{for S})] = 64 - 48 = 16$ shared electrons

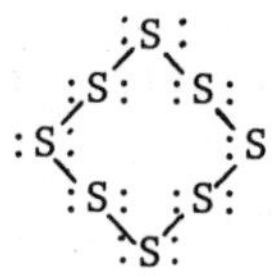

or

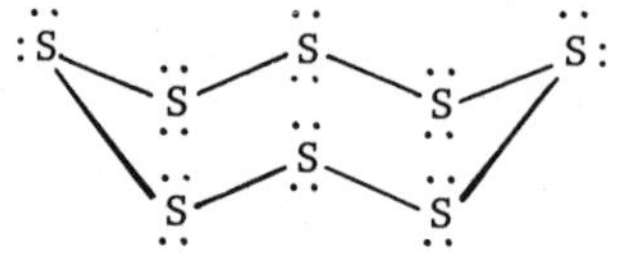

6-42. Refer to Sections 6-5, 6-6 and 6-8. . . . .

(a) BF_3 violates the octet rule (**Section 6-8**, Limitation 2).

(b) PF_3 obeys the octet rule.

$S = N - A = [1 \times 8(\text{for P}) + 3 \times 8(\text{for F})] - [1 \times 5(\text{for P}) + 3 \times 7(\text{for F})]$

$= 32 - 26$

$= 6$ shared electrons

(c) AsF_5 violates the octet rule (**Section 6-8**, Limitation 4).

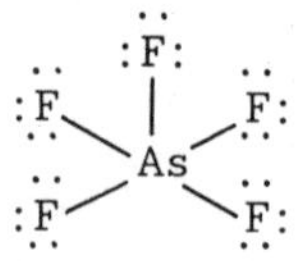

(d) $POBr_3$ obeys the octet rule. P is the least electronegative element and therefore is the central atom.

$S = N - A$

$= [1 \times 8(\text{for P}) + 1 \times 8(\text{for O}) + 3 \times 8(\text{for Br})]$

$\quad - [1 \times 5(\text{for P}) + 1 \times 6(\text{for O}) + 3 \times 7(\text{for Br})]$

$= 40 - 32$

$= 8$ shared electrons

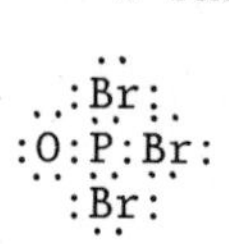

(e) HNO_2 obeys the octet rule. The hydrogen atom is attached to an oxygen atom.

$$\begin{aligned} S &= N - A \\ &= [1 \times 2(\text{for H}) + 1 \times 8(\text{for N}) + 2 \times 8(\text{for O})] \\ &\qquad - [1 \times 1(\text{for H}) + 1 \times 5(\text{for N}) + 2 \times 6(\text{for O})] \\ &= 26 - 18 \\ &= 8 \text{ shared electrons} \end{aligned}$$

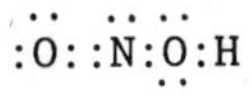

(f) SeF_4 violates the octet rule (**Section 6-8**, Limitation 4).

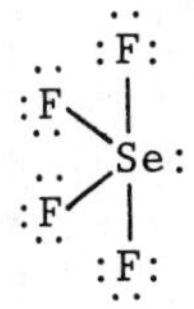

6-44. **Refer to Sections 6-5 through 6-8.** . . . .

(a) 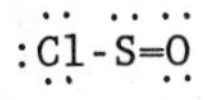

The number of valence electrons in this molecule should be = [1 × 7(for Cl) + 1 × 6(for S) + 1 × 6(for O)] = 19 electrons. However, this structure only shows 18 electrons. This molecule cannot exist as written.

(b) 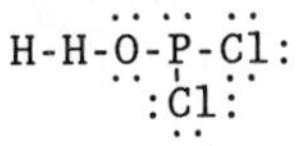

This compound will not exist because (1) a hydrogen atom cannot bond to 2 different atoms through the sharing of 4 electrons and (2) the structure as drawn requires 28 valence electrons and only 27 valence electrons are available.

(c) 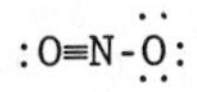

The total number of valence electrons in this polyatomic ion should be = [2 × 6(for O) + 1 × 5(for N) + 1 e^-] = 18 valence electrons. However this structure only contains 16 electrons. So, this ion cannot exist as written.

(d) 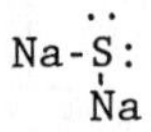

This covalent compound will not exist because Na only forms ionic compounds.

6-46. **Refer to Section 6-6.** . . . .

These compounds obey the octet rule, $S = N - A$. We will use this equation to solve for the periodic group (the number of valence electrons) of El.

(a) 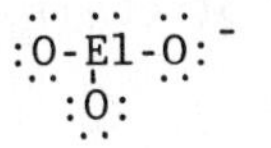

$$\begin{aligned} S &= N - A \\ 6 &= [1 \times 8(\text{for El}) + 3 \times 8(\text{for O})] \\ &\qquad - [1 \times ?(\text{for El}) + 3 \times 6(\text{for O}) + 1\ e^-] \\ 6 &= 32 - (? + 19) \\ ? &= 7 \end{aligned}$$

Therefore, El is a VIIA element. The polyatomic ion could be ClO_3^-.

(b) 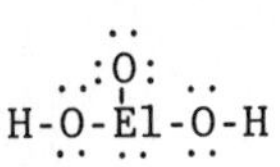

$$\begin{aligned} S &= N - A \\ 10 &= [2 \times 2(\text{for H}) + 3 \times 8(\text{for O}) + 1 \times 8(\text{for El})] \\ &\qquad - [2 \times 1(\text{for H}) + 3 \times 6(\text{for O}) + 1 \times ?(\text{for El})] \\ 10 &= 36 - (20 + ?) \\ ? &= 6 \end{aligned}$$

Therefore, El is a VIA element. This compound could be H_2SO_3.

(c) 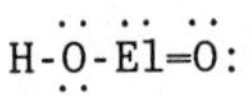

$$\begin{aligned} S &= N - A \\ 8 &= [1 \times 2(\text{for H}) + 2 \times 8(\text{for O}) + 1 \times 8(\text{for El})] \\ &\qquad - [1 \times 1(\text{for H}) + 2 \times 6(\text{for O}) + 1 \times ?(\text{for El})] \\ 8 &= 26 - (13 + ?) \\ ? &= 5 \end{aligned}$$

Therefore, El is a VA element. This compound could be HNO_2.

6-48. **Refer to Sections 6-3 and 6-5.** . . . .

The compound C_2Cl_4 is covalent because it is composed of the nonmetals, C and Cl.

Its Lewis dot structure is

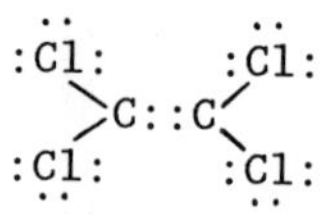

$$S = N - A$$
$$= [2 \times 8(\text{for C}) + 4 \times 8(\text{for Cl})] - [2 \times 4(\text{for C}) + 4 \times 7(\text{for Cl})]$$
$$= 12 \text{ shared electrons}$$

6-50. **Refer to Sections 6-5 through 6-8.** . . . .

(a) SO_2 exhibits resonance (**Exercise 6-36c Solution**).

(b) NO_2 exhibits resonance, but it violates the octet rule because the compound contains an odd number of valence electrons, 17 (**Section 6-8**, Limitation 3).

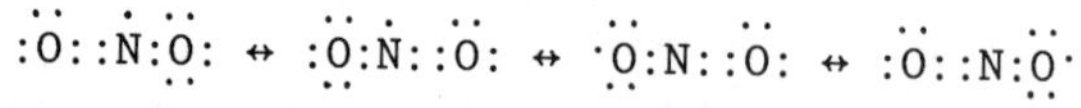

(c) CO exhibits resonance. It is known from experiments that the C-O bond in CO is intermediate between a typical double and triple bond length.

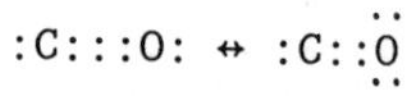

$$S = N - A$$
$$= [1 \times 8(\text{for C}) + 1 \times 8(\text{for O})] - [1 \times 4(\text{for C}) + 1 \times 6(\text{for O})]$$
$$= 16 - 10$$
$$= 6 \text{ shared electrons}$$

(d) O_3 exhibits resonance.

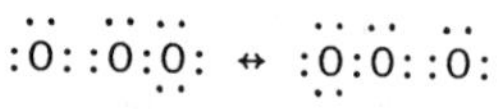

$$S = N - A$$
$$= [3 \times 8(\text{for O})] - [3 \times 6(\text{for O})]$$
$$= 24 - 18$$
$$= 6 \text{ shared electrons}$$

(e) SO_3 exhibits resonance (**Exercise 6-36d Solution**).

(f) $(NH_4)_2SO_4$ is an ionic solid composed of covalently bonded polyatomic ions:

NH_4^+

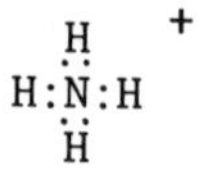

$$S = N - A$$
$$= [4 \times 2(\text{for H}) + 1 \times 8(\text{for N})] - [4 \times 1(\text{for H}) + 1 \times 5(\text{for N}) - 1\ e^-]$$
$$= 16 - 8$$
$$= 8 \text{ shared electrons}$$

SO_4^{2-}

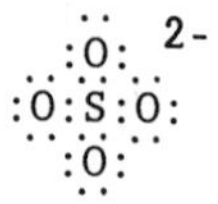

$$S = N - A$$
$$= [4 \times 8(\text{for O}) + 1 \times 8(\text{for S})] - [4 \times 6(\text{for O}) + 1 \times 6(\text{for S}) + 2\ e^-]$$
$$= 40 - 32$$
$$= 8 \text{ shared electrons}$$

6-52. **Refer to Section 6-7.** . . . .

The formal charge, FC = (Group No.) - [(No. of bonds) + (No. of unshared e^-)]

(a)

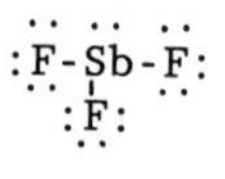

for Sb, FC = 5 - (3 + 2) = 0
for F, FC = 7 - (1 + 6) = 0

(b)

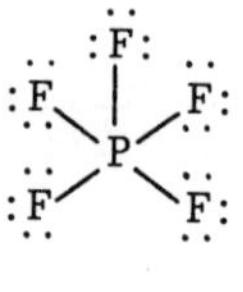

for P, FC = 5 - (5 + 0) = 0
for F, FC = 7 - (1 + 6) = 0

(c) :O=C=O: for C, FC = 4 - (4 + 0) = 0
for O, FC = 6 - (2 + 4) = 0

6-54. Refer to Section 6-7.

The formal charge, FC = (Group No.) - [(No. of bonds) + (No. of unshared e^-)]

(a) $:N{\equiv}N\text{-}N:^-$ (1 2 3)
for N(1), FC = 5 - (3 + 2) = 0
for N(2), FC = 5 - (4 + 0) = +1
for N(3), FC = 5 - (1 + 6) = -2

(b) $:O{=}N{=}O:^+$
for N, FC = 5 - (4 + 0) = +1
for O, FC = 6 - (2 + 4) = 0

(c) :Cl-Al-Cl: (with :Cl: above and below Al) $^-$
for Al, FC = 3 - (4 + 0) = -1
for Cl, FC = 7 - (1 + 6) = 0

6-56. Refer to Section 6-7.

The formal charge, FC = (Group No.) - [(No. of bonds) + (No. of unshared e^-)]

(a) :F-N=O: 2 (with :O: 1 below N)
for F, FC = 7 - (1 + 6) = 0
for N, FC = 5 - (4 + 0) = +1
for O(1), FC = 6 - (1 + 6) = -1
for O(2), FC = 6 - (2 + 4) = 0

The sum of the formal charges of the elements is zero. Therefore, this dash formula is likely (Rule 1a).

(b) :F=N-O: (with :O: below N)
for F, FC = 7 - (2 + 4) = +1
for N, FC = 5 - (4 + 0) = +1
for O, FC = 6 - (1 + 6) = -1

The sum of the elemental formal charges for the molecule is 0. However, the adjacent atoms, F and N, were both assigned formal charges of +1. According to Rule 6, this representation of the molecule is not likely.

(c) :F, :F bonded to N(1); N(1)-N(2); N(2)=O:1 and N(2)-O:2
for F, FC = 7 - (1 + 6) = 0
for N(1), FC = 5 - (3 + 0) = +2
for N(2), FC = 5 - (4 + 0) = +1
for O(1), FC = 6 - (2 + 4) = 0
for O(2), FC = 6 - (1 + 6) = -1

This is an unfavorable representation since the sum of the elemental formal charges is not zero and the adjacent nitrogen atoms were both assigned positive formal charges.

6-58. Refer to Section 6-3 and Table 5-6.

An HCl molecule is a heteronuclear diatomic molecule composed of H (EN = 2.1) and Cl (EN = 3.0). Because the electronegativities of the elements are different, the pull on the electrons in the covalent bond between them is unequal. Hence HCl is a polar molecule. A homonuclear diatomic molecule contains a nonpolar bond, since the electron pair between the two atoms is shared equally. Cl_2 is an example of a homonuclear diatomic molecule.

6-60. **Refer to Sections 6-3 and 6-4, and Table 6-2.** . . . .

The polarities decrease in the order HF > HCl > HBr > HI paralleling a decrease in the difference in electronegativity values. Remember that F has the highest value of electronegativity of all the elements.

6-62. **Refer to Section 6-4.** . . . .

The dipole moment is an indication of the polarity of a diatomic molecule on a numerical scale. It is defined as the product of the distance separating charges of equal magnitude and opposite sign and the magnitude of the charge. Generally, as the polarity of a diatomic molecule increases so does its dipole moment.

6-64. **Refer to Table 5-6 and Section 6-3.** . . . .

We know that bond polarity increases with increasing Δ(EN), the difference in electronegativity between 2 atoms that are bonded together.

bond	Br – Cl	Br – Br	Br – F	Br – I
EN	2.8 3.0	2.8 2.8	2.8 4.0	2.8 2.5
Δ(EN)	0.2	0	1.2	0.3

Therefore, in order of increasing bond polarity: Br-Br < Br-Cl < Br-I < Br-F.

6-66. **Refer to the Introduction to Covalent Bonding and Table 5-6.** . . . .

The covalent character of the bonds in a compound increases as the electronegativity differences between its component elements decreases.

(a) Al_2O_3 Δ(EN) for Al(EN = 1.5) and O(EN = 3.5) = 2.0

B_2O_3 Δ(EN) for B(EN = 2.0) and O(EN = 3.5) = 1.5

Therefore, B_2O_3 has the greater covalent character.

(b) Al_2O_3 Δ(EN) for Al(EN = 1.5) and O(EN = 3.5) = 2.0

Ga_2O_3 Δ(EN) for Ga(EN = 1.7) and O(EN = 3.5) = 1.8

Therefore, Ga_2O_3 has the greater covalent character.

(c) BeF_2 Δ(EN) for Be(EN = 1.5) and F(EN = 4.0) = 2.5

BF_3 Δ(EN) for B(EN = 2.0) and F(EN = 4.0) = 2.0

Therefore, BF_3 has the greater covalent character.

(d) SiF_4 Δ(EN) for Si(EN = 1.8) and F(EN = 4.0) = 2.2

SnF_4 Δ(EN) for Sn(EN = 1.8) and F(EN = 4.0) = 2.2

Therefore, SiF_4 and SnF_4 have similar degrees of covalent character. SiF_4 is probably more covalent due to a better overlap of orbitals of similar size and energy.

6-68. **Refer to Section 6-9.** . . . .

Oxidation numbers can be useful as mechanical aids in writing formulas and in balancing equations.

6-70. **Refer to Section 6-9 and Table 6-3.** . . . .

In simple binary ionic compounds, IA, IIA and IIIA elements exhibit +1, +2 and +3 oxidation states, respectively. The VA, VIA and VIIA elements exhibit -3, -2 and -1 oxidation states, respectively.

Therefore, we have

Element	Group Number	Oxidation Number
Cl	VIIA	-1
O	VIA	-2
Sr	IIA	+2
K	IA	+1
Al	IIIA	+3
Se	VIA	-2

6-72. **Refer to Section 6-9 and Example 6-9 and Table 6-3.** . . . .

For a compound, the sum of the oxidation numbers of the component elements must be equal to zero. If we let x = oxidation number of Cr,

(a) $Cr(OH)_3$ $0 = x + 3(\text{ox. no. O}) + 3(\text{ox. no. H}) = x + 3(-2) + 3(+1) = x - 3$
$x = +3$

(b) $CrCl_2$ $0 = x + 2(\text{ox. no. Cl}) = x + 2(-1) = x - 2$
$x = +2$

(c) $Na_2Cr_2O_7$ $0 = 2(\text{ox. no. Na}) + 2x + 7(\text{ox. no. O}) = 2(+1) + 2x + 7(-2)$
$0 = 2x - 12$
$x = +6$

(d) H_2CrO_4 $0 = 2(\text{ox. no. H}) + x + 4(\text{ox. no. O}) = 2(+1) + x + 4(-2)$
$0 = x - 6$
$x = +6$

(e) $Cr_2(SO_4)_3$ $0 = 2x + 3(\text{ox. no. S}) + 12(\text{ox. no. O}) = 2x + 3(+6) + 12(-2)$
$0 = 2x - 6$
$x = +3$

(f) Cr_2O_3 $0 = 2x + 3(\text{ox. no. O}) = 2x + 3(-2) = 2x - 6$
$x = +3$

(g) CrO $0 = x + 1(\text{ox. no. O}) = x + 1(-2) = x - 2$
$x = +2$

(h) $KCrO_2$ $0 = 1(\text{ox. no. K}) + x + 2(\text{ox. no. O}) = 1(+1) + x + 2(-2) = x - 3$
$x = +3$

(i) CrO_3 $0 = x + 3(\text{ox. no. O}) = x + 3(-2) = x - 6$
$x = +6$

6-74. **Refer to Section 6-9 and Table 6-3.** . . . .

For a compound, the sum of the oxidation numbers of the component elements must be zero.

(a) Let x = oxidation state of S

S_8 $0 = 8x$
$x = 0$

K_2S $0 = 2(\text{ox. no. K}) + x = 2(+1) + x = x + 2$
$x = -2$

$NaHSO_4$ $0 = 1(\text{ox. no. Na}) + 1(\text{ox. no. H}) + x + 4(\text{ox. no. O})$
$0 = 1(+1) + 1(+1) + x + 4(-2) = x - 6$
$x = +6$

CS_2 $\quad 0 = 1(\text{ox. no. C}) + 2x = 1(+4) + 2x = 2x + 4$
$x = -2$

Note: See explanation at the end of this exercise.

$SOCl_2$ $\quad 0 = x + 1(\text{ox. no. O}) + 2(\text{ox. no. Cl}) = x + 1(-2) + 2(-1)$
$= x - 4$
$x = +4$

(b) Let x = oxidation state of N

$NaNO_3$ $\quad 0 = 1(\text{ox. no. Na}) + x + 3(\text{ox. no. O}) = 1(+1) + x + 3(-2) = x - 5$
$x = +5$

HNO_2 $\quad 0 = 1(\text{ox. no. H}) + x + 2(\text{ox. no. O}) = 1(+1) + x + 2(-2) = x - 3$
$x = +3$

NF_3 $\quad 0 = x + 3(\text{ox. no. F}) = x + 3(-1) = x - 3$
$x = +3$

N_2O_5 $\quad 0 = 2x + 5(\text{ox. no. O}) = 2x + 5(-2) = 2x - 10$
$x = +5$

NO $\quad 0 = x + 1(\text{ox. no. O}) = x + 1(-2) = x - 2$
$x = +2$

NO_2 $\quad 0 = x + 2(\text{ox. no. O}) = x + 2(-2) = x - 4$
$x = +4$

(c) Let x = oxidation state of Mn

MnO $\quad 0 = x + 1(\text{ox. no. O}) = x + 1(-2) = x - 2$
$x = +2$

MnO_2 $\quad 0 = x + 2(\text{ox. no. O}) = x + 2(-2) = x - 4$
$x = +4$

$Mn(OH)_2$ $\quad 0 = x + 2(\text{ox. no. O}) + 2(\text{ox. no. H}) = x + 2(-2) + 2(+1) = x - 2$
$x = +2$

K_2MnO_4 $\quad 0 = 2(\text{ox. no. K}) + x + 4(\text{ox. no. O}) = 2(+1) + x + 4(-2) = x - 6$
$x = +6$

$KMnO_4$ $\quad 0 = 1(\text{ox. no. K}) + x + 4(\text{ox. no. O}) = 1(+1) + x + 4(-2) = x - 7$
$x = +7$

Mn_2O_7 $\quad 0 = 2x + 7(\text{ox. no. O}) = 2x + 7(-2) = 2x - 14$
$x = +7$

Note: The following describes a fundamental method for determining the oxidation numbers of elements in a covalent compound. First, draw the Lewis dot structure. Electrons shared between two unlike atoms are counted with the more electronegative atom. Electrons shared between two like atoms are divided equally between the sharing atoms. Then

$$\text{oxidation number} = \left(\begin{array}{c}\text{No. of Valence Electrons}\\ \text{in the element}\end{array}\right) - \left(\begin{array}{c}\text{No. of Electrons Counted}\\ \text{with the element}\end{array}\right)$$

For example, let us determine the oxidation numbers of carbon (4 valence e^-) and sulfur (6 valence e^-) in CS_2, which has the Lewis dot structure,

$$:\ddot{S}::C::\ddot{S}:$$

S is more electronegative than C, and so in CS_2, 8 electrons are counted with each S atom and zero electrons are counted with the C atom. Therefore,

oxidation number of S = 6 - 8 = -2
oxidation number of C = 4 - 0 = +4

6-76. **Refer to Section 6-9 and Table 6-3.** . . . .

For an ion, the sum of the oxidation numbers of the component elements must equal the charge on the ion.

(a) Let x = the oxidation state of S

S^{2-} x = -2

SO_4^{2-} -2 = x + 4(ox. no. O) = x + 4(-2) = x - 8
x = +6

$S_2O_3^{2-}$ -2 = 2x + 3(ox. no. O) = 2x + 3(-2) = 2x - 6
2x = 4
x = +2

$S_4O_6^{2-}$ -2 = 4x + 6(ox. no. O) = 4x + 6(-2) = 4x - 12
4x = 10
x = +2.5

(b) Let x = oxidation state of Cr

CrO_2^- -1 = x + 2(ox. no. O) = x + 2(-2) = x - 4
x = +3

$Cr(OH)_4^-$ -1 = x + 4(ox. no. O) + 4(ox. no. H) = x + 4(-2) + 4(+1) = x - 4
x = +3

CrO_4^{2-} -2 = x + 4(ox. no. O) = x + 4(-2) = x - 8
x = +6

$Cr_2O_7^{2-}$ -2 = 2x + 7(ox. no. O) = 2x + 7(-2) = 2x - 14
2x = 12
x = +6

(c) Let x = oxidation state of B

BO_2^- -1 = x + 2(ox. no. O) = x + 2(-2) = x - 4
x = +3

BO_3^{3-} -3 = x + 3(ox. no. O) = x + 3(-2) = x - 6
x = +3

$B_4O_7^{2-}$ -2 = 4x + 7(ox. no. O) = 4x + 7(-2) = 4x - 14
4x = 12
x = +3

6-78. **Refer to Section 6-10 and Table 6-4.** . . . .

(a) diarsenic pentoxide As_2O_5 (b) cadmium acetate $Cd(CH_3COO)_2$

(c) ferric carbonate $Fe_2(CO_3)_3$ (d) copper(I) iodide CuI

(e) dinitrogen pentoxide N_2O_5 (f) barium fluoride BaF_2

(g) aluminum perchlorate $Al(ClO_4)_3$ (h) cobalt(II) cyanide $Co(CN)_2$

6-80. **Refer to Section 6-10 and Table 6-4.** . . . .

(a) nitrogen dioxide NO_2

(b) selenium trioxide SeO_3

(c) barium hydrogen carbonate $Ba(HCO_3)_2$

(d) chromium(III) sulfite $Cr_2(SO_3)_3$

(e) zinc bromate $Zn(BrO_3)_2$

(f) sodium bromite $NaBrO_2$

(g) copper(II) sulfide CuS

(h) iron(II) arsenate $Fe_3(AsO_4)_2$

(i) stannous iodide SnI_2

6-82. **Refer to Section 6-10 and Table 6-4.** . . . .

(a) $Ca_3(PO_4)_2$ calcium phosphate

(b) Cr_2O_3 chromium(III) oxide

(c) $Fe_2(SO_4)_3$ iron(III) sulfate or ferric sulfate

(d) HBr (aq) hydrobromic acid

(e) $Zn(OH)_2$ zinc hydroxide

(f) $Mg(MnO_4)_2$ magnesium permanganate

(g) NH_4ClO_2 ammonium chlorite

(h) FeS iron(II) sulfide or ferrous sulfide

6-84. **Refer to Section 6-10.** . . . .

(a) H_2S hydrogen sulfide

(b) SnO_2 tin(IV) oxide or stannic oxide

(c) $Cr_2(CO_3)_3$ chromium(III) carbonate

(d) Hg_2I_2 mercury(I) iodide or mercurous iodide

(e) $Ca(OH)_2$ calcium hydroxide

(f) $Fe(OH)_2$ iron(II) hydroxide or ferrous hydroxide

(g) $NiHPO_4$ nickel(II) hydrogen phosphate

(h) Li_2CrO_4 lithium chromate

6-86. **Refer to Section 6-10.** . . . .

(a)	$MgBr_2$	magnesium bromide	KBr	potassium bromide
	$NiBr_2$	nickel(II) bromide	NH_4Br	ammonium bromide
	$FeBr_2$	iron(II) bromide or ferrous bromide	$FeBr_3$	iron(III) bromide or ferric bromide

(b)	$Mg(CN)_2$	magnesium cyanide	KCN	potassium cyanide
	$Ni(CN)_2$	nickel(II) cyanide	NH_4CN	ammonium cyanide
	$Fe(CN)_2$	iron(II) cyanide or ferrous cyanide	$Fe(CN)_3$	iron(III) cyanide or ferric cyanide
(c)	$MgSO_4$	magnesium sulfate	K_2SO_4	potassium sulfate
	$NiSO_4$	nickel(II) sulfate	$(NH_4)_2SO_4$	ammonium sulfate
	$FeSO_4$	iron(II) sulfate or ferrous sulfate	$Fe_2(SO_4)_3$	iron(III) sulfate or ferric sulfate
(d)	$Mg(HCO_3)_2$	magnesium bicarbonate	$KHCO_3$	potassium bicarbonate
	$Ni(HCO_3)_2$	nickel(II) bicarbonate	NH_4HCO_3	ammonium bicarbonate
	$Fe(HCO_3)_2$	iron(II) bicarbonate or ferrous bicarbonate	$Fe(HCO_3)_3$	iron(III) bicarbonate or ferric bicarbonate
(e)	$Mg(OH)_2$	magnesium hydroxide	KOH	potassium hydroxide
	$Ni(OH)_2$	nickel(II) hydroxide	NH_4OH	(ammonium hydroxide)
	$Fe(OH)_2$	iron(II) hydroxide or ferrous hydroxide	$Fe(OH)_3$	iron(III) hydroxide or ferric hydroxide
(f)	$Mg(NO_3)_2$	magnesium nitrate	KNO_3	potassium nitrate
	$Ni(NO_3)_2$	nickel(II) nitrate	NH_4NO_3	ammonium nitrate
	$Fe(NO_3)_2$	iron(II) nitrate or ferrous nitrate	$Fe(NO_3)_3$	iron(III) nitrate or ferric nitrate

7 Molecular Structure and Covalent Bonding Theories

7-2. **Refer to Sections 6-6 and 7-6.**

(a) "Bonding pair" is a term that refers to a pair of electrons that is shared between two nuclei in a covalent bond, while the term "lone pair" refers to an unshared pair of electrons that is associated with a single nucleus.

(b) Lone pairs of electrons occupy more space than bonding pairs. This fact was determined experimentally from measurements of bond angles of many molecules and polyatomic ions. An explanation for this is the fact that a lone pair has only one atom exerting strong attractive forces on it, and it exists closer to the nucleus than bond pairs.

(c) The relative magnitudes of the repulsive forces between pairs of electrons on an atom are as follows: *lp - lp* >> *lp - bp* > *bp - bp* where *lp* refers to lone pairs and bp refers to bonding pairs of valence shell electrons.

7-4. **Refer to Section 6-6 and 7-5.**

Consider the molecule with formula, AB_4.

If its molecular geometry is square planar, it has octahedral electronic geometry with two lone pairs, 180° apart. All molecules with octahedral electronic geometry must disobey the octet rule since they have 12 electrons, not 8, around the central atom.

If its molecular geometry is tetrahedral, it has tetrahedral electronic geometry as well, as there are no lone pairs of electrons on the central atom. The molecule does obey the octet rule since the molecule has 8 electrons around the central atom.

7-6. **Refer to Sections 6-6 and Sections 7-5, 7-7, 7-10 and 7-14.**

(a) H_2O

$S = N - A = [2 \times 2(\text{for H}) + 1 \times 8(\text{for O})] - [2 \times 1(\text{for H}) + 1 \times 6(\text{for O})] = 4$ shared electrons

```
 ..
H:O:
  ..
  H
```

The Lewis dot formula predicts 4 regions of high electron density around the central O atom and a tetrahedral electronic geometry. The number of lone pairs on the O atom is 2, so the molecular geometry of the molecule is bent (angular).

(b) $SnCl_4$

Sn is a IVA element and has 4 valence electrons.

$S = N - A = [4 \times 8(\text{for Cl}) + 1 \times 8(\text{for Sn})] - [4 \times 7(\text{for Cl}) + 1 \times 4(\text{for Sn})] = 8$ shared electrons

```
     ..
    :Cl:
 ..  ..  ..
:Cl:Sn:Cl:
 ..  ..  ..
    :Cl:
     ..
```

The Lewis dot formula predicts 4 regions of high electron density around the central Sn atom and a tetrahedral electronic geometry. Since there are no lone pairs on Sn, the molecular geometry is also tetrahedral.

(c) SF_6

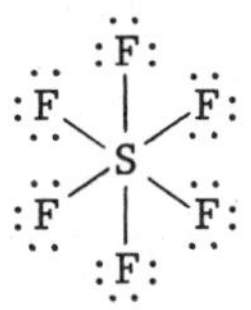

This molecule disobeys the octet rule since 12 electrons must be shared to form 6 S-F bonds. The Lewis dot structure predicts 6 regions of high electron density around the central S atom and an octahedral electronic geometry. There are no lone pairs on the S atom, so the molecular geometry is the same as the electronic geometry.

(d) SbF_6^-

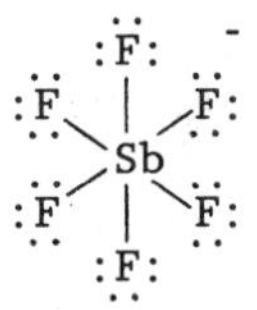

This polyatomic ion, like (c), must also violate the octet rule Since 12 electrons must be shared to form 6 Sb-F bonds. Sb is a VA element, but the charge on the ion gives an extra electron which participates in bonding. The Lewis dot structure predicts 6 regions of high electron density around the central Sb atom and an octahedral electronic geometry. There are no lone pairs on the Sb atom, so the ionic geometry is the same as the electronic geometry.

7-8. **Refer to Exercise 7-6 Solution and Section 7-7.** . . . .

(a) H_2O — The ideal bond angles would be those for a perfect tetrahedral structure, 109.5^o.

$SnCl_4$ — The ideal bond angles would be 109.5^o since the structure is tetrahedral.

SF_6 — The ideal bond angles would be those for an octahedral geometry, 90°, 180°.

SbF_6^- — The ideal bond angles would be those for an octahedron, 90° and 180°.

(b) These bond angles differ from the actual bond angles only for H_2O, since this is the only species with lone pairs of electrons on the central atom. Lone pairs of electrons require more space than bonding pairs of electrons and reduce the H-O-H bond angle from 109.5^o to 104.5^o.

7-10. **Refer to the Sections as stated.** . . . .

(a) $BeBr_2$

:Br:Be:Br:

This molecule does not obey the octet rule. The Lewis dot formula predicts 2 regions of high electron density around the central atom, Be, and a linear electronic geometry. Since there are no lone pairs on Be, the molecular geometry is the same as the electronic geometry **(Section 7-3)**.

(b) H_2Se

$$S = N - A = [2 \times 2(\text{for H}) + 1 \times 8(\text{for Se})] - [2 \times 1(\text{for H}) + 1 \times 6(\text{for Se})] = 4 \text{ shared electrons}$$

H:Se:H

The Lewis dot formula predicts 4 regions of high electron density around the central Se atom and a tetrahedral electronic geometry. The molecular geometry is bent or angular due to the presence of 2 lone pairs on Se **(Section 7-7)**.

(c) AsF_3

$$S = N - A = [3 \times 8(\text{for F}) + 1 \times 8(\text{for As})] - [3 \times 7(\text{for F}) + 1 \times 5(\text{for As})] = 6 \text{ shared electrons}$$

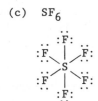

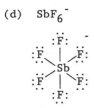

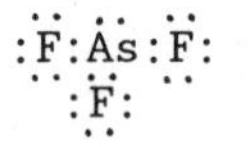

The Lewis dot formula predicts 4 regions of high electron density around the central As atom and a tetrahedral electronic geometry. The molecular geometry is pyramidal due to 1 lone pair on the central atom **(Section 7-6)**.

(d) GaI_3

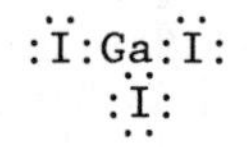

This molecule does not obey the octet rule, since Ga is a IIIA element. The Lewis dot formula predicts 3 regions of high electron density around the central Ga atom and a <u>trigonal planar</u> electronic geometry. Since there are no lone pairs on Ga, the molecular geometry is also <u>trigonal planar</u> **(Section 7-4)**.

7-12. **Refer to Exercise 7-10 Solution.** . . . .

7-14. **Refer to the Sections as stated.** . . . .

(a) SF_2

$S = N - A = [2 \times 8(\text{for F}) + 1 \times 8(\text{for S})] - [2 \times 7(\text{for F}) + 1 \times 6(\text{for S})]$
$= 4$ shared electrons

The Lewis dot formula predicts 4 regions of high electron density around the S atom and a tetrahedral electronic geometry. The ideal F-S-F bond angle is 109.5^o. The actual bond angle will be slightly smaller due to the presence of 2 lone pairs on the S **(Section 7-7)**.

SO_2

$S = N - A = [2 \times 8(\text{for O}) + 1 \times 8(\text{for S})] - [2 \times 6(\text{for O}) + 1 \times 6(\text{for S})]$
$= 6$ shared electrons

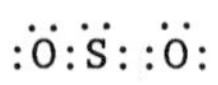

The Lewis dot formula predicts 3 regions of high electron density around the S atom and a trigonal planar electronic geometry. The ideal O-S-O bond angle is 120°. The actual bond angle, 119.50^o, is slightly smaller due to the presence of 1 lone pair on the S **(Section 7-7 and Table 7-3)**.

Therefore, SF_2 has a smaller bond angle than SO_2.

(b) BF_3 and BCl_3 both violate the octet rule. Their Lewis dot formulas predict 3 regions of high electron density around the central B atom and trigonal planar electronic geometries with ideal X-B-X bond angles of 120° for both molecules **(Section 7-4 and Table 7-3)**.

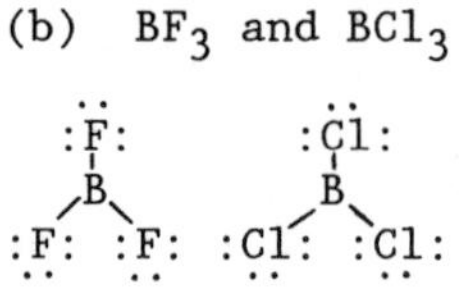

(c) CF_4

$S = N - A = [4 \times 8(\text{for F}) + 1 \times 8(\text{for C})] - [4 \times 7(\text{for F}) + 1 \times 4(\text{for C})]$
$= 8$ shared electrons

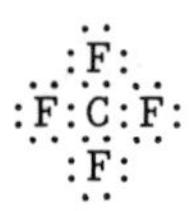

The Lewis dot formula predicts 4 regions of high electron density around the central atom and a tetrahedral electronic and molecular geometry with F-C-F bond angles of 109.5^o **(Section 7-5)**.

SF_4

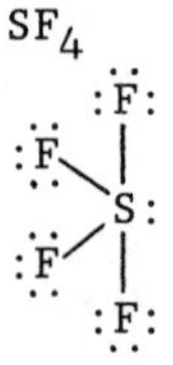

This molecule violates the octet rule. Its Lewis dot formula predicts 5 regions of high electron density and a trigonal bipyramidal electronic geometry with ideal F-S-F bond angles of 90°, 120° and 180°. The presence of 1 lone pair will make the angles slightly smaller **(Table 7-3)**.

Therefore, SF_4 has some smaller and some larger bond angles than CF_4.

(d) NH_3

$S = N - A = [3 \times 2(\text{for H}) + 1 \times 8(\text{for N})] - [3 \times 1(\text{for H}) + 1 \times 5(\text{for N})]$
$= 6$ shared electrons

H:N:H
H

The Lewis dot formula predicts 4 regions of high electron density and a tetrahedral electronic geometry. Its actual H-N-H bond angle is 107°, slightly less than the ideal bond angle of 109.5°, due to the presence of 1 lone pair **(Section 7-6)**.

H_2O $\quad S = N - A = [2 \times 2(\text{for H}) + 1 \times 8(\text{for O})] - [2 \times 1(\text{for H}) + [1 \times 6(\text{for O})] = 4$ shared electrons

H:O:
H

The Lewis dot structure predicts 4 regions of high electron density and a tetrahedral electronic geometry. The actual H-O-H bond angle is 104.5°. It is significantly smaller than the ideal bond angle of 109.5° due to the presence of 2 lone pairs of electrons on the O atom **(Section 7-7)**.

Therefore, the H-O-H bond angle is smaller than an H-N-H bond angle due to presence of 2 lone pairs in H_2O versus 1 lone pair in NH_3.

7-16. **Refer to the Sections as stated.**

(a) PF_3 $\quad S = N - A = [3 \times 8(\text{for F}) + 1 \times 8(\text{for P})] - [3 \times 7(\text{for F}) + 1 \times 5(\text{for P})] = 6$ shared electrons

:F:P:F:
:F:

The Lewis dot structure predicts 4 regions of high electron density and tetrahedral electronic geometry. One of the regions is a lone pair of electrons; the molecular geometry is pyramidal **(Section 7-6)**.

(b) AsF_5

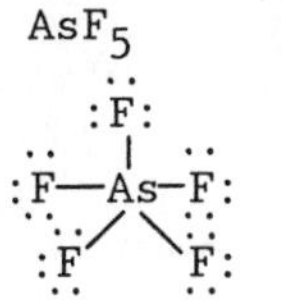

This compound violates the octet rule. As a VA element, each valence electron is shared with a F atom. The Lewis dot formula predicts 5 regions of high electron density around the central As atom and no lone pairs of electrons. The electronic geometry is the same as the molecular geometry: trigonal bipyramidal **(Section 7-9)**.

(c) $GeCl_2$

:Cl:Ge:Cl:

This compound violates the octet rule. Ge is a IVA element; 2 of the 4 valence electrons are shared with Cl atoms, the others are a lone pair. The Lewis dot structure predicts 3 regions of high electron density around Ge and an electronic geometry that is trigonal planar. The molecular geometry is bent due to the presence of a lone pair of electrons **(Section 7-4)**.

(d) IF_3

:F:
:F:I:F:

This compound violates the octet rule. Iodine is a VIIA element; 3 of its 7 valence electrons are shared with F atoms, the other 4 constitute 2 lone pairs. The Lewis dot formula predicts 5 regions of high electron density around the central atom and a trigonal bipyramidal electronic geometry. The molecular geometry is T-shaped due to the presence of 2 lone pairs **(Table 7-3)**.

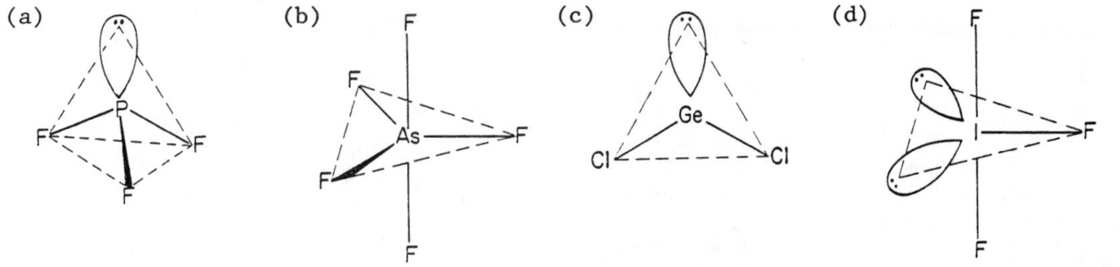

7-18.   **Refer to Section 7-14 and Tables 7-3 and 7-4.** • • • • • • • • • • • •

(a) IO_4^-

$S = N - A = [4 \times 8(\text{for } O) + 1 \times 8(\text{for } I)]$
$- [4 \times 6(\text{for } O) + 1 \times 7(\text{for } I) + 1\ e^-]$
$= 8$ shared electrons

$$\begin{array}{c} :\!\ddot{O}\!: \\ :\!\ddot{O}\!:\!\ddot{I}\!:\!\ddot{O}\!: \\ :\!\ddot{O}\!: \end{array}^-$$

The Lewis dot formula predicts 4 regions of high electron density around the central I atom with no lone pairs of electrons. The ionic geometry is the same as the electronic geometry: tetrahedral.

(b) BH_4^-

$S = N - A = [4 \times 2(\text{for } H) + 1 \times 8(\text{for } B)]$
$- [4 \times 1(\text{for } H) + 1 \times 3(\text{for } B) + 1\ e^-]$
$= 8$ shared electrons

$$\begin{array}{c} H \\ H\!:\!\ddot{B}\!:\!H \\ H \end{array}^-$$

The Lewis dot formula predicts 4 regions of high electron density around the central B atom with no lone pairs. The ionic geometry is the same as the electronic geometry: tetrahedral.

(c) H_3O^+

$S = N - A = [3 \times 2(\text{for } H) + 1 \times 8(\text{for } O)]$
$- [3 \times 1(\text{for } H) + 1 \times 6(\text{for } O) - 1\ e^-]$
$= 6$ shared electrons

$$\begin{array}{c} \ddot{} \\ H\!:\!\ddot{O}\!:\!H \\ H \end{array}^+$$

The Lewis dot formula predicts 4 regions of high electron density and a tetrahedral electronic geometry. The ionic geometry is pyramidal due to the presence of 1 lone pair.

(d) SF_5^-

This ion violates the octet rule. S is a VIA element with 6 valence electrons. There are 7 electrons which may participate in bonding due to the -1 charge on the ion. Each F atom requires 1 additional electron to complete its valence shell, leaving 2 electrons as a lone pair. The Lewis dot formula predicts a central S atom with 1 lone pair bonded to 5 F atoms. There are 6 regions of high electron density; the electronic geometry is octahedral. The ionic geometry is square pyramidal.

$$\begin{array}{c} :\!\ddot{F}\!: \quad \ddot{F}\!: \\ :\!F\!-\!\ddot{S}\!: \\ :\!\ddot{F}\!: \quad \ddot{F}\!: \end{array}^-$$

(a) ... (b) ... (c) ... (d)

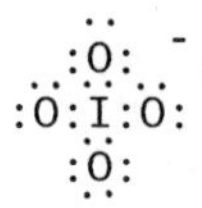

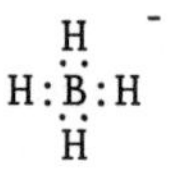

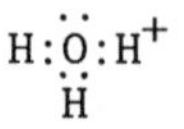

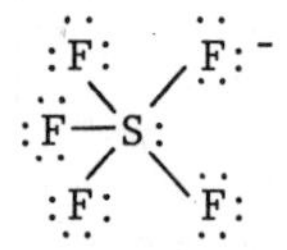

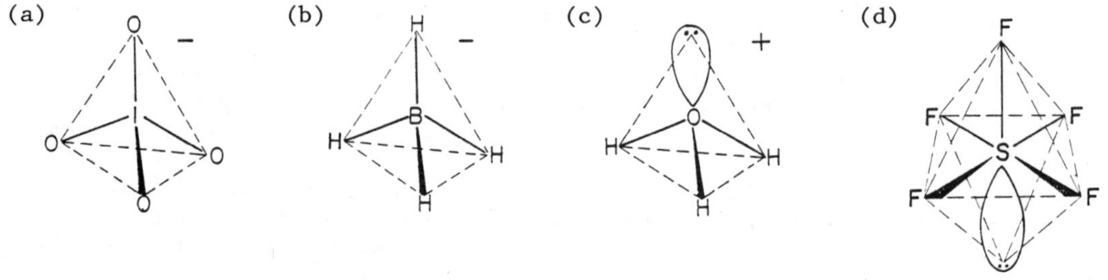

7-20.   **Refer to Section 9-9.** • • • • • • • • • • • • • • • • • •

O_3

$S = N - A = [3 \times 8(\text{for } O)] - [3 \times 6(\text{for } O)]$
$= 6$ shared electrons

$$:\!\ddot{O}\!:\!:\!\ddot{O}\!:\!\ddot{O}\!: \quad \leftrightarrow \quad :\!\ddot{O}\!:\!\ddot{O}\!:\!:\!\ddot{O}\!:$$

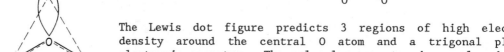

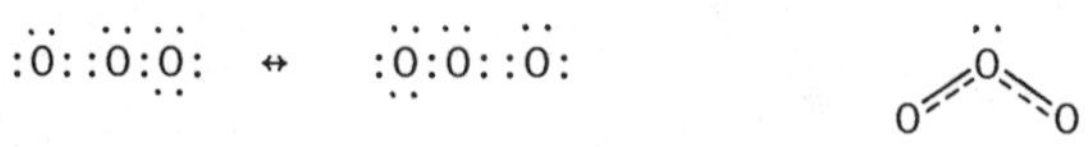

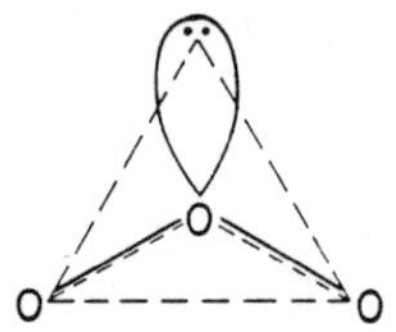

The Lewis dot figure predicts 3 regions of high electron density around the central O atom and a trigonal planar electronic geometry. The molecular geometry is angular due to the presence of 1 lone pair.

7-22. **Refer to Table 7-2.** •

(a) XeF_2

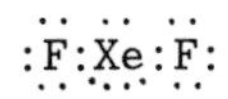

This compound does not obey the octet rule. Xe is a noble gas with 8 valence electrons, 2 of which are shared by the fluorine atoms, leaving 3 lone pairs. The Lewis dot figure predicts 5 regions of high electron density around Xe and a trigonal bipyramidal electronic geometry. Due to the presence of 3 lone pairs, the molecular geometry is linear.

(b) XeF_4

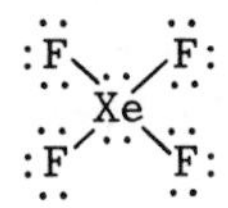

This compound does not obey the octet rule. Four of the 8 valence electrons belonging to Xe are shared with the 4 F atoms, leaving 2 lone pairs of electrons. The Lewis dot figure predicts 6 regions of high electron density around Xe and an octahedral electronic geometry. Due to the 2 lone pairs, the molecular geometry is square planar.

(c) XeO_3

$$S = N - A = [3 \times 8(\text{for O}) + 1 \times 8(\text{for Xe})] - [3 \times 6(\text{for O}) + 1 \times 8(\text{for Xe})] = 6 \text{ shared electrons}$$

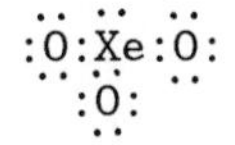

The Lewis dot structure predicts 4 regions of high electron density around Xe and a tetrahedral electronic geometry. Since 1 region is a lone pair, the molecular geometry is pyramidal.

(d) XeO_4

$$S = N - A = [4 \times 8(\text{for O}) + 1 \times 8(\text{for Xe})] - [4 \times 6(\text{for O}) + 1 \times 8(\text{for Xe})] = 8 \text{ shared electrons}$$

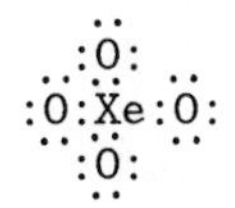

The Lewis dot structure predicts 4 regions of high electron density around Xe and a tetrahedral electronic geometry. Since there are no lone pairs, the molecular geometry is also tetrahedral.

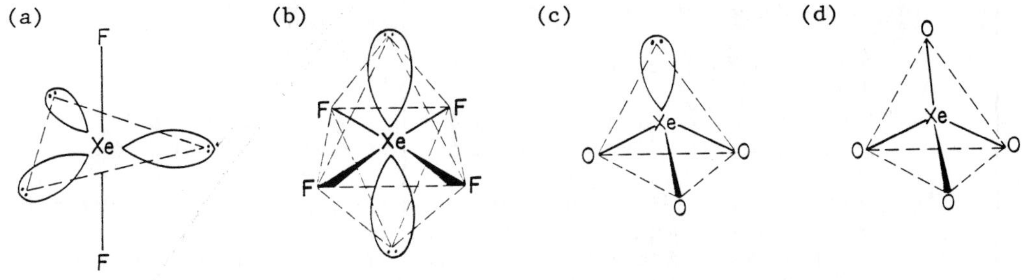

7-24. **Refer to Section 7-14 and Table 7-2.** • • • • • • • • • • • • • • •

None of the following interhalogen polyatomic ions obey the octet rule.

(a) IF_4^+

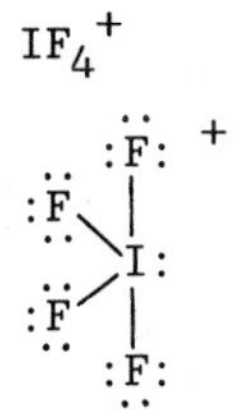

Iodine is a VIIA element with 7 valence electrons; 6 electrons may participate in bonding due to the +1 charge. Each of the 4 F atoms requires an electron to complete its octet, leaving 2 electrons as a lone pair on the I. The Lewis dot formula predicts 5 regions of high electron density around I and a trigonal bipyramidal electronic geometry. Due to the presence of a lone pair, the ionic geometry is probably a see-saw.

(b) ICl_2^-

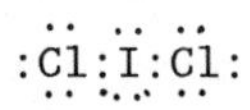

Iodine is a VIIA element with 7 valence electrons; 8 electrons may participate in bonding due to the -1 charge. Each Cl atom requires an electron to complete its octet, leaving 6 electrons as 3 lone pairs on I. The Lewis dot figure predicts 5 regions of high electron density around the central I atom and a trigonal bipyramidal electronic geometry. This ionic geometry is probably linear.

(c) BrF_4^-

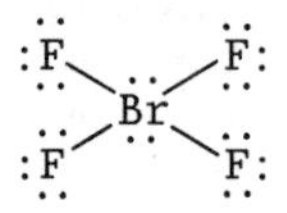

Br is a VIIA element with 7 valence electrons; 8 electrons may participate in bonding due to the -1 charge. Each of the 4 F atoms requires an electron to complete its octet, leaving 4 electrons as 2 lone pairs on Br. The Lewis dot figure predicts 6 regions of high electron density and an octahedral electronic geometry. The ionic geometry is square planar, due to the presence of 2 lone pairs.

(a)

(b)

(c)

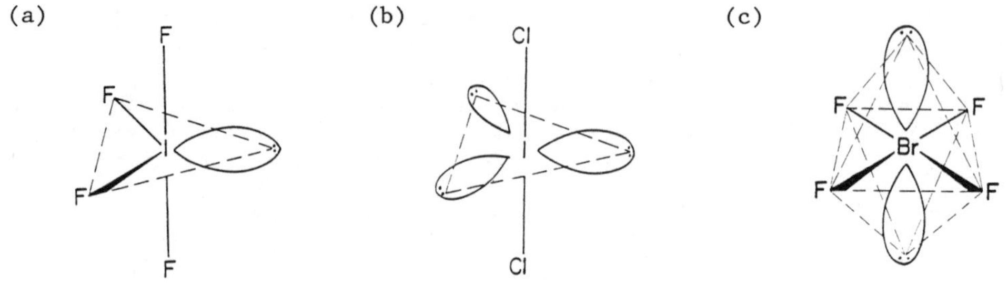

7-26. **Refer to Section 7-4.** •

The NO_2 molecule has an unpaired electron and does not obey the octet rule:

There are 4 resonance structures:

:Ö::N:Ö: ↔ :Ö:N::Ö: ↔ :Ö:N::Ö: ↔ :Ö::N:Ö:

The Lewis dot structure predicts 3 regions of high electron density around the N atom, a trigonal planar electronic geometry and a bent molecular geometry.

Nitrogen dioxide molecules easily dimerize to form dinitrogen tetroxide in a temperature-dependent equilibrium. The ease of dimerization is mainly due to the pairing of the unpaired electron in NO_2. The dimerization reaction releases energy, so at low temperatures, the forward reaction is favored.

$$2NO_2 \text{ (g)} \rightleftharpoons N_2O_4 \text{ (g)}$$

A dot formula for N_2O_4 is:

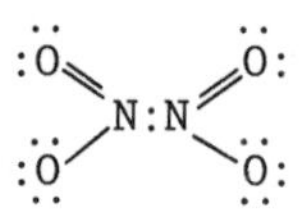

There are other resonance structures.

The molecular geometry around each N atom is the same as the electronic geometry due to the absence of lone pairs: trigonal planar.

7-28. **Refer to Table 7-3.** •

C_3O_2 $S = N - A = [2 \times 8(\text{for O}) + 3 \times 8(\text{for C})] - [2 \times 6(\text{for O}) + 3 \times 4(\text{for C})]$
              $= 16$ shared electrons

:Ö::C::C::C::Ö: or   :Ö=C=C=C=Ö:

The Lewis dot figure predicts 2 regions of high electron density around each C atom, causing the linear structure of C_3O_2.

7-30. **Refer to the stated Exercise Solutions.** • • • • • • • • • • • • • • •

		No. Lone Pairs on the Central Atom	Molecular Geometry
(a) 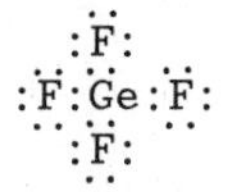	GeF_4	None	Tetrahedral (**Exercise 7-14c** with Ge similar to C)
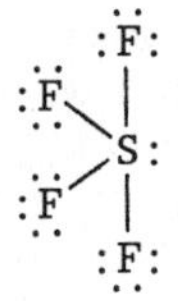	SF_4	1	See-saw (**Exercise 7-14c**)
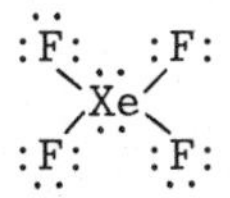	XeF_4	2	Square planar (**Exercise 7-22b**)

(b) It is obvious from the above table that the differences in molecular geometries are due to the different number of valence electrons around the central atom that must be accommodated in the molecule.

7-32. **Refer to the Sections as stated.** • • • • • • • • • • • • • • • •

(a) CdI_2

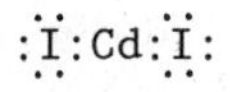

This molecule has a linear electronic and molecular geometry. The Cd-I bonds are polar. Since the molecule is symmetric, the bond dipoles cancel to give a nonpolar molecule **(Section 7-3)**.

(b) BCl_3

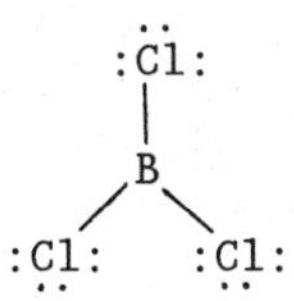

This molecule has a trigonal planar electronic geometry and trigonal planar molecular geometry. The B-Cl bonds are polar, but since the molecule is symmetrical, the bond dipoles cancel to give a nonpolar molecule **(Section 7-4 and Exercise 7-14b)**.

(c) PCl_3

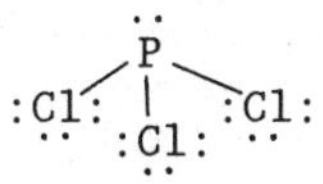

This molecule has a tetrahedral electronic geometry and a pyramidal molecular geometry. Cl (EN = 3.0) is more electronegative than P (EN = 2.1). The polar P-Cl bond dipoles oppose the effect of the lone pair. The molecule is only slightly polar **(Section 7-6)**.

(d) H_2O

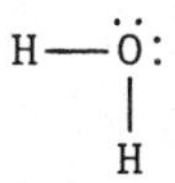

This molecule has a tetrahedral electronic geometry and an angular molecular geometry. Oxygen (EN = 3.5) is more electronegative than H (EN = 2.1). The O-H bond dipole reinforces the effect of the 2 lone pairs of electrons and so, H_2O is very polar **(Section 7-7)**.

(e) SF_6

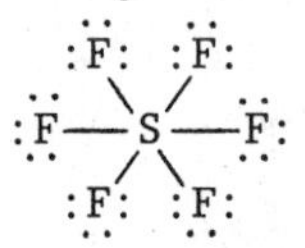

This molecule has an octahedral electronic and molecular geometry. The S-F bonds are polar, but the molecule is symmetrical. The S-F bond dipoles cancel to give a nonpolar molecule **(Section 7-10)**.

7-34. **Refer to Section 7-6.** •

Deductive logic is a method of reasoning by which concrete applications or consequences are deduced from general principles. When we predict the polarity of molecules, we use the general principles of Lewis dot formulas and VSEPR

Theory to deduce the shape of the molecule. Statement (a) is an example of this kind of reasoning; Statement (b) is not. In fact, the last portion of (b) is not a firm deduction because not all angular triatomic molecules possess three sets of electrons around the central atom. H_2O is a polar angular triatomic molecule with four sets of electrons around the O atom.

7-36. **Refer to Section 7-9.** . . . .

PF_3Cl_2

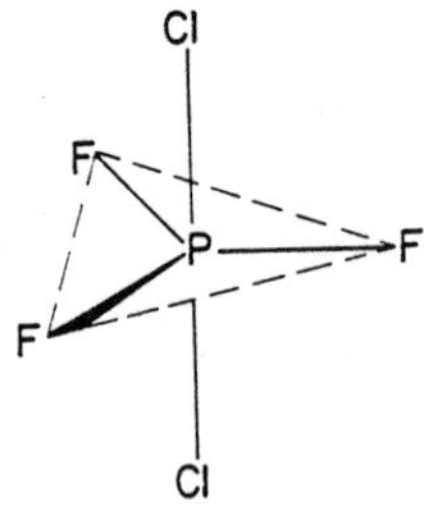

The Lewis dot formula for this molecule predicts 5 regions of high electron density with no lone pairs of electrons on the P atoms. The molecular geometry is the same as the electronic geometry: trigonal bipyramidal. For the molecule to be nonpolar (dipole moment = 0), the P-F and P-Cl bonds must be symmetrically arranged, i.e., the three F atoms must be at the equatorial positions and the two Cl atoms must be at the axial positions.

7-38. **Refer to Sections 7-6 and 7-7.** . . . .

The presence of lone pairs of valence electrons affects the polarity of a molecule in two ways. (1) VSEPR theory predicts the distribution of electron pairs. If 1 or more electron pairs are lone pairs, the arrangement of bonding pairs will be grossly affected, the molecule will adopt certain fundamental electronic geometries, and will generally be polar. (2) Both the direction and the magnitude of the molecule's dipole moment are determined by the counter balance between the dipoles of the lone pairs and the dipoles of the bonded pairs with the central atom. The direction and magnitude are also mildly modified due to the larger lone pair repulsive forces in comparison with those of the bonded pairs.

Two examples of polar molecules with lone pairs are H_2O and NH_3. In H_2O, the two lone pairs cause the molecule to adopt an angular rather than linear molecular geometry, while in NH_3, the one lone pair causes the molecule to adopt a pyramidal rather than trigonal planar molecular geometry.

One nonpolar molecule that has lone pairs is XeF_4 **(Exercise 7-22b).**

7-40. **Refer to the Key Terms for Chapter 7.** . . . .

A hybrid orbital is one of a set of equivalent orbitals formed by mixing and redistributing atomic orbitals during the formation of covalent bonds. It has properties and energies intermediate between those of the original unhybridized orbitals.

7-42. **Refer to Table 7-3.** . . . .

(a) *sp* 180° (b) sp^2 120° (c) sp^3 109.5°

(d) sp^3d 90°, 120°, 180° (e) sp^3d^2 90°, 180°

7-44. **Refer to Section 7-6.** . . . .

Actual experimental data is the primary factor used to decide whether overlap of simple atomic orbitals or hybridized orbitals should be used. If hybridization of orbitals is unnecessary to explain measured molecular angles, hybridization is not invoked.

7-46. **Refer to the Sections as stated.** . . . .

(a) $BeCl_2$

:Cl:Be:Cl: (with lone pairs above and below each Cl)

The Lewis dot formula predicts 2 regions of high electron density and a linear electronic and molecular geometry. The Be atom has *sp* hybridization. The valence electrons available for bonding from Cl are in 3*p* orbitals. The valence bond structure is shown below **(Section 7-3)**.

(b) CCl_4

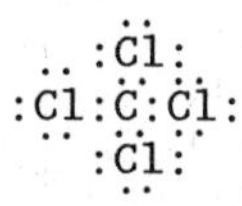

The Lewis dot formula predicts 4 regions of high electron density and a tetrahedral electronic and molecular geometry. The C atom has sp^3 hybridization. The valence electrons available for bonding from Cl are in 3*p* orbitals and the valence bond structure is shown below **(Section 7-5)**.

(c) H_2S

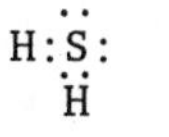

The Lewis dot formula predicts 4 regions of high electron density which corresponds to a tetrahedral electronic geometry. If hybridized, the S atom should have a sp^3 hybridization. However, the experimental H-S-H bond angle is only 92°. Since the bond angle is very different from the tetrahedral bond angle of 109.5^o, it is therefore unnecessary to invoke hybridization in this molecule. The 1*s* orbitals of H overlap with unhybridized 3*p* orbitals of S as shown in the valence bond structure **(Section 7-7)**.

(d) SeF_6

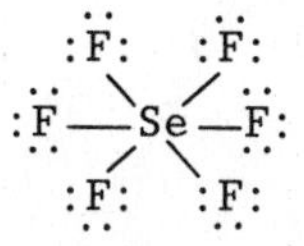

The Lewis dot formula predicts 6 regions of high electron density and an octahedral electronic and molecular geometry. The Se atom has sp^3d^2 hybridization. The valence electrons available for bonding from F are in 2*p* orbitals and the valence bond structure is shown below **(Section 7-10)**.

(a) $BeCl_2$

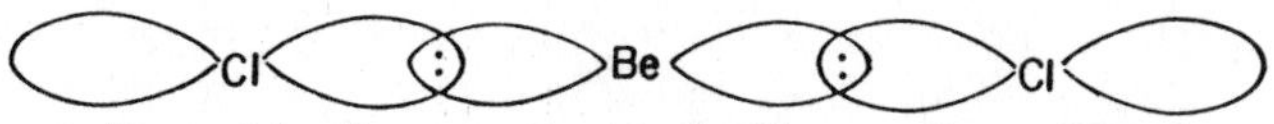

(b) CCl_4

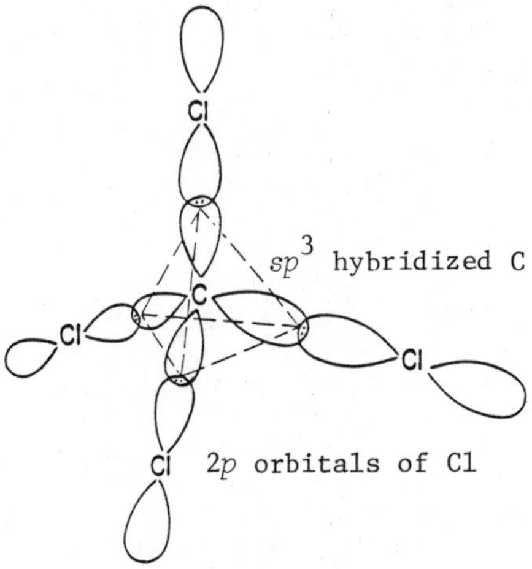

(d) SeF_6

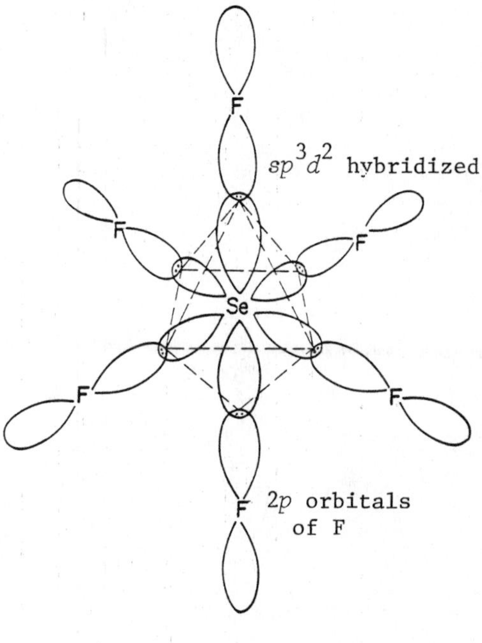

(c) H_2S

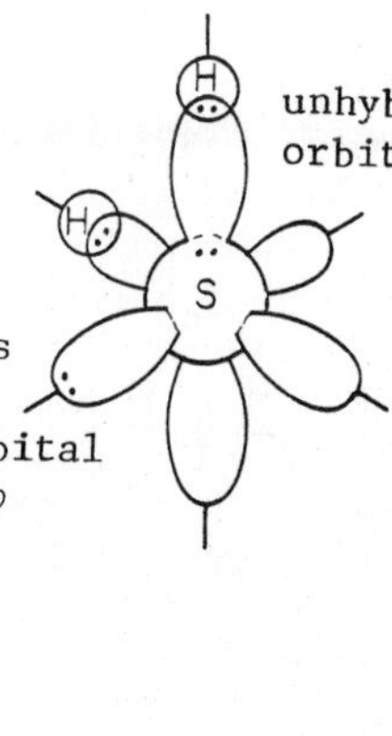

7-48. **Refer to Sections 7-5, 7-7 and 7-14.** . . . .

(a) ClO_2

```
 .. .. ..
:O:Cl:O:
 .. .. ..
```

The Lewis dot formula predicts 4 regions of high electron density, tetrahedral electronic geometry, bent molecular geometry and sp^3 hybridization for Cl. The valence electrons available for bonding from the O atoms are in $2p$ orbitals; the valence bond structure is shown below.

(b) SiH_2Cl_2

```
 ..  H  ..
:Cl:Si:Cl:
 ..  H  ..
```

The Lewis dot formula predicts 4 regions of high electron density, tetrahedral electronic and molecular geometry and sp^3 hybridization for Si. The valence electrons available for bonding from the H atoms and the Cl atoms are in $1s$ orbitals and $3p$ orbitals, respectively. The valence bond structure is given below.

(c) ClO_2^-

```
 .. .. ..
:O:Cl:O: -
 .. ..  ..
```

The Lewis dot formula predicts 4 regions of high electron density, tetrahedral electronic geometry, bent ionic geometry and sp^3 hybridization for the Cl atom. The valence electrons available for bonding from the O atoms are in $2p$ orbitals and the valence bond structure for the ion is shown below.

(d) $HClO_3$

```
 .. .. ..
:O:Cl:O:H
 .. .. ..
   :O:
    ..
```

The Lewis dot formula predicts 4 regions of high electron density around the Cl atom, a tetrahedral electronic geometry, a pyramidal molecular geometry and sp^3 hybridization for the Cl atom. A tetrahedral electronic geometry and sp^3 hybridization is also predicted for the O atom bonded to the H atom. The valence electron available for bonding from the H atom is in a $1s$ orbital. The valence bond structure is shown below.

(a) ClO_2

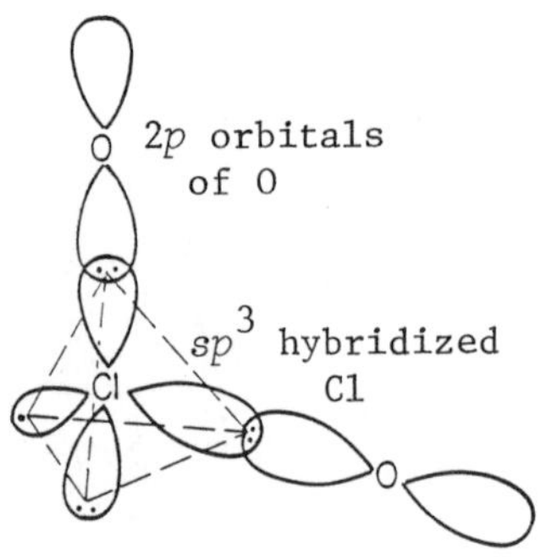

(b) SiH_2Cl_2

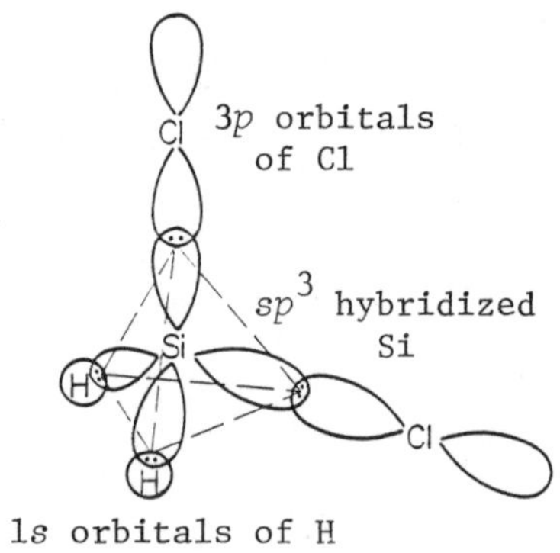

(c) ClO_2^-

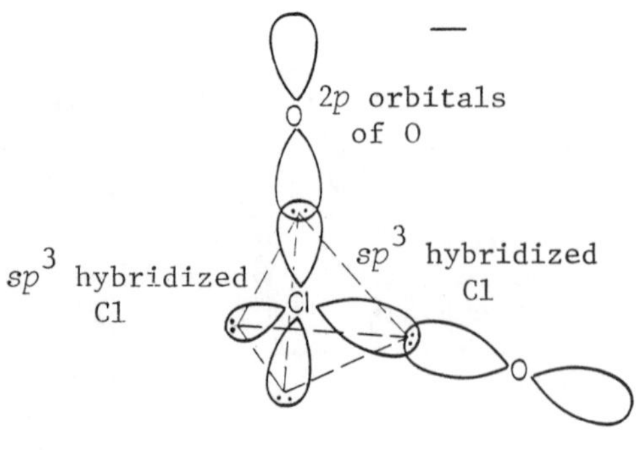

(d) $HClO_3$

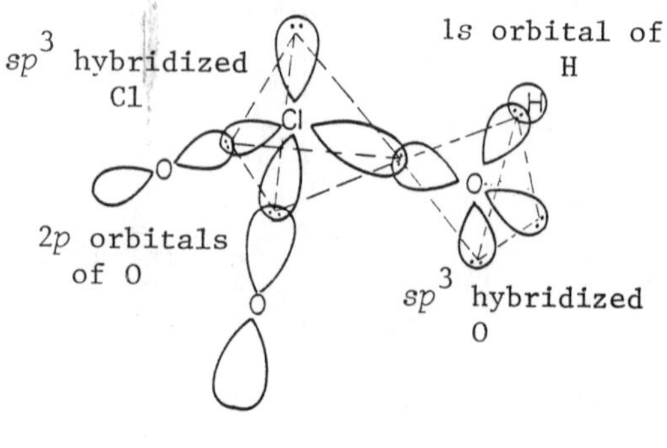

7-50. **Refer to Sections 7-3, 7-4, 7-7 and 7-12.** . . . .

Based on the observed bond angles for HN_3, we can predict that

(1) hybridization at N_a is sp^3 (ideal bond angle: 109.5°)

(2) hybridization at N_b is sp (ideal bond angle: 180°)

The Lewis dash formula for HN_3:

H-N̈-N≡N: $S = N - A$ = [1 × 2(for H) + 3 × 8(for N)]
- [1 × 1(for H) + 3 × 5(for N)]
= 10 shared electrons

This structure is consistent with this conclusion.

7-52. **Refer to Section 7-4.** . . . .

The general representation for NOF, NOCl and NOBr is NOX.

NOX $S = N - A$ = [1 × 8(for N) + 1 × 8(for O) + 1 × 8(for X)]
- [1 × 5(for N) + 1 × 6(for O) + 1 × 7(for X)]
:Ö=N̈-Ẍ: = 6 shared electrons

The Lewis dot formula predicts 3 regions of high electron density a trigonal planar electronic geometry and sp^2 hybridization on the N atom. The O-N-X bond angle increased for X = F, Cl, Br due to the increasing number of electrons in these atoms. This leads to the increasing repulsive force of the X atom against the oxygen atom and the lone pair.

7-54. **Refer to Section 7-14.** . . . .

SO_3^{2-} $S = N - A$ = [3 × 8(for O) + 1 × 8(for S)]
- [3 × 6(for O) + 1 × 6(for S) + 2 e^-)]
= 6 shared electrons

:Ö:S̈:Ö:$^{2-}$
:Ö:

The Lewis dot formula predicts 4 regions of high electron density and sp^3 hybridization at the S atom. The electronic geometry is tetrahedral and the ionic geometry is pyramidal due to the presence of 1 lone pair on the S. The ideal O-S-O bond angle should be 109.5°. However, the lone pair will push the oxygen atoms slightly closer to give a bond angle slightly less than 109°.

SiO_3^{2-} $S = N - A$ = [3 × 8(for O) + 1 × 8(for Si)]
- [3 × 6(for O) + 1 × 4(for Si) + 2 e^-]
= 8 shared electrons

:Ö::Si:Ö:$^{2-}$
:Ö:

The Lewis dot formula predicts 3 regions of high electron density and sp^2 hybridization at the Si atom. The electronic and ionic geometry are the same: trigonal planar. The ideal O-Si-O bond angle should be 120°. The dot formula shown is 1 of 3 resonance structures.

BO_3^{3-} $S = N - A$ = [3 × 8(for O) + 1 × 6(for B)]
- [3 × 6(for O) + 1 × 3(for B) + 3 e^-]
= 6 shared electrons since B does not obey the octet rule.

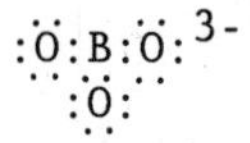

:Ö:B:Ö:³⁻
:Ö:
    The Lewis dot formula predicts 3 regions of high electron density and sp^2 hybridization at the B atom. The electronic geometry is the ionic geometry: trigonal planar. The ideal O-B-O bond angle should be 120°.

$PO_3{}^{3-}$

$$S = N - A = [3 \times 8(\text{for O}) + 1 \times 8(\text{for P})]$$
$$- [3 \times 6(\text{for O}) + 1 \times 5(\text{for P}) + 3\ e^-]$$
$$= 6 \text{ shared electrons}$$

:Ö:P:Ö:³⁻
:Ö:

    The Lewis dot formula predicts 4 regions of high electron density and sp^3 hybridization at the P atom. The electronic geometry is tetrahedral, but the ionic geometry is pyramidal due to the presence of 1 lone pair on P. The expected O-P-O bond angle should be slightly less than 109° due to the higher *lp-bp* repulsions than *bp-bp* repulsions.

7-56. **Refer to Tables 7-3 and 7-4.** • • • • • • • • • • • • • • • • • • •

(a) PF_5

    The Lewis dot formulas predict sp^3d hybridization of P in PF_5 (5 regions of high electron density) changing to sp^3d^2 hybridization of P in $PF_6{}^-$ (6 regions of high electron density.)

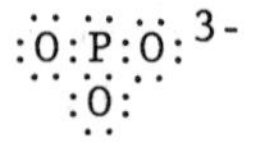

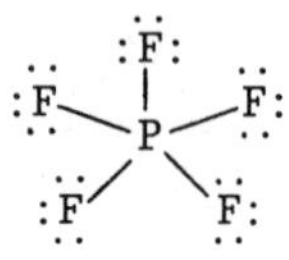

$PF_6{}^-$

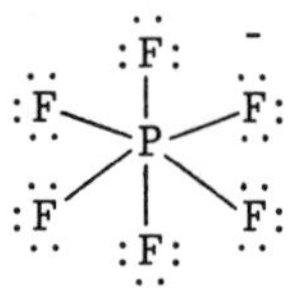

(b) CO   :C:::O:

    CO_2 :Ö::C::Ö:

    The Lewis dot formulas predict sp hybridization of C in both molecules since both C atoms have 2 regions of high electron density around them.

(c) AlI_3 :Ï:Al:Ï:
                 :Ï:

    The Lewis dot formulas predict sp^2 hybridization of Al in AlI_3 (3 regions of high electron density) changing to sp^3 hybridization of Al in $AlI_4{}^-$ (4 regions of high electron density).

$AlI_4{}^-$:Ï:⁻
          :Ï:Al:Ï:
             :Ï:

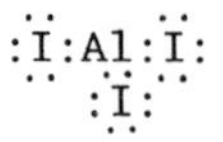

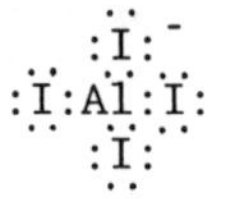

7-58. **Refer to Table 7-3.** •

S_8

    The Lewis dot formula predicts sp^3 hybridization and a S-S-S bond angle of about 109° (**Exercise 6-40 Solution**).

SO_2 :Ö:S::Ö:

    The Lewis dot formula predicts sp^2 hybridization for the S atom.

SO_3 :Ö:S::Ö:
          :Ö:

    The Lewis dot formula predicts sp^2 hybridization for the S atom.

H_2SO_4 :Ö:
       H:Ö:S:Ö:H
          :Ö:

    The Lewis dot formula predicts sp^3 hybridization for the S atom.

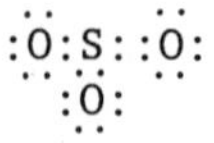

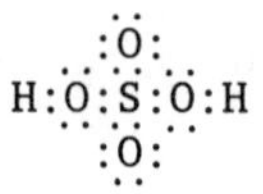

7-60. **Refer to Table 7-3 and Exercise 7-22 Solution.** • • • • • • • • • • • • • •

Using the Lewis dot formulas obtained in Exercise 7-22 and Table 7-3, we can predict the following hybridizations for Xe:

(a) XeF_2 sp^3d (b) XeF_4 sp^3d^2

(c) XeO_3 sp^3 (d) XeO_4 sp^3

Since the valence orbitals for bonding from F and O atoms are in $2p$ orbitals, the valence bond structures are shown below.

(a) XeF_2
F-Xe-F bond angle = 180°

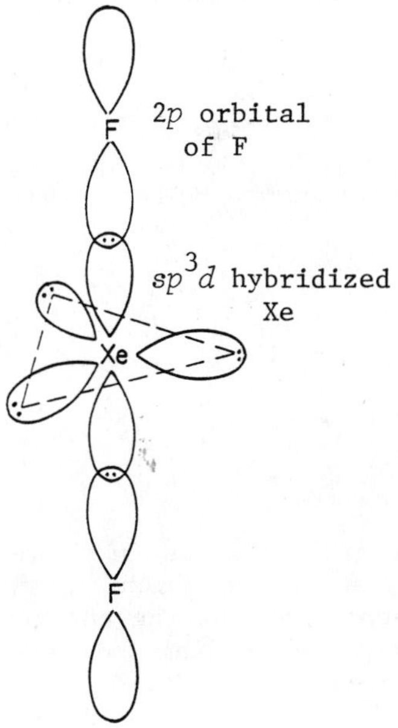

(b) XeF_4
F-Xe-F bond angles = 90° and 180°

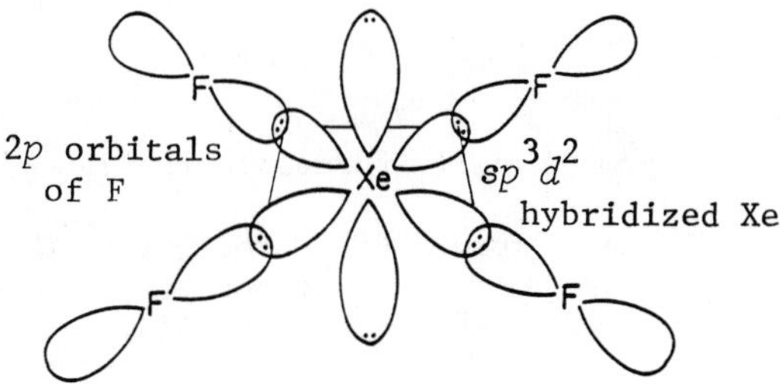

(c) XeO_3
O-Xe-O bond angle = 109°

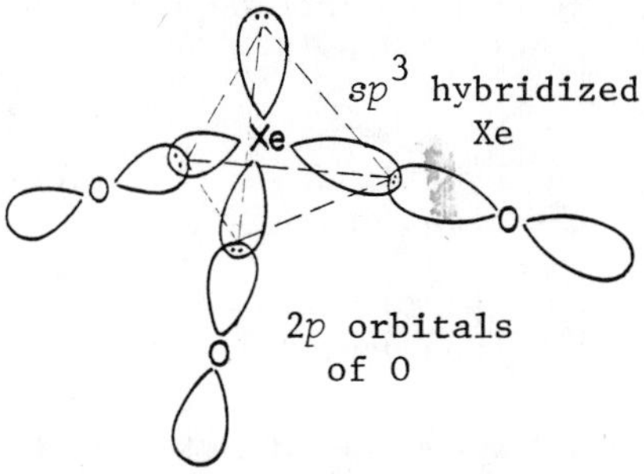

(d) XeO_4
O-Xe-O bond angle = 109°

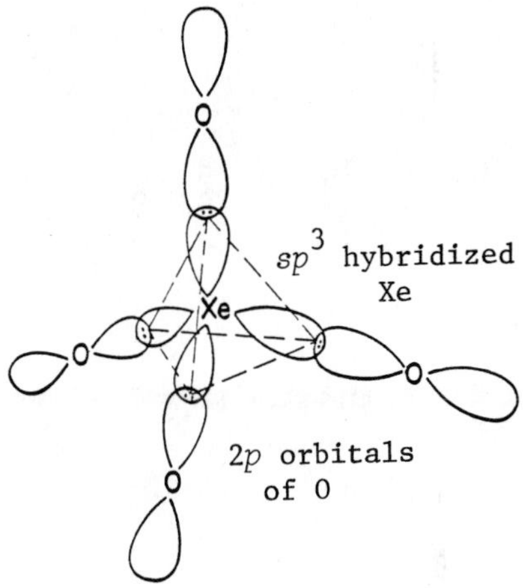

7-62. **Refer to Table 7-3 and Section 6-7.** . . . .

(a) 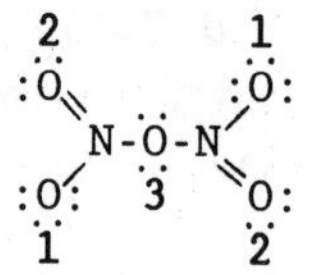

N - sp^2 hybridization (3 regions of high electron density)

O(3) - sp^3 hybridization (4 regions of high electron density, 2 bonding pairs and 2 lone pairs of electrons).

The O-N-O bond angle should equal about 120°.
The N-O-N bond angle should equal about 109°.

(b) The proposed dash formula is shown below.

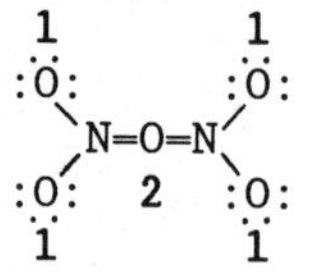

N - sp^2 hybridization (3 regions of high electron density)

O(2) - sp hybridization (2 regions of high electron density)

(c) Recall, formal charge, FC = (Group No.) - [(No. bonds) + (No. unshared e^-)]

Formula (a): for O(1), FC = 6 - (1 + 6) = -1
for O(2), FC = 6 - (2 + 4) = 0
for O(3), FC = 6 - (2 + 4) = 0
for N, FC = 5 - (4 + 0) = +1

Formula (b): for O(1), FC = 6 - (1 + 6) = -1
for O(2), FC = 6 - (4 + 0) = +2
for N, FC = 5 - (4 + 0) = +1

According to the rules for assigning formal charges in a compound, the most energetically favorable structure is the one in which the formal charge on each atom is 0 or at least has adjacent atoms having charges with opposite sign. Formula (a) should be more stable according to these rules, but it does not predict the observed bond angles.

7-64. **Refer to Sections 7-5, 7-11 and 7-12.** . . . .

(a) 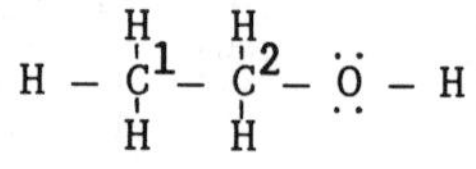

C(1) - sp^3 hybridization (4 regions of high electron density)

C(2) - sp^3 hybridization (4 regions of high electron density)

(b) 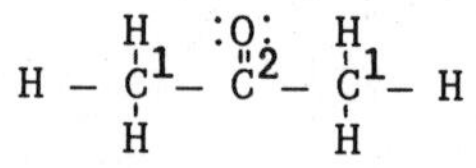

C(1) - sp^3 hybridization (4 regions of high electron density)

C(2) - sp^2 hybridization (3 regions of high electron density)

(c) 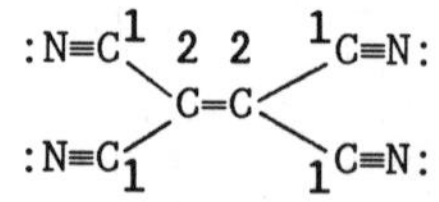

C(1) - sp hybridization (2 regions of high electron density)

C(2) - sp^2 hybridization (3 regions of high electron density)

(d) 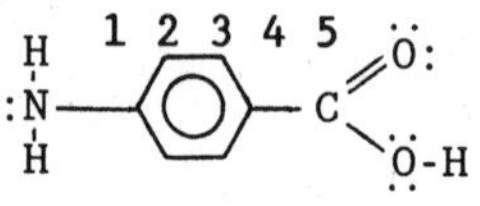

C(1,2,3,4,5) - sp^2 hybridization (3 regions of high electron density)

(e)
```
    :O:  ..  H
     ‖       |2
 H — C — N — C — H
     1   |   H
         |2
     H — C — H
         |
         H
```

C(1) - sp^2 hybridization (3 regions of high electron density)

C(2) - sp^3 hybridization (4 regions of high electron density)

7-66. **Refer to Sections 7-11 and 7-12.** . . . .

(a) The molecule contains 7 single bonds (7 sigma bonds) and 1 double bond (1 sigma bond, 1 pi bond) for a total of 8 sigma bonds and 1 pi bond.

(b) The molecule contains 4 single bonds (4 sigma bonds) and 2 double bonds (2 sigma bonds, 2 pi bonds) for a total of 6 sigma bonds and 2 pi bonds.

(c) The molecule contains 9 single bonds (9 sigma bonds) and 1 double bond (1 sigma bond, 1 pi bond) for a total of 10 sigma bonds and 1 pi bond.

(d) The molecule contains 5 single bonds (5 sigma bonds), 1 double bond (1 sigma bond, 1 pi bond) and 1 triple bond (1 sigma bond, 2 pi bonds) for a total of 7 sigma bonds and 3 pi bonds.

7-68. **Refer to Section 7-12.** . . . .

The acetylide anion is C_2H^-.

:C:::C:H$^-$

$$S = N - A = [1 \times 2(\text{for H}) + 2 \times 8(\text{for C})] - [1 \times 1(\text{for H}) + 2 \times 4(\text{for C}) + 1\ e^-] = 8 \text{ shared electrons}$$

The structure is linear with both carbons having *sp* hybridization. The C-C-H bond angle is 180°.

7-70. **Refer to Section 7-11 and 7-12.** . . . .

(a) C_4H_{10}

```
  H H H H
  | | | |
H-C-C-C-C-H
  | | | |
  H H H H
```

Each C atom is sp^3 hybridized with bond angles of 109.5°.

(b) $H_2C{=}CHCH_2CH_3$ (1 2 3 4)

```
H  H H H
 \ | | |
  C=C-C-C-H
 /  | | |
H   H H H
```

C(1), C(2) - sp^2 hybridization with bond angles of 120°

C(3), C(4) - sp^3 hybridization with bond angles of 109.5°

(c) $HC{\equiv}CCH_2CH_3$ (1 23 4)

```
        H H
        | |
H-C≡C-C-C-H
        | |
        H H
```

C(1), C(2) - *sp* hybridization with bond angles of 180°

C(3), C(4) - sp^3 hybridization with bond angles of 109.5°

(d) CH_3CHO
1 2

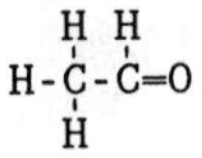

C(1) - sp^3 hybridization with bond angle of 109.5°

C(2) - sp^2 hybridization with bond angle of 120°

Three-dimensional Structures:

hydrogen atom
carbon atom
oxygen atom
single bond
double bond
triple bond

(a) butane

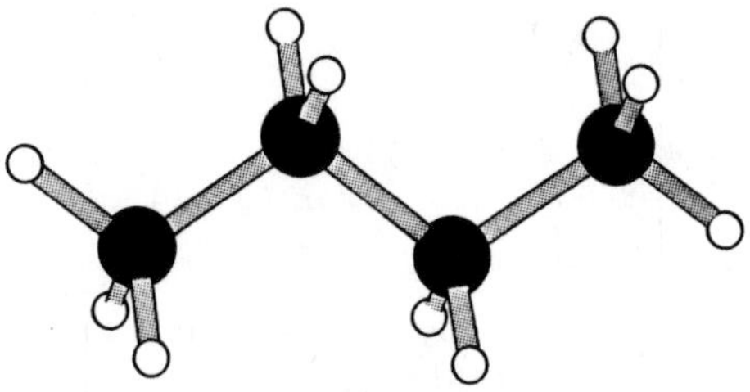

(b) 1-butene

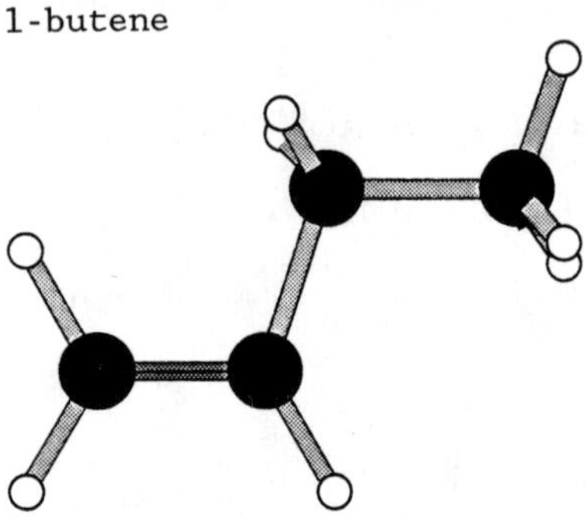

(c) 1-butyne

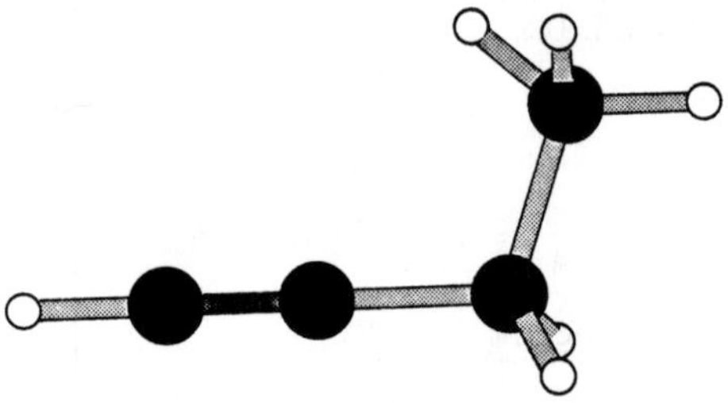

(d) acetaldehyde

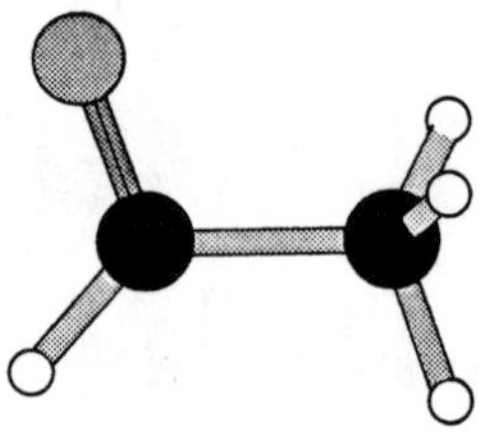

8 Molecular Orbitals in Chemical Bonding

8-2. **Refer to Section 8-1.**

Atomic orbitals, localized hybridized atomic orbitals and molecular orbitals are all referring to regions in space where the probability of finding electrons is highest in atoms or molecules. Each of these three types of orbitals can contain a maximum of two electrons, providing the electrons are of opposite spin.

The main differences among these three kinds of orbitals are as follows. Atomic orbitals are pure orbitals with no mixing with other orbitals in the same atom or molecule. Hybridized orbitals are derived from mixing pure atomic orbitals from the same atom to form a new set of atomic orbitals with the same total electron capacity and with properties and energies intermediate between those of the original unhybridized orbitals. On the other hand, molecular orbitals are orbitals resulting from overlapping and mixing of atomic orbitals of all the atoms in a molecule. A molecular orbital belongs to the molecule as a whole.

8-4. **Refer to Section 7-2.**

A set of hybridized atomic orbitals holds the same maximum number of electrons as the set of atomic orbitals from which the hybridized atomic orbitals were formed. A hybridized atomic orbital can hold a maximum of 2 electrons having opposite spin.

8-6. **Refer to Sections 8-2 and 8-4.**

Hund's rule states that electrons enter degenerate orbitals (ones with identical energy) one at a time with like (parallel) spins before pairing occurs in any of the orbitals. For example, for molecules with electrons in degenerate π orbitals such as B_2, the highest occupied MO's have an electronic configuration of $\pi_{2p_y}{}^1$ and $\pi_{2p_z}{}^1$ instead of $\pi_{2p_y}{}^2$ and $\pi_{2p_z}{}^0$.

8-8. **Refer to Figure 8-4a.**

8-10. **Refer to Sections 8-3 and Example 8-1.**

In Exercise 8-9, we obtained the electron configurations of the following molecules and ions.

$$\text{Bond order} = \frac{\text{No. Bonding Electrons - No. Antibonding Electrons}}{2}$$

			No. Bonding Electrons	No. Antibonding Electrons	Bond Order
(a)	Be_2	$\sigma_{1s}{}^2\ \sigma^*_{1s}{}^2\ \sigma_{2s}{}^2\ \sigma^*_{2s}{}^2$	4	4	0
	$Be_2{}^+$	$\sigma_{1s}{}^2\ \sigma^*_{1s}{}^2\ \sigma_{2s}{}^2\ \sigma^*_{2s}{}^1$	4	3	0.5
	$Be_2{}^-$	$\sigma_{1s}{}^2\ \sigma^*_{1s}{}^2\ \sigma_{2s}{}^2\ \sigma^*_{2s}{}^2\ \pi_{2p_y}{}^1$	5	4	0.5

			No. Bonding Electrons	No. Antibonding Electrons	Bond Order
(b)	B_2	$\sigma_{1s}^2\ \sigma_{1s}^{*2}\ \sigma_{2s}^2\ \sigma_{2s}^{*2}\ \pi_{2p_y}^1\ \pi_{2p_z}^1$	6	4	1
	B_2^+	$\sigma_{1s}^2\ \sigma_{1s}^{*2}\ \sigma_{2s}^2\ \sigma_{2s}^{*2}\ \pi_{2p_y}^1$	5	4	0.5
	B_2^-	$\sigma_{1s}^2\ \sigma_{1s}^{*2}\ \sigma_{2s}^2\ \sigma_{2s}^{*2}\ \pi_{2p_y}^2\ \pi_{2p_z}^1$	7	4	1.5
(c)	C_2^+	$\sigma_{1s}^2\ \sigma_{1s}^{*2}\ \sigma_{2s}^2\ \sigma_{2s}^{*2}\ \pi_{2p_y}^2\ \pi_{2p_z}^1$	7	4	1.5

8-12. **Refer to Section 8-3 and Exercise 8-10.** . . . .

In Molecular Orbital Theory, the greater the bond order, the more stable is the molecule or ion. Therefore, we predict:

Unstable: Be_2

Somewhat stable: Be_2^+, Be_2^-, B_2^+

Stable: B_2, B_2^-, C_2^+

This means that although Be_2 is unstable, both its + and - ions are somewhat stable. It also shows that the stability of the boron species is in the order: $B_2^+ < B_2 < B_2^-$. All these predictions are generally correct. The fact that many of these supposed stable species are not observed in nature is because they are chemically very reactive. Therefore, the chemical reactivity of a species is also instrumental when considering the survival probability of a species with stable chemical bonding.

8-14. **Refer to Section 8-5 and Figure 8-7.** . . . .

The atomic orbitals of the more electronegative element of a heteronuclear diatomic molecule are lower in energy than the corresponding orbitals of the less electronegative element. In such a molecule, a bonding molecular orbital has more of the character of the atomic orbital in the more electronegative element, and resembles the energy of that atomic orbital more closely. An antibonding molecular orbital has more of the character of the atomic orbital in the less electronegative element, and resembles the energy of the atomic orbital of the less electronegative atom. Therefore, electrons occupying bonding molecular orbitals of heteronuclear diatomic molecules are more closely associated with the more electronegative element. The greater the energy difference between the overlapping orbitals, the more polar is the bond joining the atoms and the greater its ionic character.

8-16. **Refer to Section 8-3.** . . . .

Recall, bond order $= \dfrac{\text{No. Bonding Electrons - No. Antibonding Electrons}}{2}$

(a) X_2 bond order $= \dfrac{8 - 4}{2} = 2$

(b) X_2^+ bond order $= \dfrac{10 - 7}{2} = 1.5$

(c) X_2^- bond order $= \dfrac{10 - 5}{2} = 2.5$

8-18. **Refer to Section 8-3 and Example 8-1.** . . . .

In Exercise 8-17, the following electron configurations were obtained:

O_2 $\sigma_{1s}^2\ \sigma_{1s}^{*2}\ \sigma_{2s}^2\ \sigma_{2s}^{*2}\ \sigma_{2p}^2\ \pi_{2p_y}^2\ \pi_{2p_z}^2\ \pi_{2p_y}^{*1}\ \pi_{2p_z}^{*1}$

O_2^- $\sigma_{1s}^2\ \sigma_{1s}^{*2}\ \sigma_{2s}^2\ \sigma_{2s}^{*2}\ \sigma_{2p}^2\ \pi_{2p_y}^2\ \pi_{2p_z}^2\ \pi_{2p_y}^{*2}\ \pi_{2p_z}^{*1}$

O_2^{2-} $\sigma_{1s}^2\ \sigma_{1s}^{*2}\ \sigma_{2s}^2\ \sigma_{2s}^{*2}\ \sigma_{2p}^2\ \pi_{2p_y}^2\ \pi_{2p_z}^2\ \pi_{2p_y}^{*2}\ \pi_{2p_z}^{*2}$

F_2 $\sigma_{1s}^2\ \sigma_{1s}^{*2}\ \sigma_{2s}^2\ \sigma_{2s}^{*2}\ \sigma_{2p}^2\ \pi_{2p_y}^2\ \pi_{2p_z}^2\ \pi_{2p_y}^{*2}\ \pi_{2p_z}^{*2}$

F_2^+ $\sigma_{1s}^2\ \sigma_{1s}^{*2}\ \sigma_{2s}^2\ \sigma_{2s}^{*2}\ \sigma_{2p}^2\ \pi_{2p_y}^2\ \pi_{2p_z}^2\ \pi_{2p_y}^{*2}\ \pi_{2p_z}^{*1}$

F_2^- $\sigma_{1s}^2\ \sigma_{1s}^{*2}\ \sigma_{2s}^2\ \sigma_{2s}^{*2}\ \sigma_{2p}^2\ \pi_{2p_y}^2\ \pi_{2p_z}^2\ \pi_{2p_y}^{*2}\ \pi_{2p_z}^{*2}\ \sigma_{2p}^{*1}$

Ne_2 $\sigma_{1s}^2\ \sigma_{1s}^{*2}\ \sigma_{2s}^2\ \sigma_{2s}^{*2}\ \sigma_{2p}^2\ \pi_{2p_y}^2\ \pi_{2p_z}^2\ \pi_{2p_y}^{*2}\ \pi_{2p_z}^{*2}\ \sigma_{2p}^{*2}$

Ne_2^+ $\sigma_{1s}^2\ \sigma_{1s}^{*2}\ \sigma_{2s}^2\ \sigma_{2s}^{*2}\ \sigma_{2p}^2\ \pi_{2p_y}^2\ \pi_{2p_z}^2\ \pi_{2p_y}^{*2}\ \pi_{2p_z}^{*2}\ \sigma_{2p}^{*1}$

(a) Recall, bond order = $\dfrac{\text{No. Bonding Electrons - No. Antibonding Electrons}}{2}$

	O_2	O_2^-	O_2^{2-}	F_2	F_2^+	F_2^-	Ne_2	Ne_2^+
Bonding e^-	10	10	10	10	10	10	10	10
Antibonding e^-	6	7	8	8	7	9	10	9
Bond order	2	1.5	1	1	1.5	0.5	0	0.5
(b) Unpaired e^-	2	1	0	0	1	1	0	1
Paramagnetic or Diamagnetic	P	P	D	D	P	P	D	P

(c) Based on bond orders, we have the following order of stability:

Unstable: Ne_2

Somewhat stable: F_2^- and Ne_2^+

Stable: O_2, O_2^-, O_2^{2-}, F_2 and F_2^+

8-20. **Refer to Section 8-5 and Figure 8-7.** . . . .

The MO diagram for NO:

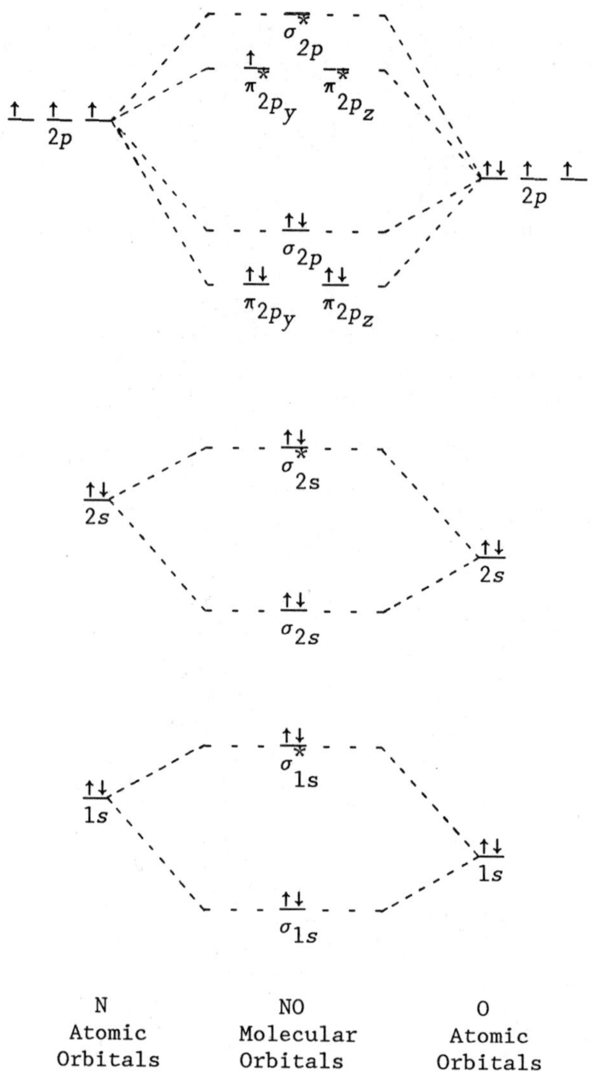

Therefore, the electron configuration of NO is

$$\sigma_{1s}^2\ \sigma_{1s}^{*2}\ \sigma_{2s}^2\ \sigma_{2s}^{*2}\ \pi_{2p_y}^2\ \pi_{2p_z}^2\ \sigma_{2p}^2\ \pi_{2p_y}^{*1}$$

$$\text{Bond order} = \frac{\text{No. Bonding Electrons - No. Antibonding Electrons}}{2} = \frac{10 - 5}{2} = 2.5$$

Therefore, the molecule is very stable.

The NO molecule is paramagnetic with 1 unpaired electron; the molecule is reactive.

8-22. **Refer to Sections 8-3 and 8-4, and Figure 8-4.** . . . .

(a) N_2^- $\sigma_{1s}^2\ \sigma_{1s}^{*2}\ \sigma_{2s}^2\ \sigma_{2s}^{*2}\ \pi_{2p_y}^2\ \pi_{2p_z}^2\ \sigma_{2p}^2\ \pi_{2p_y}^{*1}$

bond order $= \frac{10 - 5}{2} = 2.5$ and the ion is very stable

(b) N_2^+ $\quad \sigma_{1s}^2\ \sigma_{1s}^{*2}\ \sigma_{2s}^2\ \sigma_{2s}^{*2}\ \pi_{2p_y}^2\ \pi_{2p_z}^2\ \sigma_{2p}^1$

bond order $= \dfrac{9 - 4}{2} = 2.5$ and the ion is very stable

(c) Cl_2 $\quad \sigma_{1s}^2\ \sigma_{1s}^{*2}\ \sigma_{2s}^2\ \sigma_{2s}^{*2}\ \sigma_{2p}^2\ \pi_{2p_y}^2\ \pi_{2p_z}^2\ \pi_{2p_y}^{*2}\ \pi_{2p_z}^{*2}\ \sigma_{2p}^{*2}\ \sigma_{3s}^2\ \sigma_{3s}^{*2}\ \sigma_{3p}^2$

$\pi_{3p_y}^2\ \pi_{3p_z}^2\ \pi_{3p_y}^{*2}\ \pi_{3p_z}^{*2}$

bond order $= \dfrac{18 - 16}{2} = 1$ and the molecule is stable

8-24. **Refer to Sections 8-3, 8-4 and 8-5; Figures 8-4 and 8-7.** . . . .

(a) O_2^{2+} $\quad \sigma_{1s}^2\ \sigma_{1s}^{*2}\ \sigma_{2s}^2\ \sigma_{2s}^{*2}\ \sigma_{2p}^2\ \pi_{2p_y}^2\ \pi_{2p_z}^2$ $\qquad$ bond order $= \dfrac{10 - 4}{2} = 3$

The ion is very stable.

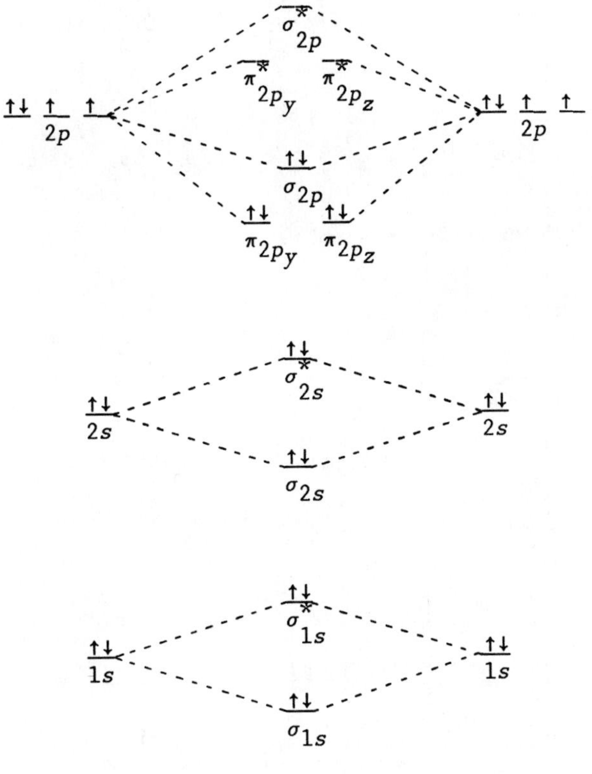

O $\qquad$ O_2^{2+} $\qquad$ O

Atomic Orbitals $\qquad$ Molecular Orbitals $\qquad$ Atomic Orbitals

(b) HO^- $1s^2\ 2s^2\ \sigma_{sp}{}^2\ 2p^2\ 2p^2$ bond order $= \frac{2 - 0}{2} = 1$

The ion is stable.

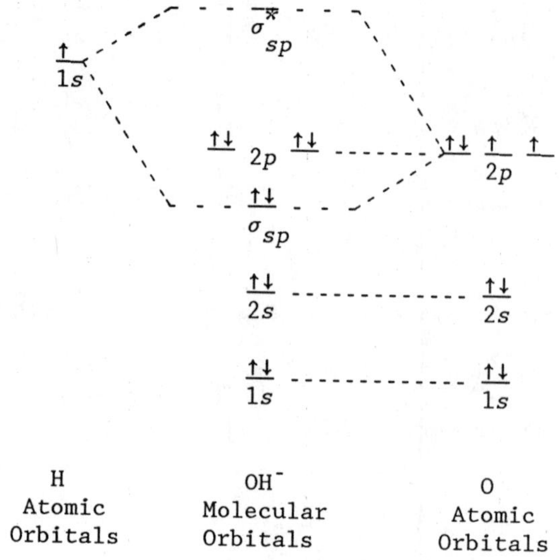

(c) HCl $1s^2\ 2s^2\ 2p_x{}^2\ 2p_y{}^2\ 2p_z{}^2\ 3s^2\ \sigma_{sp}{}^2\ 3p_x{}^2\ 3p_y{}^2$ bond order $= \frac{2 - 0}{2} = 1$

The HCl molecule is stable.

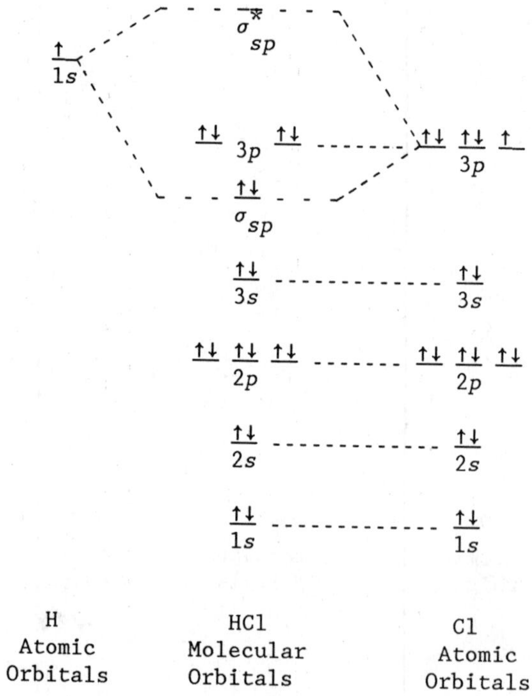

8-26. **Refer to Section 8-5 and Exercise 8-20.** • • • • • • • • • • • • • •

(a) The diagrams will be similar to that in Exercise 8-20 except that σ_{2p} has lower energy than π_{2p_y} and π_{2p_z}.

(b) NF $\quad \sigma_{1s}^2 \; \sigma_{1s}^{*2} \; \sigma_{2s}^2 \; \sigma_{2s}^{*2} \; \sigma_{2p}^2 \; \pi_{2p_y}^2 \; \pi_{2p_z}^2 \; \pi_{2p_y}^{*1} \; \pi_{2p_z}^{*1}$

NF⁻ $\quad \sigma_{1s}^2 \; \sigma_{1s}^{*2} \; \sigma_{2s}^2 \; \sigma_{2s}^{*2} \; \sigma_{2p}^2 \; \pi_{2p_y}^2 \; \pi_{2p_z}^2 \; \pi_{2p_y}^{*2} \; \pi_{2p_z}^{*1}$

(c) Recall, bond order = $\dfrac{\text{No. Bonding Electrons} - \text{No. Antibonding Electrons}}{2}$

	NF	NF⁻
Bonding Electrons	10	10
Antibonding Electrons	6	7
Bond Order	2	1.5
(d) Unpaired Electrons	2	1
Paramagnetic (P) or Diamagnetic (D)	P	P

Both NF and NF⁻ are stable, but NF is slightly more stable due to its higher bond order.

8-28. **Refer to Section 8-4.** •

(a) N and P are both VA elements but N_2 is much more stable than P_2 because N is a smaller atom than P. The $3p$ orbitals of a P atom do not overlap side-on in a pi bond with the corresponding $3p$ orbitals of another P atom nearly as effectively as the corresponding $2p$ orbitals of the much smaller nitrogen atoms do. So, as explained by valence bond theory, P forms 3 sigma bonds and acquires an octet as it uses sp^3 hybridization to form P_4 molecules.

(b) O and S are both VIA elements. However, O_2 is much more stable than S_2 because O is a smaller atom than S. The $3p$ orbitals of a S atom do not overlap side-on with the corresponding $3p$ orbitals of another S atom nearly as well as the corresponding $2p$ orbitals of the smaller O atoms . So, S forms 2 sigma bonds and obtains its octet as it uses sp^3 hybridization to form S_8 molecules.

8-30. **Refer to Section 8-6 and Figures 8-9 and 8-10.** • • • • • • • • • • • • •

The valence bond structures of the species are below. Refer to Figure 8-9 for molecular orbital representations for the delocalized pi systems.

Electronic Geometry
Around Central Atom

(a) SO_2 :Ö::S:Ö: ↔ :Ö:S::Ö: trigonal planar

(b) HCO_3^- H:Ö:C:Ö:⁻ ↔ H:Ö:C::Ö:⁻   trigonal planar
   :Ö: :Ö:

103

			Electronic Geometry Around Central Atom
(c)	SO_3	:Ö::S:Ö: ↔ :Ö:S:Ö: ↔ :Ö:S::Ö: :Ö: :Ö: :Ö:	trigonal planar
(d)	O_3	:Ö::Ö:Ö: ↔ :Ö:Ö::Ö:	trigonal planar

Molecular orbital descriptions:

(a) SO_2

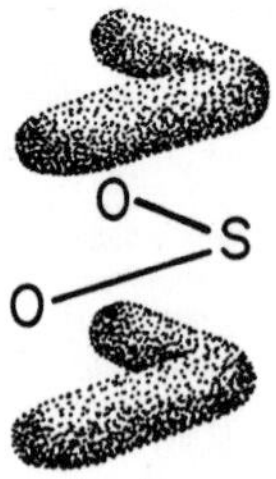

(b) HCO_3^-

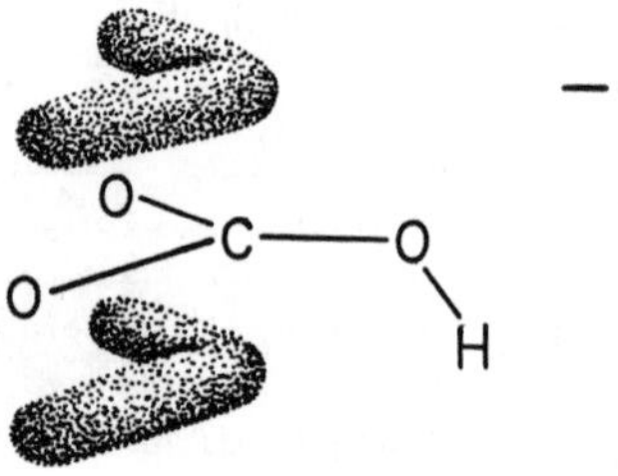

(c) SO_3

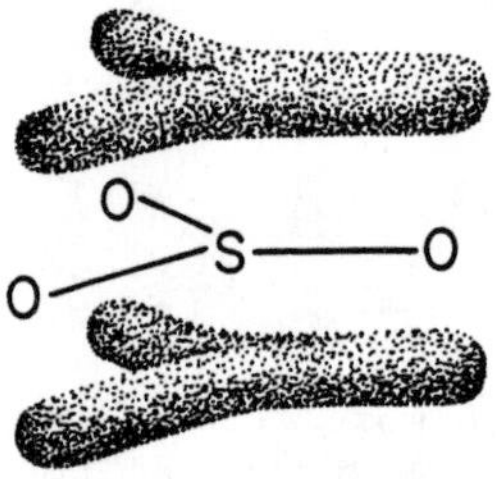

(d) O_3

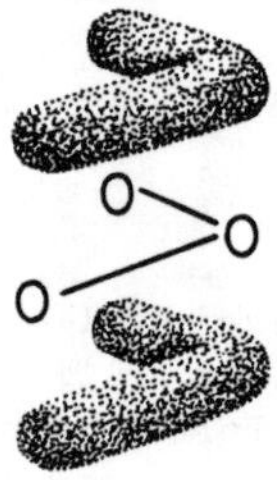

9 Chemical Reactions: A Systematic Study

9-2. **Refer to Section 9-1.** . . . .

Three major classes of compounds are electrolytes:

		Strong Electrolytes	Weak Electrolytes
(1)	acids	HCl, $HClO_4$	CH_3COOH, HF
(2)	soluble bases	NaOH, $Ba(OH)_2$	NH_3, $(CH_3)_3N$
(3)	soluble salts	NaCl, KNO_3	—

Therefore, the three classes of compounds which are strong electrolytes are strong acids, strong soluble bases and soluble salts.

9-4. **Refer to Sections 9-1 and 9-6.** . . . .

A salt is a compound that contains a cation other than H^+ and an anion other than the hydroxide ion, OH^-, or the oxide ion, O^{2-}. A salt is a product of the reaction between a particular acid and base and consists of the cation of the base and the anion of the acid. For example,

$$\underset{\textbf{base}}{NaOH} + \underset{\textbf{acid}}{HCl} \rightarrow \underset{\textbf{salt}}{NaCl} + H_2O$$

9-6. **Refer to Section 9-1.** . . . .

(a) $HCl\ (aq) \rightarrow H^+\ (aq) + Cl^-\ (aq)$

(b) $HNO_3\ (aq) \rightarrow H^+\ (aq) + NO_3^-\ (aq)$

(c) $HClO_4\ (aq) \rightarrow H^+\ (aq) + ClO_4^-\ (aq)$

9-8. **Refer to Section 9-1.** . . . .

Reactions that can occur in both directions are reversible reactions. Examples include:

$$HF\ (aq) \rightleftharpoons H^+\ (aq) + F^-\ (aq)$$

$$HCN\ (aq) \rightleftharpoons H^+\ (aq) + CN^-\ (aq)$$

$$NH_3\ (aq) + H_2O\ (\ell) \rightleftharpoons NH_4^+\ (aq) + OH^-\ (aq)$$

9-10. **Refer to Table 9-3.** . . . .

9-12. **Refer to Section 9-1 and Solubility Rule 6.** . . . .

The common metal hydroxides are insoluble in water except for those of Group IA metals and the heavier Group IIA metals beginning with $Ca(OH)_2$. Examples include

$Mg(OH)_2$	magnesium hydroxide
$Cr(OH)_3$	chromium(III) hydroxide
$Fe(OH)_2$	iron(II) hydroxide or ferrous hydroxide
$Fe(OH)_3$	iron(III) hydroxide or ferric hydroxide
$Cu(OH)_2$	copper(II) hydroxide or cupric hydroxide

9-14. **Refer to Section 9-1.** . . . .

Acidic household "chemicals": vinegar (dilute acetic acid)
vitamin C (ascorbic acid)
lemon juice (citric acid)

Basic household "chemicals": ammonia
drain cleaner (sodium hydroxide)
Milk of Magnesia (magnesium hydroxide)

9-16. **Refer to Section 9-1 and the Solubility Rules.** . . . .

(a) $BaSO_4$ insoluble in water (Rule 5)

(b) $Al(NO_3)_3$ soluble in water (Rule 3)

(c) CuS insoluble in water (Rule 8)

(d) Na_2S soluble in water (Rules 2 and 8)

(e) $Ca(CH_3COO)_2$ soluble in water (Rule 3)

9-18. **Refer to Section 9-1 and the Solubility Rules.** . . . .

(a) $KClO_3$ soluble in water (Rules 2 and 3)

(b) NH_4Br soluble in water (Rule 2 and 4)

(c) NH_3 soluble in water (Section 9-1, part 4)

(d) HNO_2 soluble in water (Rule 1)

(e) PbS insoluble in water (Rule 8)

9-20. **Refer to Section 9-1.** . . . .

(a) $NaNO_3$ is a soluble salt and is a strong electrolyte.

(b) $Ba(OH)_2$ is a strong soluble base and is a strong electrolyte.

(c) CH_3OH, methanol, is a nonelectrolyte.

(d) HCN is a weak acid and is a weak electrolyte.

(e) $Al(NO_3)_3$ is a soluble salt and is a strong electrolyte.

9-22. **Refer to Sections 9-2 and 6-1.** . . . .

(a) $Be + F_2 \rightarrow BeF_2$

Be $\underline{\uparrow\downarrow}_{1s}$ $\underline{\uparrow\downarrow}_{2s}$ $\rightarrow$ Be^{2+} $\underline{\uparrow\downarrow}_{1s}$ $\underline{\quad}_{2s}$ lost 2 e^-

F $\underline{\uparrow\downarrow}_{1s}$ $\underline{\uparrow\downarrow}_{2s}$ $\underline{\uparrow\downarrow}\ \underline{\uparrow\downarrow}\ \underline{\uparrow\ }_{2p}$ $\rightarrow$ F^- $\underline{\uparrow\downarrow}_{1s}$ $\underline{\uparrow\downarrow}_{2s}$ $\underline{\uparrow\downarrow}\ \underline{\uparrow\downarrow}\ \underline{\uparrow\downarrow}_{2p}$ gained 1 e^-

(b) $Ca + Br_2 \rightarrow CaBr_2$

Ca	[Ar] $\frac{\uparrow\downarrow}{4s}$	$\rightarrow$	Ca^{2+} [Ar] $\frac{\quad}{4s}$	lost 2 e^-
Br	[Ar] $3d^{10}$ $\frac{\uparrow\downarrow}{4s}$ $\frac{\uparrow\downarrow\ \uparrow\downarrow\ \uparrow\ }{4p}$	$\rightarrow$	Br^- [Ar] $3d^{10}$ $\frac{\uparrow\downarrow}{4s}$ $\frac{\uparrow\downarrow\ \uparrow\downarrow\ \uparrow\downarrow}{4p}$	gained 1 e^-

(c) $Ba + Cl_2 \rightarrow BaCl_2$

Ba	[Xe] $\frac{\uparrow\downarrow}{6s}$	$\rightarrow$	Ba^{2+} [Xe] $\frac{\quad}{6s}$	lost 2 e^-
Cl	[Ne] $\frac{\uparrow\downarrow}{3s}$ $\frac{\uparrow\downarrow\ \uparrow\downarrow\ \uparrow\ }{3p}$	$\rightarrow$	Cl^- [Ne] $\frac{\uparrow\downarrow}{3s}$ $\frac{\uparrow\downarrow\ \uparrow\downarrow\ \uparrow\downarrow}{3p}$	gained 1 e^-

9-24. **Refer to Section 9-2.** . . . .

Arsenic trichloride, $AsCl_3$, forms when As combines with a limited amount of Cl_2.

Arsenic pentachloride, $AsCl_5$, forms when As combines with excess Cl_2.

9-26. **Refer to Section 9-2.** . . . .

(a) $\overset{+6}{3SO_3} + \overset{+3}{Al_2O_3} \rightarrow \overset{+3\ +6}{Al_2(SO_4)_3}$

(b) $\overset{+7}{Cl_2O_7} + H_2O \rightarrow \overset{+7}{2HClO_4}$

(c) $\overset{+2}{CaO} + \overset{+4}{SiO_2} \rightarrow \overset{+2\ +4}{CaSiO_3}$

There was no change in oxidation number for any of the elements in the above combination reactions.

9-28. **Refer to Section 9-4 and Table 9-6.** . . . .

Zn and Fe are more active metals than Cu and will displace Cu from an aqueous solution of $CuSO_4$.

$Hg\ (\ell) + CuSO_4\ (aq) \rightarrow$ no reaction

$Zn\ (s) + Cu^{2+}\ (aq) \rightarrow Zn^{2+}\ (aq) + Cu\ (s)$

$Fe\ (s) + Cu^{2+}\ (aq) \rightarrow Fe^{2+}\ (aq) + Cu\ (s)$

$Ag\ (s) + CuSO_4\ (aq) \rightarrow$ no reaction

9-30. **Refer to Section 9-4, Table 9-6 and Exercise 9-28.** . . . .

In order of increasing activity: Ag < Hg < Cu < Fe < Zn

9-32. **Refer to Section 9-4, Table 9-6 and Exercise 9-31.** . . . .

In order of increasing activity: Fe < Zn < Na < Ca

9-34. **Refer to Section 9-4.** . . . .

Each halogen will displace less electronegative (heavier) halogens from their binary salts. Hence, reactions (b), (c) and (d) will occur and reaction (a) will not occur.

9-36. **Refer to Section 9-5.** . . . .

(a) $\overset{-3\ +1}{(NH_4)_2}\overset{+6\ -2}{Cr_2O_7}$ (s) → $\overset{0}{N_2}$ (g) + $\overset{+3\ -2}{Cr_2O_3}$ (s) + $\overset{+1\ -2}{4H_2O}$ (g)

(b) $\overset{+1\ +5\ -2}{2NaNO_3}$ (s) → $\overset{+1\ +3\ -2}{2NaNO_2}$(s) + $\overset{0}{O_2}$ (g)

9-38. **Refer to Section 9-6.** . . . .

(a) molecular: $3CaCl_2$ (aq) + $2K_3PO_4$ (aq) → $Ca_3(PO_4)_2$ (s) + 6KCl (aq)

total ionic: $3Ca^{2+}$ (aq) + $6Cl^-$ (aq) + $6K^+$ (aq) + $2PO_4^{3-}$ (aq)

→ Ca_3PO_4 (s) + $6K^+$ (aq) + $6Cl^-$ (aq)

net ionic: $3Ca^{2+}$ (aq) + $2PO_4^{3-}$ (aq) → $Ca_3(PO_4)_2$ (s)

(b) molecular: $Hg(NO_3)_2$ (aq) + Na_2S (aq) → HgS (s) + $2NaNO_3$ (aq)

total ionic: Hg^{2+} (aq) + $2NO_3^-$ (aq) + $2Na^+$ (aq) + S^{2-} (aq)

→ HgS (s) + $2Na^+$ (aq) + 2 NO_3^- (aq)

net ionic: Hg^{2+} (aq) + S^{2-} (aq) → HgS (s)

(c) molecular: $2CrCl_3$ (aq) + $3Ca(OH)_2$ (aq) → $2Cr(OH)_3$ (s) + $3CaCl_2$ (aq)

total ionic: $2Cr^{3+}$ (aq) + $6Cl^-$ (aq) + $3Ca^{2+}$ (aq) + $6OH^-$ (aq)

→ $2Cr(OH)_3$ (s) + $3Ca^{2+}$ (aq) + $6Cl^-$ (aq)

net ionic: $2Cr^{3+}$ (aq) + $6OH^-$ (aq) → $2Cr(OH)_3$ (s)

therefore, Cr^{3+} (aq) + $3OH^-$ (aq) → $Cr(OH)_3$ (s)

9-40. **Refer to Section 9-6.** . . . .

(a) molecular: $Cu(NO_3)_2$ (aq) + Na_2S (aq) → CuS (s) + $2NaNO_3$ (aq)

total ionic: Cu^{2+} (aq) + $2NO_3^-$ (aq) + $2Na^+$ (aq) + S^{2-} (aq)

→ CuS (s) + $2Na^+$ (aq) + $2NO_3^-$ (aq)

net ionic: Cu^{2+} (aq) + S^{2-} (aq) → CuS (s)

(b) molecular: $CdSO_4$ (aq) + H_2S (aq) → CdS (s) + H_2SO_4 (aq)

total ionic: Cd^{2+} (aq) + SO_4^{2-} (aq) + H_2S (aq)

→ CdS (s) + $2H^+$ (aq) + SO_4^{2-} (aq)

net ionic: Cd^{2+} (aq) + H_2S (aq) → CdS (s) + $2H^+$ (aq)

(c) molecular: $Bi_2(SO_4)_3$ (aq) + $3(NH_4)_2S$ (aq) → Bi_2S_3 (s) + $3(NH_4)_2SO_4$ (aq)

total ionic: $2Bi^{3+}$ (aq) + $3SO_4^{2-}$ (aq) + $6NH_4^+$ (aq) + $3S^{2-}$ (aq)

→ Bi_2S_3 (s) + $6NH_4^+$ (aq) + $3SO_4^{2-}$ (aq)

net ionic: $2Bi^{3+}$ (aq) + $3S^{2-}$ (aq) → Bi_2S_3 (s)

9-42. **Refer to Sections 9-1 and 9-6.** . . . .

(a) molecular: $2CH_3COOH$ (aq) + $Ca(OH)_2$ (aq) → $Ca(CH_3COO)_2$ (aq) + $2H_2O$ (ℓ)

total ionic: $2CH_3COOH$ (aq) + Ca^{2+} (aq) + $2OH^-$

→ Ca^{2+} (aq) + $2CH_3COO^-$ (aq) + $2H_2O$ (ℓ)

net ionic: $2CH_3COOH$ (aq) + $2OH^-$ (aq) → $2CH_3COO^-$ (aq) + $2H_2O$ (ℓ)

therefore, CH_3COOH (aq) + OH^- (aq) → CH_3COO^- (aq) + H_2O (ℓ)

(b) molecular: H_2SO_3 (aq) + 2NaOH (aq) → Na_2SO_3 (aq) + $2H_2O$ (ℓ)

total ionic: H_2SO_3 (aq) + $2Na^+$ (aq) + $2OH^-$ (aq)

→ $2Na^+$ (aq) + SO_3^{2-} (aq) + $2H_2O$ (ℓ)

net ionic: H_2SO_3 (aq) + $2OH^-$ (aq) → SO_3^{2-} (aq) + $2H_2O$ (ℓ)

(c) molecular: HF (aq) + LiOH (aq) → LiF (aq) + H_2O (ℓ)

total ionic: HF (aq) + Li^+ (aq) + OH^- (aq) → Li^+ (aq) + F^- (aq) + H_2O (ℓ)

net ionic: HF (aq) + OH^- (aq) → F^- (aq) + H_2O (ℓ)

9-44. **Refer to Sections 9-1 and 9-6.** . . . .

(a) molecular: 2NaOH (aq) + H_2SO_4 (aq) → Na_2SO_4 (aq) + $2H_2O$ (ℓ)

total ionic: $2Na^+$ (aq) + $2OH^-$ (aq) + $2H^+$ (aq) + SO_4^{2-} (aq)

→ $2Na^+$ (aq) + SO_4^{2-} (aq) + $2H_2O$ (ℓ)

net ionic: $2OH^-$ (aq) + $2H^+$ (aq) → $2H_2O$ (ℓ)

therefore, OH^- (aq) + H^+ (aq) → H_2O (ℓ)

(b) molecular: $3Ca(OH)_2$ (aq) + $2H_3PO_4$ (aq) → $Ca_3(PO_4)_2$ (s) + $6H_2O$ (ℓ)

total ionic: $3Ca^{2+}$ (aq) + $6OH^-$ (aq) + $2H_3PO_4$ (aq)

→ $Ca_3(PO_4)_2$ (s) + $6H_2O$ (ℓ)

net ionic: same as total ionic

(c) molecular: $Cu(OH)_2$ (s) + $2HNO_3$ (aq) → $Cu(NO_3)_2$ (aq) + $2H_2O$ (ℓ)

total ionic: $Cu(OH)_2$ (s) + $2H^+$ (aq) + $2NO_3^-$ (aq)

→ Cu^{2+} (aq) + $2NO_3^-$ (aq) + $2H_2O$ (ℓ)

net ionic: $Cu(OH)_2$ (s) + $2H^+$ (aq) → Cu^{2+} (aq) + $2H_2O$ (ℓ)

9-46. **Refer to Sections 9-1 and 9-6.** . . . .

(a) molecular: $2HClO_4$ (aq) + $Ca(OH)_2$ (aq) → $Ca(ClO_4)_2$ (aq) + $2H_2O$ (ℓ)

total ionic: $2H^+$ (aq) + $2ClO_4^-$ (aq) + Ca^{2+} (aq) + $2OH^-$ (aq)

→ Ca^{2+} (aq) + $2ClO_4^-$ (aq) + $2H_2O$ (ℓ)

net ionic: $2H^+$ (aq) + $2OH^-$ (aq) → $2H_2O$ (ℓ)

therefore, H^+ (aq) + OH^- (aq) → H_2O (ℓ)

(b) molecular: H_2SO_4 (aq) + $2NH_3$ (aq) → $(NH_4)_2SO_4$ (aq)

total ionic: $2H^+$ (aq) + SO_4^{2-} (aq) + $2NH_3$ (aq) → $2NH_4^+$ (aq) + SO_4^{2-} (aq)

net ionic: $2H^+$ (aq) + $2NH_3$ (aq) → $2NH_4^+$ (aq)

therefore, H^+ (aq) + NH_3 (aq) → NH_4^+ (aq)

(c) molecular: H_2S (aq) + $Cu(OH)_2$ (s) → CuS (s) + $2H_2O$ (ℓ)

total ionic: same as molecular

net ionic: same as molecular

9-48. **Refer to Sections 9-1 and 9-6.** . . . .

(a) molecular: H_2S (aq) + 2NaOH (aq) → Na_2S (aq) + $2H_2O$ (ℓ)

total ionic: H_2S (aq) + $2Na^+$ (aq) + $2OH^-$ (aq)

→ $2Na^+$ (aq) + S^{2-} (aq) + $2H_2O$ (ℓ)

net ionic: H_2S (aq) + $2OH^-$ (aq) → S^{2-} (aq) + $2H_2O$ (ℓ)

(b) molecular: $2H_3PO_4$ (aq) + $3Ba(OH)_2$ (aq) → $Ba_3(PO_4)_2$ (s) + $6H_2O$ (ℓ)

total ionic: $2H_3PO_4$ (aq) + $3Ba^{2+}$ (aq) + $6OH^-$ (aq)

→ $Ba_3(PO_4)_2$ (s) + $6H_2O$ (ℓ)

net ionic: same as total ionic

(c) molecular: $2H_3AsO_4$ (aq) + $3Pb(OH)_2$ (s) → $Pb_3(AsO_4)_2$ (s) + $6H_2O$ (ℓ)

total ionic: same as molecular

net ionic: same as molecular

9-50. **Refer to Section 9-7.** . . . .

Due to the Law of Conservation of Matter, the electrons that cause the reduction of one substance must be produced from the oxidation of another substance.

9-52. **Refer to Section 9-7 and Exercise 9-51.** . . . .

	Oxidizing Agent		Reducing Agent		
(a)	$3H_2SO_4$	+	2Al	→	$Al_2(SO_4)_3$ + $3H_2$
(b)	N_2	+	$3H_2$	→	$2NH_3$
(c)	$3O_2$	+	2ZnS	→	2ZnO + $2SO_2$
(d)	$4HNO_3$	+	C	→	$4NO_2$ + CO_2 + $2H_2O$
(e)	H_2SO_4	+	2HI	→	SO_2 + I_2 + $2H_2O$

9-54. **Refer to Section 9-7.** . . . .

The oxygen gas we inhale to keep ourselves alive has an oxidation number of 0. The gases we exhale, carbon dioxide and water vapor, contain oxygen with an oxidation number of -2. Oxygen gas is being reduced in a reaction involving our bodily processes. Therefore, these processes must involve oxidation.

9-56. **Refer to Section 9-6.** . . . .

The only metathesis reaction involving no change in oxidation number that is also a precipitation reaction is (h).

9-58. **Refer to Section 9-7.** . . . .

Oxidation-reduction reactions include (b), (c), (e), (g), (i), (j), (m) and (o). Note that some of these can also be classified as combination, decomposition or displacement reactions.

9-60. **Refer to Sections 9-4 and 9-7.** . . . .

Reactions in which one element displaces another from a compound are called displacement reactions. Those oxidation-reduction reactions that are also displacement reactions are (c) and (i). Reaction (g) may be considered by some to be a redox displacement reaction.

9-62. **Refer to Section 9-2.** . . . .

Reactions in which two or more substances combine to form a compound are called combination reactions. They may or may not involve redox reactions.

redox combination reactions: (b), (e)
non-redox combination reactions: (d), (l), (n)

9-64. **Refer to Section 9-8.** . . . .

Elemental hydrogen exists as a colorless, odorless, tasteless, diatomic gas with the lowest atomic weight and density of any known substance. It melts at -259.14°C and boils at -252.8°C.

9-66. **Refer to Section 9-8.** . . . .

(a) Hydrogen gas reacts with the alkali metals and heavier alkaline earth metals to form ionic hydrides:

$2Li\ (molten) + H_2\ (g) \rightarrow 2LiH\ (s)$

(b) Hydrogen gas reacts with other nonmetals to form covalent hydrides:

$H_2\ (g) + Cl_2\ (g) \rightarrow 2HCl\ (g)$

9-68. **Refer to Section 9-8.** . . . .

NaH, sodium hydride, is the product of hydrogen gas reacting with an active metal, sodium. A compound consisting of a metal and a nonmetal is ionic, hence this is an ionic hydride.

H_2S, hydrogen sulfide, is the product of hydrogen gas reacting with a nonmetal, sulfur. A compound consisting of two nonmetals is covalent, and therefore this is a covalent hydride.

9-70. **Refer to Sections 9-8 and 6-10 on Nomenclature.** . . . .

(a) H_2S hydrogen sulfide

(b) HF hydrogen fluoride

(c) KH potassium hydride

(d) NH_3 ammonia

(e) H_2Se hydrogen selenide

(f) MgH_2 magnesium hydride

9-72. **Refer to Sections 9-8 and 9-9.** . . . .

H_2, hydrogen, is a colorless, odorless, tasteless, nonpolar, diamagnetic, diatomic gas with the lowest atomic weight and density of any known substance. It has low solubility in water and is very flammable. Hydrogen is prepared by reactions of metals with water, steam or various acids, electrolysis of water, the water gas reaction and thermal cracking of hydrocarbons. It combines with metals and nonmetals to form hydrides.

O_2, oxygen, is nearly colorless, odorless, tasteless, nonpolar, paramagnetic, diatomic gas. It is nonflammable but participates in all combustion reactions. It is prepared by fractional distillation of liquid air, electrolysis of water and thermal decomposition of certain oxygen-containing salts. Oxygen combines with almost all other elements to form oxides and can be converted to an allotropic form, ozone, O_3.

9-74. **Refer to Section 9-9 and Table 9-7.** . . . .

The elements that react with oxygen to form mainly normal oxides include (a) Li, (d) Mg, (e) Zn and (f) Al.

9-76. **Refer to Section 9-9.** . . . .

When O_2 is limited in a reaction, the element is oxidized to its lowest oxidation number.

(a) $2Sr\ (s) + O_2\ (g) \rightarrow 2SrO\ (s)$ (O_2 is limited)

(b) $2Fe\ (s) + O_2\ (g) \rightarrow 2FeO\ (s)$ (O_2 is limited)

(c) $2Mn\ (s) + O_2\ (g) \rightarrow 2MnO\ (s)$ (O_2 is limited)

(d) $4Cu\ (s) + O_2\ (g) \rightarrow 2Cu_2O\ (s)$ (O_2 is limited)

9-78. **Refer to Section 9-9 and Exercise 9-76 Solution.** . . . .

(a) $2C\ (s) + O_2\ (g) \rightarrow 2CO\ (g)$ (O_2 is limited)

(b) $As_4\ (s) + 3O_2\ (g) \rightarrow As_4O_6\ (s)$ (O_2 is limited)

(c) $2Ge\ (s) + O_2\ (g) \rightarrow 2GeO\ (s)$ (O_2 is limited)

9-80. **Refer to Section 9-9 and Table 9-7.** . . . .

A normal oxide is a binary (two element) compound containing oxygen in the -2 oxidation state. BaO is an example of an ionic oxide and SO_2 is an example of a covalent oxide.

A peroxide can be a binary ionic compound, such as Na_2O_2, or a covalent compound, such as H_2O_2, containing the O_2^{2-} ion with oxygen in the -1 oxidation state.

A superoxide is a binary ionic compound containing the O_2^- ion with oxygen in the -1/2 oxidation state, such as KO_2.

9-82. **Refer to Section 9-9 and Example 9-7.** . . . .

(a) $SO_2\ (g) + H_2O\ (\ell) \rightarrow H_2SO_3\ (aq)$ sulfurous acid

(b) $SO_3\ (\ell) + H_2O\ (\ell) \rightarrow H_2SO_4\ (aq)$ sulfuric acid

(c) $SeO_3\ (s) + H_2O\ (\ell) \rightarrow H_2SeO_4\ (aq)$ selenic acid

(d) $N_2O_5\ (s) + H_2O\ (\ell) \rightarrow 2HNO_3\ (aq)$ nitric acid

(e) $Cl_2O_7\ (\ell) + H_2O\ (\ell) \rightarrow 2HClO_4\ (aq)$ perchloric acid

9-84. **Refer to Section 9-9.** . . . .

The acid anhydrides are: (a) SO_3, (b) CO_2, (c) SO_2, (d) As_2O_5, (e) N_2O_3

9-86. **Refer to Section 9-9.** . . . .

Combustion is an oxidation-reduction reaction in which oxygen gas combines rapidly with oxidizable materials in highly exothermic reactions with a visible flame. A combustion reaction is a redox reaction since the oxidation number of oxygen is changed from 0 in O_2 to -2 in the products, usually CO_2 and H_2O.

9-88. **Refer to Section 9-9.** . . . .

(a) $2C_2H_6$ (g) + $5O_2$ (g) → 4CO (g) + $6H_2O$ (g) (O_2 is limited)

(b) $2C_3H_8$ (g) + $7O_2$ (g) → 6CO (g) + $8H_2O$ (g) (O_2 is limited)

9-90. **Refer to Section 9-9.** . . . .

(a) $\overset{-2.5\ +1}{2C_4H_{10}}$ (g) + $\overset{0}{13O_2}$ (g) → $\overset{+4\ -2}{8CO_2}$ (g) + $\overset{+1\ -2}{10H_2O}$ (g) (O_2 is in excess)

(b) $\overset{-2.5\ +1}{2C_4H_{10}}$ (g) + $\overset{0}{9O_2}$ (g) → $\overset{+2\ -2}{8CO}$ (g) + $\overset{+1\ -2}{10H_2O}$ (g) (O_2 is limited)

(c) $\overset{-2.5\ +1}{C_4H_{10}}$ (g) + $\overset{0}{O_2}$ (g) → $\overset{-1\ +1}{C_2H_2}$ (g) + $\overset{+2\ -2}{2CO}$ (g) + $\overset{0}{4H_2}$ (g) (O_2 is very limited)

9-92. **Refer to Section 9-9.** . . . .

(a) $4C_6H_5N$ (ℓ) + $31O_2$ (g) → $24CO_2$ (g) + 10 H_2O (ℓ) + 4NO (g)

(b) $2C_2H_5SH$ (ℓ) + $9O_2$ (g) → $4CO_2$ (g) + $6H_2O$ (g) + $2SO_2$ (g)

(c) $C_7H_{10}NO_2S$ (ℓ) + $10O_2$ (g) → $7CO_2$ (g) + $5H_2O$ (g) + NO (g) + SO_2 (g)

9-94. **Refer to Section 9-9.** . . . .

"Acid rain" is the result of industrially produced acid anhydrides, particularly sulfur oxides, dissolving in atmospheric moisture. The source of the acid anhydrides is largely the burning of fossil fuels. Fossil fuels are generally thought to result from the decay of animal and vegetable matter under high pressure and temperature conditions. They contain sulfur and nitrogen compounds as well as hydrocarbons. When fossil fuels are burned, oxides of sulfur and nitrogen (acid anhydrides) are released into the atmosphere and dissolve in atmospheric moisture producing "acid rain." The reactions converting sulfur to acid rain are shown below:

$2SO_2$ (g) + O_2 (g) → $2SO_3$ (ℓ) (occurs slowly in air)

SO_2 (g) + H_2O (ℓ, moisture in air) → H_2SO_3 (ℓ)

SO_3 (ℓ) + H_2O (ℓ, moisture in air) → H_2SO_4 (ℓ)

10 Gases and the Kinetic Molecular Theory

10-2. **Refer to Section 10-1 and Table 10-1.** . . . .

From Table 10-1, we see that the densities of gases are very much (10^2-10^3 times) less than the densities of liquids or solids. We also know experimentally that gases are easily compressed. These facts let us conclude that the molecules in gases are very far apart compared to molecules in liquids or solids.

10-4. **Refer to Section 10-3 and Figure 10-1.** . . . .

A manometer is a device employing the change in liquid levels to measure gas pressure differences between a standard and an unknown system. For example, a typical mercury manometer consists of a glass tube partially filled with mercury. One arm is open to the atmosphere and the other is connected to a container of gas. When the pressure of the gas in the container is greater than atmospheric pressure, the level of the mercury in the open side will be higher and

$$P_{gas} = P_{atm} + \Delta h$$

where Δh is the difference in mercury levels (Figure 10-1). However, when the pressure of the gas is less than atmospheric pressure, the level of the mercury in the closed side will be higher and

$$P_{gas} = P_{atm} - \Delta h.$$

10-6. **Refer to Section 10-3 and Figure 10-1.** . . . .

A tire gauge does not measure the absolute tire pressure:

$$\text{gauge reading} = P_{\text{gas in tire}} - P_{atm}.$$

The gas inside the tire exerts a pressure that is opposed by the atmospheric pressure acting on the gauge.

10-8. **Refer to Section 10-3.** . . . .

Atmospheric pressure varies with distance above sea level. It is greater at lower elevations, e.g., in Chicago as opposed to Denver, since there is more air above the low elevation air which compresses it and therefore increases its pressure.

10-10. **Refer to Section 10-4 and Figure 10-2.** . . . .

Boyle studied the effect of changing pressure on a volume of a known mass of gas at constant temperature. Boyle's Law states: at a given temperature, the product of pressure and volume of a definite mass of gas is constant.

10-12. **Refer to Section 10-4 and Figure 10-3.** . . . .

When the mathematical relationship, XY = constant, is plotted on the X-Y axes, a hyperbola results. Boyle's Law can be stated as

$$\text{pressure} \times \text{volume} = \text{constant}$$

resulting in the graph shown in Figure 10-3. Since pressure and volume can never have negative values, the other branch of the hyperbola is omitted.

10-14. **Refer to Section 10-4 and Examples 10-1 and 10-2.** . . . .

(a) This is a Boyle's Law calculation since the system is at constant n and T with changes in P and V.

Boyle's Law: $P_1V_1 = P_2V_2$ at constant n and T

$$? \ V_2 \ (\text{mL}) = \frac{P_1V_1}{P_2} = \frac{732 \text{ torr} \times 549 \text{ mL}}{100 \text{ torr}} = \mathbf{4020 \ mL}$$

(b) $$? \ \% \text{ volume change} = \frac{\text{volume change}}{\text{initial volume}} \times 100 = \frac{V_2 - V_1}{V_1} \times 100$$

$$= \frac{4020 - 549}{549} \times 100$$

$= \mathbf{632\% \ increase \ in \ volume}$

(c) $$? \ \% \text{ pressure change} = \frac{\text{pressure change}}{\text{initial pressure}} \times 100 = \frac{P_2 - P_1}{P_1} \times 100$$

$$= \frac{100 - 732}{732} \times 100$$

$= -86.3\%$

$= \mathbf{86.3\% \ decrease \ in \ pressure}$

(d) The answers in (b) and (c) are very different. Boyle's Law states that (pressure × volume) is a constant at constant temperature, not (% change in pressure × % change in volume) is a constant.

10-16. **Refer to Section 10-4 and Exercise 10-14 Solution.** . . . .

Boyle's Law states: $P_1V_1 = P_2V_2$ at constant n and T. If the pressure is tripled, then $P_2 = 3P_1$.

Substituting and solving for V_2,

$$V_2 \ (\text{L}) = \frac{P_1V_1}{P_2} = \frac{P_1 \times 100 \text{ L}}{3P_1} = 33 \text{ L}$$

Note: When the pressure is tripled, the volume is reduced by a factor of 3, i.e., the volume becomes 1/3 of the original volume.

$$? \text{ fraction volume change} = \frac{V_2 - V_1}{V_1} = \frac{33 - 100}{100} = -0.67 = \mathbf{0.67 \ decrease \ in \ volume}$$

$$? \ \% \text{ volume change} = \frac{V_2 - V_1}{V_1} \times 100 = -67 \ \% = \mathbf{67\% \ decrease \ in \ volume}$$

10-18. **Refer to Section 10-5 and Figure 10-4.** . . . .

(a) An "absolute temperature scale" is a scale in which properties such as gas volume change linearly with temperature while the origin of the scale is set at absolute zero. The Kelvin scale is a typical example of it.

(b) Boyle, in his experiments, noticed that temperature affected volume. About 1800, Charles and Guy-Lussac found that the rate of expansion with increased temperature was constant at constant pressure. Later Lord Kelvin noticed that for a series of constant pressure systems, volume decreased as temperature decreased and the extrapolation of these different *T-V* lines back to zero volume yielded a common intercept, $-273.15^{o}C$ on the temperature axis. He defined this temperature as absolute zero. The relationship between the Celsius and Kelvin temperature scales is

$$K = {}^{o}C + 273.15^{o}.$$

(c) Absolute zero may be thought of as the limit of thermal contraction for an ideal gas. In other words, an ideal gas would have zero volume at absolute zero temperature. Theoretically, it is also the temperature at which molecular motion ceases.

10-20. **Refer to Section 10-5 and Figure 10-4.** . . . .

Experiments have shown that at constant pressure, the volume of a definite mass of gas is directly proportional to its absolute temperature (in K). This is known as Charles' Law and is expressed as V/T = constant at constant n and P. Therefore, for a sample of gas when volume is plotted against temperature (in K) a straight line results.

10-22. **Refer to Section 10-5 and Example 10-3.** . . . .

This is a Charles' Law calculation: $\frac{V_1}{T_1} = \frac{V_2}{T_2}$ at constant n and P.

Plan: (1) For T_1, convert ^{o}F to K.

(2) Use Charles' Law to solve for T_2 (K).

(3) For T_2, convert K to ^{o}F.

(1) $T_1\ (^{o}C) = \frac{1.0^{o}C}{1.8^{o}F}(45^{o}F - 32^{o}F) = 7.2^{o}C$

$T_1\ (K) = {}^{o}C + 273.15 = 7.2^{o}C + 273.15^{o} = 280.4\ K$

(2) $T_2\ (K) = \frac{V_2T_1}{V_1} = \frac{485\ mL \times 280.4\ K}{316\ mL} = 430\ K$

(3) $T_2\ (^{o}C) = K - 273.15^{o} = 430\ K - 273.15^{o} = 157^{o}C$

$T_2\ (^{o}F) = (157^{o}C \times \frac{1.8^{o}F}{1.0^{o}C}) + 32^{o}F = \mathbf{315^{o}F}$

10-24. **Refer to Section 10-5.** . . . .

Recall Charles' Law: $\frac{V_1}{T_1} = \frac{V_2}{T_2}$ at constant n and P.

Given: $V_1 = 400\ mL$ $\quad T_1\ (^{o}C) = 50.0^{o}C$ $\quad T_1\ (K) = 50.0^{o}C + 273.15\ K = 323.2\ K$

$V_2 = ?$ $\quad T_2\ (^{o}C) = 100^{o}C$ $\quad T_2\ (K) = 100^{o}C + 273.15\ K = 373\ K$

Note that to use Charles' Law, all temperature values must be in Kelvin.

Substituting,

$$? \; V_2 \text{ (mL)} = \frac{T_2V_1}{T_1} = \frac{373 \text{ K} \times 400 \text{ mL}}{323.2 \text{ K}} = \mathbf{462 \text{ mL}}$$

(a) On the Celsius scale:

$$? \text{ fractional increase in temperature} = \frac{T_2 - T_1}{T_1} = \frac{100^\circ\text{C} - 50.0^\circ\text{C}}{50.0^\circ\text{C}} = \mathbf{1}$$

$$? \text{ \% increase in temperature} = \frac{T_2 - T_1}{T_1} \times 100 = \mathbf{100\%}$$

On the Kelvin scale:

$$? \text{ fractional increase in temperature} = \frac{T_2 - T_1}{T_1} = \frac{373 - 323}{323} = \mathbf{0.155}$$

$$? \text{ \% increase in temperature} = \frac{T_2 - T_1}{T_1} \times 100 = \mathbf{15.5\%}$$

(b) The fractional increase in the volume of the hydrogen sample should be 0.155, i.e. the volume should increase by 15.5%, since Charles' Law states that the volume of a gas sample is directly proportional to the absolute temperature at constant pressure.

$$? \text{ \% increase in volume} = \frac{V_2 - V_1}{V_1} \times 100 = \frac{462 - 400}{400} \times 100 = \mathbf{15.5\%}$$

10-26. **Refer to Section 10-5.** . . . .

Recall Charles' Law: $\frac{V_1}{T_1} = \frac{V_2}{T_2}$ at constant n and P

Given: $V_1 = 75.0$ mL T_1 (K) $= 15^\circ\text{C} + 273^\circ = 288$ K

$V_2 = 25.0$ mL $T_2 = ?$

$$? \; T_2 \text{ (K)} = \frac{V_2T_1}{V_1} = \frac{25.0 \text{ mL} \times 288 \text{ K}}{75.0 \text{ mL}} = 96.0 \text{ K}$$

$$? \; t_2 \text{ (}^\circ\text{C)} = 96.0 \text{ K} - 273.15^\circ = \mathbf{-177.2^\circ C}$$

10-28. **Refer to Section 10-6.** . . . .

(a) Standard conditions of temperature and pressure (STP) are a reference point for describing the state of matter in a system.

(b) Standard temperature: 0°C, 273.15 K, 32°F

(c) Standard pressure: 760 torr, 760 mm Hg, 1 atm, 1.013×10^5 Pa, 101.3 kPa

10-30. **Refer to Section 10-7.** . . . .

(a) The "combined gas law" is a single equation combining Boyle's Law and Charles' Law.

(b) $\frac{P_1V_1}{T_1} = \frac{P_2V_2}{T_2}$ at constant n

(c) The mathematical expression of Boyle's Law is $PV = k$ (at constant T and n) and that of Charles' Law is $V/T = k'$ (at constant P and n) where k and k' are different proportionality constants. Therefore, PV/T is equal to another proportionality constant and since PV/T for all systems at constant n is equal to the new proportionality constant:

$$\frac{P_1V_1}{T_1} = \frac{P_2V_2}{T_2} \quad \text{at constant } n$$

10-32. **Refer to Section 10-7.** . . . .

Recall: at STP, $T = 273.15$ K and $P = 1$ atm

Given: $T_1 = 273.15$ K $\quad V_1 = 375$ mL $\quad P_1 = 1$ atm

$T_2 = 819^{\circ}C + 273.15^{\circ} = 1092$ K $\quad V_2 = 375$ mL $\quad P_2 = ?$

Combined Gas Law: $$\frac{P_1V_1}{T_1} = \frac{P_2V_2}{T_2} \quad \text{at constant } n$$

$$\frac{1 \text{ atm} \times 375 \text{ mL}}{273.15 \text{ K}} = \frac{P_2 \times 375 \text{ mL}}{1092 \text{ K}}$$

Solving, P_2 = **4.00 atm**

10-34. **Refer to Section 10-7.** . . . .

Given: $P_1 = 830$ torr $\quad V_1 = 350$ mL $\quad T_1 = 22^{\circ}C + 273.15^{\circ} = 295$ K

$P_2 = 600$ torr $\quad V_2 = 500$ mL $\quad T_2 = ?$

Combined Gas Law: $$\frac{P_1V_1}{T_1} = \frac{P_2V_2}{T_2} \quad \text{at constant } n$$

$$\frac{830 \text{ torr} \times 350 \text{ mL}}{295 \text{ K}} = \frac{600 \text{ torr} \times 500 \text{ mL}}{T_2}$$

Solving, T_2 = **305 K or 32°C**

10-36. **Refer to Section 10-7 and Exercise 10-6 Solution.** . . . .

Given: V is constant; $V_1 = V_2$

T_1 (°F) = 40°F $\qquad$ $T_1 \text{ (K)} = \frac{1.0^{\circ}C}{1.8^{\circ}F}(40^{\circ}F - 32^{\circ}F) + 273^{\circ} = 277 \text{ K}$

T_2 (°F) = 125°F $\qquad$ $T_2 \text{ (K)} = \frac{1.0^{\circ}C}{1.8^{\circ}F}(125^{\circ}F - 32^{\circ}F) + 273^{\circ} = 325 \text{ K}$

Recall that a tire gauge measures "relative" pressure, which is the difference between the internal tire pressure and the external atmospheric pressure. Therefore, the actual pressure in the tire, P_1, is the sum of the gauge pressure and the atmospheric pressure. So, if we assume that the atmospheric pressure is 1 atm or 14.7 psi, then

$P_1 = 28$ psi + 14.7 psi = 43 psi

$P_2 = ?$

Combined Gas Law: $\frac{P_1V_1}{T_1} = \frac{P_2V_2}{T_2}$ at constant n simplifying, $(V_1 = V_2)$ $\frac{P_1}{T_1} = \frac{P_2}{T_2}$

Therefore, $P_2 = \frac{T_2P_1}{T_1} = \frac{325\text{ K} \times 43\text{ psi}}{277\text{ K}} = 50\text{ psi}$

gauge reading = 50 psi - 14.7 psi = **35 psi**

10-38. **Refer to Section 10-8.** . . . .

Avogadro's Law states that, at the same temperature and pressure, equal volumes of all gases contain the same number of molecules. This means that equal number of moles of any gas take up equal volumes as long as the temperature and pressure are the same.

10-40. **Refer to Section 10-8 and Example 10-6.** . . . .

At STP conditions:

$$D\text{ (g/L)} = \frac{1\text{ mol }F_2}{22.4\text{ L }F_2} \times \frac{38.0\text{ g }F_2}{1\text{ mol }F_2} = \mathbf{1.70\text{ g/L}}\text{ or }\mathbf{1.70 \times 10^{-3}\text{ g/mL}}$$

10-42. **Refer to Section 10-9 and Examples 10-8 and 10-9.** . . . .

(a) Use the ideal gas equation, $PV = nRT$.

$$n = \frac{PV}{RT} = \frac{[(420/760)\text{ atm}](124\text{ L})}{(0.0821\text{ L·atm/mol·K})(75°\text{C} + 273°)} = \mathbf{2.40\text{ mol }F_2}$$

(b) ? g F_2 = 2.40 mol F_2 × 38.0 g/mol = **91.2 g F_2**

10-44. **Refer to Section 10-10.** . . . .

Method 1: Recall that at STP, 1 mole of gas occupies 22.4 L.

Plan: V_{STP} gas $\xrightarrow{(1)}$ mol gas $\xrightarrow{(2)}$ FW gas $\xrightarrow{(3)}$ g of 6 mol gas

(1) $\text{mol gas} = \frac{V_{STP}}{22.4\text{ L/mol}} = \frac{1.12\text{ L}}{22.4\text{ L/mol}} = 0.0500\text{ mol}$

(2) $\text{FW gas} = \frac{4.00\text{ g gas}}{0.0500\text{ mol}} = 80.0\text{ g/mol}$

(3) g of 6.00 mol gas = 80.0 g/mol × 6.00 mol = **4.80×10^2 g**

Method 2: Dimensional Analysis

$$?\text{ g gas} = 6.00\text{ mol} \times \frac{22.4\text{ L}_{STP}}{1\text{ mol}} \times \frac{4.00\text{ g}}{1.12\text{ L}_{STP}} = \mathbf{4.80 \times 10^2\text{ g}}$$

10-46. **Refer to Section 10-10 and Example 10-11.** . . . .

We know from the density that there are 2.083 g of gas present in 1.00 L of gas.

Plan: (1) Use the ideal gas law, $PV = nRT$, to find the number of moles of gas present in 1 liter.

(2) Use mass and the number of moles of gas in 1.00 L to calculate molecular weight.

$$(1)\quad P = \frac{741 \text{ torr}}{760 \text{ torr/atm}} = 0.975 \text{ atm}$$

$$(2)\quad n = \frac{PV}{RT} = \frac{(0.975 \text{ atm})(1.00 \text{ L})}{(0.0821 \text{ L·atm/mol·K})(52°\text{C} + 273°)} = 0.0365 \text{ mol gas}$$

$$(3)\quad \text{MW (g/mol)} = \frac{2.083 \text{ g gas}}{0.0365 \text{ mol}} = \mathbf{57.1\ g/mol}$$

10-48. **Refer to Section 10-10.** •

Plan: (1) Use the ideal gas law, $PV = nRT$, to determine the moles of N_2 in 1 liter at the new conditions.
(2) Use molecular weight and moles to calculate the mass of N_2 in 1 liter. This value is equivalent to density.

$$(1)\quad n = \frac{PV}{RT} = \frac{(0.500 \text{ atm})(1.00 \text{ L})}{(0.0821 \text{ L·atm/mol·K})(100°\text{C} + 273°)} = 0.0163 \text{ mol } N_2$$

(2) ? g N_2 in 1.00 L = 0.0163 mol N_2 × 28.0 g/mol = 0.456 g N_2 in 1.00 L

Therefore, the density of N_2 at 100°C and 0.500 atm is **0.456 g/L.**

10-50. **Refer to Section 10-9.** •

An "ideal gas" is a hypothetical gas that obeys exactly all the postulates of the kinetic molecular theory.

10-52. **Refer to Section 10-9 and Example 10-7.** • • • • • • • • • • • • • •

Use the ideal gas law, $PV = nRT$, and the fact that at STP, 1 mole of an ideal gas occupies 22.4 L.

$$R = \frac{PV}{nT} = \frac{(1\text{atm})(22.4 \text{ L})}{(1 \text{ mol})(273 \text{ K})} = \mathbf{0.0821\ L·atm/mol·K}$$

(A more exact value for R is 0.08206 L·atm/mol·K)

We can now use dimensional analysis and Appendix C to convert R to different units.

$$R\ (\text{kPa·dm}^3/\text{mol·K}) = 0.08206\ \frac{\text{L·atm}}{\text{mol·K}} \times \frac{1\ \text{dm}^3}{1\ \text{L}} \times \frac{101.32\ \text{kPa}}{1\ \text{atm}} = \mathbf{8.314\ kPa·dm^3/mol·K}$$

$$R\ (\text{J/mol·K}) = 0.08206\ \frac{\text{L·atm}}{\text{mol·K}} \times \frac{101.32\ \text{J}}{1\ \text{L·atm}} = \mathbf{8.314\ J/mol·K} = \mathbf{8.314 \times 10^{-3}\ kJ/mol·K}$$

10-54. **Refer to Section 10-9.** •

Plan: (1) Calculate the moles of CH_4.
(2) Use the ideal gas law, $PV = nRT$, to determine temperature.

$$(1)\quad n = \frac{36.2 \text{ g } CH_4}{16.0 \text{ g/mol}} = 2.26 \text{ mol } CH_4$$

(2) $P = \frac{686}{760}$ atm = 0.903 atm

$$T\ (K) = \frac{PV}{nR} = \frac{(0.903\ atm)(41.0\ L)}{(2.26\ mol)(0.0821\ L\cdot atm/mol\cdot K)} = 200\ K$$

$t\ (^oC) = 200\ K - 273^o = \mathbf{-73^oC}$

10-56. Refer to Section 10-10 and Example 10-11. . . . .

Plan: (1) Use the ideal gas law, $PV = nRT$, to solve for n.
(2) Calculate the molecular weight from the mass and the number of moles.

(1) $$n = \frac{PV}{RT} = \frac{[(1672/760)\ atm](1.425\ L)}{(0.0821\ L\cdot atm/mol\cdot K)(86°C + 273°)} = 0.106\ mol\ gas$$

(2) $$MW\ (g/mol) = \frac{5.56\ g\ gas}{0.106\ mol\ gas} = \mathbf{52.5\ g/mol}$$

10-58. Refer to Section 10-9. . . . .

Plan: (1) Use the ideal gas law, $PV = nRT$, to calculate moles of gas.
(2) Use Avogadro's Number, **N**, to calculate the number of molecules of N_2 and multiply it by a factor of 2 to calculate atoms of N.

(1) $$n = \frac{PV}{RT} = \frac{[(3040/760)\ atm](0.328\ L)}{(0.0821\ L\cdot atm/mol\cdot K)(527°C + 273°)} = \mathbf{0.0200\ mol\ N_2}$$

(2) ? molecules N_2 = **N** × mol N_2 = 6.022×10^{23} molecules/mol × 0.0200 mol
= 1.20×10^{22} molecules N_2

? atoms N = molecules N_2 × 2 atoms/molecule

= 1.20×10^{22} molecules × 2 atoms/molecule = $\mathbf{2.40 \times 10^{22}}$ **atoms N**

10-60. Refer to Section 10-9. . . . .

Plan: (1) Use the ideal gas law, $PV = nRT$, to find n.
(2) Use Avogadro's Number, **N**, to calculate the number of molecules.

(1) $$n = \frac{PV}{RT} = \frac{[(1 \times 10^{-3}/760)\ atm](0.500\ L)}{(0.0821\ L\cdot atm/mol\cdot K)(20°C + 273°)} = 3 \times 10^{-8}\ mol\ gas$$

(2) ? molecules = mol gas × **N** = 3×10^{-8} mol × 6.022×10^{23} molecules/mol
= $\mathbf{2 \times 10^{16}}$ **molecules**

10-62. Refer to Section 10-10. . . . .

Plan: (1) Use the ideal gas law, $PV = nRT$, to calculate the number of moles of ethane in the container at STP.
(2) Determine the experimental molecular weight of ethane and compare it to the theoretical value.

(1) $$n = \frac{PV}{RT} = \frac{(1\ atm)(0.165\ L)}{(0.0821\ L\cdot atm/mol\cdot K)(273\ K)} = 7.36 \times 10^{-3}\ mol\ C_2H_6$$

(2) $$MW\ C_2H_6 = \frac{0.218\ g\ C_2H_6}{7.36 \times 10^{-3}\ mol} = \mathbf{29.6\ g/mol}$$

The actual molecular weight of ethane is 30.1 g/mol.

$$\text{Percent error} = \frac{\text{actual MW - experimental MW}}{\text{actual MW}} \times 100 = \frac{30.1 - 29.6}{30.1} \times 100 = \mathbf{2\%}$$

Possible sources of error which would result in a slightly low experimental molecular weight include

(a) the container volume was slightly less than 165 mL, and
(b) the mass of ethane was slightly more than 0.218 g.
(c) ethane deviates slightly from ideal behavior under STP conditions.

10-64. **Refer to Sections 2-8 and 10-10.** . . . .

The simplest (empirical) formula for a compound is the smallest whole-number ratio of atoms present.

For molecular compounds, the molecular formula indicates the actual numbers of atoms present in a molecule of the compound. It may be the same as the simplest formula or some multiple of it.

10-66. **Refer to Section 10-10 and Examples 10-13 and 10-14.** . . . .

Plan: (1) Evaluate the simplest formula.
(2) Use the ideal gas law, $PV = nRT$, to find n.
(3) Calculate the molecular weight from n and mass.
(4) Determine the molecular formula.

(1) $$?\ mol\ B = \frac{0.589\ g\ B}{10.81\ g/mol} = 0.0545\ mol \qquad \text{Ratio} = \frac{0.0545\ mol}{0.0545\ mol} = 1$$

$$?\ mol\ H = \frac{0.137\ g\ H}{1.008\ g/mol} = 0.136\ mol \qquad \text{Ratio} = \frac{0.136\ mol}{0.0545\ mol} = 2.5$$

A 1:2.5 ratio converts to a 2:5 ratio by multiplying by 2.
Therefore, the simplest formula is B_2H_5 (FW = 26.7 g/mol).

(2) $$n = \frac{PV}{RT} = \frac{[(780/760)\ atm](0.0504\ L)}{(0.0821\ L\cdot atm/mol\cdot K)(23°C + 273°)} = 2.13 \times 10^{-3}\ mol\ gas$$

(3) $$MW = \frac{0.113\ g\ gas}{2.13 \times 10^{-3}\ mol} = 53.1\ g/mol$$

(4) $$\text{factor x} = \frac{MW}{\text{simplest FW}} = \frac{53.1}{26.7} = 2$$

Therefore, the molecular formula is $(B_2H_5)_2 = \mathbf{B_4H_{10}}$.

10-68. **Refer to Section 10-11.** . . . .

(a) The change in pressure was due to the change in the number of moles of gas. From Dalton's Law of Partial Pressures, we can say that

$$(\Delta P)V = (\Delta n)RT \quad \text{where } \Delta n = \text{moles He added} = \frac{2.48\ g\ He}{4.003\ g/mol} = 0.620\ mol$$

$$\text{or} \quad V = \frac{(\Delta n)RT}{\Delta P} \qquad \Delta P = P_{final} - P_{initial} = 1863 - 742 = 1121\ torr$$

Substituting,

$$V = \frac{(0.620 \text{ mol})(0.0821 \text{ L·atm/mol·K})(20^\circ\text{C} + 273^\circ)}{(1121/760) \text{ atm}} = \mathbf{10.1\ L}$$

(b) Using the initial conditions and the ideal gas law, $PV = nRT$, we can calculate the moles of gas.

$$n = \frac{PV}{RT} = \frac{[(742/760) \text{ atm}](10.1 \text{ L})}{(0.0821 \text{ L·atm/mol·K})(20^\circ\text{C} + 273^\circ)} = \mathbf{0.410\ mol\ gas}$$

(c) $$\text{MW (g/mol)} = \frac{20.1 \text{ g gas}}{0.410 \text{ mol}} = \mathbf{49.0\ g/mol}$$

10-70. **Refer to Section 10-11 and Example 10-15.** . . . .

From Dalton's Law of Partial Pressures,

$$P_{total} = \frac{n_{total}RT}{V} \qquad \text{where } n_{total} = mol_{O_2} + mol_{CH_4} + mol_{SO_2}$$

$$= \frac{3.2 \text{ g } O_2}{32 \text{ g/mol}} + \frac{6.4 \text{ g } CH_4}{16 \text{ g/mol}} + \frac{6.4 \text{ g } SO_2}{64 \text{ g/mol}}$$

$$= 0.10 + 0.40 + 0.10$$

$$= 0.60 \text{ mol}$$

Therefore,

$$P_{total} = \frac{(0.60 \text{ mol})(0.0821 \text{ L·atm/mol·K})(127^\circ\text{C} + 273^\circ)}{40.0 \text{ L}} = \mathbf{0.49\ atm\ or\ 370\ torr}$$

10-72. **Refer to Section 10-11 and Example 10-15.** . . . .

(a) Boyle's Law states that $P_1V_1 = P_2V_2$.

For each gas, $P_2 = \frac{P_1V_1}{V_2} = \frac{1.50 \text{ atm} \times 3.50 \text{ L}}{1.00 \text{ L}} = 5.25 \text{ atm}$

Dalton's Law of Partial Pressure states that $P_{total} = P_1 + P_2 + P_3 + \ldots$

Therefore, $P_{total} = P_{O_2} + P_{N_2} + P_{He} = 5.25 + 5.25 + 5.25 = \mathbf{15.75\ atm}$

(b) partial pressure of O_2 = **5.25 atm**

(c) partial pressure of N_2 = partial pressure of He = **5.25 atm**

10-74. **Refer to Section 10-11 and Exercise 10-73.** . . . .

In Exercise 10-73, the mole fraction of each gas was calculated.

$$\text{mole fraction} = \frac{\text{mol substance}}{\text{total mol}}$$

$$\text{total mol} = mol_{H_2} + mol_{He} + mol_{N_2}$$

$$= \frac{6.0 \text{ g } H_2}{2.0 \text{ g/mol}} + \frac{12.0 \text{ g He}}{4.00 \text{ g/mol}} + \frac{24.0 \text{ g } N_2}{28.0 \text{ g/mol}}$$

$$= 3.0 \text{ mol } H_2 + 3.00 \text{ mol He} + 0.857 \text{ mol } N_2$$

$$= 6.86 \text{ mol}$$

$$\text{mole fraction } H_2 = \frac{\text{mol } H_2}{\text{total mol}} = \frac{3.0 \text{ mol}}{6.86 \text{ mol}} = 0.44$$

$$\text{mole fraction He} = \frac{\text{mol He}}{\text{total mol}} = \frac{3.00 \text{ mol}}{6.86 \text{ mol}} = 0.437$$

$$\text{mole fraction } N_2 = \frac{\text{mol } N_2}{\text{total mol}} = \frac{0.857 \text{ mol}}{6.86 \text{ mol}} = 0.125$$

(a) $P_{H_2} = P_{total} \times \text{mole fraction of } H_2 = 1488 \text{ torr} \times 0.44 = \mathbf{650 \text{ torr}}$

$P_{He} = P_{total} \times \text{mole fraction of He} = 1488 \text{ torr} \times 0.437 = \mathbf{650. \text{ torr}}$

$P_{N_2} = P_{total} \times \text{mole fraction of } N_2 = 1488 \text{ torr} \times 0.125 = \mathbf{186 \text{ torr}}$

(b) From Dalton's Law of Partial Pressure, $P_{total}V = n_{total}RT$

$$V = \frac{n_{total}RT}{P_{total}} = \frac{(6.86 \text{ mol})(0.0821 \text{ L·atm/mol·K})(25°C + 273°)}{(1488/760) \text{ atm}} = \mathbf{85.7 \text{ L}}$$

10-76. **Refer to Section 10-11 and Example 10-16.** . . . .

Plan: (1) Calculate the partial pressure of the gas in the container.
(2) Use the ideal gas law, $PV = nRT$, to calculate n.
(3) Determine the molecular weight.

(1) $P_{gas} = P_{atm} - P_{H_2O} = 718 \text{ torr} - 18.65 \text{ torr} = 699 \text{ torr}$

(2) $$n = \frac{PV}{RT} = \frac{[(699/760) \text{ atm}](3.46 \text{ L})}{(0.0821 \text{ L·atm/mol·K})(21°C + 273°)} = 0.132 \text{ mol gas}$$

(3) $$\text{MW (g/mol)} = \frac{4.20 \text{ g gas}}{0.132 \text{ mol}} = \mathbf{31.8 \text{ g/mol}}$$

10-78. **Refer to Section 10-11, Example 10-16 and Table 10-4.** . . . .

Plan: (1) Calculate the partial pressure of hydrogen in the container.
(2) From the ideal gas law, $PV = nRT$, determine n.
(3) Using n and the molecular weight of H_2, determine mass. (The value of density was not needed to solve the problem in this way.)

(1) $P_{H_2} = P_{atm} - P_{H_2O} = 755 \text{ torr} - 24 \text{ torr} = 731 \text{ torr}$

(2) $$n = \frac{PV}{RT} = \frac{[(731/760) \text{ atm}](0.750 \text{ L})}{(0.0821 \text{ L·atm/mol·K})(25°C + 273°)} = 0.0295 \text{ mol } H_2$$

(3) $? \text{ g } H_2 = 0.0295 \text{ mol } H_2 \times 2.02 \text{ g/mol} = \mathbf{0.0596 \text{ g } H_2}$

An alternate plan for solving this problem is to (1) calculate V at STP, then (2) multiply by the density at STP to determine the mass of H_2.

10-80. **Refer to Section 10-12.** . . . .

The third assumption of the Kinetic-Molecular Theory states that the average kinetic energy of gaseous molecules is directly proportional to the absolute temperature of the sample.

kinetic energy = $1/2mv^2 \propto T$ where m = mass (g)
v = average molecular speed (m/s)
T = absolute temperature (K)

It is readily seen that the average molecular speed is directly proportional to the square root of the absolute temperature.

10-82. **Refer to Section 10-12 and Exercise 10-80 Solution.** . . . .

According to the Kinetic-Molecular Theory, all gas molecules have the same average kinetic energy at the same temperature. Hence, at the same T:

average kinetic energy of silane = average kinetic energy of methane

$$1/2(m_{SiH_4})(v_{SiH_4})^2 = 1/2(m_{CH_4})(v_{CH_4})^2$$

$$(v_{CH_4})/(v_{SiH_4}) = [(m_{SiH_4})/(m_{CH_4})]^{1/2} = [(MW_{SiH_4})/(MW_{CH_4})]^{1/2}$$

$$= (32/16)^{1/2}$$

$$= 1.4$$

SiH_4 is heavier than CH_4; however, both molecules have the same average kinetic energy. This is due to the fact that methane molecules have an average speed which is 1.4 times faster than that of silane molecules.

10-84. **Refer to Section 10-11 and Exercise 10-80.** . . . .

From the Kinetic Molecular Theory, we know that the average kinetic energy of gaseous molecules is directly proportional to the absolute temperature. We can therefore write:

$$\frac{\text{rms speed of } N_2 \text{ molecules at } 100°C}{\text{rms speed of } N_2 \text{ molecules at } 0°C} = [(100°C + 273°)/(0°C + 273°)]^{1/2} = \mathbf{1.17}$$

10-86. **Refer to Section 10-13.** . . . .

Graham's Law states that the rates of effusion of gases are inversely proportional to the square roots of their molecular weights (or densities). This means that lighter gas molecules will effuse through tiny holes faster than heavier molecules because they move faster.

10-88. **Refer to Section 10-13 and Figure 10-14a.** . . . .

(a) A bell jar full of hydrogen is lowered over a porous cup full of air. Hydrogen is lighter than nitrogen and oxygen, which are the two primary components of air. And so, hydrogen diffuses into the cup faster than nitrogen and oxygen effuse out of the cup, thereby causing an increase in pressure in the cup sufficient to produce bubbles though the water in the beaker.

(b) Sulfur hexafluoride molecules are heavier than nitrogen and oxygen molecules. And so, nitrogen and oxygen will effuse out of the cup faster than sulfur hexafluoride would diffuse into the cup, and sulfur hexafluoride could not be used in this kind of bubbler.

10-90. **Refer to the Accompanying Figure and Section 10-13.** . . . .

From Graham's Law, we have

$$\frac{\text{effusion rate of HCl}}{\text{effusion rate of } NH_3} = [(MW_{NH_3})/(MW_{HCl})]^{1/2} = (17.0/36.5)^{1/2} = 0.682$$

Therefore, HCl molecules effuse 0.683 times as fast as NH_3 molecules. We can now say that

$$\frac{\text{effusion rate of HCl}}{\text{effusion rate of NH}_3} = \frac{\text{distance from HCl end}}{\text{distance from NH}_3\text{ end}} = 0.683$$

let x = distance from HCl end
$1-x$ = distance from NH_3 end

Substituting, we have

$$\frac{x}{1-x} = 0.683 \qquad \text{Solving,} \quad x = 0.406 \text{ m} = \mathbf{40.6\ cm}$$

10-92. **Refer to Section 10-13.** . . . .

(a) $$\frac{\text{effusion rate of } ^1H_2}{\text{effusion rate of } ^2H_2} = [(M \text{ of } ^2H_2)/(M \text{ of } ^1H_2)]^{1/2}$$
$$= (4.0280 \text{ amu}/2.015650 \text{ amu})^{1/2}$$
$$= \mathbf{1.4136}$$

(b) $$\frac{\text{effusion rate of } ^1H^{35}Cl}{\text{effusion rate of } ^1H^{37}Cl} = [(M \text{ of } ^1H^{37}Cl)/(M \text{ of } ^1H^{35}Cl)]^{1/2}$$
$$= (37.973725 \text{ amu}/35.976675 \text{ amu})^{1/2}$$
$$= \mathbf{1.0273800}$$

(c) $^1H_2$ and $^2H_2$ would be easier to separate since the ratio of their effusion rates is larger than that for $^1H^{35}Cl$ and $^1H^{37}Cl$.

10-94. **Refer to Section 10-14.** . . . .

An "ideal gas" is a hypothetical gas that obeys exactly the postulates of the Kinetic-Molecular Theory. In particular, the volume of the gas molecules is negligible when compared with the volume occupied by the gas, and the attractive forces between the molecules is negligible. However, in reality, gas molecules <u>do</u> occupy volume and there <u>are</u> attractive forces between molecules. This is why a real gas deviates from ideal behavior.

Under normal conditions, most gases approach ideal behavior. However, at low temperatures and/or high pressures a real gas deviates significantly from ideality. Deviations from ideality become very serious at very low temperatures and/or high pressures such as those conditions close to the liquefaction point of a gas. In such a situation, the volume of the gas molecules themselves is no longer negligible when compared to the volume occupied by the gas. Also, the forces of attraction between the molecules become significant.

10-96. **Refer to Section 10-14 and Exercise 10-94 Solution.** . . . .

10-98. **Refer to Section 10-14.** . . . .

(a) According to the Kinetic-Molecular Theory, pressure is the result of collisions between gas molecules and the walls of the container. Real gases have attractive forces acting between molecules, causing the molecules to "stick together" more, not collide as frequently with container walls, and produce a lower pressure. Hence, P_{ideal} is greater than P_{real}.

(b) Volume is defined as the total volume occupied by the gas molecules and is the sum of the volume of the gas molecules themselves plus the volume of the space between the molecules. The molecules of ideal gases are assumed to have essentially no volume, whereas real gases have a finite volume. Hence, V_{ideal} is less than V_{real}.

10-100. **Refer to Section 10-14, Example 10-19 and Table 10-5.** . . . .

Ammonia as an ideal gas:

$$P = \frac{nRT}{V} = \frac{(10.0\text{ mol})(0.0821\text{ L}\cdot\text{atm/mol}\cdot\text{K})(100°\text{C} + 273°)}{60.0\text{ L}} = \mathbf{5.10\text{ atm}}$$

Ammonia as a real gas:

Plan: Use the van der Waals equation to calculate P.

From Table 10-5 for NH_3: a = 4.17 $L^2\cdot atm/mol^2$, b = 0.0371 L/mol

van der Waals equation: $(P + n^2a/V^2)(V - nb) = nRT$

Substituting, we have

$$\left[P + \frac{(10.0\text{ mol})^2(4.17\text{ L}^2\cdot\text{atm/mol}^2)}{(60.0\text{ L})^2}\right]\left[60.0\text{ L} - (10.0\text{ mol})(0.0371\text{ L/mol})\right]$$

$$= (8.82\text{ mol})(0.0821\text{ L}\cdot\text{atm/mol}\cdot\text{K})(100°\text{C} + 273°)$$

Simplifying,

$$(P + 0.166\text{ atm})(59.6\text{ L}) = 306\text{ L}\cdot\text{atm}$$
$$P + 0.116\text{ atm} = 5.13\text{ atm}$$
$$P = \mathbf{5.02\text{ atm}}$$

$$\%\text{ difference} = \frac{P_{ideal} - P_{real}}{P_{ideal}} \times 100 = \frac{5.10 - 5.02}{5.10} \times 100 = 1.57\%\text{ or } \mathbf{2\%}$$

(1 sig. fig.)

10-102. **Refer to Section 10-9.** . . . .

Plan: (1) Calculate the moles of argon, n.
(2) Convert P (torr) to P (atm).
(3) Use the ideal gas law, $PV = nRT$, to calculate R.

(1) $n = \dfrac{4.46\text{ g Ar}}{39.95\text{ g/mol}} = 0.112\text{ mol Ar}$

(2) $P\text{ (atm)} = 760\text{ torr} \times \dfrac{1\text{ atm}}{760\text{ torr}} = 1.00\text{ atm}$

(3) $R = \dfrac{PV}{nT} = \dfrac{(1.00\text{ atm})(2.50\text{ L})}{(0.112\text{ mol})(0°\text{C} + 273°)} = \mathbf{0.0818\text{ L}\cdot\text{atm/mol}\cdot\text{K}}$

10-104. **Refer to Section 10-14.** . . . .

The pressure and volume given are P_{real} and V_{real}. Therefore,

$$\text{compressibility factor} = \frac{(P_{real})(V_{real})}{RT} = \frac{(30.0\text{ atm})(0.500\text{ L})}{(0.0821\text{ L}\cdot\text{atm/mol}\cdot\text{K})(-10°\text{C} + 273°)}$$

$$= \mathbf{0.695}$$

If we assume that NH_3 is behaving like an ideal gas, the <u>ideal</u> pressure can be calculated from the ideal gas law.

$$P = \frac{nRT}{V} = \frac{(1.00\text{ mol})(0.0821\text{ L·atm/mol·K})(-10.0°\text{C} + 273°)}{0.500\text{ L}}$$

$$= \mathbf{43.2\ atm}$$

By comparison, the real pressure is much smaller than the ideal pressure. Since the system is at relatively high pressure and low temperature, there are apparently attractive forces at work between the ammonia molecules to render the real pressure lower.

10-106. **Refer to Section 10-15.** . . . .

Gay-Lussac's Law states that at constant T and P, the volumes of reacting gases can be expressed as ratios of simple whole numbers related to the stoichiometric coefficients in the balanced equation. Therefore,

$$\frac{\text{volume of } H_2}{\text{volume of } N_2} = \frac{3}{1} \quad \text{and} \quad V_{H_2} = 3 \times V_{N_2} = 3 \times 2.45\text{ L} = \mathbf{7.35\ L}$$

$$\frac{\text{volume of } NH_3}{\text{volume of } N_2} = \frac{2}{1} \quad \text{and} \quad V_{NH_3} = 2 \times V_{N_2} = 2 \times 2.45\text{ L} = \mathbf{4.90\ L}$$

10-108. **Refer to Section 10-15 and Exercise 10-106 Solution.** . . . .

Balanced equation: $8Cl_2\ (g) + 8H_2S\ (g) \rightarrow S_8\ (s) + 16HCl\ (g)$

Using Gay-Lussac's Law of Combining Volumes,

$$\frac{\text{volume of HCl}}{\text{volume of } H_2S} = \frac{16}{8} = 2 \quad \text{and} \quad V_{HCl} = 2 \times V_{H_2S} = 2 \times 15.2\text{ L} = \mathbf{30.4\ L}$$

10-110. **Refer to Section 10-16 and Example 10-21.** . . . .

Balanced equation: $2KNO_3\ (s) \xrightarrow{\Delta} 2KNO_3\ (s) + O_2\ (g)$

Recall that 1 mole of gas at STP occupies 22.4 L.

Method 1: Plan: $V_{STP}\ O_2 \xrightarrow{(1)} \text{mol } O_2 \xrightarrow{(2)} \text{mol } KNO_3 \xrightarrow{(3)} \text{g } KNO_3$

(1) $? \text{ mol } O_2 = \dfrac{18.4\ L_{STP}\ O_2}{22.4\ L_{STP}/\text{mol}} = 0.821 \text{ mol } O_2$

(2) $? \text{ mol } KNO_3 = 0.821 \text{ mol } O_2 \times 2 = 1.64 \text{ mol } KNO_3$

(3) $? \text{ g } KNO_3 = 1.64 \text{ mol } KNO_3 \times 101 \text{ g/mol} = \mathbf{166\ g\ KNO_3}$

Method 2: Dimensional Analysis

$$? \text{ g } KNO_3 = 18.4\ L_{STP}\ O_2 \times \underbrace{\frac{1 \text{ mol } O_2}{22.4\ L_{STP}\ O_2}}_{\text{Step 1}} \times \underbrace{\frac{2 \text{ mol } KNO_3}{1 \text{ mol } O_2}}_{\text{Step 2}} \times \underbrace{\frac{101 \text{ g } KNO_3}{1 \text{ mol } KNO_3}}_{\text{Step 3}} = \mathbf{166\ g\ KNO_3}$$

10-112. **Refer to Section 10-16 and Exercise 10-110 Solution.** . . . .

Balanced equation: H_2 (g) + Cl_2 (g) → 2HCl (g)

Method 1: Plan: mol HCl $\xrightarrow{(1)}$ mol H_2 $\xrightarrow{(2)}$ V_{STP} H_2

(1) ? mol H_2 = 0.400 mol HCl × 1/2 = 0.200 mol H_2

(2) ? L_{STP} H_2 = 22.4 L_{STP}/mol × 0.200 mol H_2 = **4.48 L_{STP} H_2**

Method 2: Dimensional Analysis

$$? \ L_{STP}\ H_2 = 0.400 \text{ mol HCl} \times \frac{1 \text{ mol } H_2}{2 \text{ mol HCl}} \times \frac{22.4\ L_{STP}\ H_2}{1 \text{ mol } H_2} = \mathbf{4.48\ L_{STP}\ H_2}$$

10-114. **Refer to Section 10-16.** . . . .

Plan: V SO_2 $\xrightarrow{(1)}$ mol SO_2 $\xrightarrow{(2)}$ mol S $\xrightarrow{(3)}$ g S $\xrightarrow{(4)}$ % S by mass

(1) $? \text{ mol } SO_2 = n = \dfrac{PV}{RT} = \dfrac{[(739/760) \text{ atm}](1.053 \text{ L})}{(0.0821 \text{ L·atm/mol·K})(66°C + 273°)} = 0.0368 \text{ mol } SO_2$

(2) ? mol S = mol SO_2 = 0.0368 mol S

(3) ? g S = 0.0368 mol S × 32.06 g/mol = 1.18 g S

(4) $? \text{ \% S by mass} = \dfrac{\text{g S}}{\text{g sample}} \times 100 = \dfrac{1.18 \text{ g}}{6.862 \text{ g}} \times 100 = \mathbf{17.2\%\ S\ by\ mass}$

10-116. **Refer to Sections 10-16 and 10-11.** . . . .

Balanced equation: Fe (s) + 2HCl (aq) → $FeCl_2$ (aq) + H_2(g)

Plan: g Fe $\xrightarrow{(1)}$ mol Fe $\xrightarrow{(2)}$ mol H_2 $\xrightarrow{(3)}$ V_{STP} H_2 $\xrightarrow{(4)}$ V H_2

(1) $? \text{ mol Fe} = \dfrac{3.787 \text{ g Fe}}{55.85 \text{ g/mol}} = 0.0678 \text{ mol Fe}$

(2) ? mol H_2 = mol Fe = 0.0678 mol H_2

(3) V_{STP} H_2 = 22.4 L_{STP}/mol × 0.0678 mol H_2 = 1.52 L_{STP} H_2

(4) The temperature (°C) is calculated: $? \ °C = \dfrac{1.0°C}{1.8°C}(66°F - 32°F) = 19°C$

From Table 10-4, P_{H_2O} at 19°C = 16 torr.

Recall: $P_{H_2} = P_{atm} - P_{H_2O}$ = 710 torr - 16 torr = 694 torr.

$$V\ H_2 = \frac{nRT}{P} = \frac{(0.0678 \text{ mol})(0.0821 \text{ L·atm/mol·K})(19°C + 273°)}{(694/760) \text{ atm}} = \mathbf{1.78\ L\ H_2}$$

10-118. **Refer to Sections 10-16 and 10-11 and Exercises 10-116 and 10-117.** . . . . .

Balanced equations: Fe (s) + 2HCl (aq) → $FeCl_2$ (aq) + H_2 (g)

2Al (s) + 6HCl (aq) → $2AlCl_3$ (aq) + $3H_2$ (g)

Plan: (1) Use the ideal gas law, $PV = nRT$, to solve for the moles of H_2.
(2) Use algebra to find the mass of Fe in the alloy.

(1) From Table 10-4, P_{H_2O} at 23°C = 21 torr.

Therefore, $P_{H_2} = P_{atm} - P_{H_2O}$ = 742 torr - 21 torr = 721 torr

$$? \text{ mol } H_2 = n = \frac{PV}{RT} = \frac{[(721/760) \text{ atm}](6.436 \text{ L})}{(0.0821 \text{ L·atm/mol·K})(23°\text{C} + 273°)} = 0.251 \text{ mol } H_2$$

(2) let x = g Fe in alloy Therefore, $\frac{x}{55.85}$ = mol Fe in alloy

6.01 - x = g Al in alloy $\frac{6.01 - x}{26.98}$ = mol Al in alloy

From the balanced equations in Exercises 10-116 and 10-117, we know that 1 mole of Fe produces 1 mole of H_2 and 1 mole of Al produces 1.5 moles of H_2. So,

mol H_2 = mol Fe + 1.5 × mol Al

Substituting, we have

$$0.251 = \frac{x}{55.85} + \frac{1.5(6.01 - x)}{26.98}$$

Solve for x by multiplying both sides by (55.85 × 26.98)

$$378 = 26.98x + 503 - 83.8x$$
$$56.8x = 125$$
x = **2.20 g Fe** % Fe in alloy = $\frac{2.20 \text{ g}}{6.01 \text{ g}} \times 100$ = **36.6% Fe**

6.01 - x = 3.81 g Al % Al in alloy = $\frac{3.81 \text{ g}}{6.01 \text{ g}} \times 100$ = **63.4% Al**

10-120. **Refer to Section 10-16 and Exercise 10-119.** . . . .

(a) Balanced equation: $C_{12}H_{22}O_{11} + 12O_2 \rightarrow 12CO_2 + 11H_2O$

(b) Plan: g glucose $\xrightarrow{(1)}$ mol glucose $\xrightarrow{(2)}$ mol O_2 $\xrightarrow{(3)}$ V O_2

(1) ? mol glucose = $\frac{5.62 \text{ g glucose}}{342 \text{ g/mol}}$ = 0.0164 mol glucose

(2) ? mol O_2 = 12 × mol glucose = 0.197 mol glucose

(3) $V\ O_2 = \frac{nRT}{P} = \frac{(0.197 \text{ mol})(0.0821 \text{ L·atm/mol·K})(37°\text{C} + 273°)}{1.00 \text{ atm}}$ = **5.01 L**

(c) From the balanced equation, we see that the moles of CO_2 produced is equal to the moles of O_2 consumed: **5.01 L.**

11 Liquids and Solids

11-2. **Refer to the Introduction to Chapter 11.** . . . .

In liquids and solids, the particles are close together and there are strong interactions between them. These interactions are very difficult to describe mathematically.

11-4. **Refer to Sections 11-2, 6-3 and Chapter 7.** . . . .

Permanent dipole-dipole forces are found in substances composed of polar covalent molecules. Examples include (c), (e), (f) and (i).

(a) CO_2 - linear, symmetrical, and therefore nonpolar

(b) Ar - monatomic molecules, hence nonpolar

(c) SO_2 - angular, nonsymmetrical due to lone pairs, and therefore polar

(d) PF_5 - trigonal bipyramidal, symmetrical, and therefore nonpolar

(e) HBr - polar heteronuclear diatomic molecules

(f) ClF_3 - distorted "T"-shaped, nonsymmetrical due to lone pairs, and therefore polar

(g) I_2 - homonuclear diatomic molecule, hence nonpolar

(h) SiH_4 - tetrahedral, symmetrical, and therefore nonpolar

(i) SiH_3Cl - distorted tetrahedral, nonsymmetrical, and therefore polar

11-6. **Refer to Section 11-2.** . . . .

Hydrogen bonding is a strong dipole-dipole interaction between molecules containing hydrogen directly bonded to a small, highly electronegative atom, such as nitrogen, oxygen or fluorine.

Strong hydrogen bonding occurs in (a) CH_3OH, (f) NH_3, (h) HF and (j) CH_3NH_2.

11-8. **Refer to Section 11-2.** . . . .

The normal boiling points of compounds increase as the intermolecular forces between the molecules in the compounds increase. If we arrange the molecules in order of increasing boiling points, we have $CH_4 < CH_3F < CH_3Br < CH_3OH$.

CH_4 is a nonpolar covalent molecule and has only London forces acting between the molecules.

CH_3F and CH_3Br are polar covalent molecules having both London forces and permanent dipole-dipole interactions as intermolecular forces. CH_3Br is a larger molecule than CH_3F; its London forces are thereby stronger and CH_3Br has a higher boiling point.

CH_3OH is a polar covalent molecule and has London forces, permanent dipole-dipole interactions and hydrogen bonding acting between its molecules, giving it the highest boiling point.

11-10. **Refer to Section 11-1.** . . . .

Particles in a gas are so far apart that the volume occupied by them is negligible under ordinary conditions. As the sample of gas is cooled and compressed sufficiently, the particles approach each other closely and move slowly enough for intermolecular forces of attraction to overcome the kinetic energies and cause condensation or liquefaction to occur. Therefore, the average separations among liquid particles are far less than in gases, and the densities of liquids are much higher than densities of gases.

11-12. **Refer to Section 11-5.** . . . .

Plant roots take up water and dissolved nutrients from fertilizers from the soil by capillary action. The roots exhibit strong adhesive forces for water and aqueous solutions.

11-14. **Refer to Sections 11-7 and 11-8.** . . . .

The normal boiling point of a liquid is the temperature at which the vapor pressure of a liquid is equal to one atmosphere (760 torr) pressure. The vapor pressure of liquids always increases with increasing temperature. Substances with high vapor pressures, at room temperature, require less heating in order for their vapor pressure to become equal to one atmosphere, than do substances with low vapor pressures at room temperature. This is expected, since substances with low vapor pressures have relatively high intermolecular forces of attraction. So, the statement that liquids with high normal boiling points have low vapor pressure at 25°C is true.

Diethyl ether, gasoline and methyl alcohol have relatively high vapor pressures at 25°C, whereas mercury, baby oil and motor oil have relatively low vapor pressures at 25°C.

11-16. **Refer to Section 11-2.** . . . .

(a) CF_4 is much smaller than CI_4. Therefore it is more difficult to polarize and has much smaller London forces. CF_4 boils easily because there is less attraction between molecules, whereas CI_4 is a solid.

(b) CF_4 is a nonpolar covalent molecule, whereas CH_3F is polar. Both substances have London forces, but CH_3F also has permanent dipole-dipole interactions. Therefore the boiling point of CF_4 is lower than that of CH_3F.

(c) One might expect the polar molecule, CH_3I, to boil at a higher temperature than CI_4, a nonpolar molecule, since CH_3I has permanent dipole-dipole interactions. However, the larger London forces present in CI_4 due to its size, is the overriding factor.

(d) Due to increased London forces, the larger molecule, CH_3I boils at a higher temperature than CH_3F.

11-18. **Refer to Sections 11-7 and 11-8.** . . . .

(a) true

(b) false - The normal boiling point of a liquid is defined as the boiling point at 760 torr.

(c) false - As long as the temperature remains constant, the vapor pressure of a liquid will remain the same. The escaping power of liquid molecules possessing sufficient kinetic energy to go into the gas phase stays the same regardless of the quantity of the liquid.

11-20. **Refer to Table 11-3 and Figure 11-12.** . . . .

From the graph in Figure 11-12, the normal boiling points of methyl alcohol and benzene are estimated to be approximately 67°C and 83°C, respectively. *The Handbook of Chemistry and Physics* (67th Ed.) gives the values as 64.96°C and 80.1°C, respectively. Benzene boils at a higher temperature due to its larger molecular size and hence larger London forces.

11-22. **Refer to Section 11-9 and Example 11-4.** . . . .

(a) In order of increasing melting points:
$CH_4 < CH_3NH_2 < CH_3CH_2NH_2 < NH_3 < CdF_2$

(b) In order of increasing boiling points:
$CH_4 < NH_3 < CH_3NH_2 < CH_3CH_2NH_2 < CdF_2$

CH_4 is a small nonpolar molecule with only London forces and it has the lowest melting and boiling points.

CdF_2 is an ionic solid at room temperature and pressure with strong ion-ion forces of attraction and it has the highest melting and boiling points.

NH_3, CH_3NH_2 and $CH_3CH_2NH_2$ are polar molecules with London forces and hydrogen bonding. The trend for increasing boiling points is due to increasing molecular size and polarizability of the molecules. The melting point trend is different due to the relative ease in packing NH_3 molecules in the solid state.

11-24. **Refer to Section 11-9 and Exercise 11-14 Solution.** . . . .

Normal boiling points generally increase with decreasing vapor pressure at a set temperature. In order of increasing normal boiling points:

CS_2 (309 torr at 20°C) < acetone (185 torr at 20°C) < ethanol (44 torr at 20°C)

11-26. **Refer to Section 11-9.** . . . .

(a) at 37°C, the heat of vaporization of water = 2.41 kJ/g

$$\text{at } 37^\circ\text{C},\ \Delta H^\circ_{vap}\ (\text{kJ/mol}) = 2.41\ \frac{\text{kJ}}{\text{g}} \times \frac{18.0\ \text{g}}{1\ \text{mol}} = 43.4\ \text{kJ/mol}$$

(b) The heat of vaporization at a certain temperature is the amount of heat required to change 1 gram of liquid to 1 gram of vapor at that temperature. The heat of vaporization is greater at 37°C than at 100°C because the average kinetic energy of the molecules is lower at lower temperatures. Therefore, more energy must be added per unit mass of the liquid to break the intermolecular forces between the molecules at the lower temperature.

11-28. **Refer to Section 11-8 and Appendix E.** . . . .

Water will boil when the vapor pressure of water equals the external or barometric pressure. If the barometric pressure is 612 torr in Denver, water will boil at approximately 94°C.

11-30. **Refer to Section 11-8.** . . . .

The order of increasing boiling points corresponds with the order of increasing temperatures when their vapor pressures are constant, e.g. 100 torr:

normal butane < diethyl ether < 1-butanol

11-32. **Refer to Section 11-8.** . . . .

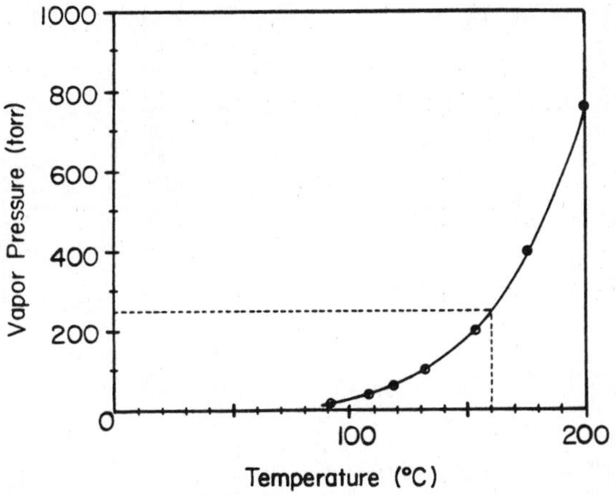

From the graph of vapor pressure (torr) versus temperature (°C), the boiling point of $GaCl_3$ is estimated to be **160° C** at 250 torr.

11-34. **Refer to Section 11-9 and Example 11-3.** . . . .

We must use the Clausius-Clapeyron equation:

$$\log\left(\frac{P_2}{P_1}\right) = \frac{\Delta H_{vap}}{2.303R}\left(\frac{1}{T_1} - \frac{1}{T_2}\right)$$

where ΔH_{vap} = molar enthalpy of vaporization 40.7 kJ/mol
P_1 = 1 atm
P_2 = 0.100 atm
R = 8.314×10^{-3} kJ/mol·K
T_1 = 113.5°C + 273° = 386.5 K
T_2 = ?

Substituting,

$$\log\left(\frac{0.100 \text{ atm}}{1 \text{ atm}}\right) = \frac{40.7 \text{ kJ/mol}}{2.303(8.314 \times 10^{-3} \text{ kJ/mol·K})}\left(\frac{1}{386.5 \text{ K}} - \frac{1}{T_2}\right)$$

$$-1.00 = 2.13 \times 10^3 \text{ K}(2.59 \times 10^{-3} \text{ K}^{-1} - \frac{1}{T_2})$$

$$-4.69 \times 10^{-4} \text{ K}^{-1} = 2.59 \times 10^{-3} \text{ K}^{-1} - \frac{1}{T_2}$$

$$\frac{1}{T_2} = 3.06 \times 10^{-3} \text{ K}^{-1}$$

T_2 = 327 K or **54°C**

11-36. **Refer to Section 11-9, Example 11-3 and Appendix E.** . . . .

From Appendix E, the molar enthalpy of vaporization of mercury at the normal boiling point is 58.6 kJ/mol.

Using the Clausius-Clapeyron equation, we have

$$\log\left(\frac{P_2}{P_1}\right) = \frac{\Delta H_{vap}}{2.303R}\left(\frac{1}{T_1} - \frac{1}{T_2}\right)$$

where ΔH_{vap} = 58.6 kJ/mol
P_1 = 760 torr
P_2 = ?
R = 8.314×10^{-3} kJ/mol·K
T_1 = 357°C + 273° = 630 K
T_2 = 25°C + 273° = 298 K

Substituting,

$$\log\left(\frac{P_2}{760\ \text{torr}}\right) = \frac{58.6\ \text{kJ/mol}}{2.303(8.314 \times 10^{-3}\ \text{kJ/mol}\cdot\text{K})}\left(\frac{1}{630\ \text{K}} - \frac{1}{298\ \text{K}}\right)$$

$$\log\left(\frac{P_2}{760\ \text{torr}}\right) = -5.41$$

$$\frac{P_2}{760\ \text{torr}} = 3.89 \times 10^{-6}$$

$$P_2 = \mathbf{2.96 \times 10^{-3}\ torr}$$

11-38. **Refer to Section 11-9.** . . . .

(a) When log P is plotted against $1/T$, the slope of the line is $-\Delta H_{vap}/2.303R$.

(b) Data Table:

t (°C)	T (K)	$1/T$ (K^{-1})	P (torr)	log P
-43.4	229.8	4.35×10^{-3}	1	0.00
-23.5	249.7	4.00×10^{-3}	5	0.70
-13.5	259.7	3.85×10^{-3}	10	1.00
- 3.0	270.2	3.70×10^{-3}	20	1.30
9.1	282.3	3.54×10^{-3}	40	1.60
16.1	289.3	3.46×10^{-3}	60	1.78
27.0	300.2	3.33×10^{-3}	100	2.00
42.0	315.2	3.17×10^{-3}	200	2.30
59.3	332.5	3.01×10^{-3}	400	2.60
?	?	?	760	2.88

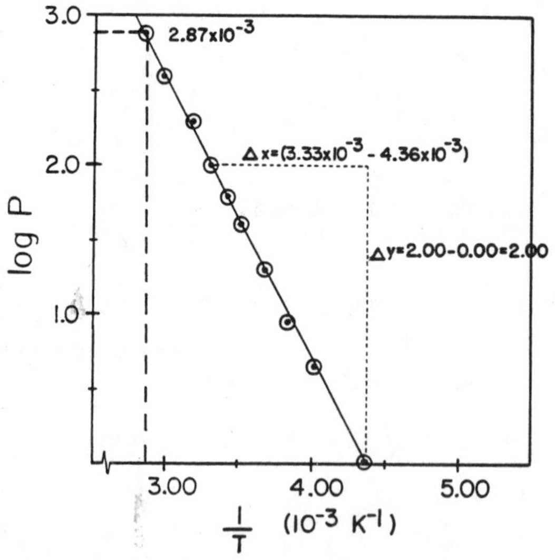

(c) $\text{Slope} = \frac{\Delta y}{\Delta x} = \frac{-\Delta H_{vap}}{2.303R} = \frac{(2.00 - 0.00)}{(3.33 \times 10^{-3} - 4.35 \times 10^{-3})} = -1.96 \times 10^3\ \text{K}$

Therefore,

$$\Delta H_{vap} = -(-1.96 \times 10^3 \text{ K})(2.303)(8.314 \text{ J/mol·K}) = \mathbf{3.75 \times 10^4 \text{ J/mol}}$$

Note that a slightly different slope to the line would give a slightly different value for ΔH_{vap}.

(d) The normal boiling point is the temperature of which the vapor pressure of acetone is 760 torr. From the graph, when log(760 torr) = 2.88,

$$1/T = 2.87 \times 10^{-3} \text{ K}^{-1} \qquad \text{and solving, } T = 3.48 \times 10^2 \text{ K or } \mathbf{75^oC}$$

11-40. **Refer to Section 11-9, Example 11-3 and Exercise 11-39.** . . . .

The solution to Exercise 11-39, applying the Clausius-Clapeyron equation, gives

$$\Delta H_{vap} = 4.37 \times 10^4 \text{ J/mol}$$

Using one set of data, e.g., vapor pressure of isopropyl alcohol = 400 torr at 67.8°C, and the molar heat of vaporization in the Clausius-Clapeyron equation, we can solve for the normal boiling point.

$$\log\left(\frac{P_2}{P_1}\right) = \frac{\Delta H_{vap}}{2.303R}\left(\frac{1}{T_1} - \frac{1}{T_2}\right)$$

where
$\Delta H_{vap} = 4.37 \times 10^4$ J/mol
$P_1 = 400$ torr
$P_2 = 760$ torr
$R = 8.314$ J/mol·K
$T_1 = 67.8^oC + 273.2^o = 341.0$ K
$T_2 = ?$

$$\log\left(\frac{760}{400}\right) = \frac{4.37 \times 10^4 \text{ J/mol}}{2.303(8.314 \text{ J/mol·K})}\left(\frac{1}{341.0 \text{ K}} - \frac{1}{T_2}\right)$$

$$0.279 = 2.28 \times 10^3 \text{ K}(2.933 \times 10^{-3} \text{ K}^{-1} - \frac{1}{T_2})$$

$$1.22 \times 10^{-4} \text{ K}^{-1} = 2.933 \times 10^{-3} \text{ K}^{-1} - \frac{1}{T_2}$$

$$\frac{1}{T_2} = 2.811 \times 10^{-3} \text{ K}^{-1}$$

$$T_2 = 355.8 \text{ K or } \mathbf{82.6^oC}$$

11-42. **Refer to Section 11-1.** . . . .

Ice, i.e. solid water, floats in liquid water because the solid state is less dense that the liquid state. However, like most other substances, solid mercury is more dense than liquid mercury and therefore, solid mercury sinks when placed in liquid mercury.

11-44. **Refer to Section 11-16 and Table 11-9.** . . . .

The characteristics and examples of solid types are given in Table 11-9.

11-46. **Refer to Section 11-16.** . . . .

MoF_6 - molecular solid

Se_8 - molecular solid

RbI - ionic solid

BN - covalent (network) solid

Pt - metallic solid

11-48. **Refer to Section 11-16.** . . . .

(a) SO_2F - molecular solid

(b) MgF_2 - ionic solid

(c) W - metallic solid

(d) Pb - metallic solid

(e) PF_5 - molecular solid

11-50. **Refer to Section 11-2.** . . . .

Melting points of ionic compounds increase with increasing ion-ion interactions which are functions of *d*, the distance between the ions, and *q*, the charge on the ions:

$$F \propto \frac{q^+q^-}{d^2}$$

Due to the differences in the number of charges on the cations (Na^+, Mg^{2+} and Al^{3+}), the following order of increasing melting points is predicted:

$$NaF < MgF_2 < AlF_3$$

11-52. **Refer to Section 11-2 and Exercise 11-50.** . . . .

The Ba^{2+} cation is larger in size than the Sr^{2+} cation. The distance between Ba^{2+} and F^- in its solid is greater than the distance between Sr^{2+} and F^- in its solid. The resulting ion-ion interaction is weaker due to Coulomb's Law. Therefore, SrF_2 melts at a higher temperature than BaF_2.

11-54. **Refer to Section 11-15 and Figure 11-26.** . . . .

(a) CsCl (simple cubic): The unit cell contains

in the corners	8 Cl^- ions × 1/8 = 1 Cl^- ion
in the center	1 Cs^+ ion × 1 = 1 Cs^+ ion

This agrees with the stoichiometry of CsCl.

(b) NaCl (face-centered cubic): The unit cell contains

in the corners	8 Cl^- ions × 1/8 = 1 Cl^- ion
on the faces	6 Cl^- ions × 1/2 = 3 Cl^- ions
in the center	1 Na^+ ion × 1 = 1 Na^+ ion
on the face edges	12 Na^+ ions × 1/4 = 3 Na^+ ions

Therefore, there are a total of 4 Na^+ ions and 4 Cl^- ions in a unit cell of NaCl. This agrees with the stoichiometry of NaCl.

(c) ZnS (face-centered cubic): The unit cell contains

inside the cube	4 Zn^{2+} ions × 1 = 4 Zn^{2+} ions
on the faces	6 S^{2-} ions × 1/2 = 3 S^{2-} ions
in the corners	8 S^{2-} ions × 1/8 = 1 S^{2-} ions

Therefore, there are a total of 4 Zn^{2+} ions and 4 S^{2-} ions in a unit cell of ZnS. This agrees with the stoichiometry of ZnS.

11-56. **Refer to Section 11-15, Table 11-8 and Figure 11-22.** . . . .

A unit cell is the smallest unit of volume of a crystal showing all the structural characteristics of a crystal. The unit cell is repeated in three dimensions to form the overall lattice structure.

A simple cubic unit cell:

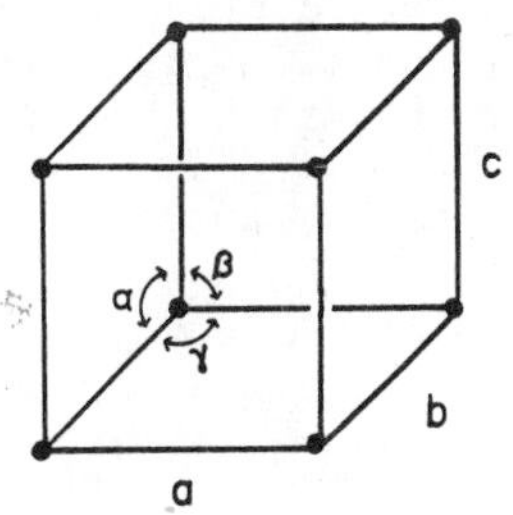

A cubic unit cell has sides of equal length (a=b=c) and all angles equal to 90° ($\alpha=\beta=\gamma=90^{o}$). A tetragonal unit cell has two sides of equal length, a third side of different length ($a=b\neq c$), and all angles equal to 90°. A hexagonal unit cell also has two sides of equal length and a third side of different length ($a=b\neq c$) but with two angles equal to 90° and a third angle equal to 120° ($\alpha=\beta=90^{o}$ and $\gamma=120^{o}$).

11-58. **Refer to Section 11-16 and Figures 11-28 and 11-29.** . . . .

The closest-packed structures of metals consist of layers of atoms which are packed together as closely as possible. As seen in Figure 11-26, there are two closest-packed crystal structures. In the hexagonal closest-packed structure, the first and third layers are oriented in the same direction. In the cubic closest packed structure, the first and third layers are oriented in opposite directions. Both have a coordination number of 12 for each atom. In a given volume, 24% is empty space in the ideal situation. All other structures would involve a less efficient arrangement with more empty space.

11-60. **Refer to Section 11-16.** . . . .

The edge length, a = 3.62 Å, is the hypotenuse of a right triangle. Each of the other two sides equal 2r, where r is the radius of a Cu atom. Using the Pythagorean Theorem,

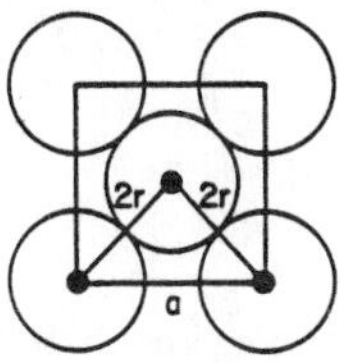

$$a^2 = (2r)^2 + (2r)^2 = 2(2r)^2$$

Solving, $(3.62\ \text{Å})^2 = 8r^2$

$$r = \mathbf{1.28\ \text{Å}}$$

11-62. **Refer to Section 11-16, Example 11-9 and Example 11-10.** . . . .

Plan: (1) Calculate the mass of an Ir unit cell.
(2) Calculate the volume of an Ir unit cell.
(3) Calculate the density of Ir using

$$\text{Density} = \frac{\text{mass of unit cell, } M}{\text{volume of unit cell, } V}$$

(1) $M_{\text{face centered unit cell}}$ = mass of 4 Ir atoms

$$= 4\ \text{Ir atoms} \times \frac{1\ \text{mol Ir}}{6.022 \times 10^{23}\ \text{Ir atoms}} \times \frac{192.2\ \text{g}}{1\ \text{mol Ir}}$$

$$= 1.277 \times 10^{-21}\ \text{g}$$

(2) The volume of a cubic unit cell = (length of unit cell edge, a)^3

For a face centered cubic unit cell, the diagonal of a face is the hypotenuse of a right triangle and equals 4r where r is the radius of an Ir atom. From the Pythagorean Theorem,

$$(4r)^2 = a^2 + a^2$$

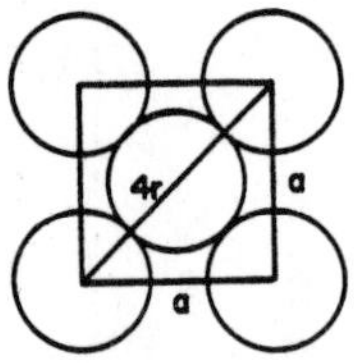

Solving, $(4 \times 1.36 \text{ Å})^2 = 2a^2$

$a\ (\text{Å}) = 3.85 \text{ Å}$

$a\ (\text{cm}) = 3.85 \times 10^{-8} \text{ cm}$

$$V_{\text{unit cell}} = a^3 = (3.85 \times 10^{-8} \text{ cm})^3$$

$$= 5.71 \times 10^{-23} \text{ cm}^3$$

(3) Therefore,

$$D = \frac{M}{V} = \frac{1.277 \times 10^{-21} \text{ g}}{5.71 \times 10^{-23} \text{ cm}^3} = \mathbf{22.4 \ g/cm^3}$$

The Handbook of Chemistry and Physics (67th Ed.) gives the density of Ir as 22.421 g/cm^2.

11-64. **Refer to Section 11-16 and Example 11-9.** . . . .

Plan: (1) Calculate the volume of metal in the unit cell.
(2) Calculate the volume of the unit cell.
(3) Using the density and volume of the unit cell, calculate the atomic weight of the metal.

(1) $V_{\text{metal atom}} = (4/3)\pi r^3$ where the radius of the metal atom, r = 1.46 Å

$$= (4/3)\pi[1.46 \text{ Å} \times (1 \times 10^{-8} \text{ cm}^3)/1 \text{ Å}]^3 = 1.30 \times 10^{-23} \text{ cm}^3$$

$$V_{\text{metal in unit cell}} = 4 \text{ atoms} \times 1.30 \times 10^{-23} \text{ cm}^3 = 5.20 \times 10^{-23} \text{ cm}^3$$

(2) $$V_{\text{unit cell}} = V_{\text{metal in unit cell}} \times \frac{1 \text{ unit cell}}{(1 - 0.24) \text{ metal in unit cell}}$$

$$= 5.20 \times 10^{-23} \text{ cm}^3 \times 1/0.76$$

$$= 6.84 \times 10^{-23} \text{ cm}^3$$

(3) We know: $$D = \frac{\text{mass of unit cell, } M}{\text{volume of unit cell, } V}$$

$$M_{\text{unit cell}} = D \times V_{\text{unit cell}} = (4.50 \text{ g/cm}^3)(6.84 \times 10^{-23} \text{ cm}^3)$$

$$= 3.08 \times 10^{-22} \text{ g}$$

Since the cell contains 4 metal atoms, then

$$? \text{ g metal atom} = 1/4 \times M_{\text{unit cell}} = (1/4)(3.08 \times 10^{-22} \text{ g}) = 7.70 \times 10^{-23} \text{ g}$$

$$AW = (7.70 \times 10^{-23} \text{ g/atom})(6.022 \times 10^{23} \text{ atoms/mol}) = 46.4 \text{ g/mol}$$

This value is close to either the atomic weight of Sc (AW = 44.96 g/mol) or the atomic weight of Ti (AW = 47.88 g/mol). And so, within experimental error, the element could be **Sc** or **Ti**.

11-66. **Refer to Section 11-16 and Example 11-9.** . . . .

Plan: (1) Calculate the mass, *M*, of the unit cell.
(2) From the density and the mass of the unit cell, calculate the volume of the unit cell, *V*.
(3) Calculate the volume of metal in the unit cell.
(4) Calculate % empty space.

(1) $M_{\text{unit cell}} = 4 \text{ Zn atoms} \times \dfrac{1 \text{ mol Zn}}{6.022 \times 10^{23} \text{ Zn atoms}} \times \dfrac{65.39 \text{ g}}{1 \text{ mol Zn}}$

$= 4.34 \times 10^{-22}$ g

(2) $V_{\text{unit cell}} = \dfrac{M_{\text{unit cell}}}{D} = \dfrac{4.34 \times 10^{-22} \text{ g}}{7.14 \text{ g/cm}^3} = 6.08 \times 10^{-23} \text{ cm}^3$

(3) $V_{\text{atoms in unit cell}} = 4 \text{ atoms} \times V_{\text{Zn atom}}$

$= 4 \times (4/3)\pi r^3$ where r is the radius of a Zn atom

$= 4 \times (4/3)\pi[1.34 \text{ Å} \times (1 \times 10^{-8} \text{ cm/Å})]^3$

$= 4.03 \times 10^{-23} \text{ cm}^3$

(4) $\% \text{ empty space} = \dfrac{V_{\text{unit cell}} - V_{\text{atoms in unit cell}}}{V_{\text{unit cell}}} \times 100$

$= \dfrac{6.08 \times 10^{-23} \text{ cm}^3 - 4.03 \times 10^{-23} \text{ cm}^3}{6.08 \times 10^{-23} \text{ cm}^3} \times 100$

$= 33.7\%$

11-68. **Refer to Section 11-16 and Example 11-9.** . . . .

Plan: (1) Calculate the volume of the unit cell, V.
(2) Calculate the volume of Ba atoms in the unit cell.
(3) Calculate the fraction of empty space.

(1) let the length of the cube edge = a = 5.025 Å

$V_{\text{unit cell}} = a^3 = [5.025 \text{ Å} \times (1 \times 10^{-8} \text{ cm/Å})]^3 = 1.269 \times 10^{-22} \text{ cm}^3$

(2) A body centered cubic unit cell contains (8 × 1/8) + (1 × 1) = 2 Ba atoms. To calculate the volume of 2 Ba atoms, we must determine the radius of a Ba atom. In a body-centered cubic unit cell, the atom at the center of the cube is touching atoms at opposite corners of the cube, but atoms do not touch along the cube edges as shown in Figure 11-23. The diagonal that runs from corner to corner through the center is the hypotenuse of a right angle triangle having an edge and a face diagonal as other sides. Applying the Pythagorean Theorem as in Example 11-7, we obtain the following.

Let h = length of the hypotenuse

$(a^2 + a^2)^{1/2}$ = length of face diagonal

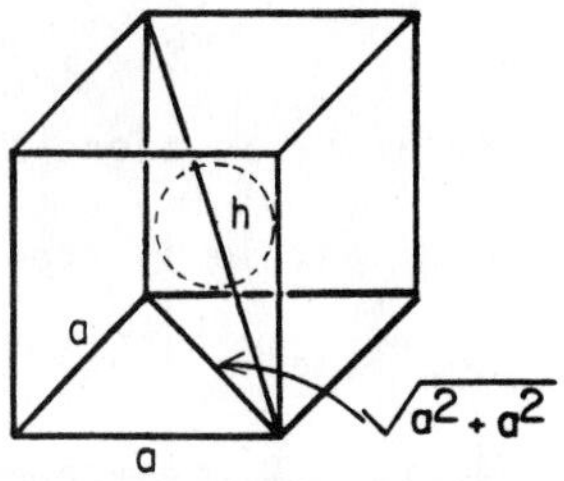

Then,

$h^2 = a^2 + [(a^2 + a^2)^{1/2}]^2 = a^2 + 2a^2$

$= 3(5.025 \times 10^{-8} \text{ cm})^2$

$h = 8.704 \times 10^{-8}$ cm

However, h = 4 × radius of a Ba atom, r

Therefore, $r = h/4 = 2.176 \times 10^{-8}$ cm

$V_{\text{atoms in unit cell}} = 2 \text{ atoms} \times (4/3)\pi r^3$

$= 2 \times (4/3)\pi(2.176 \times 10^{-8} \text{ cm})^3$

$= 8.632 \times 10^{-23} \text{ cm}^3$

(3) fraction of empty space $= \dfrac{V_{\text{unit cell}} - V_{\text{atoms in unit cell}}}{V_{\text{unit cell}}}$

$= \dfrac{1.269 \times 10^{-22}\ \text{cm}^3 - 8.632 \times 10^{-23}\ \text{cm}^3}{1.269 \times 10^{-22}\ \text{cm}^3}$

$= \mathbf{0.320}$

This value is higher than 0.24, the fraction of empty space in a close-packed lattice.

11-70. **Refer to Section 11-16 and Example 11-7.** . . . .

Plan: (1) Determine the length of unit cell edge, a, from the volume.
(2) Calculate the atomic radius of nickel, r.
(3) Calculate the atomic volume of nickel, *V*.

(1) $V_{\text{unit cell}} = 43.763\ \text{Å}^3 = a^3$ Solving, a = 3.5240 Å

(2) For a face-centered cubic unit cell, the diagonal of a face is the hypotenuse of a right isosceles triangle and equals 4r. From the Pythagorean Theorum,

$$(4r)^2 = a^2 + a^2 = 2a^2 = 24.837\ \text{Å}^3$$

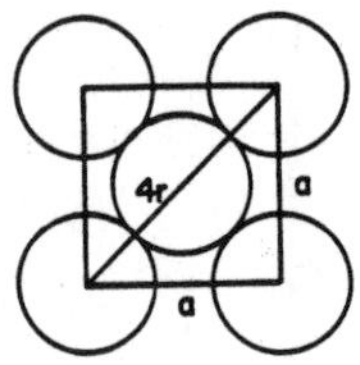

Solving, r = 1.2459 Å

(3) $V_{\text{Ni atom}} = (4/3)\pi r^3 = (4/3)\pi(1.2459\ \text{Å})^3$

$= \mathbf{8.1010\ Å^3}$

11-72. **Refer to Section 11-14.** . . . .

In X-ray diffraction, the Bragg equation is used:

$n\lambda = 2d\sin\theta$ where n = 1 for the minimum diffraction angle
λ = wavelength, 0.70926 Å
θ = angle of incidence, 8.683°
d = spacing between parallel layers of Au atoms

Solving for d,

$$d = \frac{n\lambda}{2\sin\theta} = \frac{1 \times 0.70926\ \text{Å}}{2 \times \sin(8.683^\circ)} = 2.349\ \text{Å}$$

11-74. **Refer to Sections 1-11, 11-9, 11-11 and Exercise 1-86 Solution.** . . . . . .

heat gained by freon-12 = $|(\text{mass})(\text{Heat of Vap.})|_{\text{freon-12}}$

heat lost by water = $|(\text{mass})(\text{Sp. ht.})(\text{temp. change}) + (\text{mass})(-\Delta H^{\circ}_{\text{Fusion}})|_{\text{water}}$

We recall that the heat gained by freon-12 = heat lost by water.

Therefore,

$$(\text{mass}_{\text{freon-12}})(289\ \text{J/g}) = (450\ \text{g})(4.18\ \text{J/g}^\circ\text{C})(20^\circ\text{C} - 0^\circ\text{C}) + (450\ \text{g})(334\ \text{J/g})$$

$\text{mass}_{\text{freon-12}} = \mathbf{650\ g}$

11-76. Refer to Section 11-9. . . . .

$$? \text{ heat required} = |(\text{mass})(\text{Heat of Vap.})| = (1.00 \text{ mol} \times 84 \text{ g/mol})(390 \text{ J/g}) = 3.28 \times 10^4 \text{ J}$$

Then,

$$? \text{ rate of heating (J/s)} \times \text{t (s)} = \text{heat required (J)}$$

$$\text{t (s)} = \frac{3.28 \times 10^4 \text{ J}}{10.0 \text{ J/s}} = 3.28 \times 10^3 \text{ s}$$

$$\text{t (min)} = \frac{3.28 \times 10^3 \text{ s}}{60.0 \text{ s/min}} = \mathbf{54.7 \text{ min}}$$

11-78. Refer to Section 1-13 and Exercise 1-86 Solution. . . . .

heat gained by cold water = heat lost by hot water

$$|(\text{mass})(\text{Sp. ht.})(\text{temp. change})|_{cold} = |(\text{mass})(\text{Sp. ht.})(\text{temp. change})|_{hot}$$

$$(500 \text{ g})(4.18 \text{ J/g}^\circ\text{C})(t_{final} - 30^\circ\text{C}) = (300 \text{ g})(4.18 \text{ J/g}^\circ\text{C})(100^\circ\text{C} - t_{final})$$

$$500 \times t_{final} - 15000 = 30000 - 300 \times t_{final}$$

$$800 \times t_{final} = 45000$$

$$t_{final} = \mathbf{56.2^\circ C}$$

11-80. Refer to Sections 11-9 and 11-11. . . . .

This exercise involves 5 separate calculations:

$$\underset{-12.0^\circ\text{C}}{\text{ice}} \xrightarrow{(1)} \underset{0.0^\circ\text{C}}{\text{ice}} \xrightarrow{(2)} \underset{0.0^\circ\text{C}}{\text{water}} \xrightarrow{(3)} \underset{100.0^\circ\text{C}}{\text{water}} \xrightarrow{(4)} \underset{100.0^\circ\text{C}}{\text{steam}} \xrightarrow{(5)} \underset{115.0^\circ\text{C}}{\text{steam}}$$

(1) heat required = $(\text{mass})(\text{Sp. ht.}_{ice})(\text{temp. change})$
$= (40.0 \text{ g})(2.09 \text{ J/g}^\circ\text{C})(0.0^\circ\text{C} - (-12.0^\circ\text{C})) = 1.00 \times 10^3 \text{ J}$

(2) heat required = $(\text{mass})(\text{Heat of Fusion}) = (40.0 \text{ g})(334 \text{ J/g}) = 1.34 \times 10^4 \text{ J}$

(3) heat required = $(\text{mass})(\text{Sp. ht.}_{water})(\text{temp. change})$
$= (40.0 \text{ g})(4.18 \text{ J/g}^\circ\text{C})(100.0^\circ\text{C} - 0.0^\circ\text{C}) = 1.67 \times 10^4 \text{ J}$

(4) heat required = $(\text{mass})(\text{Heat of Vap.}) = (40.0 \text{ g})(2260 \text{ J/g}) = 9.04 \times 10^4 \text{ J}$

(5) heat required = $(\text{mass})(\text{Sp. ht.}_{steam})(\text{temp. change})$
$= (40.0 \text{ g})(2.03 \text{ J/g}^\circ\text{C})(115.0^\circ\text{C} - 100.0^\circ\text{C}) = 1.22 \times 10^3 \text{ J}$

Therefore, the total heat required = (1) + (2) + (3) + (4) + (5)
$= 1.227 \times 10^5$ J or **122.7 kJ**

11-82. Refer to Sections 11-9 and 1-13. . . . .

Plan: (1) Determine if the final phase will be liquid or gas.
(2) Calculate the final temperature, t_f (°C).

(1) The amount of heat required to change the liquid water to steam at 100°C is

$$\text{heat required} = |(\text{mass})(\text{Sp. ht.})(\text{temp change})|_{water} + |(\text{mass})(\text{Heat of Vap})|$$
$$= (105 \text{ g})(4.18 \text{ J/g}^\circ\text{C})(100.0^\circ\text{C} - 0.0^\circ\text{C}) + (105 \text{ g})(2.26 \times 10^3 \text{ J/g}) = 2.81 \times 10^5 \text{ J}$$

The amount of heat released when the steam changes to water at 100°C is

$$\text{heat released} = |(\text{mass})(\text{Sp. ht.})(\text{temp change})|_{\text{steam}} + |(\text{mass})(\text{Heat of Vap})|$$
$$= (10.5\ g)(2.03\ J/g^{o}C)(110.0^{o}C - 100.0^{o}C) + (10.5\ g)(2.26 \times 10^{3}\ J/g) = 2.39 \times 10^{4}\ J$$

The heat required to convert water to steam is greater than the heat released when the steam condenses to water. Therefore, when the two systems are mixed, the liquid water will cause all of the steam to condense and the final temperature will be between 0°C and 100°C.

(2) $$|\text{amount of heat gained by water}| = |\text{amount of heat lost by steam}|$$

$$|(\text{mass})(\text{Sp. ht.})(\text{temp. change})|_{\text{water}} = |(\text{mass})(\text{Sp. ht.})(\text{temp. change})|_{\text{steam}} + |(\text{mass})(\text{Heat of Vap.})| + |(\text{mass})(\text{Sp. ht.})(\text{temp. change})|_{\text{water from steam}}$$

$$(105\ g)(4.18\ J/g^{o}C)(t_f - 0.0^{o}C) = (10.5\ g)(2.03\ J/g^{o}C)(110.0^{o}C - 100.0^{o}C) + (10.5\ g)(2.26 \times 10^{3}\ J/g) + (10.5\ g)(4.18\ J/g^{o}C)(100.0^{o}C - t_f)$$

$$(439)t_f = 213\ J + 2.37 \times 10^{4}\ J + 4.39 \times 10^{3}\ J - (43.9)t_f$$
$$t_f = \mathbf{58.6^{o}C}$$

11-84. **Refer to Sections 1-13 and 11-9.** . . . .

$$\text{heat lost by copper} = \text{heat gained by water}$$
$$|(\text{mass})(\text{Sp. ht.})(\text{temp. change})|_{\text{copper}} = |(\text{mass})(\text{Sp. ht.})(\text{temp. change})|_{\text{water}}$$

$$(300\ g)(0.385\ J/g^{o}C)(100.0^{o}C - t_f) = (500\ g)(4.18\ J/g^{o}C)(t_f - 30.0^{o}C)$$
$$1.16 \times 10^{4} - (116)t_f = (2090)t_f - 6.27 \times 10^{4}$$
$$t_f = \mathbf{33.7^{o}C}$$

The final temperature is much lower than in Exercise 11-78 since 1 gram of copper releases only 0.385 J per 1°C of cooling, whereas 1 gram of water requires 4.18 J per 1°C of heating.

11-86. **Refer to Section 11-9.** . . . .

Heating 100 g of water from 0°C to 100°C requires 4.18×10^{4} J of heat since

$$\text{heat required} = (\text{mass})(\text{Sp. ht.})(\text{temp. change})$$
$$= (100\ g)(4.18\ J/g^{o}C)(100^{o}C - 0^{o}C) = 4.18 \times 10^{4}\ J$$

Vaporizing 100 g of water at 100°C requires 2.26×10^{5} J since

$$\text{heat required} = (\text{mass})(\text{Heat of Vap.})$$
$$= (100\ g)(2.26 \times 10^{3}\ J) = 2.26 \times 10^{5}\ J$$

If heat is supplied at an identical constant rate, it would therefore take longer to vaporize 100 g of water at 100°C since more heat is required.

11-88. **Refer to Sections 11-9, 11-11 and Example 11-2.** . . . .

The amount of heat required to change 400 g of ice at 0°C to 400 g of water at 37°C is

$$\text{heat required} = (\text{mass})(\text{Heat of Fusion}) + (\text{mass})(\text{Sp. ht.})(\text{temp. change})$$
$$= (400\ g)(334\ J/g) + (400\ g)(4.18\ J/g^{o}C)(37^{o}C - 0^{o}C)$$
$$= 1.95 \times 10^{5}\ J$$

The amount of heat required to evaporate 400 g of water (since the density of water = 1.00 g/mL) at 37°C is

$$\text{heat required} = (\text{mass})(\text{Heat of Vap.})$$
$$= (400\ \text{g})(2.41 \times 10^3\ \text{J/g})$$
$$= 9.64 \times 10^5\ \text{J}$$

Therefore, approximately 5 times more "cooling" is obtained by "sweating out" 400 mL of water at body temperature than by eating 400 g of ice.

11-90. **Refer to Section 11-12.** . . . .

The heat of sublimation is the heat required to change 1 gram of solid directly into a gas.

Heat of sublimation of Eu (s) at 298 K

$$= 1170\ \frac{\text{cal}}{\text{g}} \times \frac{4.184\ \text{J}}{1.000\ \text{cal}} \times \frac{1\ \text{kJ}}{1000\ \text{J}} \times \frac{152.0\ \text{g}}{1\ \text{mol}}$$
$$= 744.1\ \text{kJ/mol}$$

11-92. **Refer to Section 11-13 and Figure 11-17b.** . . . .

(a) This point lies on the sublimation curve where the solid and gas phases are in equilibrium. So, both the solid and gaseous phases are present.

(b) This point is called the triple point where all three phases are in equilibrium with each other. So, the solid, liquid and gaseous phases are all present.

11-94. **Refer to Section 11-13 and Figure 11-17b.** . . . .

The melting point of carbon dioxide increases with increasing pressure, since the solid-liquid equilibrium line on its phase diagram slopes up and to the right. If the pressure on a sample of liquid carbon dioxide is increased at constant temperature, causing the molecules to get closer together, the liquid will solidify. This indicates that solid carbon dioxide has a higher density than the liquid phase. This is true for most substances. The notable exception is water.

11-96. **Refer to Section 11-13.** . . . .

(a) liquid
(b) solid (rhombic)
(c) solid (rhombic)
(d) solid (monoclinic)
(e) vapor
(f) liquid

11-98. **Refer to Section 11-13.** . . . .

(a) Sulfur is a vapor at 160°C and 10^{-4} atm. When the system is compressed at constant temperature to a pressure of $\sim 10^{-3.25}$ or 6×10^{-4} atm on curve E-F, the sulfur vapor begins to condense. At this temperature and pressure, vapor ⇌ liquid. When the pressure is greater than its value on the vapor pressure curve, all the sulfur is in the condensed state. Sulfur will remain in liquid form as the pressure is increased to $10^{3.2}$ or $+1.6 \times 10^3$ atm.

(b) When liquid sulfur at 160°C and 1 atm is compressed at constant temperature, to a pressure of $10^{3.2}$ or 1.6×10^3 atm on curve C-G, part of the melting curve for the rhombic solid, the liquid begins to solidify into the rhombic form. At this temperature and pressure, liquid ⇌ solid (rhombic). When the pressure is greater than its value on the melting curve, all the sulfur has crystallized into the rhombic form. Sulfur will remain as a rhombic solid as the pressure increases to 5×10^3 or $10^{3.7}$ atm.

(c) When liquid sulfur at 130°C and 1 atm is compressed at constant temperature to a pressure of ~$10^{2.3}$ or 200 atm on curve E-D, the melting curve for the monoclinic solid, the liquid begins to solidify into the monoclinic form. Under these conditions, liquid ⇌ solid (monoclinic). When the pressure is greater than its value on the melting curve, all the sulfur has crystallized into the monoclinic form. Sulfur will remain as a monoclinic solid as the pressure increases to 500 or $10^{2.7}$ atm.

11-100. **Refer to Section 11-16.** . . . .

When a metal is distorted (e.g., rolled into sheets or drawn into wire), new metallic bonds are formed and the environment around each atom is essentially unchanged. This can happen because the valence electrons of bonded metal atoms are only loosely associated with individual atoms, as though metal cations exist in a "cloud of electrons." In ionic solids, the lattice arrangements of cations and anions are more rigid. When an ionic solid is distorted, it is possible for cation-cation and anion-anion alignments to occur. But this will cause the solid to shatter due to electrostatic repulsions between ions of like charge.

12 Solutions

12-2. **Refer to Sections 9-1 and 12-1.**

The solubility rules apply only to aqueous solutions under normal temperature and pressure conditions.

12-4. **Refer to Section 12-1 and Figure 12-1.**

Dissolution is favored when (a) solute-solute attractions and (b) solvent-solvent attractions are relatively small and (c) solvent-solute attractions are relatively large. Both steps (a) and (b) require energy input, first to separate the solute particles from each other, then to separate the solvent molecules from each other. Step (c) releases energy as solute particles and solvent molecules interact. If the absolute value of heat absorbed in steps (a) and (b) is less than the absolute value of heat released in step (c), then the dissolving process is favored and releases heat.

12-6. **Refer to Section 12-1, 12-2 and 12-3.**

The dissolution of many solids in liquids is endothermic (requires heat) due to the large solute-solute attractions that must be overcome relative to the solvent-solute attractions.

The mixing of two miscible liquids is exothermic (releases heat) since solute-solute attractions are less than the solvent-solute attractions.

12-8. **Refer to Section 12-1.**

There are two factors which control the spontaneity of a dissolution process: (1) the amount of heat absorbed or released and (2) the amount of increase in the disorder, or randomness of the system. All dissolution processes are accompanied by an increase in the disorder of both solvent and solute. Thus, their disorder factor is invariably favorable to solubility. Dissolution will always occur if the dissolution process is exothermic and the disorder term increases. Dissolution will occur when the dissolution process is endothermic if the disorder term is large enough to overcome the endothermicity, which opposes dissolution.

12-10. **Refer to Section 12-3.**

Bromine is a nonpolar covalent molecule. The "Rule of Thumb" says that "like dissolves like," i.e. substances with similar structures are soluble in each other. Bromine liquid would be more soluble in carbon tetrachloride since it also is a nonpolar covalent molecule, whereas water is a very polar solvent with hydrogen bonding.

12-12. **Refer to Sections 12-2, 12-3 and 12-4.**

(a) HCl in H_2O - high solubility since both are polar covalent molecules.

(b) HF in H_2O - high solubility since both are polar covalent molecules with hydrogen bonding.

(c) Al_2O_3 in H_2O - low solubility since both Al^{3+} and O^{2-} in the ionic solid have high charge-to-size ratios. The dissolution of Al_2O_3 is very endothermic and is not favored.

(d) S_8 in H_2O - low solubility since S_8 is a nonpolar covalent molecule and H_2O is very polar.

12-14. **Refer to Section 9-1 and Exercise 12-12.** . . . .

(a) HCl in H_2O - strong electrolyte

(b) HF in H_2O - weak electrolyte

(c) Al_2O_3 in H_2O - cannot be prepared in "reasonable" concentration

(d) S_8 in H_2O - cannot be prepared in "reasonable" concentration

12-16. **Refer to Section 12-5.** . . . .

A saturated solution of KCl is guaranteed by adding solid KCl to the beaker and mixing. As long as solid KCl is present, the solution is saturated.

12-18. **Refer to Section 12-2 and Table 12-1.** . . . .

The hydration energy of ions generally increases with increasing charge and decreasing size. The greater the hydration energy for an ion, the more strongly hydrated it is. In order of decreasing hydration energy:

(a) $Na^+ > K^+$ (b) $F^- > Cl^-$

(c) $Fe^{3+} > Fe^{2+}$ (d) $Mg^{2+} > Na^+$

12-20. **Refer to Section 12-2.** . . . .

For a solution at infinite dilution at 25°C,

the overall heat of solution = hydration energy + $(-\Delta H_{xtal})$

The reason for using $-\Delta H_{xtal}$ is as follows.

The crystal lattice energy, ΔH_{xtal} = -825.9 kJ/mol, means that 825.9 kJ of heat is released in the reaction:

$K^+ (g) + F^- (g) \rightarrow KF (s) + 825.9 \text{ kJ}$

However, we require the reverse reaction: $KF (s) + 825.9 \text{ kJ} \rightarrow K^+ (g) + F^- (g)$ and $\Delta H_{rxn} = -\Delta H_{xtal}$ = +825.9 kJ/mol. We will discuss sign conventions more thoroughly in Chapter 15.

Rearranging the above equation, we have

hydration energy of KF = heat of solution + ΔH_{xtal}

= -17.7 kJ/mol - 825.9 kJ/mol

= **-843.6 kJ/mol**

12-22. **Refer to Sections 12-4 and 12-6.** . . . .

Most gases undergo exothermic dissolution in water: gas $\rightleftharpoons$ dissolved gas + heat. Applying LeChatelier's Principle to this process, we see that at higher temperatures, more heat is available; the dissolution process will reverse itself to minimize the heat gain. Therefore at higher temperatures, the solubilities of most gases decrease.

12-24. **Refer to Section 12-8.** . . . .

We know: $m = \frac{\text{mol solute}}{\text{kg solvent}}$

? mol sucrose, $C_{12}H_{22}O_{11} = \frac{211 \text{ g sucrose}}{342 \text{ g/mol}} = 0.617$ mol sucrose

Therefore,

$$m = \frac{0.617 \text{ mol sucrose}}{0.325 \text{ kg water}} = \mathbf{1.90}\ \boldsymbol{m}$$

12-26. **Refer to Section 12-8.** . . . .

? mol $C_6H_5COOH = \frac{90.0 \text{ g } C_6H_5COOH}{122 \text{ g/mol}} = 0.738$ mol C_6H_5COOH

? kg C_2H_5OH = 500 mL $C_2H_5OH \times \frac{0.789 \text{ g } C_2H_5OH}{1 \text{ mL } C_2H_5OH} \times \frac{1 \text{ kg}}{1000 \text{ g}} = 0.394$ kg C_2H_5OH

Therefore,

$$m = \frac{\text{mol } C_6H_5COOH}{\text{kg } C_2H_5OH} = \frac{0.738 \text{ mol}}{0.394 \text{ kg}} = \mathbf{1.87}\ \boldsymbol{m}\ \mathbf{C_6H_5COOH \text{ in } C_2H_5OH}$$

12-28. **Refer to Section 12-8 and Example 12-2.** . . . .

Plan: m sucrose, kg $H_2O \xrightarrow{(1)}$ mol sucrose $\xrightarrow{(2)}$ g sucrose

(1) Recall: m = mol solute/kg solvent

Rearranging,

? mol sucrose = m sucrose × kg H_2O = 0.250 m × 0.200 kg = 0.0500 mol

(2) ? g sucrose = 0.0500 mol sucrose × 342 g/mol = **17.1 g**

12-30. **Refer to Section 12-8.** . . . .

Plan: (1) Assuming 100 g of solution, calculate the moles of HF.
(2) Calculate the mass of water.
(3) Determine the molality of HF.

(1) Recall: % by mass $= \frac{\text{g solute}}{\text{g solution}} \times 100$

? g HF in 100 g soln $= \frac{55 \text{ \% HF} \times 100 \text{ g soln}}{100} = 55$ g HF

? mol HF $= \frac{55 \text{ g HF}}{20. \text{ g/mol}} = 2.8$ mol HF

(2) ? g H_2O = g soln - g HF = 100 g - 55 g = 45 g H_2O

(3) $m = \frac{\text{mol HF}}{\text{kg } H_2O} = \frac{2.8 \text{ mol}}{0.045 \text{ kg}} = \mathbf{62}\ \boldsymbol{m}\ \mathbf{HF}$

12-32. **Refer to Section 12-8 and Exercise 12-30.** . . . .

(a) (1) Assume 100 g of solution.

$$? \text{ g } C_6H_{12}O_6 = \frac{32.0 \text{ \% } C_6H_{12}O_6 \times 100 \text{ g soln}}{100} = 32.0 \text{ g } C_6H_{12}O_6$$

$$? \text{ mol } C_6H_{12}O_6 = \frac{32 \text{ g } C_6H_{12}O_6}{180 \text{ g/mol}} = 0.178 \text{ mol } C_6H_{12}O_6$$

(2) $? \text{ g } H_2O = \text{g soln} - \text{g } C_6H_{12}O_6 = 100.0 \text{ g} - 32.0 \text{ g} = 68.0 \text{ g } H_2O$

(3) $$m = \frac{\text{mol } C_6H_{12}O_6}{\text{kg } H_2O} = \frac{0.178 \text{ mol}}{0.0680 \text{ kg}} = \mathbf{2.62\ \mathit{m}\ C_6H_{12}O_6}$$

(b) Molality is independent of density since it is a measure of the number of moles of substance dissolved in 1 kilogram of solvent. Only concentration units that involve volume, e.g. molarity, are dependent on density.

12-34. **Refer to Section 12-8 and Example 12-3.** . . . .

Plan: (1) Calculate the moles of each component.
(2) Calculate the mole fraction.

(1) $$? \text{ mol } CCl_4 = \frac{64.0 \text{ g } CCl_4}{154 \text{ g/mol}} = 0.416 \text{ mol } CCl_4$$

$$? \text{ mol } C_6H_6 = \frac{36.0 \text{ g } C_6H_6}{78.1 \text{ g/mol}} = 0.461 \text{ mol } C_6H_6$$

(2) $$? \text{ mole fraction } CCl_4 = \frac{\text{mol } CCl_4}{\text{mol } CCl_4 + \text{mol } C_6H_6} = \frac{0.416 \text{ mol}}{0.416 \text{ mol} + 0.461 \text{ mol}} = 0.474$$

12-36. **Refer to Section 12-8.** . . . .

We know: $$\text{mole fraction } C_6H_{12}O_6 = \frac{\text{mol } C_6H_{12}O_6}{\text{mol } C_6H_{12}O_6 + \text{mol } H_2O}$$

$$? \text{ mol } H_2O = \frac{150.0 \text{ g } H_2O}{18.0 \text{ g/mol}} = 8.33 \text{ mol } H_2O \qquad \text{since the density of } H_2O = 1.00 \text{ g/mL}$$

Substituting,

$$0.125 = \frac{\text{mol } C_6H_{12}O_6}{\text{mol } C_6H_{12}O_6 + 8.33 \text{ mol}}$$

Rearranging and solving for the moles of $C_6H_{12}O_6$,

$$(0.125 \times \text{mol } C_6H_{12}O_6) + (0.125 \times 8.33 \text{ mol}) = \text{mol } C_6H_{12}O_6$$

$$0.875 \times \text{mol } C_6H_{12}O_6 = 1.04 \text{ mol}$$

$$\text{mol } C_6H_{12}O_6 = 1.19 \text{ mol}$$

$$? \text{ g } C_6H_{12}O_6 = 1.19 \text{ mol } C_6H_{12}O_6 \times 180 \text{ g/mol} = \mathbf{214 \text{ g } C_6H_{12}O_6}$$

12-38. **Refer to Section 12-8, Example 12-3 and Exercise 12-34.** . . . .

$$? \text{ mol } C_2H_5OH = \frac{50.0 \text{ g } C_2H_5OH}{46.1 \text{ g/mol}} = 1.08 \text{ mol } C_2H_5OH$$

$$? \text{ mol } H_2O = \frac{50.0 \text{ g } H_2O}{18.0 \text{ g/mol}} = 2.78 \text{ mol } H_2O$$

Therefore,

$$? \text{ mole fraction } C_2H_5OH = \frac{\text{mol } C_2H_5OH}{\text{mol } C_2H_5OH + \text{mol } H_2O} = \frac{1.09 \text{ mol}}{1.09 \text{ mol} + 2.78 \text{ mol}} = \mathbf{0.282}$$

$$? \text{ mole fraction } H_2O = 1 - \text{mole fraction } C_2H_5OH = 1.000 - 0.282 = \mathbf{0.718}$$

12-40. **Refer to Section 12-8 and Exercise 12-30.** . . . .

Plan: (1) Calculate the masses and moles of HF and H_2O in 100 g of solution.
(2) Calculate the mole fraction of HF.

(1) Assuming 100 g of solution,

$? \text{ g HF} = 55 \text{ g HF}$ $\qquad$ $? \text{ g } H_2O = \text{g soln} - 55 \text{ g} = 45 \text{ g } H_2O$

$$? \text{ mol HF} = \frac{55 \text{ g HF}}{20. \text{ g/mol}} = 2.8 \text{ mol HF} \qquad ? \text{ mol } H_2O = \frac{45 \text{ g } H_2O}{18.0 \text{ g/mol}} = 2.5 \text{ mol } H_2O$$

(2) $$? \text{ mole fraction HF} = \frac{\text{mol HF}}{\text{mol HF} + \text{mol } H_2O} = \frac{2.8 \text{ mol}}{2.8 \text{ mol} + 2.5 \text{ mol}} = \mathbf{0.53}$$

12-42. **Refer to Section 12-9.** . . . .

The vapor pressure of a liquid depends upon the ease with which the molecules are able to escape from the surface of the liquid. The vapor pressure of a liquid always decreases when nonvolatile solutes (ions or molecules) are dissolved in it, since after dissolution there are fewer solvent molecules at the surface to vaporize.

12-44. **Refer to Section 12-9 and Example 12-4.** . . . .

(a) From Raoult's Law, $\Delta P_{benzene} = X_{naphthalene} P^o_{benzene}$

where $\Delta P_{benzene}$ = vapor pressure lowering of benzene, C_6H_6

$P^o_{benzene}$ = vapor pressure of pure benzene, 74.6 torr at 20°C

$X_{naphthalene}$ = mole fraction of naphthalene, $C_{10}H_8$

$$? \text{ mol } C_6H_6 = \frac{120.0 \text{ g } C_6H_6}{78.1 \text{ g/mol}} = 1.54 \text{ mol } C_6H_6$$

$$? \text{ mol } C_{10}H_8 = \frac{40.0 \text{ g } C_{10}H_8}{128 \text{ g/mol}} = 0.312 \text{ mol } C_{10}H_8$$

$$X_{naphthalene} = \frac{\text{mol } C_{12}H_8}{\text{mol } C_{10}H_8 + \text{mol } C_6H_6} = \frac{0.312 \text{ mol}}{0.312 \text{ mol} + 1.54 \text{ mol}} = 0.168$$

Substituting into Raoult's Law, $\Delta P_{benzene} = 0.168 \times 74.6 \text{ torr} = \mathbf{12.5 \text{ torr}}$

(b) $$P_{benzene} = P^o_{benzene} - \Delta P_{benzene}$$
$$= 74.6 \text{ torr} - 12.5 \text{ torr}$$
$$= \mathbf{62.1 \text{ torr}}$$

12-46. **Refer to Section 12-9 and Exercise 12-45.** . . . .

From Raoult's Law, $P_{water} = X_{water}P^o_{water}$

$$X_{water} = \frac{\text{mol } H_2O}{\text{mol } H_2O + \text{mol } C_{12}H_{22}O_{11}}$$

$$= \frac{(400 \text{ g } H_2O/18.0 \text{ g/mol})}{(400 \text{ g } H_2O/18.0 \text{ g/mol}) + (60.0 \text{ g } C_{12}H_{22}O_{11}/342 \text{ g/mol})} = 0.992$$

We know: since 100°C is the boiling point of water at 760 torr,

P^o_{water} = 760 torr at 100°C

Substituting into Raoult's Law, P_{water} = 0.992 × 760 torr = **754 torr**

12-48. **Refer to Sections 12-11, 12-12 and Table 12-2.** . . . .

(a) $\Delta T_f = K_f m$. Therefore, a 0.100 *m* solution of a nonvolatile, nonelectrolyte solute in the solvent, phenol, which has the highest value of K_f would have the greatest freezing point depression.

(b) We know that $T_{f(soln)} = T_{f(solvent)} - \Delta T_f = T_{f(solvent)} - (0.100\ m)K_f$.

A 0.100 *m* in water would have the lowest freezing point.

(c) $\Delta T_b = K_b m$. Therefore, a 0.100 *m* solution of nonvolatile nonelectrolyte solute in the solvent, nitrobenzene which has the highest value of K_b would have the greatest boiling elevation.

(d) We know that $T_{b(soln)} = T_{b(solvent)} - \Delta T_b = T_{b(solvent)} - (0.100\ m)K_b$.

A 0.100 *m* solution in nitrobenzene would have the highest boiling point.

12-50. **Refer to Sections 12-11, 12-12 and Exercise 12-54 Solution.** . . . .

(a) The molal freezing point depression constant, K_f (°C/*m*) for a particular solvent, represents the number of degrees Celsius the freezing point of that particular solvent is lowered for a 1 *m* solution of a nonvolatile nonelectrolyte.

(b) The molal boiling point elevation constant, K_b (°C/*m*) for a particular solvent, represents the number of degrees Celsius the boiling point of that particular solvent is raised for a 1 *m* solution of a nonvolatile nonelectrolyte.

K_f and K_b for a particular solvent can be determined directly in the laboratory by measuring the new freezing point and boiling point of a solution of a nonvolatile nonelectrolyte with known molality and by substituting the obtained ΔT_f and ΔT_b values into the $\Delta T_f = K_f m$ and $\Delta K_b = K_b m$ equations to obtain the K_f and K_b values.

12-52. **Refer to Section 12-11.** . . . .

An antifreeze is a substance that is added to a liquid, usually water, to lower its freezing point. Nearly all antifreezes on the market are either ethylene glycol or methanol. Ethylene glycol, because of its high boiling point, is less volatile than water, stays in the radiator longer and is considered to be a "permanent" antifreeze. Methanol, on the other hand, is a "temporary" antifreeze since it has a lower boiling point than water and is therefore more volatile.

12-54. **Refer to Section 12-11.**

The boiling point elevation, $\Delta T_b = K_b m$

$\Delta T_b = 84.27^oC - 77.06^oC = 7.21^oC$

$$m = \frac{\text{mol } C_{10}H_8}{\text{kg ethylacetate}} = \frac{50.0 \text{ g } C_{10}H_8/128 \text{ g/mol}}{0.150 \text{ kg}} = 2.60\ m\ C_{10}H_8$$

Substituting and solving for K_b, $K_b = \frac{\Delta T_b}{m} = \frac{7.21^oC}{2.60\ m} = 2.77\ ^oC/m$

12-56. **Refer to Sections 12-11, 12-12 and Table 12-2.**

From Table 12-2, for benzene: $T_b = 80.1^oC$ $K_b = 2.53\ ^oC/m$
$T_f = 5.48^oC$ $K_f = 5.12\ ^oC/m$

$$m = \frac{\text{mol } C_{12}H_{10}}{\text{kg } C_6H_6} = \frac{18.0 \text{ g/154 g/mol}}{0.0650 \text{ kg}} = 1.80\ m\ C_{12}H_{10}$$

Freezing point calculation:

$\Delta T_f = K_f m = (5.12\ ^oC/m)(1.80\ m) = 9.22^oC$

$T_{f(soln)} = T_{f(benzene)} - \Delta T_f = 5.48^oC - 9.22^oC = \mathbf{-3.74^oC}$

Boiling point calculation:

$\Delta T_b = K_b m = (2.53\ ^oC/m)(1.80\ m) = 4.55^oC$

$T_{b(soln)} = T_{b(benzene)} + \Delta T_b = 80.1^oC + 4.6^oC = \mathbf{84.7^oC}$

12-58. **Refer to Sections 12-11, 12-12 and Table 12-2.**

From Table 12-2, for water: $T_b = 100^oC$ $K_b = 0.512\ ^oC/m$
$T_f = 0^oC$ $K_f = 1.86\ ^oC/m$

$$m = \frac{\text{mol } N_2H_4CO}{\text{kg } H_2O} = \frac{25.0 \text{ g/60.1 g/mol}}{0.250 \text{ g}} = 1.66\ m\ N_2H_4CO$$

Freezing point calculation:

$\Delta T_f = K_f m = (1.86\ ^oC/m)(1.66\ m) = \mathbf{3.09^oC}$

$T_{f(soln)} = T_{f(water)} - \Delta T_f = 0.00^oC - 3.09^oC = -3.09^oC$

Boiling point calculation:

$\Delta T_b = K_b m = (0.512^oC/m)(1.66\ m) = 0.850^oC$

$T_{b(soln)} = T_{b(water)} + \Delta T_b = 100.000^oC + 0.850^oC = \mathbf{100.850^oC}$

12-60. **Refer to Section 12-13, Table 12-2 and Example 12-8.**

From Table 12-2, for benzene: $T_f = 5.48^oC$, $K_f = 5.12\ ^oC/m$

Plan: (1) Calculate the molality of the unknown solution using $\Delta T_f = K_f m$.
(2) Calculate the moles and molecular weight.

(1) $m = \frac{\Delta T_f}{K_f} = \frac{5.48^\circ C - 2.06^\circ C}{5.12\ ^\circ C/m} = 0.668\ m$

(2) Recall, $m = \frac{\text{mol solute}}{\text{kg solvent}} = \frac{\text{g solute/MW solute}}{\text{kg solvent}}$

Solving,

MW solute $= \frac{\text{g solute}}{m \times \text{kg solvent}} = \frac{0.500\ g}{0.668\ m \times 0.015\ kg} =$ **49.9 g/mol**

12-62. **Refer to Section 12-11 and Table 12-2.** . . . .

From Table 12-2, for water: $T_b = 100^\circ C$, $K_b = 0.512\ ^\circ C/m$

(a) $\Delta T_b = K_b m = (0.512\ ^\circ C/m)(1.00\ m) = 0.512^\circ C$

$T_{b(soln)} = T_{b(water)} + \Delta T_b = 100.000^\circ C + 0.512^\circ C =$ **100.512°C**

(b) As the water boils away, the solution becomes more concentrated. The boiling point of the solution increases since $\Delta T_b \propto m$.

(c) When half of the water has boiled away, the concentration of the new, more concentrated solution is 2.00 *m*. Therefore,

$\Delta T_b = K_b m = (0.512\ ^\circ C/m)(2.00\ m) = 1.02^\circ C$

$T_{b(soln)} = 100.000^\circ C + 1.02^\circ C =$ **101.02°C**

12-64. **Refer to Section 12-11 and Table 12-2.** . . . .

Plan: (1) Solve for the molality of the solution using $\Delta T_b = K_b m$.
(2) Calculate the mass of $C_{10}H_8$.

From Table 12-2, for nitrobenzene: $T_b = 210.88^\circ C$ $K_b = 5.24\ ^\circ C/m$

(1) $m = \frac{\Delta T_b}{K_b} = \frac{213.50^\circ C - 210.88^\circ C}{5.24^\circ C/m} = 0.500\ m\ C_{10}H_8$

(2) Recall, $m = \frac{\text{mol solute}}{\text{kg solvent}} = \frac{\text{g solute/MW solute}}{\text{kg solvent}}$

Solving,

? g $C_{10}H_8$ = $(m)(\text{MW } C_{10}H_8)(\text{kg } C_6H_5NO_2)$ = (0.500 *m*)(128 g/mol)(0.250 kg)
= **16.0 g**

12-66. **Refer to Section 12-11 , Example 12-5 and Exercise 12-65.** . . . .

Plan: (1) Assume a sample of antifreeze consists of 100 mL of water (solvent) and 100 mL of ethylene glycol (solute). Calculate the moles and molality of ethylene glycol in the antifreeze.
(2) Calculate $T_{b(soln)}$ for the antifreeze.

(1) ? mol ethylene glycol $= \frac{110.9\ \text{g ethylene glycol}}{62.07\ \text{g/mol}} = 1.787$ mol

$m = \frac{\text{mol ethylene glycol}}{\text{kg water}} = \frac{1.787\ \text{mol}}{0.1000\ \text{kg}} = 17.87\ m$ ethylene glycol

(2) From Table 12-2, for water: $T_b = 100.00°C$, $K_b = 0.512°C/m$

$$\Delta T_b = K_b m = (0.512\ °C/m)(17.87\ m) = 9.15°C$$

$$T_{b(soln)} = T_{b(water)} + \Delta T_b = 100.00°C + 9.15°C = \mathbf{109.15°C}$$

12-68. **Refer to Section 12-11 and Exercise 12-67.** . . . .

The molality of the ethylene glycol solution in Exercise 12-67 is 12.5 m.

From Table 12-2, for water: $T_b = 100°C$, $K_b = 0.512\ °C/m$

$$\Delta T_b = K_b m = (0.512\ °C/m)(12.5\ m) = 6.40°C$$

$$T_{b(soln)}\ (°C) = T_{b(water)} + \Delta T_b = 100.00°C + 6.40°C = 106.40°C$$

$$T_{b(soln)}\ (°F) = (106.40°C \times \frac{1.8°F}{1.0°C}) + 32.00°F = \mathbf{223.52°F}$$

12-70. **Refer to Section 12-12 and Fundamental Algebra.** . . . .

Plan: (1) Calculate the molality of the solution using $\Delta T_f = K_f m$.
(2) Calculate the total number of moles of solutes.
(3) Calculate the masses of $C_{10}H_8$ and $C_{14}H_{10}$.

(1) From Table 12-2, for benzene: $T_f = 5.48°C$, $K_f = 5.12\ °C/m$

$$m = \frac{\Delta T_f}{K_f} = \frac{5.48°C - 4.85°C}{5.12\ °C/m} = 0.12\ m \text{ solutes}$$

(2) Recall, $m = \frac{\text{mol solute}}{\text{kg solvent}}$

$$?\ \text{mol solutes} = 0.12\ m \times 0.300\ \text{kg} = 0.036\ \text{mol solutes}$$

(3) let $x = \text{g } C_{10}H_8$ and $\frac{x}{128\ \text{g/mol}} = \text{mol } C_{10}H_8$

$5.00\ \text{g} - x = \text{g } C_{14}H_{10}$ $\frac{5.00\ \text{g} - x}{178\ \text{g/mol}} = \text{mol } C_{14}H_{10}$

Therefore,
the total moles of solute = mol $C_{10}H_8$ + mol $C_{14}H_{10}$ $= \frac{x}{128} + \frac{5.00 - x}{178}$

$$\text{total moles} = 0.036\ \text{mol} = \frac{x}{128} + \frac{5.00 - x}{178}$$

$$820 = 178x + 640 - 128x$$

$$x = 3.6\ \text{g } C_{10}H_8$$

$$5.00 - x = 1.4\ \text{g } C_{14}H_{10}$$

And $$\%\ C_{10}H_8 = \frac{3.6\ \text{g } C_{10}H_8}{5.00\ \text{g sample}} \times 100 = \mathbf{72\%}$$

$$\%\ C_{14}H_{10} = \frac{1.4\ \text{g } C_{14}H_{10}}{5.00\ \text{g sample}} \times 100 = \mathbf{28\%}$$

12-72. **Refer to Section 12-12.** . . . .

Since both ethylene glycol and sucrose are nonvolatile nonelectrolytes, the number of moles of sucrose required to lower the freezing point of water to -10.0°F with 6.00 gallons of water equals the number of moles of ethylene glycol required to do the same.

Therefore,

mol ethylene glycol, $C_2H_6O_2$ = mol sucrose, $C_{12}H_{22}O_{11}$

$$\frac{17.6 \times 10^3 \text{ g}}{62.1 \text{ g/mol}} = \frac{\text{g sucrose}}{342 \text{ g/mol}}$$

$$? \text{ g sucrose} = 9.69 \times 10^4 \text{ g or } \mathbf{96.9 \text{ kg}}$$

12-74. **Refer to Section 12-14.** . . . .

In an ionic solution, the electrical interactions between dissolved ions increase with (1) increasing charge on the ions and (2) increasing concentration. The weaker the electrical attraction between the ions of a dissolved salt, the more completely dissociated is the salt. A 0.10 *M* $LiNO_3$ solution is more completely dissociated than either a 0.10 *M* $Ca(NO_3)_2$ or a 0.10 *M* $Al(NO_3)_3$ solution since the charge on the lithium ion is +1, compared to the +2 charge on the calcium ion and the +3 charge on the aluminum ion. On the other hand, the solution with the greatest number of ions will conduct electricity the strongest. If completely dissociated, the total number of ions expected from $LiNO_3$, $Ca(NO_3)_2$ and $Al(NO_3)_3$ would be 2, 3 and 4, respectively. Therefore, $Al(NO_3)_3$ is expected to be the best conductor of electricity among the three.

12-76. **Refer to Section 12-14.** . . . .

The ideal value for the van't Hoff factor, *i*, for strong electrolytes at infinite dilution is the total number of ions present in a formula unit.

		ideal value of I
(a)	HCl	2
(b)	$Ca(OH)_2$	3
(c)	$Al_2(SO_4)_3$	5
(d)	$Fe(NO_3)_3$	4

12-78. **Refer to Section 12-14.** . . . .

HF (aq) is a weak electrolyte and therefore dissociates only slightly in aqueous solution. Its van't Hoff factor is much less than the ideal value, 2, due to extensive ion association.

12-80. **Refer to Section 12-14 and Table 12-2.** . . . .

From Table 12-2, for water: T_b = 100.00°C, K_b = 0.512 °C/*m*

For $Ca(NO_3)_2$, i_{Ideal} = 3

$$\Delta T_b = iK_bm = (3)(0.512\ ^\circ\text{C}/m)(0.092\ m) = 0.14^\circ\text{C}$$

$$T_{b(soln)} = T_{b(water)} + \Delta T_b = 100.00^\circ\text{C} + 0.14^\circ\text{C} = \mathbf{100.14^\circ C}$$

The true boiling point will be higher than 100.00°C but slightly below 100.14°C since the actual value of *i* should be slightly less than 3.

12-82. **Refer to Section 12-14 and Example 12-9.** . . . .

Plan: (1) Calculate $m_{effective}$.
(2) Determine the % ionization.

(1) From Table 12-2, for water: $T_f = 0^oC$, $K_f = 1.86\ ^oC/m$

$$m_{effective} = \frac{\Delta T_f}{K_f} = \frac{0.0000^oC - (-0.1884^oC)}{1.86\ ^oC/m} = 0.101\ m$$

(2) Let x = molality of CH_3COOH that ionizes
Then x = molality of H^+ and CH_3COO^- that formed.

Consider:	CH_3COOH	→	H^+	+	CH_3COO^-
Initial	0.100 m		0		0
Final	(0.100 - x)m		x m		x m

$$m_{effective} = m_{CH_3COOH} + m_{H^+} + m_{CH_3COO^-}$$
$$0.101 = (0.100 - x) + x + x$$
$$0.101 = (0.100 + x)$$
$$x = 0.001\ m$$

$$\%\ ionization = \frac{m_{ionized}}{m_{original}} \times 100 = \frac{0.001\ m}{0.100\ m} \times 100 = \mathbf{1\%}$$

12-84. **Refer to Section 12-14.** . . . .

Plan: (1) Determine $m_{effective}$ and the van't Hoff factor, i.
(2) Calculate the freezing point of the solution.

(1) $$m_{ionized} = \frac{\%\ ionized}{100\ \%} \times m_{original} = \frac{6.1\ \%}{100\ \%} \times 0.00100\ m = 6.1 \times 10^{-5}\ m$$

Consider the ionization of a weak base, such as R_3N.

	R_3N	+	H_2O	→	R_3NH^+	+	OH^-
Initial	0.00100 m				0		0
Final	$(1.00 \times 10^{-3} - 6.1 \times 10^{-5})m$				$6.1 \times 10^{-5}\ m$		$6.1 \times 10^{-5}\ m$

$$m_{effective} = m_{R_3N} + m_{R_3NH^+} + m_{OH^-}$$
$$= (1.00 \times 10^{-3} - 6.1 \times 10^{-5}) + 6.1 \times 10^{-5} + 6.1 \times 10^{-5}$$
$$= 1.06 \times 10^{-3}\ m$$

$$\text{van't Hoff factor, } i = \frac{m_{effective}}{m_{stated}} = \frac{1.06 \times 10^{-3}\ m}{1.00 \times 10^{-3}\ m} = 1.06$$

(2) For water, $K_f = 1.86\ ^oC/m$

$$\Delta T_f = K_f m_{effective}\ (= iK_f m_{stated})$$
$$= (1.86\ ^oC/m)(1.06 \times 10^{-3}\ m) = 1.97 \times 10^{-3}\ ^oC$$

$$T_{f(soln)} = T_{f(water)} - \Delta T_f = 0.000^oC - 1.97 \times 10^{-3}\ ^oC = \mathbf{-1.97 \times 10^{-3}\ ^oC}$$

12-86. **Refer to Section 12-14.**

Plan: (1) Find ΔT_f for the solution if CsCl had been a nonelectrolyte.
(2) Determine the van't Hoff factor, i.
(3) Calculate $m_{effective}$ for CsCl.
(4) Calculate the apparent percentage dissociation.

(1) If CsCl were a nonelectrolyte

$\Delta T_f = K_f m = (1.86\ ^oC/m)(0.121\ m) = 0.225^oC$ (for water, $K_f = 1.86\ ^oC/m$)

(2) The van't Hoff factor, $i = \dfrac{\Delta T_{f(actual)}}{\Delta T_{f(nonelectrolyte)}} = \dfrac{0^oC - (-0.403^oC)}{0.225^oC} = \mathbf{1.79}$

(3) $m_{effective} = i \times m_{stated} = 1.79 \times 0.121\ m = 0.217\ m$

(4) Let x = molality of CsCl that apparently dissociated
Then x = molality of Cs^+ and Cl^- that formed

Consider:	$CsCl$	$\rightarrow$	Cs^+	+	Cl^-
Initial	$0.121\ m$		0		0
Final	$(0.121 - x)m$		$x\ m$		$x\ m$

$$m_{effective} = 0.217\ m = m_{CsCl} + m_{Cs^+} + m_{Cl^-}$$
$$= (0.121 - x) + x + x$$
$$= (0.121 + x)$$
$$x = 0.096\ m$$

$$\%\ \text{apparent dissociation} = \frac{m_{apparently\ dissociated}}{m_{stated}} \times 100 = \frac{0.096\ m}{0.121\ m} \times 100$$
$$= \mathbf{79\%}$$

12-88. **Refer to Section 12-4 and Exercise 12-86 Solution.**

(1) If $K_3[Fe(CN)_6]$ were a nonelectrolyte,

$\Delta T_f = K_f m = (1.86\ ^oC/m)(0.126\ m) = 0.234^oC$ (for water, $K_f = 1.86\ ^oC/m$)

(2) van't Hoff factor, $i = \dfrac{\Delta T_{f(actual)}}{\Delta T_{f(nonelectrolyte)}} = \dfrac{0^oC - (-0.649^oC)}{0.234^oC} = \mathbf{2.77}$

(3) $m_{effective} = i \times m_{stated} = 2.77 \times 0.126\ m = 0.349\ m$

(4) Let x = molality of $K_3[Fe(CN)_6]$ that apparently dissociated
3x = molality of K^+ formed
x = molality of $[Fe(CN)_6]^{3-}$ formed

Consider:	$K_3[Fe(CN)_6]$	$\rightarrow$	$3K^+$	+	$[Fe(CN)_6]^{3-}$
Initial	$0.126\ m$		0		0
Final	$(0.126 - x)m$		$(3x)m$		$x\ m$

$$m_{effective} = 0.349\ m = m_{K_3[Fe(CN)_6]} + m_{K^+} + m_{[Fe(CN)_6]^{3-}}$$
$$= (0.126 - x) + (3x) + x$$
$$= (0.126 + 3x)$$
$$x = 0.0743\ m$$

$$\%\ \text{apparent dissociation} = \frac{m_{apparently\ dissociated}}{m_{stated}} \times 100 = \frac{0.0743\ m}{0.126\ m} \times 100$$
$$= \mathbf{59.0\%}$$

12-90. **Refer to Section 12-14.** . . . .

Solutions of compounds which have van't Hoff factor, *i*, values less than 1 are not dissociating at all, but in fact are undergoing molecular association. In this case, benzoic acid form hydrogen bonds in benzene with other benzoic acid molecules. Since *i* is approximately 0.5, we can deduce that on the average, 2 molecules of benzoic acid undergo molecular association to give a dimer unit.

12-92. **Refer to Section 12-14.** . . . .

Plan: (1) Convert 15.0% NaCl to molality.
(2) Determine the freezing point depression for this solution if NaCl were a nonelectrolyte.
(3) Calculate the van't Hoff factor, *i*.

(1) Assume we have 100.0 g of solution.
Therefore, it contains 15.0 g NaCl and 85.0 g H_2O.

$$m = \frac{\text{mol NaCl}}{\text{kg H}_2\text{O}} = \frac{15.0\text{ g NaCl}/58.4\text{ g/mol}}{0.0850\text{ kg H}_2\text{O}} = 3.02\ m\text{ NaCl}$$

(2) $\Delta T_f = K_f m = (1.86\ ^\circ\text{C}/m)(3.02\ m) = 5.62^\circ\text{C}$ (for water, $K_f = 1.86\ ^\circ\text{C}/m$)

(3) $$\text{van't Hoff factor, } i = \frac{\Delta T_{f(\text{actual})}}{\Delta T_{f(\text{nonelectrolyte})}} = \frac{0^\circ\text{C} - (-10.888^\circ\text{C})}{5.62^\circ\text{C}} = 1.94$$

12-94. **Refer to Section 12-15.** . . . .

Another experiment which could be used to measure the osmotic pressure of a solution is the following. Place the solution in an inverted thistle tube which has a membrane across the bottom. Immerse it in a container of solvent so that the levels of both solutions are the same. Measure the applied pressure to the thistle tube which is necessary to keep the solution levels the same. That pressure is related to the osmotic pressure.

12-96. **Refer to Section 12-8.** . . . .

Assume 1 liter of solution.

$$?\text{ g NaCl in 1 L soln} = 1.00 \times 10^{-4}\text{ mol NaCl} \times 58.4\text{ g/mol} = 5.84 \times 10^{-3}\text{ g}$$

The contribution of NaCl to the mass of 1 liter of water is nearly negligible. Since the density of water (1.00 g/mL) is essentially the same as the density of the solution:

$$M = \frac{1.00 \times 10^{-4}\text{ mol NaCl}}{1\text{ L soln}} \simeq \frac{1.00 \times 10^{-4}\text{ mol NaCl}}{1\text{ kg H}_2\text{O}} = m$$

However, if acetonitrile (density at 20°C = 0.786 g/mL) were the solvent, 1 liter of solution which would be essentially pure solvent would have a mass of only 786 g. We can see that molarity would <u>not</u> be equivalent to molality.

12-98. **Refer to Section 12-15 and Example 12-10.** . . . .

The osmotic pressure equation states:

$$\pi = MRT = \frac{nRT}{V}$$

where π = osmotic pressure
n = moles of enzyme
R = universal gas constant, 0.0821 L·atm/mol·K
T = temperature (K)
V = volume (L)

Substituting,

$$\pi = \frac{(20.0\text{ g}/3.8 \times 10^6\text{ g/mol})(0.0821\text{ L·atm/mol·K})(38^\circ\text{C} + 273^\circ)}{1.560\text{ L}}$$

$$= \mathbf{8.61 \times 10^{-5}\ atm}$$

12-100. **Refer to Section 12-15.** . . . .

Because the solution is dilute, an approximation of the osmotic pressure equation can be used:

$$\pi \simeq mRT = \left(\frac{n}{\text{kg solvent}}\right)RT$$

where π = *osmotic pressure*
n = moles of enzyme = (g sample)/MW
R = universal gas constant, 0.0821 L·atm/mol·K
T = temperature (K)

Rearranging,

$$\text{MW} = \frac{(\text{g sample})RT}{\pi(\text{kg solvent})} = \frac{(0.481\text{ g})(0.0821\text{ L·atm/mol·K})(21.4^\circ\text{C} + 273.2\text{ K})}{(5.97/760)\text{atm }(0.0500\text{ kg})}$$

$$= \mathbf{2.96 \times 10^4\ g/mol}$$

12-102. **Refer to Section 12-15 and Table 12-3.** . . . .

From Table 12-3, the van't Hoff factor, i, for 0.10 m K_2CO_3 is 2.45.

$$m_{\text{effective}} = i \times m_{\text{stated}} = 2.45 \times 0.100\ m = 0.245\ m$$

Therefore, using the approximation equation for dilute solutions, $M \simeq m$:

$$\text{osmotic pressure, } \pi = m_{\text{effective}}RT = (0.245\ m)(0.0821\text{ L·atm/mol·K})(25^\circ\text{C} + 273^\circ)$$

$$= \mathbf{5.99\ atm}$$

12-104. **Refer to Sections 12-4 and 12-15.** . . . .

(a) $$m = \frac{\text{mol substance}}{\text{kg solvent}} = \frac{0.0100\text{ g}/2.00 \times 10^4\text{ g/mol}}{0.0100\text{ kg}} = 5.00 \times 10^{-5}\ m$$

$$\Delta T_f = K_{f(\text{water})}m = (1.86\ ^\circ\text{C}/m)(5.00 \times 10^{-5}\ m) = 9.30 \times 10^{-5}\ ^\circ\text{C}$$

$$T_{f(\text{soln})} = T_{f(\text{water})} - \Delta T_f = 0^\circ\text{C} - 9.30 \times 10^{-5}\ ^\circ\text{C} = \mathbf{-9.30 \times 10^{-5}\ ^\circ C}$$

(b) $T = 25^\circ\text{C} = 298\text{ K}$.

$$\text{Osmotic pressure, } \pi = mRT = (5.00 \times 10^{-5}\ m)(0.0821\text{ L·atm/mol·K})(298\text{ K})$$
$$= \mathbf{1.22 \times 10^{-3}\ atm}\text{ or }\mathbf{0.930\ torr}$$

(c) From the equation given in (a),

% error in $T_{f(\text{soln})}$ = % error in ΔT_f = % error in MW

$$?\ \%\text{ error in } T_{f(\text{soln})} = \frac{\text{error in } T_f}{T_f} \times 100 = \frac{0.001^\circ\text{C}}{9.30 \times 10^{-5}\ ^\circ\text{C}} \times 100 = \mathbf{1000\%}$$

Therefore, an error of only 0.001°C in the freezing point temperature corresponds to a 1000% error in the macromolecule's molecular weight.

(d) Since the osmotic pressure, $\pi = \frac{nRT}{V} = (\text{mol/MW})\frac{RT}{V}$,

% error in π = % error in MW

$$\% \text{ error in } \pi = \frac{\text{error in } \pi}{\pi} \times 100 = \frac{0.1 \text{ torr}}{0.930 \text{ torr}} \times 100 = \mathbf{10\% \text{ error}}$$

Therefore, an error of 0.1 torr in osmotic pressure gives only a 10% error in molecular weight.

12-106. **Refer to Table 12-4 and the Key Terms for Chapter 12.** . . . .

(a) sol - a colloidal suspension of a solid dispersed in a liquid, e.g. detergents in water.

(b) gel - a collidal suspension of a solid dispersed in a liquid; a semirigid sol, e.g. jelly.

(c) emulsion - a colloidal suspension of a liquid in a liquid, e.g. some cough medicines.

(d) foam - a colloidal suspension of a gas in a liquid, e.g. bubbles in a bubble bath.

(e) solid sol - a colloidal suspension of a solid in a solid, e.g. dirty ice.

(f) solid emulsion - a colloidal suspension of a liquid dispersed in a solid, e.g. some kinds of sea ice containing pockets of brine.

(g) solid foam - a colloidal suspension of a gas dispersed in a solid, e.g. marshmallows.

(h) solid aerosol - a colloidal suspension of a solid in a gas, e.g. fine dust.

(i) liquid aerosol - a colloidal suspension of a liquid in gas, e.g. insect spray.

12-108. **Refer to Section 12-19 and the Key Terms for Chapter 12.** . . . .

Hydrophilic colloids are colloidal particles that attract water molecules, whereas hydrophobic colloids are colloidal particles that repel water molecules.

12-110. **Refer to Section 10-23.** . . . .

Soaps and detergents are both emulsifying agents. Solid soaps are usually sodium salts of long chain organic acids called fatty acids with the general formula, $R\text{-}COO^-Na^+$. On the other hand, synthetic detergents contain sulfonate, $-SO_3^-$, sulfate, $-SO_4^-$, or phosphate groups instead of carboxylate groups, $-COO^-$.

"Hard" water contains Fe^{3+}, Ca^{2+} and/or Mg^{2+} ions, all of which displace Na^+ from soap molecules and give an undesirable precipitate coating. However, detergents do not form precipitates with the ions of "hard" water.

A typical equation between soap and hard water that contains Ca^{2+} ions is shown below.

$$Ca^{2+}(aq) + 2RCOO^-Na^+(aq) \rightarrow (RCOO^-)_2Ca^{2+}(s) + 2Na^+(aq)$$

13 Acids, Bases, and Salts

13-2. **Refer to Section 13-1.** . . . .

Gay-Lussac, in 1814, concluded that acids and bases should be defined in terms of their reactions with each other, i.e., an acid is a substance that neutralizes a base.

13-4. **Refer to Section 13-1.** . . . .

(a) According to Arrhenius, an acid is a substance that contains hydrogen and produces hydrogen ions in aqueous solution. A base is a substance that contains the OH group and produces hydroxide ions in aqueous solution. Neutralization is the reaction between hydrogen ions and hydroxide ions yielding water molecules.

(b) acid: $HBr\ (aq) \rightarrow H^+\ (aq) + Br^-\ (aq)$

base: $Ba(OH)_2\ (aq) \rightarrow Ba^{2+}\ (aq) + 2OH^-\ (aq)$

neutralization: $2HBr\ (aq) + Ba(OH)_2\ (aq) \rightarrow BaCl_2\ (aq) + 2H_2O\ (\ell)$

13-6. **Refer to Section 9-1.** . . . .

(a) A strong acid is an acid that ionizes completely into its ions or nearly so in dilute aqueous solution, e.g. HCl (aq). A weak acid ionizes partially into its ions in dilute aqueous solution, e.g. HF (aq) and CH_3COOH (aq).

(b) A strong soluble base is a hydroxide of a Group IA or IIA metal that is soluble in water and is completely or nearly completely dissociated into its ions in dilute aqueous solutions, e.g. NaOH (aq). A weak base is a molecular base that is slightly ionized in dilute aqueous solution, e.g., NH_3 (aq), CH_3NH_2 (aq).

(c) The definition of a strong soluble base is given in (b). An insoluble base is an insoluble metal hydroxide, e.g. $Cu(OH)_2$ (s), $Fe(OH)_3$ (s).

13-8. **Refer to Section 13-2.** . . . .

In aqueous solution, the hydrogen ion does not exist as a bare proton, H^+, but as a hydrated ion, $H^+(H_2O)_n$, where n is some small integer. We sometimes represent this fact by writing H_3O^+ instead of H^+, by letting n = 1.

13-10. **Refer to Section 13-2 and Exercise 13-8.** . . . .

The statement, "The hydrated hydrogen ion should always be represented as H_3O^+," has two main flaws. First, the true extent of hydration of the H^+ in many solutions is unknown. Secondly, when balancing equations, it is generally much easier to use H^+ rather than H_3O^+.

13-12. **Refer to Section 13-3.** . . . .

(a) acid: a proton donor, e.g. HCl, NH_4^+, H_2O, H_3O^+

(b) conjugate base: a species that is produced when an acid donates a proton, e.g. Cl^- is the conjugate base of HCl OH^- is the conjugate base of H_2O

NH_3 is the conjugate base of NH_4^+ H_2O is the conjugate base of H_3O^+

(c) base: a proton acceptor, e.g. NH_3, H_2O, OH^-

(d) conjugate acid: a species that is produced when a base accepts a proton,

e.g. HCl is the conjugate acid of Cl^- H_2O is the conjugate acid of OH^-

NH_4^+ is the conjugate acid of NH_3 H_3O^+ is the conjugate acid of H_2O

(e) conjugate acid-base pair: two species with formulas that differ only by a proton, e.g. HCl and Cl^-, NH_4^+ and NH_3, $H_2PO_4^-$ and HPO_4^{2-}. The species with the extra proton is the conjugate acid, whereas the other is the conjugate base.

13-14. **Refer to Section 13-3.** . . . .

(a) In dilute aqueous solution, ammonia is a Bronsted-Lowry base and water is an acid. The reaction produces an ammonium ion and a hydroxide ion.

$$NH_3\ (aq) + H_2O\ (\ell) \rightleftharpoons NH_4^+\ (aq) + OH^-\ (aq)$$

(b) In the gaseous state, ammonia also behaves as a Bronsted-Lowry base when reacting with gaseous hydrogen chloride. It takes a proton from the gaseous acid and produces the ionic salt, ammonium chloride.

$$NH_3\ (g) + HCl\ (g) \rightarrow NH_4Cl\ (s)$$

13-16. **Refer to Section 13-3.** . . . .

Water is amphiprotic because a water molecule can either accept or donate a proton.

(a) We can describe water as amphoteric because it has the ability to react either as an acid or as a base. It can act as a Bronsted-Lowry acid by donating a proton to form a hydroxide ion or it can act as a Bronsted-Lowry base by accepting a proton to form a hydronium ion.

(b) Water as a Bronsted-Lowry acid: $H_2O\ (\ell) + CN^-\ (aq) \rightarrow HCN\ (aq) + OH^-\ (aq)$

Water as a Bronsted-Lowry base: $H_2O\ (\ell) + HCN\ (aq) \rightarrow H_3O^+\ (aq) + CN^-\ (aq)$

13-18. **Refer to Section 13-4.** . . . .

Aqueous solutions of strong soluble bases (1) have a bitter taste, (2) have a slippery feeling, (3) change the colors of many acid-base indicators, (4) react with protonic acids to form salts and water, and (5) conduct an electrical current since they contain ions.

Aqueous ammonia reacts with protonic acids to form salts, but exhibits all the other traits to a lesser degree since it is a weak base exhibiting limited ionization and providing a lower OH^- concentration.

13-20. **Refer to Section 13-5.** . . . .

Base strength refers to the relative tendency to produce OH^- ions in aqueous solution by (1) the dissociation of soluble metal hydroxides or (2) by ionization reactions with water. A more general definition, applying Bronsted-Lowry theory, is that base strength is a measure of the relative tendency to accept a proton from any acid.

13-22. **Refer to Section 13-5.** . . . .

(a) A binary protonic acid is a covalent compound consisting of a hydrogen atom and one other element. The compound can act as a proton donor.

(b) hydrofluoric acid - HF (aq) hydrobromic acid - HBr (aq)
hydrosulfuric acid - H_2S (aq) hydroselenic acid - H_2Se (aq)

13-24. **Refer to Section 13-5 and the Key Terms for Chapter 13.** . . . .

The leveling effect is the effect by which all bases stronger than the base formed by the autoionization of the solvent reacts with the solvent to produce that base. The same statement applies to acids. In other words, the strongest base (or acid) that can exist in a given solvent is the base (or acid) that is characteristic of that solvent. (a) In H_2O, the strongest acid that can persistantly exist is H_3O^+ and the strongest base that can persistantly exist is OH^-. (b) In liquid NH_3, the corresponding pair is NH_4^+ and NH_2^-. (c) In liquid HF, the corresponding pair is H_2F^+ and F^-.

13-26. **Refer to Section 13-6.** . . . .

Solubility refers to the extent to which a solid substance will dissolve in a solvent. Substances that are dissolved in water may or may not dissociate into ions. NaCl, a soluble salt, dissociates almost totally into its ions, whereas glucose, a soluble molecule, does not dissociate at all. Weak acids, such as HF, are soluble in water but dissociate only slightly into their ions.

13-28. **Refer to Section 13-6.** . . . .

Acid-base reactions are called neutralization reactions because the reaction of an acid with a base generally produces a salt with little, if any, acid-base character and water.

13-30. **Refer to Sections 9-6, 13-6 and Exercise 9-42 Solution.** . . . .

(a) molecular: HNO_3 (aq) + KOH (aq) → KNO_3 (aq) + H_2O (ℓ)
nitric acid; potassium hydroxide; potassium nitrate

total ionic: H^+ (aq) + NO_3^- (aq) + K^+ (aq) + OH^- (aq)
→ K^+ (aq) + NO_3^- (aq) + H_2O (ℓ)

net ionic: H^+ (aq) + OH^- (aq) → H_2O (ℓ)

(b) molecular: H_2SO_4 (aq) + 2NaOH (aq) → Na_2SO_4 (aq) + $2H_2O$ (ℓ)
sulfuric acid; sodium hydroxide; sodium sulfate

total ionic: $2H^+$ (aq) + SO_4^{2-} (aq) + $2Na^+$ (aq) + $2OH^-$ (aq)
→ $2Na^+$ (aq) + SO_4^{2-} (aq) + $2H_2O$ (ℓ)

net ionic: $2H^+$ (aq) + $2OH^-$ (aq) → $2H_2O$ (ℓ)

therefore, H^+ (aq) + OH^- (aq) → H_2O (ℓ)

(c) molecular: $2HCl (aq) + Ca(OH)_2 (aq) \rightarrow CaCl_2 (aq) + 2H_2O (\ell)$
hydrochloric acid — calcium hydroxide — calcium chloride

total ionic: $2H^+ (aq) + 2Cl^- (aq) + Ca^{2+} (aq) + 2OH^- (aq)$

$\rightarrow Ca^{2+} (aq) + 2Cl^- (aq) + 2H_2O (\ell)$

net ionic: $2H^+ (aq) + 2OH^- (aq) \rightarrow 2H_2O (\ell)$

therefore, $H^+ (aq) + OH^- (aq) \rightarrow H_2O (\ell)$

(d) molecular: $CH_3COOH (aq) + KOH (aq) \rightarrow KCH_3COO (aq) + H_2O (\ell)$
acetic acid — potassium hydroxide — potassium acetate

total ionic: $CH_3COOH (aq) + K^+ (aq) + OH^- (aq)$

$\rightarrow K^+ (aq) + CH_3COO^- (aq) + H_2O (\ell)$

net ionic: $CH_3COOH (aq) + OH^- (aq) \rightarrow CH_3COO^- (aq) + H_2O (\ell)$

13-32. **Refer to Sections 9-6, 13-6 and Exercise 9-42 Solution.** . . . .

(a) molecular: $2HClO_4 (aq) + Ba(OH)_2 (aq) \rightarrow Ba(ClO_4)_2 (aq) + 2H_2O (\ell)$
perchloric acid — barium hydroxide — barium perchlorate

total ionic: $2H^+ (aq) + 2ClO_4^- (aq) + Ba^{2+} (aq) + 2OH^- (aq)$

$\rightarrow Ba^{2+} (aq) + 2ClO_4^- (aq) + 2H_2O (\ell)$

net ionic: $2H^+ (aq) + 2OH^- (aq) \rightarrow 2H_2O (\ell)$

therefore, $H^+ (aq) + OH^- (aq) \rightarrow H_2O (\ell)$

(b) molecular: $HCl (aq) + NH_3 (aq) \rightarrow NH_4Cl (aq)$
hydrochloric acid — ammonia — ammonium chloride

total ionic: $H^+ (aq) + Cl^- (aq) + NH_3 (aq) \rightarrow NH_4^+ (aq) + Cl^- (aq)$

net ionic: $H^+ (aq) + NH_3 (aq) \rightarrow NH_4^+ (aq)$

(c) molecular: $HNO_3 (aq) + NH_3 (aq) \rightarrow NH_4NO_3 (aq)$
nitric acid — ammonia — ammonium nitrate

total ionic: $H^+ (aq) + NO_3^- (aq) + NH_3 (aq) \rightarrow NH_4^+ (aq) + NO_3^- (aq)$

net ionic: $H^+ (aq) + NH_3 (aq) \rightarrow NH_4^+ (aq)$

(d) molecular: $3H_2SO_4 (aq) + 2Fe(OH)_3 (s) \rightarrow Fe_2(SO_4)_3 (aq) + 6H_2O (\ell)$
sulfuric acid — iron(III) hydroxide — iron(III) sulfate

total ionic: $6H^+ (aq) + 3SO_4^{2-} (aq) + 2Fe(OH)_3 (s)$

$\rightarrow 2Fe^{3+} (aq) + 3SO_4^{2-} (aq) + 6H_2O (\ell)$

net ionic: $6H^+ (aq) + 2Fe(OH)_3 (s) \rightarrow 2Fe^{3+} (aq) + 6H_2O (\ell)$

therefore, $3H^+ (aq) + Fe(OH)_3 (s) \rightarrow Fe^{3+} (aq) + 3H_2O (\ell)$

(e) molecular: $2H_3PO_4 (aq)$ (phosphoric acid) + $3Ba(OH)_2 (aq)$ (barium hydroxide) $\rightarrow$ $Ba_3(PO_4)_2 (s)$ (barium phosphate) + $6H_2O (\ell)$

total ionic: $2H_3PO_4 (aq) + 3Ba^{2+} (aq) + 6OH^- (aq)$

$\rightarrow Ba_3(PO_4)_2 (s) + 6H_2O (\ell)$

net ionic: same as total ionic

13-34. **Refer to Sections 9-6, 13-6 and Exercise 9-46 Solution.** . . . .

(a) molecular: $2HNO_3 (aq) + Pb(OH)_2 (s) \rightarrow Pb(NO_3)_2 (aq) + 2H_2O (\ell)$

total ionic: $2H^+ (aq) + 2NO_3^- (aq) + Pb(OH)_2 (s)$

$\rightarrow Pb^{2+} (aq) + 2NO_3^- (aq) + 2H_2O (\ell)$

net ionic: $2H^+ (aq) + Pb(OH)_2 (s) \rightarrow Pb^{2+} (aq) + 2H_2O (\ell)$

(b) molecular: $3HCl (aq) + Fe(OH)_3 (s) \rightarrow FeCl_3 (aq) + 3H_2O (\ell)$

total ionic: $3H^+ (aq) + 3Cl^- (aq) + Fe(OH)_3 (s)$

$\rightarrow Fe^{3+} (aq) + 3Cl^- (aq) + 3H_2O (\ell)$

net ionic: $3H^+ (aq) + Fe(OH)_3 (s) \rightarrow Fe^{3+} (aq) + 3H_2O (\ell)$

(c) molecular: $H_2CO_3 (aq) + 2NH_3 (aq) \rightarrow (NH_4)_2CO_3 (aq)$

total ionic: $H_2CO_3 (aq) + 2NH_3 (aq) \rightarrow 2NH_4^+ (aq) + CO_3^{2-} (aq)$

net ionic: same as total ionic

(d) molecular: $2HClO_4 (aq) + Ca(OH)_2 (aq) \rightarrow Ca(ClO_4)_2 (aq) + 2H_2O (\ell)$

total ionic: $2H^+ (aq) + 2ClO_4 (aq) + Ca^{2+} (aq) + 2OH^- (aq)$

$\rightarrow Ca^{2+} (aq) + 2ClO_4^- (aq) + 2H_2O (\ell)$

net ionic: $2H^+ (aq) + 2OH^- (aq) \rightarrow 2H_2O (\ell)$

therefore, $H^+ (aq) + OH^- (aq) \rightarrow H_2O (\ell)$

(e) molecular: $3H_2SO_4 (aq) + 2Al(OH)_3 (s) \rightarrow Al_2(SO_4)_3 (aq) + 6H_2O (\ell)$

total ionic: $6H^+ (aq) + 3SO_4^{2-} (aq) + 2Al(OH)_3 (s)$

$\rightarrow 2Al^{3+} (aq) + 3SO_4^{2-} (aq) + 6H_2O (\ell)$

net ionic: $6H^+ (aq) + 2Al(OH)_3 (s) \rightarrow 2Al^{3+} (aq) + 6H_2O (\ell)$

therefore, $3H^+ (aq) + Al(OH)_3 (s) \rightarrow Al^{3+} (aq) + 3H_2O (\ell)$

13-36. **Refer to Section 13-7.** . . . .

An acidic salt is a salt that contains an ionizable hydrogen atom. It is the product which results from reacting less than a stoichiometric amount of base with a polyprotic acid:

H_2SO_3 (aq) + NaOH (aq) → $NaHSO_3$ (aq) + H_2O (ℓ)

H_2CO_3 (aq) + NaOH (aq) → $NaHCO_3$ (aq) + H_2O (ℓ)

H_3PO_4 (aq) + NaOH (aq) → NaH_2PO_4 (aq) + H_2O (ℓ)

H_3PO_4 (aq) + 2NaOH (aq) → Na_2HPO_4 (aq) + $2H_2O$ (ℓ)

13-38. **Refer to Section 13-7.** . . . .

(a) HNO_3 (aq) + NH_3 (aq) → NH_4NO_3 (aq)

(b) H_3PO_4 (aq) + NH_3 (aq) → $NH_4H_2PO_4$ (aq)

(c) H_3PO_4 (aq) + $2NH_3$ (aq) → $(NH_4)_2HPO_4$ (aq)

(d) H_3PO_4 (aq) + $3NH_3$ (aq) → $(NH_4)_3PO_4$ (aq)

13-40. **Refer to Section 13-7.** . . . .

(a) CH_3COOH (aq) + $NaHCO_3$ (aq) → $NaCH_3COO$ (aq) + CO_2 (g) + H_2O (ℓ)

The "fizz" is caused by the gaseous product, carbon dioxide.

(b) $CH_3CH(OH)COOH$ (aq) + $NaHCO_3$ (aq) → $NaCH_3CH(OH)COO$ (aq) + CO_2 (g) + H_2O (ℓ)

"Quick" bread "rises" during baking due to the reaction between baking soda and lactic acid found in the added milk. The resulting carbon dioxide gas bubbles are caught in the bread dough, giving the bread more volume. Yeast breads "rise" due to carbon dioxide bubbles released in the fermentation of sugar by yeast.

13-42. **Refer to Section 13-7.** . . . .

A basic salt is a salt containing an ionizable OH group and can therefore neutralize acids.

(a),(b) HCl (aq) + $Ba(OH)_2$ (aq) → Ba(OH)Cl (aq) + H_2O (ℓ)
1 mol   1 mol

HCl (aq) + $Al(OH)_3$ (aq) → $Al(OH)_2Cl$ (aq) + H_2O (ℓ)
1 mol   1 mol

2HCl (aq) + $Al(OH)_3$ (aq) → $Al(OH)Cl_2$ (aq) + $2H_2O$ (ℓ)
2 mol   1 mol

13-44. **Refer to Section 13-9.** . . . .

Ternary acids, including nitric and sulfuric acids, can be described as hydroxides of nonmetals since they contain 1 or more -O-H groups attached to the central nonmetal atom. For example,

$HNO_3 \equiv NO_2(OH)$ $H_2SO_4 \equiv SO_2(OH)_2$

13-46. **Refer to Section 13-9.** . . . .

(a) Acid strengths of most ternary acids containing different elements in the same oxidation state from the same group in the periodic table increase with increasing electronegativity of the central element.

(b) In order of increasing acid strength:

(1) $H_3PO_4 < HNO_3$ (2) $H_3AsO_4 < H_3PO_4$

(3) $H_2SeO_4 < H_2SO_4$ (4) $HBrO_3 < HClO_3$

13-48. **Refer to Section 13-9 and Table 13-3.** . . . .

(a) <u>for $Cr(OH)_3$</u>

molecular: $Cr(OH)_3\ (s) + 3HNO_3\ (aq) \rightarrow Cr(NO_3)_3\ (aq) + 3H_2O\ (\ell)$

total ionic: $Cr(OH)_3\ (s) + 3H^+\ (aq) + 3NO_3^-\ (aq)$

$\rightarrow Cr^{3+}\ (aq) + 3NO_3^-\ (aq) + 3H_2O\ (\ell)$

net ionic: $Cr(OH)_3\ (s) + 3H^+\ (aq) \rightarrow Cr^{3+}\ (aq) + 3H_2O\ (\ell)$

<u>for $Pb(OH)_2$</u>

molecular: $Pb(OH)_2\ (s) + 2HNO_3\ (aq) \rightarrow Pb(NO_3)_2\ (aq) + 2H_2O\ (\ell)$

total ionic: $Pb(OH)_2\ (s) + 2H^+\ (aq) + 2NO_3^-\ (aq)$

$\rightarrow Pb^{2+}\ (aq) + 2NO_3^-\ (aq) + 2H_2O\ (\ell)$

net ionic: $Pb(OH)_2\ (s) + 2H^+\ (aq) \rightarrow Pb^{2+}\ (aq) + 2H_2O\ (\ell)$

(b) <u>for $Cr(OH)_3$</u>

molecular: $Cr(OH)_3\ (s) + KOH\ (aq) \rightarrow K[Cr(OH)_4]\ (aq)$

total ionic: $Cr(OH)_3\ (s) + K^+\ (aq) + OH^-\ (aq) \rightarrow K^+\ (aq) + [Cr(OH)_4]^-\ (aq)$

net ionic: $Cr(OH)_3\ (s) + OH^-\ (aq) \rightarrow [Cr(OH)_4]^-\ (aq)$

<u>for $Pb(OH)_2$</u>

molecular: $Pb(OH)_2\ (s) + 2KOH\ (aq) \rightarrow K_2[Pb(OH)_4]\ (aq)$

total ionic: $Pb(OH)_2\ (s) + 2K^+\ (aq) + 2OH^-\ (aq)$

$\rightarrow 2K^+\ (aq) + [Pb(OH)_4]^{2-}\ (aq)$

net ionic: $Pb(OH)_2\ (s) + 2OH^-\ (aq) \rightarrow [Pb(OH)_4]^{2-}\ (aq)$

13-50. **Refer to Sections 9-8 and 13-10.** . . . .

(a) Hydrogen sulfide, H_2S (g), can be prepared by combining elemental sulfur with hydrogen gas.

$$S_8\ (s) + 8H_2\ (g) \rightarrow 8H_2S\ (g)$$

(b) Hydrogen chloride, HCl (g), can be prepared in small quantities by dropping concentrated nonvolatile acids such as sulfuric acid, onto an appropriate salt such as NaCl (s).

$$H_2SO_4\ (\ell) + NaCl\ (s) \rightarrow HCl\ (g) + NaHSO_4\ (s)$$

(c) Dinitrogen pentoxide, N_2O_5 (g), is the acid anhydride of nitric acid, HNO_3 (aq). When N_2O_5 (g) is dissolved in water, nitric acid is the product.

$$N_2O_5\ (g) + H_2O\ (\ell) \rightarrow 2HNO_3\ (aq)$$

13-52. **Refer to Section 13-11.** . . . .

Strong soluble bases can be prepared by dissolving the appropriate metal oxide, called a basic anhydride, in water.

(a) $K_2O\ (s) + H_2O\ (\ell) \rightarrow 2KOH\ (aq)$

(b) $BaO\ (s) + H_2O\ (\ell) \rightarrow Ba(OH)_2\ (aq)$

(c) $Na_2O\ (s) + H_2O\ (\ell) \rightarrow 2NaOH\ (aq)$

$2Na_2O_2\ (s) + 2H_2O\ (\ell) \rightarrow 4NaOH\ (aq) + O_2\ (g)$

$2Na\ (s) + 2H_2O\ (\ell) \rightarrow 2NaOH\ (aq) + H_2\ (g)$

$2NaCl\ (aq) + 2H_2O\ (\ell) \xrightarrow{\text{electrolysis}} 2NaOH\ (aq) + Cl_2\ (g) + H_2\ (g)$

13-54. **Refer to Section 13-3.** . . . .

The advantages of Bronsted-Lowry theory over Arrhenius theory are twofold. (1) Bronsted-Lowry theory extends the definitions of acids and bases so that they can be used in nonaqueous systems and (2) this theory also recognizes that other constituents besides the hydroxide ion can have basic properties.

A limitation of the Bronsted-Lowry theory is that it does not permit the acid-base classification to include those systems which have neither H^+ nor OH^- ions present. This limitation is met by the Lewis theory.

13-56. **Refer to Section 13-12.** . . . .

```
      ..       ..       ..  +       ..  -
(a) H:O:  +  H:O:  →  H:O:H    +  H:O:
      ..       ..       ..          ..
      H        H        H
    base     acid

      ..       ..       ..  -       ..  +
(b) H:Cl:  +  H:O:  →  :Cl:    +  H:O:H
      ..       ..       ..          ..
               H                    H
    acid     base

                         H  +
      ..       ..        ..         ..  -
(c) H:N:H  +  H:O:  →  H:N:H    +  H:O:
      ..       ..        ..         ..
      H        H         H
    base     acid

                          H
      ..       ..         ..  +     ..  -
(d) H:N:H  +  H:F:  →  [H:N:H  ][:F:  ]
      ..       ..         ..        ..
      H                   H
    base     acid
```

13-58. **Refer to Section 3-5.** . . . .

Molarity is defined as the number of moles of solute per 1 liter of solution and has units of mol/L. If we multiply molarity by unity = $10^{-3}/10^{-3}$,

$$M\left(\frac{\text{mol}}{\text{L}}\right) = \frac{\text{mol solute}}{\text{L soln}} \times \frac{10^{-3}}{10^{-3}} = \frac{\text{mmol solute}}{\text{mL soln}}$$

we find that M (mol/L) ≡ M (mmol/mL).

13-60. **Refer to Section 3-5.** . . . .

(a) $? \ M \ K_2SO_4 = \dfrac{(17.5 \text{ g } K_2SO_4)/(174 \text{ g/mol})}{0.300 \text{ L}} = \mathbf{0.335} \ M$

(b) $? \ M \ Al_2(SO_4)_3 = \dfrac{(143 \text{ g } Al_2(SO_4)_3)/(342 \text{ g/mol})}{3.00 \text{ L}} = \mathbf{0.139} \ M$

(c) $? \ M \ Al_2(SO_4)_3{\cdot}18H_2O = \dfrac{(143 \text{ g } Al_2(SO_4)_3{\cdot}18H_2O)/(666 \text{ g/mol})}{3.00 \text{ L}} = \mathbf{0.0716} \ M$

13-62. **Refer to Section 3-5.** . . . .

Plan: (1) Calculate the moles of $(NH_4)_2SO_4$ present in the solution.
(2) Calculate the mass required.

(1) ? mol $(NH_4)_2SO_4$ = 0.288 M $(NH_4)_2SO_4$ × 3.75 L = 1.08 mol

(2) ? g $(NH_4)_2SO_4$ = 1.08 mol $(NH_4)_2SO_4$ × 132 g/mol = **143 g**

13-64. **Refer to Section 3-5 and Exercise 3-88 Solution.** . . . .

Assume a 1 liter solution of 39.77 % H_2SO_4 with a density of 1.305 g/mL.

$? \text{ g } H_2SO_4 \text{ soln in 1 L soln} = \dfrac{1.305 \text{ g soln}}{1 \text{ mL soln}} \times 1000 \text{ mL soln} = 1305 \text{ g soln}$

$? \text{ g } H_2SO_4 \text{ in 1 L soln} = 1305 \text{ g soln} \times \dfrac{39.77 \text{ g } H_2SO_4}{100 \text{ g soln}} = 519.0 \text{ g } H_2SO_4$

$? \text{ mol } H_2SO_4 \text{ in 1 L soln} = \dfrac{519.0 \text{ g } H_2SO_4}{98.08 \text{ g/mol}} = 5.292 \text{ mol } H_2SO_4$

Therefore,

? M H_2SO_4 = **5.292** M

13-66. **Refer to Section 13-13 and Example 13-3.** . . . .

This is a possible limiting reagent problem.

Plan: (1) Calculate the number of moles of HCl and NaOH.
(2) Determine the limiting reagent, if there is one.
(3) Calculate the moles of NaCl formed.
(4) Determine the molarity of NaCl in the solution.

Balanced equation: HCl (aq) + NaOH (aq) → NaCl (aq) + H_2O (ℓ)

(1) ? mol HCl = 4.32 *M* HCl × 0.200 L = 0.864 mol HCl
? mol NaOH = 2.16 *M* NaOH × 0.400 L = 0.864 mol NaOH

(2) This is not a problem with a single limiting reagent since we have stoichiometric amounts of both HCl and NaOH. Our final solution is a salt solution with no excess acid or base.

(3) ? mol NaCl = mol HCl = mol NaOH = 0.864 mol NaCl

(4) $$? \; M \text{ NaCl} = \frac{\text{mol NaCl}}{\text{total volume}} = \frac{0.864 \text{ mol}}{(0.200 \text{ L} + 0.400 \text{ L})} = 1.44 \; M \text{ NaCl}$$

13-68. **Refer to Section 13-13 and Exercise 13-66 Solution.** . . . .

Balanced equation: H_3PO_4 (aq) + 3NaOH (aq) → Na_3PO_4 (aq) + $3H_2O$ (ℓ)

Plan: (1) Calculate the moles of H_3PO_4 and NaOH.
(2) Determine the limiting reagent, if there is one.
(3) Calculate the moles of Na_3PO_4 formed.
(4) Determine the molarity of the salt in the solution.

(1) ? mol H_3PO_4 = 3.68 *M* H_3PO_4 × 0.225 L = 0.828 mol H_3PO_4

? mol NaOH = 3.68 *M* NaOH × 0.675 L = 2.484 mol NaOH

(2) In the balanced equation, H_3PO_4 reacts with NaOH in a 1:3 mole ratio.

mol H_3PO_4: mol NaOH = 0.828 mol:2.484 mol = 1:3

Since we have stoichiometric amounts of the reactants, this is not a problem with a single limiting reagent. The final salt solution contains no excess acid or base.

(3) ? mol Na_3PO_4 = mol H_3PO_4 = 1/3 × mol NaOH = 0.828 mol Na_3PO_4

(4) $$? \; M \; Na_3PO_4 = \frac{0.828 \text{ mol } Na_3PO_4}{(0.225 \text{ L} + 0.675 \text{ L})} = \mathbf{0.920 \; M \; Na_3PO_4}$$

13-70. **Refer to Section 13-13.** . . . .

Balanced equation: 3NaOH (aq) + H_3PO_4 (aq) → Na_3PO_4 (aq) + $3H_2O$ (ℓ)

Plan: (1) Calculate the moles of NaOH and H_3PO_4 required to form 1 mole of Na_3PO_4.
(2) Find the volumes of each solution.

(1) ? mol NaOH = 3 × mol Na_3PO_4 = 3 × 1.00 mol = 3.00 mol NaOH

? mol H_3PO_4 = mol Na_3PO_4 = 1.00 mol H_3PO_4

(2) $$? \text{ L NaOH soln} = \frac{3.00 \text{ mol NaOH}}{1.00 \; M \text{ NaOH}} = \mathbf{3.00 \text{ L NaOH soln}}$$

$$? \text{ L } H_3PO_4 \text{ soln} = \frac{1.00 \text{ mol } H_3PO_4}{1.50 \; M \; H_3PO_4} = \mathbf{0.667 \text{ L } H_3PO_4 \text{ soln}}$$

13-72. **Refer to Section 13-13.** . . . .

Balanced equation: $2NH_3 + H_2SO_4 \rightarrow (NH_4)_2SO_4$

Plan: $\text{lb }(NH_4)_2SO_4 \xrightarrow{(1)} \text{g }(NH_4)_2SO_4 \xrightarrow{(2)} \text{mol }(NH_4)_2SO_4 \xrightarrow{(3)} \text{mol }H_2SO_4$
$\xrightarrow{(4)} \text{g }H_2SO_4 \xrightarrow{(5)} \text{g }H_2SO_4\text{ soln} \xrightarrow{(6)} \text{mL }H_2SO_4\text{ soln}$

$$? \text{ mL } H_2SO_4 \text{ soln} = 50.0 \text{ lb } (NH_4)_2SO_4 \times \underbrace{\frac{454 \text{ g } (NH_4)_2SO_4}{1 \text{ lb } (NH_4)_2SO_4}}_{\textbf{Step 1}} \times \underbrace{\frac{1 \text{ mol } (NH_4)_2SO_4}{132 \text{ g } (NH_4)_2SO_4}}_{\textbf{Step 2}}$$

$$\times \underbrace{\frac{1 \text{ mol } H_2SO_4}{1 \text{ mol } (NH_4)_2SO_4}}_{\textbf{Step 3}} \times \underbrace{\frac{98.1 \text{ g } H_2SO_4}{1 \text{ mol } H_2SO_4}}_{\textbf{Step 4}} \times \underbrace{\frac{100 \text{ g } H_2SO_4 \text{ soln}}{98.20 \text{ g } H_2SO_4}}_{\textbf{Step 5}}$$

$$\times \underbrace{\frac{1 \text{ mL } H_2SO_4 \text{ soln}}{1.841 \text{ g } H_2SO_4 \text{ soln}}}_{\textbf{Step 6}} = \mathbf{9.33 \times 10^3} \textbf{ mL } \mathbf{H_2SO_4} \textbf{ soln}$$

13-74. **Refer to Section 3-5 and Exercise 3-88 Solution.** . . . .

Assume a 1 liter solution of 5.11% CH_3COOH.

$$? \text{ g } CH_3COOH \text{ soln in 1 L soln} = \frac{1.007 \text{ g soln}}{1 \text{ mL soln}} \times 1000 \text{ mL soln} = 1007 \text{ g soln}$$

$$? \text{ g } CH_3COOH \text{ in 1 L soln} = 1007 \text{ g soln} \times \frac{5.11 \text{ g } CH_3COOH}{100 \text{ g soln}} = 51.5 \text{ g } CH_3COOH$$

$$? \text{ mol } CH_3COOH \text{ in 1 L soln} = \frac{51.5 \text{ g } CH_3COOH}{60.1 \text{ g/mol}} = 0.857 \text{ mol } CH_3COOH$$

Therefore,

$? \ M \ CH_3COOH = \mathbf{0.857} \ M$

13-76. **Refer to Section 13-13, Example 13-5, Exercises 13-73 and 13-74 Solutions.** . .

In Exercises 13-73 and 13-74, the solution concentrations were calculated to be 2.90 M NH_3 and 0.858 M CH_3COOH.

Balanced equation: $CH_3COOH + NH_3 \rightarrow NH_4CH_3COO$

At neutralization, the number of moles of CH_3COOH is exactly equal to the number of moles of NH_3.

$$\text{mol } CH_3COOH = \text{mol } NH_3$$
$$0.857 \ M \times 1.00 \text{ L } CH_3COOH \text{ soln} = 2.90 \ M \times ? \text{ L } NH_3 \text{ soln}$$
$$? \text{ L } NH_3 \text{ soln} = \mathbf{0.296 \text{ L soln}}$$

13-78. **Refer to Sections 13-14 and 13-15.** . . . .

A standard solution of NaOH cannot be prepared directly because the solid is hydroscopic and absorbs moisture and CO_2 from the air.

Step 1: Weigh out an amount of solid NaOH and dissolve it in water to obtain a solution with an approximate concentration.

Step 2: Weigh out an appropriate amount of an acidic material, suitable for use as a primary standard, such as potassium hydrogen phthalate (KHP).

Step 3: Titrate the KHP sample with the NaOH solution and calculate the molarity of the NaOH solution.

The NaOH solution thusly prepared is called a secondary standard because its concentration is determined by titration against a primary standard.

13-80. **Refer to Section 13-14.** • • • • • • • • • • • • • • • • • • •

(a) An ideal primary standard:
(1) does not react with or absorb water vapor, oxygen or carbon dioxide,
(2) reacts according to a single known reaction,
(3) is available in high purity,
(4) has a high formula weight,
(5) is soluble in the solvent of interest, and
(6) is nontoxic.

(b) The significance of each factor is given below.
(1) The compound must be weighed accurately and must not undergo composition change due to reaction with atmospheric components.
(2) The reaction must be one of known stoichiometry with no side reactions.
(3) Solutions of precisely known concentration must be prepared by directly weighing the primary standard.
(4) The high formula weight is necessary to minimize the effect of weighing errors.
(5),(6) The significance is self-explanatory.

13-82. **Refer to Section 13-15.** • • • • • • • • • • • • • • • • • • •

(a) Potassium hydrogen phthalate (KHP) is the acidic salt, $KC_6H_4(COO)(COOH)$.

(b) KHP is used as a primary standard for the standardization of strong bases.

13-84. **Refer to Section 13-15.** • • • • • • • • • • • • • • • • • • •

Balanced equation: $2NaOH + H_2SO_4 \rightarrow Na_2SO_4 + 2H_2O$

Plan: M, mL H_2SO_4 soln $\xrightarrow{(1)}$ mol H_2SO_4 $\xrightarrow{(2)}$ mol NaOH $\xrightarrow{(3)}$ L NaOH soln

(1) ? mol H_2SO_4 = 0.400 M H_2SO_4 × 0.0253 L = 0.0101 mol H_2SO_4

(2) ? mol NaOH = 2 × mol H_2SO_4 = 2 × 0.0101 mol = 0.0202 mol NaOH

(3) ? L NaOH soln = $\dfrac{0.0202 \text{ mol NaOH}}{0.112\ M \text{ NaOH}}$ = **0.180 L or 180 mL NaOH soln**

13-86. **Refer to Section 13-5 and Example 13-9.** • • • • • • • • • • • • • • •

Balanced equation: $(COOH)_2 + 2NaOH \rightarrow Na_2(COO)_2 + 2H_2O$

Plan: M, mL NaOH $\xrightarrow{(1)}$ mol NaOH $\xrightarrow{(2)}$ mol $(COOH)_2$ $\xrightarrow{(3)}$ mol $(COOH)_2 \cdot 2H_2O$

$\xrightarrow{(4)}$ g $(COOH)_2 \cdot 2H_2O$ $\xrightarrow{(5)}$ % purity

(1) ? mol NaOH = 0.198 M NaOH × 0.04032 L = 0.00798 mol

(2) ? mol $(COOH)_2$ = 1/2 × mol NaOH = 1/2 × 0.00798 mol = 0.00399 mol

(3) ? mol $(COOH)_2 \cdot 2H_2O$ = mol $(COOH)_2$ = 0.00399 mol

(4) ? g $(COOH)_2 \cdot 2H_2O$ = 0.00399 mol $(COOH)_2 \cdot 2H_2O \times 126$ g/mol = 0.503 g

(5) $? \ \% \ (COOH)_2 \cdot 2H_2O = \dfrac{g \ (COOH)_2 \cdot 2H_2O}{g \ sample} \times 100 = \dfrac{0.503 \ g}{2.00 \ g} \times 100 = \mathbf{25.2 \ \%}$

13-88. **Refer to Section 13-15 and Example 13-8.** . . . .

Balanced equation: $NaOH + KHP \rightarrow NaKP + H_2O$

Plan: g KHP $\xrightarrow{(1)}$ mmol KHP $\xrightarrow{(2)}$ mmol NaOH $\xrightarrow{(3)}$ *M* NaOH

(1) $? \ mmol \ KHP = \dfrac{0.8407 \ g \ KHP}{204.2 \ g/mol} \times \dfrac{1000 \ mmol}{1 \ mol} = 4.117 \ mmol \ KHP$

(2) ? mmol NaOH = mmol KHP = 4.117 mmol NaOH

(3) $? \ M \ NaOH = \dfrac{4.117 \ mmol \ NaOH}{(46.16 \ mL - 0.23 \ mL)} = \mathbf{0.08964} \ M \ \mathbf{NaOH}$

13-90. **Refer to Section 13-15 and Exercise 13-88 Solution.** . . . .

Balanced equation: $H_2SO_4 + 2NaOH \rightarrow Na_2SO_4 + 2H_2O$

The NaOH solution concentration from Exercise 13-88 is 0.08964 *M*.

Plan: *M*, mL NaOH $\xrightarrow{(1)}$ mmol NaOH $\xrightarrow{(2)}$ mmol H_2SO_4 $\xrightarrow{(3)}$ *M* H_2SO_4

(1) ? mmol NaOH = 0.08964 *M* NaOH × 27.86 mL = 2.497 mmol NaOH

(2) ? mmol H_2SO_4 = 1/2 × mmol NaOH = 1/2 × 2.497 mmol = 1.249 mmol H_2SO_4

(3) $? \ M \ H_2SO_4 = \dfrac{1.249 \ mmol \ H_2SO_4}{34.53 \ mL} = \mathbf{0.03617} \ M \ \mathbf{H_2SO_4}$

13-92. **Refer to Section 13-15.** . . . .

Balanced equation: $2HCl + CaCO_3 \rightarrow CaCl_2 + CO_2 + H_2O$

Plan: *M*, mL HCl $\xrightarrow{(1)}$ mol HCl $\xrightarrow{(2)}$ mol $CaCO_3$ $\xrightarrow{(3)}$ g $CaCO_3$

(1) ? mol HCl = 0.0932 *M* HCl × 0.0226 L = 0.00211 mol HCl

(2) ? mol $CaCO_3$ = 1/2 × mol HCl = 0.00106 mol $CaCO_3$

(3) ? g $CaCO_3$ = 0.00106 mol $CaCO_3$ × 100. g/mol = **0.106 g $CaCO_3$**

13-94. **Refer to Section 13-15.** . . . .

Balanced equation: $NaAl(OH)_2CO_3 + 4HCl \rightarrow NaCl + AlCl_3 + CO_2 + 3H_2O$

Plan: (1) Calculate the mmoles of HCl in your stomach acid.
(2) Calculate the mmoles of HCl that can be neutralized by the antacid tablet.

(1) ? mmol HCl in stomach = 0.10 *M* HCl × 800 mL = 80 mmol HCl

(2) $? \text{ mmol NaAl(OH)}_2\text{CO}_3 = \dfrac{334 \text{ mg NaAl(OH)}_2\text{CO}_3}{144 \text{ mg/mmol}} = 2.32 \text{ mmol NaAl(OH)}_2\text{CO}_3$

$? \text{ mmol neutralized HCl} = 4 \times \text{mmol NaAl(OH)}_2\text{CO}_3 = 4 \times 2.32 \text{ mmol} = \mathbf{9.28 \text{ mmol}}$

The number of mmoles of HCl in your stomach is roughly nine times larger than the number of mmoles of HCl that can be neutralized by a single antacid tablet. However, about 2 tablets are sufficient to neutralize the excess HCl in the stomach by reducing its concentration down to the normal 8.0×10^{-2} *M* level.

13-96. **Refer to Section 13-15 and Example 13-12.** . . . .

For complete neutralization, 1 mole of H_3PO_4 yields 3 moles of H^+ ions. Therefore, the equivalent weight of H_3PO_4 = molecular weight/(3 eq/mol)
= (98.0 g/mol)/(3 eq/mol)
= 32.7 g/eq

$$? \ N \ H_3PO_4 = \frac{\text{eq } H_3PO_4}{\text{L soln}} = \frac{(7.08 \text{ g } H_3PO_4)/(32.7 \text{ g/eq})}{0.185 \text{ L}} = \mathbf{1.17 \ \mathit{N} \ H_3PO_4}$$

13-98. **Refer to Section 13-15 and Example 13-12.** . . . .

$$? \ M \ H_3AsO_4 = \frac{(19.6 \text{ g } H_3AsO_4)/(142 \text{ g/mol})}{0.600 \text{ L}} = \mathbf{0.230 \ \mathit{M} \ H_3AsO_4}$$

For complete neutralization, 1 mole of H_3AsO_4 yields 3 moles of H^+ ions. Therefore,

$? \ N \ H_3AsO_4 = 3 \times 0.230 \ M \ H_3AsO_4 = \mathbf{0.690 \ \mathit{N} \ H_3AsO_4}$

13-100. **Refer to Section 13-15.** . . . .

At neutralization, $N_{acid} \times V_{acid} = N_{base} \times V_{base}$

Substituting and solving for N_{base},

$$? \ N \ Ba(OH)_2 = \frac{0.206 \ N \ HNO_3 \times 25.0 \text{ mL } HNO_3}{39.3 \text{ mL } Ba(OH)_2} = 0.131 \ N \ Ba(OH)_2$$

For complete neutralization, 1 mole of $Ba(OH)_2$ will yield 2 moles of OH^- ions. Therefore,

$? \ M \ Ba(OH)_2 = 1/2 \times 0.131 \ N \ Ba(OH)_2 = \mathbf{0.0655 \ \mathit{M} \ Ba(OH)_2}$

13-102. **Refer to Section 13-15.** . . . .

Plan: (1) Calculate the normality of the NaOH solution.
(2) Calculate the normality of the HCl solution.

(1) At neutralization, $N_{acid} \times V_{acid} = N_{base} \times V_{base}$
Therefore,

$$? \ N \ NaOH = \frac{0.100 \ N \ H_2SO_4 \times 41.8 \text{ mL } H_2SO_4}{44.4 \text{ mL NaOH}} = 0.0941 \ N \ NaOH$$

(2) Using the same general formula,

$$? \ N \ HCl = \frac{0.0941 \ N \ NaOH \times 47.2 \text{ mL NaOH}}{36.0 \text{ mL HCl}} = \mathbf{0.123 \ \mathit{N} \ HCl}$$

13-104. **Refer to Section 13-15.** . . . .

Balanced equation: $2HCl + Na_2CO_3 \rightarrow 2NaCl + CO_2 + H_2O$

Plan: $g\ Na_2CO_3 \xrightarrow{(1)} eq\ Na_2CO_3 \xrightarrow{(2)} eq\ HCl \xrightarrow{(3)} N\ HCl \xrightarrow{(4)} M\ HCl$

(1) ? eq wt Na_2CO_3 = MW (g/mol)/(2 eq/mol) = (106.0 g/mol)/(2 eq/mol)
= 53.0 g/eq

(1) $? \text{ eq } Na_2CO_3 = \dfrac{0.318 \text{ g } Na_2CO_3}{53.0 \text{ g/eq}} = 6.00 \times 10^{-3} \text{ eq } Na_2CO_3$

(2) $? \text{ eq HCl} = \text{eq } Na_2CO_3 = 6.00 \times 10^{-3} \text{ eq HCl}$

(3) $? \ N \text{ HCl} = \dfrac{6.00 \times 10^{-3} \text{ eq HCl}}{0.0331 \text{ L}} = \mathbf{0.181}\ N\ \textbf{HCl}$

(4) ? M HCl = N HCl = **0.181 *M* HCl**

14 Oxidation-Reduction Reactions

14-2. **Refer to Section 6-9 and Exercise 6-72 Solution.** • • • • • • • • • • •

(a) compound	Oxidation No. for N	(b) compound	Oxidation No. for C	(c) compound	Oxidation No. for S
NO	+2	CO	+2	S_8	0
N_2O_3	+3	CO_2	+4	H_2S	-2
N_2O_4	+4	CH_2O	0	SO_2	+4
NH_3	-3	CH_4O	-2	SO_3	+6
N_2H_4	-2	C_2H_6O	-2	Na_2SO_3	+4
NH_2OH	-1	$(COOH)_2$	+3	H_2SO_4	+6
HNO_3	+5	Na_2CO_3	+4	K_2SO_4	+6

14-4. **Refer to Section 6-9 and Exercise 6-72 Solution.** • • • • • • • • • • •

(a) ion	Oxidation No. for S	(b) ion	Oxidation No. for Cr	(c) ion	Oxidation No. for B
S^{2-}	-2	CrO_2^-	+3	BO_2^-	+3
SO_3^{2-}	+4	$Cr(OH)_4^-$	+3	BO_3^{3-}	+3
SO_4^{2-}	+6	CrO_4^{2-}	+6	$B_4O_7^{2-}$	+3
$S_2O_3^{2-}$	+2	$Cr_2O_7^{2-}$	+6		
$S_4O_6^{2-}$	+2.5				

14-6. **Refer to Sections 9-7, 14-1 and Exercise 9-58 Solution.** • • • • • • • • •

Reaction (b) is the only oxidation-reduction reaction. In reactions (a), (c) and (d), there are no elements that are changing oxidation number.

Reducing Agent $4NH_3(g)$ + **Oxidizing Agent** $3O_2(g)$ → $2N_2(g)$ + $6H_2O(g)$

Oxidation No. (-3) (0) (0) (-2)

+ 3

- 2

14-8. **Refer to Sections 9-7, 14-1 and Example 14-1.** • • • • • • • • • • • •

(a) molecular: $Zn(s) + 2HClO_4(aq) \rightarrow Zn(ClO_4)_2(aq) + H_2(g)$

net ionic: $Zn(s) + 2H^+(aq) \rightarrow Zn^{2+}(aq) + H_2(g)$

This is a redox equation: $\overset{0}{Zn} \rightarrow \overset{+2}{Zn}$ and $\overset{+1}{H} \rightarrow \overset{0}{H}$

oxidizing agent: $HClO_4$ (H^+) substance reduced: $HClO_4$ (H^+)
reducing agent: Zn substance oxidized: Zn

(b) molecular: $2K(s) + 2H_2O(\ell) \rightarrow 2KOH(aq) + H_2(g)$

net ionic: $2K(s) + 2H_2O(\ell) \rightarrow 2K^+(aq) + 2OH^-(aq) + H_2(g)$

This is a redox equation: $\overset{0}{K} \rightarrow \overset{+1}{K}$ and $\overset{+1}{H} \rightarrow \overset{0}{H}$

oxidizing agent: H_2O substance reduced: H_2O
reducing agent: K substance oxidized: K

(c) molecular: $2NaClO_3(s) \xrightarrow{\Delta} 2NaCl(s) + 3O_2(g)$

This is a redox equation: $\overset{+5}{Cl} \rightarrow \overset{-1}{Cl}$ and $\overset{-2}{O} \rightarrow \overset{0}{O}$

oxidizing and reducing agent: $NaClO_3$
substance reduced and oxidized: $NaClO_3$

14-10. **Refer to Section 14-2 and Example 14-2.** . . . .

(a) $Fe(s) + 2HCl(aq) \rightarrow FeCl_2(aq) + H_2(g)$

(b) $2Cr(s) + 3H_2SO_4(aq) \rightarrow Cr_2(SO_4)_3(aq) + 3H_2(g)$

(c) $Sn(s) + 4HNO_3(aq) \rightarrow SnO_2(s) + 4NO_2(g) + 2H_2O(\ell)$

14-12. **Refer to Section 14-2 and Example 14-3.** . . . .

Remember: a balanced equation must have both mass balance and charge balance. Using the Change-in-Oxidation Number Method:

(a) $2MnO_4^-(aq) + 16H^+(aq) + 10Br^-(aq) \rightarrow 2Mn^{2+}(aq) + 5Br_2(\ell) + 8H_2O(\ell)$

(+7) (-1) (+2) (0) Oxidation No.

- 5 (Mn: +7 → +2)

+ 1 (Br: -1 → 0)

Oxidation Numbers	Change/Atom	Equalizing Charges Gives
Mn = +7 → Mn = +2	-5	1(-5) = -5
Br = -1 → Br = 0	+1	5(+1) = +5

Each change is multiplied by 2 since there are 2 Br atoms in Br_2.

(b) $Cr_2O_7^{2-}(aq) + 14H^+(aq) + 6I^-(aq) \rightarrow 2Cr^{3+}(aq) + 3I_2(s) + 7H_2O(\ell)$

Oxidation Numbers	Change/Atom	Equalizing Charges Gives
Cr = +6 → Cr = +3	-3	1(-3) = -3
I = -1 → I = 0	+1	3(+1) = +3

Each change is multiplied by 2 since there are 2 Cr atoms in $Cr_2O_7^{2-}$.

(c) $2MnO_4^-(aq) + 5SO_3^{2-}(aq) + 6H^+(aq) \rightarrow 2Mn^{2+}(aq) + 5SO_4^{2-}(aq) + 3H_2O(\ell)$

Oxidation Numbers	Change/Atom	Equalizing Charges Gives
Mn = +7 → Mn = +2	-5	2(-5) = -10
S = +4 → S = +6	+2	5(+2) = +10

(d) $Cr_2O_7^{2-}(aq) + 6Fe^{2+}(aq) + 14H^+(aq) \rightarrow 2Cr^{3+}(aq) + 6Fe^{3+}(aq) + 7H_2O(\ell)$

Oxidation Numbers	Change/Atom	Equalizing Charges Gives
Cr = +6 → Cr = +3	-3	1(-3) = -3
Fe = +2 → Fe = +3	+1	3(+1) = +3

Each change is multiplied by 2 since there are 2 Cr atoms in $Cr_2O_7^{2-}$.

14-14. **Refer to Section 14-2.** . . . .

Remember: a balanced equation must have charge balance as well as mass balance.

(a) $8Al(s) + 3NO_3^-(aq) + 5OH^-(aq) + 18H_2O(\ell) \rightarrow 8Al(OH)_4^-(aq) + 3NH_3(g)$

(0) (+5) (+3) (-3) Ox. No.

+ 3

- 8

Oxidation Numbers	Change/Atom	Equalizing Charges Gives
Al = 0 → Al = +3	+3	8(+3) = +24
N = +5 → N = -3	-8	3(-8) = -24

(b) $2NO_2(g) + 2OH^-(aq) \rightarrow NO_3^-(aq) + NO_2^-(aq) + H_2O(\ell)$

Oxidation Numbers	Change/Atom	Equalizing Charges Gives
N = +4 → N = +5	+1	-
N = +4 → N = +3	-1	-

(c) $2MnO_4^-(aq) + H_2O(\ell) + 3NO_2^-(aq) \rightarrow 2MnO_2(s) + 3NO_3^-(aq) + 2OH^-(aq)$

Oxidation Numbers	Change/Atom	Equalizing Charges Gives
Mn = +7 → Mn = +4	-3	2(-3) = -6
N = +3 → N = +5	+2	3(+2) = +6

(d) $2I^-(aq) + 4H^+(aq) + 2NO_2^-(aq) \rightarrow 2NO(g) + 2H_2O(\ell) + I_2(s)$

Oxidation Numbers	Change/Atom	Equalizing Charges Gives
I = -1 → I = 0	+1	-
N = +3 → N = +2	-1	-

Each change is multiplied by 2 since there are 2 I atooms in I_2.

(e) $Hg_2Cl_2(s) + 2NH_3(aq) \rightarrow Hg(\ell) + HgNH_2Cl(s) + NH_4^+(aq) + Cl^-(aq)$

Oxidation Numbers	Change/Atom	Equalizing Charges Gives
Hg = +1 → Hg = 0	-1	-
Hg = +1 → Hg = +2	+1	-

14-16. **Refer to Sections 14-3, 14-4 and Example 14-6.** . . . .

Using the Ion-Electron Method:

(a) skeletal eqn: $Fe^{2+}(aq) + MnO_4^-(aq) \rightarrow Fe^{3+}(aq) + Mn^{2+}(aq)$

ox. half-rxn: $Fe^{2+}(aq) \rightarrow Fe^{3+}(aq)$

balanced ox. half-rxn: $Fe^{2+}(aq) \rightarrow Fe^{3+}(aq) + e^-$

red. half-rxn: $MnO_4^-(aq) \rightarrow Mn^{2+}(aq)$

$8H^+(aq) + MnO_4^-(aq) \rightarrow Mn^{2+}(aq) + 4H_2O$

balanced red. half-rxn: $5e^- + 8H^+(aq) + MnO_4^-(aq) \rightarrow Mn^{2+}(aq) + 4H_2O$

Now, we balance the electron transfer and add the half-reactions term-by-term and cancel electrons:

oxidation: $5[Fe^{2+}(aq) \rightarrow Fe^{3+}(aq) + e^-]$

reduction: $5e^- + MnO_4^-(aq) + 8H^+(aq) \rightarrow Mn^{2+}(aq) + 4H_2O(\ell)$

balanced: $5Fe^{2+}(aq) + MnO_4^-(aq) + 8H^+(aq) \rightarrow 5Fe^{3+}(aq) + Mn^{2+}(aq) + 4H_2O(\ell)$

(b) skeletal eqn: $Br_2(\ell) + SO_2(g) \rightarrow Br^-(aq) + SO_4^{2-}(aq)$

ox. half-rxn: $SO_2(g) \rightarrow SO_4^{2-}(aq)$

$2H_2O(\ell) + SO_2(g) \rightarrow SO_4^{2-}(aq) + 4H^+$

balanced ox. half-rxn: $2H_2O(\ell) + SO_2(g) \rightarrow SO_4^{2-}(aq) + 4H^+ + 2e^-$

red. half-rxn: $Br_2(\ell) \rightarrow Br^-(aq)$

balanced red. half-rxn: $2e^- + Br_2(\ell) \rightarrow 2Br^-(aq)$

The electron transfer is already balanced and we can write:

oxidation: $2H_2O\ (\ell) + SO_2(g) \rightarrow SO_4^{2-}(aq) + 4H^+(aq) + 2e^-$

reduction: $2e^- + Br_2(\ell) \rightarrow 2Br^-(aq)$

balanced: $Br_2(\ell) + SO_2(g) + 2H_2O(\ell) \rightarrow 2Br^-(aq) + SO_4^{2-}(aq) + 4H^+(aq)$

(c) oxidation: $Cu(s) \rightarrow Cu^{2+}(aq) + 2e^-$

reduction: $2[1e^- + 2H^+(aq) + NO_3^-(aq) \rightarrow NO_2(g) + H_2O(\ell)]$

balanced: $Cu(s) + 2NO_3^-(aq) + 4H^+(aq) \rightarrow Cu^{2+}(aq) + 2NO_2(g) + 2H_2O\ (\ell)$

(d) oxidation: $2Cl^-(aq) \rightarrow Cl_2(g) + 2e^-$

reduction: $2e^- + 4H^+(aq) + 2Cl^-(aq) + PbO_2(s) \rightarrow PbCl_2(s) + 2H_2O(\ell)$

balanced: $PbO_2(s) + 4Cl^-(aq) + 4H^+(aq) \rightarrow PbCl_2(s) + Cl_2(g) + 2H_2O(\ell)$

(e) oxidation: $5[Zn(s) \rightarrow Zn^{2+}(aq) + 2e^-]$

reduction: $10e^- + 12H^+(aq) + 2NO_3^-(aq) \rightarrow N_2(g) + 6H_2O(\ell)$

balanced: $5Zn(s) + 2NO_3^-(aq) + 12H^+(aq) \rightarrow 5Zn^{2+}(aq) + N_2(g) + 6H_2O(\ell)$

14-18. **Refer to Sections 14-3, 14-4 and Example 14-7.** . . . .

Using the Ion-Electron Method:

(a) skeletal eqn: $MnO_4^-(aq) + NO_2^-(aq) \rightarrow MnO_2(s) + NO_3^-(aq)$

ox. half-rxn: $NO_2^-(aq) \rightarrow NO_3^-(aq)$

$2OH^-(aq) + NO_2^-(aq) \rightarrow NO_3^-(aq) + H_2O(\ell)$

balanced ox. half-rxn: $2OH^-(aq) + NO_2^-(aq) \rightarrow NO_3^-(aq) + H_2O(\ell) + 2e^-$

red. half-rxn: $MnO_4^-(aq) \rightarrow MnO_2(s)$

$2H_2O(\ell) + MnO_4^-(aq) \rightarrow MnO_2(s) + 4OH^-(aq)$

balanced
red. half-rxn: $3e^- + 2H_2O(\ell) + MnO_4^-(aq) \rightarrow MnO_2(s) + 4OH^-(aq)$

Now, we balance the electron transfer and add the half-reactions term-by-term:

oxidation: $3[2OH^-(aq) + NO_2^-(aq) \rightarrow NO_3^-(aq) + H_2O(\ell) + 2e^-]$

reduction: $2[3e^- + 2H_2O(\ell) + MnO_4^-(aq) \rightarrow MnO_2(s) + 4OH^-(aq)]$

balanced: $2MnO_4^-(aq) + 3NO_2^-(aq) + H_2O(\ell) \rightarrow 2MnO_2(s) + 3NO_3^-(aq) + 2OH^-(aq)$

(b) oxidation: $4[4OH^-(aq) + Zn(s) \rightarrow Zn(OH)_4^{2-}(aq) + 2e^-]$

reduction: $8e^- + 6H_2O(\ell) + NO_3^-(aq) \rightarrow NH_3(aq) + 9OH^-(aq)$

balanced: $4Zn(s) + NO_3^-(aq) + 6H_2O(\ell) + 7OH^-(aq) \rightarrow NH_3(aq) + 4Zn(OH)_4^{2-}(aq)$

(c) oxidation: $4OH^-(aq) + N_2H_4(aq) \rightarrow N_2(g) + 4H_2O(\ell) + 4e^-$

reduction: $2[2e^- + Cu(OH)_2(s) \rightarrow Cu(s) + 2OH^-(aq)]$

balanced: $N_2H_4(aq) + 2Cu(OH)_2(s) \rightarrow N_2(g) + 2Cu(s) + 4H_2O(\ell)$

(d) oxidation: $3[4OH^-(aq) + Mn^{2+}(aq) \rightarrow MnO_2(s) + 2H_2O + 2e^-]$

reduction: $2[3e^- + 2H_2O(\ell) + MnO_4^-(aq) \rightarrow MnO_2(s) + 4OH^-]$

balanced: $3Mn^{2+}(aq) + 2MnO_4^-(aq) + 4OH^-(aq) \rightarrow 5MnO_2(s) + 2H_2O(\ell)$

(e) oxidation: $12OH^-(aq) + Cl_2(g) \rightarrow 2ClO_3^-(aq) + 6H_2O(\ell) + 10e^-$

reduction: $5[2e^- + Cl_2(g) \rightarrow 2Cl^-(aq)]$

balanced: $6Cl_2(g) + 12OH^-(aq) \rightarrow 2ClO_3^-(aq) + 10Cl^-(aq) + 6H_2O(\ell)$

or $3Cl_2(g) + 6OH^-(aq) \rightarrow ClO_3^-(aq) + 5Cl^-(aq) + 3H_2O(\ell)$

14-20. **Refer to Section 14-3 and Exercise 14-14.** . . . .

Using the Change-in-Oxidation Number Method:

(a) $2H_2SO_4(aq) + C(s) \rightarrow CO_2(g) + 2SO_2(g) + 2H_2O(\ell)$

Oxidation Numbers	**Change/Atom**	**Equalizing Charges Gives**
S = +6 → S = +4	-2	2(-2) = -4
C = 0 → C = +4	+4	1(+4) = +4

(b) $MnO_2(s) + 4HCl(aq) \rightarrow Cl_2(g) + MnCl_2(aq) + 2H_2O(\ell)$

Oxidation Numbers	**Change/Atom**	**Equalizing Charges Gives**
Mn = +4 → Mn = +2	-2	1(-2) = -2
Cl = -1 → Cl = 0	+1	2(+1) = +2

Note: there are Cl atoms that are not changing oxidation number.

(c) $2KMnO_4(aq) + 10NaI(aq) + 8H_2SO_4(aq)$

$$\rightarrow 5I_2(s) + 2MnSO_4(aq) + 5Na_2SO_4(aq) + K_2SO_4(aq) + 8H_2O(\ell)$$

Oxidation Numbers	**Change/Atom**	**Equalizing Charges Gives**
Mn = +7 → Mn = +2	-5	1(-5) = -5
I = -1 → I = 0	+1	5(+1) = +5

Each change is multiplied by 2 since there are 2 I atoms in I_2.

(d) $K_2Cr_2O_7(aq) + 6KBr(aq) + 7H_2SO_4(aq)$

$$\rightarrow 3Br_2(\ell) + 4K_2SO_4(aq) + Cr_2(SO_4)_3(aq) + 7H_2O(\ell)$$

Oxidation Numbers	**Change/Atom**	**Equalizing Charges Gives**
Cr = +6 → Cr = +3	-3	1(-3) = -3
Br = -1 → Br = 0	+1	3(+1) = +3

Each change is multiplied by 2 since there are 2 Br atoms in Br_2.

14-22. **Refer to Section 14-5 and Example 14-9.** . . . .

Balanced <u>net ionic</u> equation is:

$$5Fe^{2+}(aq) + MnO_4^-(aq) + 8H^+(aq) \rightarrow 5Fe^{3+}(aq) + Mn^{2+}(aq) + 4H_2O\ (\ell)$$

Note: This is the net ionic equation. This exercise uses $KMnO_4$ and $FeSO_4$. These are both soluble salts which dissociate into their ions. The K^+ and SO_4^{2-} ions are spectator ions and are omitted from the balanced net ionic equation.

Plan: M, mL $FeSO_4 \xrightarrow{(1)}$ mmol $FeSO_4 \xrightarrow{(2)}$ mmol $KMnO_4 \xrightarrow{(3)}$ mL $KMnO_4$

(1) ? mmol $FeSO_4$ = 0.10 M × 20 mL = 2.0 mmol $FeSO_4$

(2) ? mmol $KMnO_4$ = 1/5 × mmol $FeSO_4$ = 1/5 × 2.0 mmol = 0.40 mmol $KMnO_4$

(3) $? \text{ mL } KMnO_4 = \dfrac{0.40 \text{ mmol } KMnO_4}{0.10\ M\ KMnO_4} = \mathbf{4.0 \text{ mL } KMnO_4}$

14-24. **Refer to Section 14-5 and Exercise 14-20(c) Solution.** . . . .

Balanced equation:

$$2KMnO_4(aq) + 10KI(aq) + 8H_2SO_4(aq) \rightarrow 5I_2(s) + 2MnSO_4(aq) + 6K_2SO_4(aq) + 8H_2O(\ell)$$

Plan: M, L KI soln $\xrightarrow{(1)}$ mol KI $\xrightarrow{(2)}$ mol $KMnO_4 \xrightarrow{(3)}$ L $KMnO_4$ soln

(1) ? mol KI = 0.10 M × 0.050 L = 0.0050 mol KI

(2) ? mol $KMnO_4$ = 1/5 × mol KI = 0.0010 mol $KMnO_4$

(3) $? \text{ L } KMnO_4 \text{ soln} = \dfrac{0.0010 \text{ mol } KMnO_4}{0.10\ M\ KMnO_4} = 0.010 \text{ L or } \mathbf{10 \text{ mL } KMnO_4 \text{ soln}}$

14-26. **Refer to Section 14-5 and Exercise 14-16(a) Solution.** . . . .

Balanced net ionic equation:

$$5Fe^{2+}(aq) + MnO_4^-(aq) + 8H^+(aq) \rightarrow 5Fe^{3+}(aq) + Mn^{2+}(aq) + 4H_2O(\ell)$$

Note: when $KMnO_4$ dissociates, the produced K^+ ion is a spectator ion and does not appear in the balanced net ionic equation.

Plan: M, L $KMnO_4$ soln $\xrightarrow{(1)}$ mol $KMnO_4$ $\xrightarrow{(2)}$ mol Fe^{2+} $\xrightarrow{(3)}$ g Fe $\xrightarrow{(4)}$ % Fe

(1) ? mol $KMnO_4$ = 0.06402 M × 0.03068 L = 0.001964 mol $KMnO_4$

(2) ? mol Fe^{2+} = 5 × mol $KMnO_4$ = 5 × 0.001964 mol = 0.009820 mol Fe^{2+}

(3) ? g Fe = 0.009820 mol Fe^{2+} × 55.85 g/mol = **0.5484 g Fe**

(4) ? % Fe in ore = $\frac{0.5484 \text{ g Fe}}{5.026 \text{ g ore}}$ × 100 = **10.91% Fe in ore**

14-28. **Refer to Section 14-5 and Exercise 14-12(d) Solution.** . . . .

Balanced net ionic equation:

$$Cr_2O_7^{2-}(aq) + 6Fe^{2+}(aq) + 14H^+(aq) \rightarrow 2Cr^{3+}(aq) + 6Fe^{3+}(aq) + 7H_2O(\ell)$$

Plan: M, L $Na_2Cr_2O_7$ soln $\xrightarrow{(1)}$ mol $Cr_2O_7^{2-}$ $\xrightarrow{(2)}$ mol Fe^{2+} $\xrightarrow{(3)}$ g Fe $\xrightarrow{(4)}$ % Fe in limonite

(1) ? mol $Cr_2O_7^{2-}$ = 0.02130 M $Na_2Cr_2O_7$ × 0.04296 L = 9.150 × 10^{-4} mol $Cr_2O_7^{2-}$

(2) ? mol Fe^{2+} = 6 × mol $Cr_2O_7^{2-}$ = 6 × 9.150 × 10^{-4} mol = 5.490 × 10^{-3} mol Fe^{2+}

(3) ? g Fe = 5.490 × 10^{-3} mol Fe^{2+} × 55.85 g/mol = 0.3066 g Fe

(4) ? % Fe in limonite = $\frac{0.3066 \text{ g Fe}}{0.5166 \text{ g ore}}$ × 100 = **59.35% Fe in limonite**

14-30. **Refer to Section 14-6, Example 14-12 and Exercises 14-20(c),(d) Solutions.** . .

Plan: N, L reagent soln $\xrightarrow{(1)}$ eq reagent $\xrightarrow{(2)}$ mol reagent $\xrightarrow{(3)}$ g reagent

(a) (1) ? eq $KMnO_4$ = 0.137 N $KMnO_4$ × 0.475 L = 0.0651 eq $KMnO_4$

(2) The balanced half-reaction in Exercise 14-20(c) involving the reduction of MnO_4^- to Mn^{2+} requires 5 electrons:

$$5e^- + 8H^+(aq) + MnO_4^-(aq) \rightarrow Mn^{2+}(aq) + 4H_2O(\ell)$$

? mol $KMnO_4$ = 1/5 × 0.0651 eq $KMnO_4$ = 0.0130 mol $KMnO_4$

(3) ? g $KMnO_4$ = 0.0130 mol $KMnO_4$ × 158 g/mol = **2.05 g $KMnO_4$**

(b) (1) ? eq NaI = 0.267 N NaI × 0.325 L = 0.0868 eq NaI

(2) The balanced half-reaction in Exercise 14-20(c) involving the oxidation of I^- to I_2 requires 2 electrons for 2 formula units of NaI.

$$2I^-(aq) \rightarrow I_2(s) + 2e^-$$

The change is therefore 1 electron for each NaI.

? mol NaI = 0.0868 eq NaI

(3) ? g NaI = 0.0868 mol NaI × 150 g/mol = **13.0 g NaI**

(c) (1) ? eq $K_2Cr_2O_7$ = 0.183 *N* $K_2Cr_2O_7$ × 0.198 L = 0.0362 eq $K_2Cr_2O_7$

(2) The balanced half-reaction in Exercise 14-20(d) involving the reduction of $Cr_2O_7^{2-}$ to Cr^{3+} requires 6 electrons:

$$6e^- + 14H^+(aq) + Cr_2O_7^{2-}(aq) \rightarrow 2Cr^{3+}(aq) + 7H_2O(\ell)$$

? mol $K_2Cr_2O_7$ = 1/6 × 0.0362 eq $K_2Cr_2O_7$ = 0.00603 mol $K_2Cr_2O_7$

(3) ? g $K_2Cr_2O_7$ = 0.00603 mol $K_2Cr_2O_7$ × 294 g/mol = **1.77 g $K_2Cr_2O_7$**

14-32. **Refer to Section 14-6 and Exercise 14-20(c) Solution.** . . . .

Plan: g $KMnO_4$ $\xrightarrow{(1)}$ mol $KMnO_4$ $\xrightarrow{(2)}$ *M* $KMnO_4$ $\xrightarrow{(3)}$ *N* $KMnO_4$

(1) ? mol $KMnO_4$ = $\dfrac{15.8 \text{ g } KMnO_4}{158 \text{ g/mol}}$ = 0.100 mol $KMnO_4$

(2) ? *M* $KMnO_4$ = $\dfrac{0.100 \text{ mol } KMnO_4}{0.500 \text{ L}}$ = **0.200 *M* $KMnO_4$**

(3) The balanced half-reaction involving the reduction of MnO_4^- to Mn^{2+} in Exercise 14-20(c) requires 5 electrons:

$$5e^- + 8H^+(aq) + MnO_4^-(aq) \rightarrow Mn^{2+}(aq) + 4H_2O(\ell)$$

? *N* $KMnO_4$ = 5 × 0.200 *M* $KMnO_4$ = **1.00 *N* $KMnO_4$**

14-34. **Refer to Section 14-6 and Exercise 14-16(a) Solution.** . . . .

Plan: g $FeSO_4$ $\xrightarrow{(1)}$ mol $FeSO_4$ $\xrightarrow{(2)}$ *M* $FeSO_4$ $\xrightarrow{(3)}$ *N* $FeSO_4$

(1) ? mol $FeSO_4$ = $\dfrac{16.2 \text{ g } FeSO_4}{152 \text{ g/mol}}$ = 0.107 mol $FeSO_4$

(2) ? *M* $FeSO_4$ = $\dfrac{0.107 \text{ mol } FeSO_4}{0.200 \text{ L}}$ = **0.535 *M* $FeSO_4$**

(3) The balanced half-reaction involving the oxidation of Fe^{2+} to Fe^{3+} in Exercise 14-16(a) requires 1 electron:

$$Fe^{2+}(aq) \rightarrow Fe^{3+}(aq) + 1e^-$$

? *N* $FeSO_4$ = *M* $FeSO_4$ = **0.535 *N* $FeSO_4$**

15 Chemical Thermodynamics

15-2. **Refer to the Key Terms for Chapters 1 and 15.** . . . .

(a) Heat is a form of energy that flows between two samples of matter due to their differences in temperature.

(b) Temperature is a measure of the intensity of heat, i.e., the hotness or coldness of an object.

(c) The system refers to the substances of interest in a process, i.e., it is the part of the universe that is under investigation.

(d) The surroundings refer to everything in the environment of the system of interest.

(e) The thermodynamic state of a system refers to a set of conditions that completely specifies all of the thermodynamic properties of the system.

(f) Work is the application of a force through a distance. For physical or chemical changes that occur at constant pressure, the work done by the system is $P\Delta V$.

15-4. **Refer to Section 1-1.** . . . .

Potential energy is defined as the energy that matter possesses by virtue of its position, condition or composition, whereas kinetic energy is the energy that matter possesses by virtue of its motion.

(a) A marble sitting on the edge of a bowl possesses potential energy. If it is given a slight nudge, the marble will begin to roll into the bowl. Its potential energy is being converted into kinetic energy.

(b) As the marble passes through the bottom of the bowl and begins to climb up the other side, it will slow down and momentarily stop. At this point, its kinetic energy has been converted back into potential energy.

15-6. **Refer to Sections 1-1 and 15-1.** . . . .

An endothermic process absorbs heat energy from its surroundings; an exothermic process releases heat energy to its surroundings. If a reaction is endothermic in one direction, it is exothermic in the opposite direction. For example, the melting of 1 mole of ice water is an endothermic process requiring 6.02 kJ of heat:

$$H_2O(s) + 6.02\text{ kJ} \rightarrow H_2O(\ell)$$

The reverse process, the freezing of 1 mole of liquid water, releasing 6.02 kJ of heat, is an exothermic process:

$$H_2O(\ell) \rightarrow H_2O(s) + 6.02\text{ kJ}$$

15-8. **Refer to Sections 15-2 and 15-7.** . . . .

A state function is a variable that defines the state of a system. It is a function that is independent of the pathway by which a state is reached. Examples of state functions are pressure, temperature and volume. And so, if a process involves a pressure change from P_1 to P_2, then $\Delta P = P_2 - P_1$, no matter how that pressure change occurred.

Hess' Law states that the enthalpy change, ΔH, for a reaction is the same whether it occurs by a single step or a series of steps. Therefore, Hess' Law would not be a law if enthalpy, H, were not a state function.

15-10. **Refer to Sections 15-5 and 15-6.** . . . .

(a) ΔH, the enthalpy change or heat of reaction, is the heat change of a reaction occurring at some constant pressure.

ΔH^o, is the standard enthalpy change of a reaction that occurs at 1 atm pressure. Unless otherwise stated, the reaction temperature is 25°C.

(b) As stated in (a), ΔH^o is the standard enthalpy change of a reaction occurring at 1 atm pressure.

ΔH_f^o, the standard molar enthalpy of formation of some substance, is the amount of heat absorbed in a reaction in which 1 mole of substance is formed from its elements in their standard states.

15-12. **Refer to Sections 15-1 and 15-7.** . . . .

Yes, it is possible to calculate the amount of heat released when 1 liter of gasoline is combusted.

Plan: (1) From the specific composition of gasoline, determine the number of moles of each component in 1 liter of gasoline.

(2) The general unbalanced equation for the complete combustion of hydrocarbons at 25°C is

$$C_xH_y + O_2(g) \rightarrow CO_2(g) + H_2O(\ell)$$

For each component, use Hess' Law and the balanced equation to determine its heat of combustion for 1 mole.

(3) For each component, multiply the heat of combustion per mole by the number of moles of component in 1 liter of gasoline to obtain the heat released per component.

(4) Calculate the total amount of heat released by adding the values in Step (3) together.

15-14. **Refer to Section 15-1.** . . . .

Consider these reactions: (1) $CaO(s) + H_2O(g) \rightarrow Ca(OH)_2(s)$
(2) $CaO(s) + H_2O(\ell) \rightarrow Ca(OH)_2(s)$

The only difference between them is that Reaction (1) involves water as water vapor and Reaction (2) uses water in the liquid phase. To convert Reaction (1) into Reaction (2), more energy is released when $H_2O(g) \rightarrow H_2O(\ell)$. Therefore, Reaction (1) is more exothermic than Reaction (2).

15-16. **Refer to Section 15-3.** . . . .

Internal energy, a state function, is the sum of all forms of energy associated with a specific amount of substance including the kinetic energies of the molecules and the energies among all the particles in the substance.

ΔE for a physical change or a chemical reaction is equal to the difference between the internal energy of the products and the internal energy of the reactants:

$$\Delta E = E_{products} - E_{reactants}$$

15-18. **Refer to Section 15-3.** . . . .

Work has been defined as force, F, times distance, d: work = $F \times d$. On the other hand, pressure, P, times the change in volume, ΔV, can also be shown to be equal to $F \times d$. We know that pressure has units of force per unit area (F/d^2) and volume has units of d^3.
Therefore,

$$P\Delta V = \frac{F}{d^2} \times d^3 = F \times d = \text{-work}$$

The ideal gas law states that $PV = nRT$ for an ideal gas. At constant pressure and temperature, the change in volume can only be due to a change in the number of moles of gas (Gay-Lussac's Law) and

$$\text{work} = -P\Delta V = -\Delta nRT \quad \text{where } \Delta n = \begin{pmatrix}\text{total moles of}\\ \text{gas products}\end{pmatrix} - \begin{pmatrix}\text{total moles of}\\ \text{gas reactants}\end{pmatrix}$$

At constant T, work $\propto \Delta n$.

15-20. **Refer to Section 15-3.** . . . .

The vaporization process is ethanol(ℓ) → ethanol(g)

$\Delta E = q + w$ where ΔE = change in internal energy
q = heat absorbed by the system
w = work done on the system

(1) The heat absorbed by the system,

$$q = \Delta H_{vap} \times \text{g ethanol} = +854 \text{ J/g} \times 1.00 \text{ g} = \mathbf{+854\ J}$$

(2) The work done on the system in going from a liquid to a gas,

$$w = -P\Delta V = -P(V_{gas} - V_{liquid})$$

$$\text{where } V_{gas} = \frac{nRT}{P} = \frac{(1.00 \text{ g}/46.0 \text{ g/mol})(0.0821 \text{ L·atm/mol·K})(78.0^{o}\text{C} + 273^{o})}{1.00 \text{ atm}}$$
$$= 0.626 \text{ L}$$

$$V_{liquid} = 1.00 \text{ g ethanol} \times \frac{1.00 \text{ mL ethanol}}{0.789 \text{ g ethanol}} = 1.27 \text{ mL or } 0.00127 \text{ L}$$

Therefore,

$$w = -P\Delta V = -1.00 \text{ atm}(0.626 \text{ L} - 0.001 \text{ L}) = -0.625 \text{ L·atm}$$

To find a factor to convert L·atm to J, we can equate two values of the molar gas constant, R

$$0.0821 \text{ L·atm/mol·K} = 8.314 \text{ J/mol·K}$$
$$1 \text{ L·atm} = 101 \text{ J}$$

$$\text{And so, } w = -0.625 \text{ L·atm} \times \frac{101 \text{ J}}{1 \text{ L·atm}} = \mathbf{-63.1\ J}$$

(3) Finally, $\Delta E = q + w = 854 \text{ J} + (-63 \text{ J}) = \mathbf{791\ J}$

15-22. **Refer to Section 15-3.** . . . .

(a) The balanced reaction for the oxidation of 1 mole of HCl:
$HCl(g) + 1/4\ O_2(g) \rightarrow 1/2\ Cl_2(g) + 1/2\ H_2O(g)$

$$\text{work} = -P\Delta V = -\Delta n_{gas}RT = -(n_{gas\ products} - n_{gas\ reactants})RT$$
$$= -(1 \text{ mol} - 5/4 \text{ mol})(8.314 \text{ J/mol·K})(200^{o}\text{C} + 273^{o})$$
$$= \mathbf{+983\ J}$$

Work is a positive number, therefore, work is done on the system by the surroundings. As the system "shrinks" from 5/4 mole of gas to 1 mole of gas, work is done on the system by the surroundings to decrease the volume (Recall that $V \propto n$ at constant T and P).

(b) The balanced reaction for the oxidation of 1 mole of NO:

$$NO(g) \rightarrow 1/2\ N_2(g) + 1/2\ O_2(g)$$

$$\text{work} = -P\Delta V = -\Delta n_{gas}RT = -(1\text{ mol} - 1\text{ mol})RT = \mathbf{0\ J}$$

There is no work done since the number of moles of gas, and hence the volume of the system, remains constant.

15-24. **Refer to Section 15-4.** . . . .

Plan: (1) Verify that Mg is the limiting reagent.
(2) Calculate the mass of the solution.
(3) Determine the total amount of heat evolved in the reaction of 1.22 g Mg with excess HCl.
(4) Convert the heat evolved (J/1.22 g Mg) to ΔH (J/mol Mg).

(1) $$?\text{ mol Mg} = \frac{1.22\text{ g Mg}}{24.3\text{ g/mol}} = 0.0502\text{ mol Mg}$$

$$?\text{ mol HCl} = 6.02\ M\text{ HCl} \times 0.100\text{ L} = 0.602\text{ mol HCl}$$

The limiting reagent is Mg. This fact is important to verify because we needed to know that the heat evolved was due to the reaction of the entire sample of Mg.

(2) We know: mass of final solution = mass of Mg + mass of HCl solution

$$?\text{ g final soln} = 1.22\text{ g Mg} + \left(100\text{ mL HCl soln} \times \frac{1.10\text{ g HCl soln}}{1.00\text{ mL HCl soln}}\right) = 111\text{ g}$$

(3) $$\left|\begin{matrix}\text{amount of heat}\\ \text{lost by reaction}\end{matrix}\right| = \left|\begin{matrix}\text{amount of heat gained}\\ \text{by calorimeter}\end{matrix}\right| + \left|\begin{matrix}\text{amount of heat}\\ \text{gained by solution}\end{matrix}\right|$$

$$\left|\begin{matrix}\text{amount of heat gained}\\ \text{by calorimeter}\end{matrix}\right| = \left|\text{heat capacity of calorimeter (J/}^\circ\text{C)} \times \Delta T\right|$$
$$= (562\text{ J/}^\circ\text{C})(45.5^\circ\text{C} - 23.0^\circ\text{C})$$
$$= 1.26 \times 10^4\text{ J}$$

$$\left|\begin{matrix}\text{amount of heat}\\ \text{gained by solution}\end{matrix}\right| = \left|(\text{Sp. Ht.})(\text{mass})(\Delta T)\right|_{\text{soln}}$$
$$= (4.18\text{ J/g}\cdot^\circ\text{C})(111\text{ g})(45.5^\circ\text{C} - 23.0^\circ\text{C})$$
$$= 1.04 \times 10^4\text{ J}$$

Therefore, $$\left|\begin{matrix}\text{amount of heat}\\ \text{lost by reaction}\end{matrix}\right| = (1.26 \times 10^4\text{ J}) + (1.04 \times 10^4\text{ J})$$
$$= 2.30 \times 10^4\text{ J}$$

(4) $$?\ \Delta H\text{ (J/mol Mg)} = \frac{-2.30 \times 10^4\text{ J}}{1.22\text{ g Mg}} \times \frac{24.31\text{ g Mg}}{1\text{ mol Mg}} = \mathbf{-4.58 \times 10^5\ J/mol\ Mg}$$

ΔH is a negative number because heat is released in the reaction.

15-26. **Refer to Section 15-4 and Figure 15-4b.** • • • • • • • • • • • • • • •

A bomb calorimeter is a mechanical device used to measure the heat transfer between a system and its surroundings at constant volume. Generally, the systems are chemical reactions. No work is done in a bomb calorimeter since $\Delta V = 0$.

Therefore, $\Delta E = q_V + w = q_V - P\Delta V = q_V - 0 = q_V$

and we see that bomb calorimeters measure the change in internal energy, ΔE, for a reaction directly.

15-28. **Refer to Section 15-4 and Example 15-1.** • • • • • • • • • • • • • • •

(a) $2C_6H_6(\ell) + 15O_2(g) \rightarrow 12CO_2(g) + 6H_2O(\ell)$

(b) The heat absorbed by the calorimeter is given as its water equivalent. This means that we are treating the calorimeter as if it were an additional 216 g of water. In our calculation, we can add this "pretend" mass of water to the actual mass of water.

$$\left|\begin{matrix}\text{amount of heat}\\\text{lost in reaction}\end{matrix}\right| = \left|\begin{matrix}\text{amount of heat gained}\\\text{by bomb calorimeter}\end{matrix}\right| + \left|\begin{matrix}\text{amount of heat}\\\text{gained by water}\end{matrix}\right|$$

$$= |(4.184\ \text{J/g}^o\text{C})(216\ \text{g} + 826\ \text{g})(33.700^o\text{C} - 23.640^o\text{C})|$$

$$= 4.386 \times 10^4\ \text{J per } 1.048\ \text{g } C_6H_6(\ell)$$

Since heat is released in this reaction, ΔE is a negative quantity.

$$\Delta E = \frac{-4.386 \times 10^4\ \text{J}}{1.048\ \text{g } C_6H_6(\ell)} = -4.185 \times 10^4\ \text{J/g } C_6H_6(\ell) \text{ or } \mathbf{-41.85\ kJ/g\ C_6H_6(\ell)}$$

$$\Delta E = -41.85\ \text{kJ/g} \times 78.11\ \text{g/mol} = \mathbf{-3269\ kJ/mol\ C_6H_6(\ell)}$$

15-30. **Refer to Sections 15-4, 15-5 and Exercise 15-20 Solution.** • • • • • • • •

The process: $H_2O(s) + 6.02\ \text{kJ} \rightarrow H_2O(\ell)$ at constant temperature and pressure

Since heat released during the reaction was measured at constant pressure,
$\Delta H = \mathbf{+6.02\ kJ/mol\ H_2O}$

$$\begin{aligned}\Delta E &= \Delta H - P\Delta V\\ &= +6.02\ \text{kJ} - (1.00\ \text{atm})(0.0180\ \text{L} - 0.0196\ \text{L})(101\ \text{J/L·atm})(1\ \text{kJ/1000 J})\\ &= +6.02\ \text{kJ} - (-1.62 \times 10^{-4}\ \text{kJ})\\ &= \mathbf{+6.02\ kJ}\end{aligned}$$

The values of ΔH and ΔE are essentially the same. The work term for processes involving only liquids and solids is negligible.

15-32. **Refer to Section 15-6, Appendix K and Example 15-5.** • • • • • • • • • •

(a) Balanced equation: $C_6H_6(\ell) + 15/2\ O_2(g) \rightarrow 6CO_2(g) + 3H_2O(\ell)$

Note: we left the fractional coefficient in front of $O_2(g)$ because we want the reaction in terms of 1 mole of $C_6H_6(\ell)$.

Using Hess' Law,

$$\begin{aligned}\Delta H^o_{rxn} &= [6\Delta H^o_f\ CO_2(g) + 3\Delta H^o_f\ H_2O(\ell)] - [\Delta H^o_f\ C_6H_6(\ell) + (15/2)\Delta H^o_f\ O_2(g)]\\ &= [(6\ \text{mol})(-393.5\ \text{kJ/mol}) + (3\ \text{mol})(-285.8\ \text{kJ/mol})]\\ &\qquad - [(1\ \text{mol})(49.03\ \text{kJ/mol}) + (15/2\ \text{mol})(0\ \text{kJ/mol})]\\ &= (-3218\ \text{kJ/mol}) - (49.03\ \text{kJ/mol})\\ &= \mathbf{-3267\ kJ/mol\ C_6H_6(\ell)}\end{aligned}$$

(b) In Exercise 15-28, ΔE = for this reaction was -3269 kJ/mol $C_6H_6(\ell)$. The ΔE value is very similar to ΔH. We know that

$$\Delta H = \Delta E + \Delta n_{gas}RT$$

$$\text{where } \Delta n_{gas}RT = (6 \text{ mol} - 7.5 \text{ mol})(8.314 \text{ J/mol·K})(298.15 \text{ K})$$
$$= -3718 \text{ J/mol } C_6H_6(\ell) \quad \text{or} \quad -3.718 \text{ kJ/mol } C_6H_6(\ell)$$

Substituting and comparing,

$$\Delta H = -3267 \text{ kJ/mol}$$
$$\Delta H = \Delta E + \Delta nRT = -3269 - 4 = -3273 \text{ kJ/mol}$$

We see that ΔH and ΔE are similar but not the same. The small difference between ΔH and ($\Delta E + \Delta n_{gas}RT$) is because ΔH and $\Delta n_{gas}RT$ were obtained from theory, but ΔE was determined experimentally.

15-34. **Refer to Section 15-6.** . . . .

The thermochemical standard state of a substance is its most stable state under standard pressure (1 atm) and at some specific temperature (usually 25°C). Examples of elements in their standard states are the following:

Element	Standard State
hydrogen	$H_2(g)$
chlorine	$Cl_2(g)$
bromine	$Br_2(\ell)$
helium	$He(g)$
copper	$Cu(s)$

15-36. **Refer to Section 15-6 and Appendix K.** . . . .

The standard molar enthalpy of formation, ΔH^o_f, of elements in their standard states is zero. From the tabulated values of standard molar enthalpies in Appendix K, we can identify the standard states of elements.

(a) fluorine - $F_2(g)$ (b) iron - $Fe(s)$

(c) carbon - C(s,graphite) (d) iodine - $I_2(s)$

(e) oxygen - $O_2(g)$

15-38. **Refer to Section 15-6 and Example 15-2.** . . . .

(a) $1/2\ H_2(g) + 1/2\ Cl_2(g) \rightarrow HCl(g)$ (b) $2C(graphite) + 3H_2(g) \rightarrow C_2H_6(g)$

(c) $Na(s) + 1/2\ Br_2(\ell) \rightarrow NaBr(s)$ (d) $1/2\ N_2(g) + 3/2\ H_2(g) \rightarrow NH_3(g)$

(e) $C(graphite) + 1/2\ O_2(g) + Cl_2(g) \rightarrow COCl_2(g)$

(f) $6C(graphite) + 6H_2(g) + 3O_2(g) \rightarrow C_6H_{12}O_6(s)$

(g) $1/2\ H_2(g) \rightarrow H(g)$

15-40. **Refer to Section 15-6.** . . . .

The balanced equation representing the standard molar enthalpy of formation of $Li_2O(s)$ is

$$2Li(s) + 1/2\ O_2(g) \rightarrow Li_2O(s)$$

$$? \text{ kJ/mol } Li_2O(s) = \frac{420 \text{ kJ}}{10.0 \text{ g Li}} \times \frac{6.94 \text{ g Li}}{1 \text{ mol Li}} \times \frac{2 \text{ mol Li}}{1 \text{ mol } Li_2O} = 583 \text{ kJ/mol}$$

And so, $\Delta H^o_f \; Li_2O(s)$ = **-583 kJ/mol** since the reaction is exothermic.

15-42. **Refer to Sections 15-5 and 15-6.** . . . .

This statement is false. The total heat content of a system is given by enthalpy, *H*, which is the absolute measurement of all the internal energy of the system (such as electronic, atomic and molecular motions), and all the external *PV* energy of the system. On the other hand, ΔH^o of formation, is the change in enthalpy when a substance is formed from its elements in their standard states. For an element, the fact that the ΔH^o of formation value is zero means that there is no energy change when forming the element from itself. It definitely does not imply that the total heat content of an element is zero.

15-44. **Refer to Section 15-7, Example 15-5 and Appendix K.** . . . .

(a) Balanced equation: $SiO_2(s) + Na_2CO_3(s) \rightarrow Na_2SiO_3(s) + CO_2(g)$

$$\Delta H^o_{rxn} = [\Delta H^o_f \; Na_2SiO_3(s) + \Delta H^o_f \; CO_2(g)] - [\Delta H^o_f \; SiO_2(s) + \Delta H^o_f \; Na_2CO_3(s)]$$
$$= [(1 \text{ mol})(-1079 \text{ kJ/mol}) + (1 \text{ mol})(-393.5 \text{ kJ/mol})] - [(1 \text{ mol})(-910.9 \text{ kJ/mol}) + (1 \text{ mol})(-1131 \text{ kJ/mol})]$$
$$= \mathbf{569 \text{ kJ}}$$

(b) Balanced equation: $H_2SiF_6(aq) \rightarrow 2HF(aq) + SiF_4(g)$

$$\Delta H^o_{rxn} = [2\Delta H^o_f \; HF(aq) + \Delta H^o_f \; SiF_4(g)] - [\Delta H^o_f \; H_2SiF_6(aq)]$$
$$- [(2 \text{ mol})(-320.8 \text{ kJ/mol}) + (1 \text{ mol})(-1615 \text{ kJ/mol})] - [(1 \text{ mol})(-2331 \text{ kJ/mol})]$$
$$= \mathbf{74 \text{ kJ}}$$

(c) Balanced equation: $2Al(s) + Fe_2O_3(s) \rightarrow 2Fe(s) + Al_2O_3(s)$

$$\Delta H^o_{rxn} = [2\Delta H^o_f \; Fe(s) + \Delta H^o_f \; Al_2O_3(s)] - [2\Delta H^o_f \; Al(s) + \Delta H^o_f \; Fe_2O_3(s)]$$
$$= [(2 \text{ mol})(0 \text{ kJ/mol}) + (1 \text{ mol})(-1676 \text{ kJ/mol})] - [(2 \text{ mol})(0 \text{ kJ/mol}) + (1 \text{ mol})(-824.2 \text{ kJ/mol})]$$
$$= \mathbf{-852 \text{ kJ}}$$

15-46. **Refer to Section 15-7 and Appendix K.** . . . .

For a reaction to be exothermic at 25°C, ΔH^o_{rxn} must be a negative number.

Balanced equation: $CO(g) + Cl_2(g) \rightarrow COCl_2(g)$

$$\Delta H^o_{rxn} = [\Delta H^o_f \; COCl_2(g)] - [\Delta H^o_f \; CO(g) + \Delta H^o_f \; Cl_2(g)]$$
$$= [(1 \text{ mol})(-223.0 \text{ kJ/mol})] - [(1 \text{ mol})(-110.5 \text{ kJ/mol}) + (1 \text{ mol})(0 \text{ kJ/mol})]$$
$$= -112.5 \text{ kJ}$$

Therefore, the reaction is exothermic.

15-48. **Refer to Section 15-7 and Appendix K.** . . . .

Balanced equation: $SiBr_4(\ell) + 4Na(s) \rightarrow 4NaBr(s) + Si(s)$

$$\Delta H^o_{rxn} = [4\Delta H^o_f \; NaBr(s) + \Delta H^o_f \; Si(s)] - [\Delta H^o_f \; SiBr_4(\ell) + 4\Delta H^o_f \; Na(s)]$$
$$= [(4 \text{ mol})(-359.9 \text{ kJ/mol}) + (1 \text{ mol})(0 \text{ kJ/mol})] - [(1 \text{ mol})(-457.3 \text{ kJ/mol}) + (4 \text{ mol})(0 \text{ kJ/mol})]$$
$$= \mathbf{-982.3 \text{ kJ}}$$

15-50. **Refer to Section 15-7 and Appendix K.**

Balanced equation: $Ca(s) + Cl_2(g) \rightarrow CaCl_2(s)$

Plan: (1) Calculate ΔH^o_{rxn}.
(2) Determine the amount of heat released when 14.2 g of Ca reacts.

(1) In this reaction, 1 mole of $CaCl_2(s)$ is formed from its elements in their standard states. Therefore,

$$\Delta H^o_{rxn} = \Delta H^o_f\ CaCl_2(s) = -795.0\ \text{kJ/mol}\ CaCl_2(s)$$

(2) $$\Delta H^o_{rxn} = 14.2\ \text{g Ca} \times \frac{1\ \text{mol Ca}}{40.1\ \text{g Ca}} \times \frac{1\ \text{mol}\ CaCl_2}{1\ \text{mol Ca}} \times \frac{-795.0\ \text{kJ}}{1\ \text{mol}\ CaCl_2} = -282\ \text{kJ}$$

Therefore, **282 kJ** of heat is released.

15-52. **Refer to Section 15-7, Appendix K and Table 1-7.**

Balanced equation: $4FeS_2(s) + 11O_2(g) \rightarrow 2Fe_2O_3(s) + 8SO_2(g)$

The standard molar enthalpy change of reaction is

$$\begin{aligned}\Delta H^o_{rxn} &= [2\Delta H^o_f\ Fe_2O_3(s) + 8\Delta H^o_f\ SO_2(g)] - [4\Delta H^o_f\ FeS_2(s) + 11\Delta H^o_f\ O_2(g)]\\ &= [(2\ \text{mol})(-824.2\ \text{kJ/mol}) + (8\ \text{mol})(-296.8\ \text{kJ/mol})]\\ &\qquad - [4\ \text{mol})(-177.5\ \text{kJ/mol}) + (11\ \text{mol})(0\ \text{kJ/mol})]\\ &= -3312.8\ \text{kJ}\end{aligned}$$

And so, when 4 moles of $FeS_2(s)$ is roasted, 3312.8 kJ of heat is released.

(a) The amount of heat released when 4.65 g of $FeS_2(s)$ is roasted is **32.1 kJ** as shown below.

$$\Delta H_{rxn} = 4.65\ \text{g}\ FeS_2 \times \frac{1\ \text{mol}\ FeS_2}{120.0\ \text{g}\ FeS_2} \times \frac{-3312.8\ \text{kJ}}{4\ \text{mol}\ FeS_2} = -32.1\ \text{kJ}$$

(b) The amount of heat released when 4.65 tons of $FeS_2(s)$ is roasted is **2.91×10^7 kJ** as shown below.

$$\begin{aligned}\Delta H_{rxn} &= 4.65\ \text{tons}\ FeS_2 \times \frac{2000\ \text{lb}\ FeS_2}{1\ \text{ton}\ FeS_2} \times \frac{453.6\ \text{g}\ FeS_2}{1\ \text{lb}\ FeS_2} \times \frac{1\ \text{mol}\ FeS_2}{120\ \text{g}\ FeS_2}\\ &\qquad \times \frac{-3312.8\ \text{kJ}}{4\ \text{mol}\ FeS_2}\\ &= -2.91 \times 10^7\ \text{kJ}\end{aligned}$$

15-54. **Refer to Section 15-7 and Appendix K.**

Balanced equation: $SO_3(g) + H_2O(\ell) \rightarrow H_2SO_4(\ell)$

The standard molar enthalpy change for the reaction is

$$\begin{aligned}\Delta H^o_{rxn} &= [\Delta H^o_f\ H_2SO_4(\ell)] - [\Delta H^o_f\ SO_3(g) + \Delta H^o_f\ H_2O(\ell)]\\ &= [(1\ \text{mol})(-814.0\ \text{kJ/mol})]\\ &\qquad - [(1\ \text{mol})(-395.6\ \text{kJ/mol}) + (1\ \text{mol})(-285.8\ \text{kJ/mol})]\\ &= -132.6\ \text{kJ}\end{aligned}$$

Therefore, 132.6 kJ of heat are released when 1 mole of SO_3 reacts with H_2O.

The standard enthalpy change when 66.4 g of SO_3 react with H_2O is

$$\Delta H^o_{rxn} = 66.4 \text{ g } SO_3 \times \frac{1 \text{ mol } SO_3}{80.1 \text{ g } SO_3} \times \frac{-132.6 \text{ kJ}}{1 \text{ mol } SO_3} = \mathbf{-110 \text{ kJ}}$$

15-56. **Refer to Section 15-7 and Example 15-3.**

To obtain the desired equation,

(1) reverse the first equation and multiply it by 2. This places 4HBr on the reactant side. Then

(2) leave the second equation as it is.

	ΔH^o
$4HBr(g) \rightarrow 2H_2(g) + 2Br_2(\ell)$	+2(72.8) kJ
$2H_2(g) + O_2(g) \rightarrow 2H_2O(g)$	-483.7 kJ
$4HBr(g) + O_2(g) \rightarrow 2Br_2(\ell) + 2H_2O(g)$	**-338.1 kJ**

15-58. **Refer to Section 15-7 and Example 15-3.**

To obtain the desired equation,

(1) divide the first equation by 3 to give 1 mole of $Fe_2O_3(s)$ on the reactant side,

(2) multiply the second equation by 2 to give 2 moles of Fe(s) on the product side, and

(3) multiply the third equation by 2/3, so that Fe_3O_4 and FeO can be eliminated when the equations are added together.

	ΔH^o
$Fe_2O_3(s) + 1/3\ CO(g) \rightarrow 2/3\ Fe_3O_4(s) + 1/3\ CO_2(g)$	-46.4/3 kJ
$2FeO(s) + 2CO(g) \rightarrow 2Fe(s) + 2CO_2(g)$	2(9.0) kJ
$2/3\ Fe_3O_4(s) + 2/3\ CO(g) \rightarrow 2FeO(s) + 2/3\ CO_2(g)$	(2/3)(-41.0) kJ
$Fe_2O_3(s) + 3CO(g) \rightarrow 2Fe(s) + 3CO_2(g)$	**-24.8 kJ**

15-60. **Refer to Sections 15-5 and 15-6, Example 15-6 and Appendix K.**

Plan: Use Hess' Law and solve for ΔH^o_f of the organic compound.

(a) Balanced equation: $C_6H_{12}(\ell) + 9O_2(g) \rightarrow 6CO_2(g) + 6H_2O(\ell)$

$$\Delta H^o_{combustion} = [6\Delta H^o_f\ CO_2(g) + 6\Delta H^o_f\ H_2O(\ell)] - [\Delta H^o_f\ C_6H_{12}(\ell) + 9\Delta H^o_f\ O_2(g)]$$

$$-3920 \text{ kJ} = [(6 \text{ mol})(-393.5 \text{ kJ/mol}) + (6 \text{ mol})(-285.8 \text{ kJ/mol})] - [(1 \text{ mol})\Delta H^o_f\ C_6H_{12}(\ell) + (9 \text{ mol})(0 \text{ kJ/mol})]$$

$$-3920 \text{ kJ} = -4075.8 \text{ kJ} - (1 \text{ mol})\Delta H^o_f\ C_6H_{12}(\ell)$$

$$\Delta H^o_f\ C_6H_{12}(\ell) = \mathbf{-156 \text{ kJ/mol}}$$

(b) Balanced equation: $C_6H_5OH(s) + 7O_2(g) \rightarrow 6CO_2(g) + 3H_2O(\ell)$

$$\Delta H^o_{combustion} = [6\Delta H^o_f\ CO_2(g) + 3\Delta H^o_f\ H_2O(\ell)] - [\Delta H^o_f\ C_6H_5OH(s) + 7\Delta H^o_f\ O_2(g)]$$

$$-3053 \text{ kJ} = [(6 \text{ mol})(-393.5 \text{ kJ/mol}) + (3 \text{ mol})(-285.8 \text{ kJ/mol})] - [(1 \text{ mol})\Delta H^o_f\ C_6H_5OH(s) + (7 \text{ mol})(0 \text{ kJ/mol})]$$

$$-3053 \text{ kJ} = -3218.4 \text{ kJ} - (1 \text{ mol})\Delta H^o_f\ C_6H_5OH(s)$$

$$\Delta H^o_f\ C_6H_5OH(s) = \mathbf{-165 \text{ kJ/mol}}$$

(c) Balanced equation: $CH_3CHO(\ell) + 5/2\ O_2(g) \rightarrow 2CO_2(g) + 2H_2O(\ell)$

$$\Delta H^o_{combustion} = [2\Delta H^o_f\ CO_2(g) + 2\Delta H^o_f\ H_2O(\ell)] - [\Delta H^o_f\ CH_3CHO(\ell) + 5/2\Delta H^o_f\ O_2(g)]$$

$$-1166\ kJ = [(2\ mol)(-393.5\ kJ/mol) + (2\ mol)(-285.8\ kJ/mol)] - [1\ mol)\Delta H^o_f\ CH_3CHO(\ell) + (5/2\ mol)(0\ kJ/mol)]$$

$$-1166\ kJ = -1358.6\ kJ - (1\ mol)\Delta H^o_f\ CH_3CHO(\ell)$$

$$\Delta H^o_f\ CH_3CHO(\ell) = \mathbf{-193\ kJ/mol}$$

15-62. **Refer to Sections 15-5 and 15-6, and Example 15-6.** . . . .

Plan: Use Hess' Law and solve for ΔH^o_f of silicon carbide. Assume that the ΔH^o_f of coke is the same as that of graphite.

Balanced equation: $SiO_2(s) + 3C(s) \rightarrow SiC(s) + 2CO(g)$

$$\Delta H^o_{rxn} = [\Delta H^o_f\ SiC(s) + 2\Delta H^o_f\ CO(g)] - [\Delta H^o_f\ SiO_2(s) + 3\Delta H^o_f\ C(s)]$$

$$624.6\ kJ = [(1\ mol)\Delta H^o_f\ SiC(s) + (2\ mol)(-110.5\ kJ/mol)] - [(1\ mol)(-910.9\ kJ/mol) + (3\ mol)(0\ kJ/mol)]$$

$$624.6\ kJ = (1\ mol)\Delta H^o_f\ SiC(s) + 689.9\ kJ$$

$$\Delta H^o_f\ SiC(s) = \mathbf{-65.3\ kJ/mol}$$

15-64. **Refer to Section 15-7 and Appendix K.** . . . .

(a) Balanced equations: $CH_4(g) + 2O_2(g) \rightarrow CO_2(g) + 2H_2O(\ell)$

$2C_8H_{18}(\ell) + 25O_2(g) \rightarrow 16CO_2(g) + 18H_2O(\ell)$

$2C_{10}H_{22}(\ell) + 31O_2(g) \rightarrow 20CO_2(g) + 22H_2O(\ell)$

(b) The combustion of $CH_4(g)$:

$$\Delta H^o_{rxn} = [\Delta H^o_f\ CO_2(g) + 2\Delta H^o_f\ H_2O(\ell)] - [\Delta H^o_f\ CH_4(g) + 2\Delta H^o_f\ O_2(g)]$$
$$= [(1\ mol)(-393.5\ kJ/mol) + (2\ mol)(-285.8\ kJ/mol)] - [(1\ mol)(-74.81\ kJ/mol) + (2\ mol)(0\ kJ/mol)]$$
$$= \mathbf{-890.3\ kJ/mol\ CH_4(g)}$$

The combustion of $C_8H_{18}(\ell)$:

$$\Delta H^o_{rxn} = [16\Delta H^o_f\ CO_2(g) + 18\Delta H^o_f\ H_2O(\ell)] - [2\Delta H^o_f\ C_8H_{18}(\ell) + 25\Delta H^o_f\ O_2(g)]$$
$$= [(16\ mol)(-393.5\ kJ/mol) + (18\ mol)(-285.8\ kJ/mol)] - [(2\ mol)(-268.8\ kJ/mol) + (25\ mol)(0\ kJ/mol)]$$
$$= -10902.8\ kJ/rxn = -10902.8\ kJ/2\ mol\ C_8H_{18}(\ell)$$
$$= \mathbf{-5451.4\ kJ/mol\ C_8H_{18}(\ell)}$$

The combustion of $C_{10}H_{22}(\ell)$:

$$\Delta H^o_{rxn} = [20\Delta H^o_f\ CO_2(g) + 22\Delta H^o_f\ H_2O(\ell)] - [2\Delta H^o_f\ C_{10}H_{22}(\ell) + 31\Delta H^o_f\ O_2(g)]$$
$$= [(20\ mol)(-393.5\ kJ/mol) + (22\ mol)(-285.8\ kJ/mol)] - [(2\ mol)(-249.6\ kJ/mol) + (31\ mol)(0\ kJ/mol)]$$
$$= -13658.4\ kJ/rxn = -13658.4\ kJ/2\ mol\ C_{10}H_{22}(\ell)$$
$$= \mathbf{-6829.2\ kJ/mol\ C_{10}H_{22}(\ell)}$$

(c) Since ΔH^o_{rxn} (kJ/mol) is most negative for the combustion of $C_{10}H_{22}(\ell)$, the burning of $\mathbf{C_{10}H_{22}(\ell)}$ at standard conditions produces the most heat per mole.

(d) for $CH_4(g)$: $\Delta H^o_{rxn} = \frac{-890.3 \text{ kJ}}{1 \text{ mol } CH_4} \times \frac{1 \text{ mol } CH_4}{16.04 \text{ g } CH_4(g)} = -55.50 \text{ kJ/g } CH_4$

for $C_8H_{18}(\ell)$: $\Delta H^o_{rxn} = \frac{-5451.4 \text{ kJ}}{1 \text{ mol } C_8H_{18}} \times \frac{1 \text{ mol } C_8H_{16}}{114.22 \text{ g } C_8H_{18}} = -47.723 \text{ kJ/g } C_8H_{18}$

for $C_{10}H_{22}(\ell)$: $\Delta H^o_{rxn} = \frac{-6829.2 \text{ kJ}}{1 \text{ mol } C_{10}H_{22}} \times \frac{1 \text{ mol } C_{10}H_{22}}{142.28 \text{ g } C_{10}H_{22}} = -48.00 \text{ kJ/g } C_{10}H_{22}$

Since ΔH^o_{rxn} (kJ/g) is most negative for the combustion of $CH_4(g)$, the burning of CH_4**(g)** at standard conditions produces the most heat per gram.

15-66. **Refer to Section 15-8.** . . . .

For a reaction occurring in the gaseous phase, the net enthalpy change, ΔH^o_{rxn}, equals the sum of the bond energies in the reactants minus the sum of the bond energies in the products:

$$\Delta H^o_{rxn} = \Sigma \text{ B.E.}_{reactants} - \Sigma \text{ B.E.}_{products}$$

15-68. **Refer to Section 15-8.** . . . .

In ammonia, NH_3, the N-H bonds are equivalent. The same amount of energy is required to remove any one of the three H atoms by breaking an N-H bond. However, as soon as we remove a H, we no longer have an NH_3 molecule, but an NH_2 radical. It requires a different amount of energy to break an N-H bond in NH_2 and form an NH radical. Likewise, it requires a different amount of energy to remove the third H atom by breaking the N-H bond in the NH radical.

15-70. **Refer to Section 15-8 and Appendix K.** . . . .

The ΔH^o_{rxn} of this reaction: $NH_3(g) \rightarrow N(g) + 3H(g)$

is equal to 3 times the average N-H bond energy in NH_3, since this reaction involves the breaking of 3 N-H bonds.

$$\Delta H^o_{rxn} = [3\Delta H^o_f \text{ H(g)} + \Delta H^o_f \text{ N(g)}] - \Delta H^o_f \text{ NH}_3\text{(g)}$$
$$= [(3 \text{ mol})(218.0 \text{ kJ/mol}) + (1 \text{ mol})(472.704 \text{ kJ/mol})] - [(1 \text{ mol})(-46.11 \text{ kJ/mol})]$$
$$= 1172.8 \text{ kJ}$$

Therefore, the average bond energy of an N-H bond in $NH_3(g)$ is (1172.8/3) kJ or **390.94 kJ.**

15-72. **Refer to Section 15-8 and Appendix K.** . . . .

The ΔH^o_{rxn} of this reaction: $SF_6(g) \rightarrow S(g) + 6F(g)$

is equal to 6 times the average S-F bond energy in $SF_6(g)$ since this reaction involves the breaking of 6 S-F bonds.

$$\Delta H^o_{rxn} = [\Delta H^o_f \text{ S(g)} + 6\Delta H^o_f \text{ F(g)}] - \Delta H^o_f \text{ SF}_6\text{(g)}$$
$$= [(1 \text{ mol})(278.8 \text{ kJ/mol}) + (6 \text{ mol})(78.99 \text{ kJ/mol})] - [(1 \text{ mol})(-1209 \text{ kJ/mol})]$$
$$= 1962 \text{ kJ}$$

Therefore, the average bond energy of an S-F bond in $SF_6(g)$ is (1962/6) kJ or **327.0 kJ.**

15-74. **Refer to Section 15-8, Table 15-2 and Example 15-7.** . . . .

Balanced reaction: $BrF_3(g) + 2H_2(g) \rightarrow HBr(g) + 3HF(g)$

$\Delta H^o_{rxn} = \Sigma\, B.E._{reactants} - \Sigma\, B.E._{products}$ in the gas phase only

$= (3B.E._{Br-F} + 2B.E._{H-H}) - (B.E._{H-Br} + 3B.E._{H-F})$

$= [(3\ mol)(197\ kJ/mol) + (2\ mol)(435\ kJ/mol)]$
$\quad - [(1\ mol)(368\ kJ/mol) + (3\ mol)(569\ kJ/mol)]$

$=$ **-614 kJ**

15-76. **Refer to Section 15-8, Table 15-2 and Example 15-8.** . . . .

$\Delta H^o_{rxn} = [5B.E._{C-H} + B.E._{C-C} + B.E._{C-N} + 2B.E._{N-H}]$
$\quad - [4B.E._{C-H} + B.E._{C=C} + 3B.E._{N-H}]$

Substituting,

$54.68\ kJ = [(5\ mol)(414\ kJ/mol) + (1\ mol)(347\ kJ/mol) + (1\ mol)(B.E._{C-N})$
$\quad + (2\ mol)(389\ kJ/mol)]$
$\quad - [(4\ mol)(414\ kJ/mol) + (1\ mol)(611\ kJ/mol) + (3\ mol)(389\ kJ/mol)]$

$54.68\ kJ = (1\ mol)(B.E._{C-N}) - 239\ kJ$

$B.E._{C-N} =$ **294 kJ/mol**

Table 15-2 gives the bond energy for an average C-N bond as 293 kJ/mol. The two values agree very well.

15-78. **Refer to Section 15-9 and Appendix K.** . . . .

The Born-Haber cycle for the formation of KI(s) is calculated as follows:

$K(s) \rightarrow K(g)$	ΔH_{subl}	= 90 kJ
$K(g) \rightarrow K^+(g) + e^-$	ΔH_{ie}	= 419 kJ
$1/2\ I_2(s) \rightarrow 1/2\ I_2(g)$	$1/2\ \Delta H_{subl}$	= 31 kJ
$1/2\ I_2(g) \rightarrow I(g)$	$1/2\ \Delta H_{diss}$	= 76 kJ
$I(g) + e^- \rightarrow I^-(g)$	ΔH_{ea}	= -295 kJ
$K^+(g) + I^-(g) \rightarrow KI(s)$	ΔH_{xtal}	= ? kJ
$K(s) + 1/2\ I_2(s) \rightarrow KI(s)$	ΔH_{rxn}	= -328 kJ = ΔH^o_f

Therefore,

$-328\ kJ = 90\ kJ + 419\ kJ + 31\ kJ + 76\ kJ - 295\ kJ + \Delta H_{xtal}$

$\Delta H_{xtal} =$ **-649 kJ**

15-80. **Refer to Section 15-9.** . . . .

The Born-Haber cycle for the formation of $CaBr_2(s)$ is calculated as follows:

$Ca(s) \rightarrow Ca(g)$	ΔH_{subl}	= 193 kJ
$Ca(g) \rightarrow Ca^+(g) + e^-$	ΔH_{ie1}	= 590 kJ
$Ca^+(g) \rightarrow Ca^{2+}(g) + e^-$	ΔH_{ie2}	= 1145 kJ
$Br_2(\ell) \rightarrow Br_2(g)$	ΔH_{vap}	= 315 kJ
$Br_2(g) \rightarrow 2Br(g)$	ΔH_{diss}	= 193 kJ
$2Br(g) + 2e^- \rightarrow 2Br^-(g)$	$2\Delta H_{ea}$	= -648 kJ
$Ca^{2+}(g) + 2Br^-(g) \rightarrow CaBr_2(s)$	ΔH_{xtal}	= ? kJ
$Ca(s) + Br_2(\ell) \rightarrow CaBr_2(s)$	ΔH_{rxn}	= -675 kJ = ΔH^o_f

Therefore,

$$-675 \text{ kJ} = 193 \text{ kJ} + 590 \text{ kJ} + 1145 \text{ kJ} + 315 \text{ kJ} + 193 \text{ kJ} - 648 \text{ kJ} + \Delta H_{xtal}$$

$$\Delta H_{xtal} = -2463 \text{ kJ}$$

15-82. **Refer to Sections 15-1 and 15-9, and Exercises 15-78 and 15-80 Solutions.** • •

In the formation of ionic solids from their elements, the ionization of a metal atom is just one of several steps as illustrated by the Born-Haber cycle. In general, the endothermicity of the ionization step is more than compensated by the exothermicity of the crystal lattice energy of the ionic solid.

15-84. **Refer to Section 15-11.** •

The Third Law of Thermodynamics states that the entropy of a pure, perfect crystalline substance is zero at 0 K. This means that all substances have some disorder except when the substance is a pure, perfect, motionless crystal at absolute zero Kelvin. This also implies that the entropy of a substance can be expressed on an absolute basis.

15-86. **Refer to Section 15-11.** •

At temperatures above 0 K, a substance has some motion associated with it: vibrational, rotational, translational, etc. It therefore has some measure of disorder and has a positive value of entropy.

15-88. **Refer to Section 15-11 and Table 15-4.** • • • • • • • • • • • • • • • • •

(a) decrease in entropy - The pennies are placed in a definite pattern (separated and heads up) and are more ordered on the table.

(b) increase in entropy - The pennies become disordered when put back into the bag.

(c) decrease in entropy - The solid phase is always more ordered than the liquid phase of a substance.

(d) increase in entropy - The gas phase is always more disordered than the liquid phase of a substance.

(e) increase in entropy - The reaction is producing 2 moles of gas from 1 mole of gas. A system with 2 moles of gas is more disordered than one with only 1 mole of gas.

(f) decrease in entropy - The reaction is the opposite of (e).

15-90. **Refer to Section 15-11.** •

There are at least 6 processes that effect entropy increases in a system:

(1) A temperature increase causes an increase in random motion of a system's particles. This results in an increase in disorder and an increase in the system's entropy.

(2) Phase changes such as melting, vaporization and sublimation will increase the the amount of disorder and the entropy of a system.

(3) Expansion of a gas sample increases the amount of disorder and the entropy of a system.

(4) Simple mixing of gases with no chemical reaction occurring increases the disorder and the entropy of a system.

(5) Diffusion of particles of one substance into the particles of another substance increases the disorder of the system and its entropy.

(6) Processes which increase the number of particles in the system increases disorder and entropy in the system. For example,
   (a) dissociation of homonuclear diatomic molecules: $Cl_2(g) \rightarrow 2Cl(g)$
   (b) decomposition of compounds: $MgCO_3(s) \rightarrow MgO(s) + CO_2(g)$
   (c) dissolution of ionic solids in water: $NaCl(s) \rightarrow Na^+(aq) + Cl^-(aq)$
   (d) reaction causing an increase in the number of moles of gaseous substances: $2H_2O(\ell) \rightarrow 2H_2(g) + O_2(g)$

15-92. **Refer to Section 15-11 and Exercise 15-90 Solution.** . . . .

(a) entropy increases (b) entropy decreases (c) entropy decreases

(d) entropy decreases

15-94. **Refer to Section 15-12.** . . . .

Consider the boiling of a pure liquid at constant pressure.

(a) $\Delta S_{system} > 0$ (b) $\Delta H_{system} > 0$ (c) $\Delta T_{system} = 0$

15-96. **Refer to Section 15-12.** . . . .

The spontaneity of a process is favored if it involves a negative change in enthalpy and a positive change in entropy. In other words, if the process is exothermic ($\Delta H < 0$) and increases the disorder of the system ($\Delta S > 0$), spontaneity is favored.

15-98. **Refer to Section 15-12 and Exercise 15-96 Solution.**

For a reaction to be spontaneous, free energy must decrease and so, $\Delta G < 0$. The Gibbs-Helmholtz equation states that $\Delta G = \Delta H + T\Delta S$ at constant T and P. An exothermic reaction ($\Delta H < 0$) is spontaneous ($\Delta G < 0$) if ΔH is the major factor affecting spontaneity. However, an endothermic reaction ($\Delta H > 0$) can be spontaneous if there is a large enough increase in disorder ($\Delta S > 0$) at a particular temperature to offset the positive ΔH value and make ΔG negative.

15-100. **Refer to Section 15-12.** . . . .

The Gibbs-Helmholtz equation states:

$$\Delta G_{rxn} = \Delta H_{rxn} - T\Delta S_{rxn} \qquad \text{at constant } T \text{ and } P$$

Therefore, absolute values of G, H or S are not required to predict spontaneity, only the change in free energy, ΔG_{rxn}, the change in enthalpy, ΔH_{rxn}, and the change in entropy, ΔS_{rxn}.

15-102. **Refer to Section 15-12.** . . . .

Only two of the five conditions guarantee that a reaction would be spontaneous at constant temperature and pressure: (b) ΔG decreases and (e) ΔH decreases and ΔS increases.

15-104. **Refer to Section 15-12, Example 15-11 and Appendix K.** . . . .

Plan: Calculate ΔH_{rxn} and ΔS_{rxn}, then use the Gibbs-Helmholtz equation to determine ΔG_{rxn}.

(a) $3NO_2(g) + H_2O(\ell) \rightarrow 2HNO_3(\ell) + NO(g)$

$$\Delta H^o_{rxn} = [2\Delta H^o_f\ HNO_3(\ell) + \Delta H^o_f\ NO(g)] - [3\Delta H^o_f\ NO_2(g) + \Delta H^o_f\ H_2O(\ell)]$$
$$= [(2\text{ mol})(-174\text{ kJ/mol}) + (1\text{ mol})(90.25\text{ kJ/mol})] - [(3\text{ mol})(33.2\text{ kJ/mol}) + (1\text{ mol})(-285.8\text{ kJ/mol})]$$
$$= -71.75\text{ kJ}$$

$$\Delta S^o_{rxn} = [2S^o_{HNO_3(\ell)} + S^o_{NO(g)}] - [3S^o_{NO_2(g)} + S^o_{H_2O(\ell)}]$$
$$= [(2\text{ mol})(155.6\text{ J/mol·K}) + (1\text{ mol})(210.7\text{ J/mol·K})] - [(3\text{ mol})(240.0\text{ J/mol·K}) + (1\text{ mol})(69.91\text{ J/mol·K}]$$
$$= -268.0\text{ J/K}$$

$$\Delta G^o_{rxn} = \Delta H^o_{rxn} - T\Delta S^o_{rxn} = -71.75\text{ kJ} - (298.15\text{ K})(-0.2680\text{ kJ/K}) = \mathbf{8.15\text{ kJ}}$$

(b) $SnO_2(s) + 2CO(g) \rightarrow 2CO_2(g) + Sn(s,white)$

$$\Delta H^o_{rxn} = [2\Delta H^o_f\ CO_2(g) + \Delta H^o_f\ Sn(s)] - [\Delta H^o_f\ SnO_2(s) + 2\Delta H^o_f\ CO(g)]$$
$$= [(2\text{ mol})(-393.5\text{ kJ/mol}) + (1\text{ mol})(0\text{ kJ/mol})] - [(1\text{ mol})(-580.7\text{ kJ/mol}) + (2\text{ mol})(-110.5\text{ kJ/mol})]$$
$$= 14.7\text{ kJ}$$

$$\Delta S^o_{rxn} = [2S^o_{CO_2(g)} + S^o_{Sn(s)}] - [S^o_{SnO_2(s)} + 2S^o_{CO(g)}]$$
$$= [(2\text{ mol})(213.6\text{ J/mol·K}) + (1\text{ mol})(51.55\text{ J/mol·K})] - [(1\text{ mol})(52.3\text{ J/mol·K}) + (2\text{ mol})(197.6\text{ J/mol·K})]$$
$$= 31.2\text{ J/K}$$

$$\Delta G^o_{rxn} = \Delta H^o_{rxn} - T\Delta S^o_{rxn} = 14.7\text{ kJ} - (298.15\text{ K})(0.0312\text{ kJ/K}) = \mathbf{5.4\text{ kJ}}$$

(c) $2Na(s) + 2H_2O(\ell) \rightarrow 2NaOH(aq) + H_2(g)$

$$\Delta H^o_{rxn} = [2\Delta H^o_f\ NaOH(aq) + \Delta H^o_f\ H_2(g)] - [2\Delta H^o_f\ Na(s) + 2\Delta H^o_f\ H_2O(\ell)]$$
$$= [(2\text{ mol})(-469.6\text{ kJ/mol}) + (1\text{ mol})(0\text{ kJ/mol})] - [(2\text{ mol})(0\text{ kJ/mol}) + (2\text{ mol})(-285.8\text{ kJ/mol})]$$
$$= -367.6\text{ kJ}$$

$$\Delta S^o_{rxn} = [2S^o_{NaOH(aq)} + S^o_{H_2(g)}] - [2S^o_{Na(s)} + 2S^o_{H_2O(\ell)}]$$
$$= [(2\text{ mol})(49.8\text{ J/mol·K}) + (1\text{ mol})(130.6\text{ J/mol·K})] - [(2\text{ mol})(51.0\text{ J/mol·K}) + (2\text{ mol})(69.91\text{ J/mol·K})]$$
$$= -11.62\text{ J/K}$$

$$\Delta G^o_{rxn} = \Delta H^o_{rxn} - T\Delta S^o_{rxn} = -367.6\text{ kJ} - (298.15\text{ K})(-0.01162\text{ kJ/K}) = \mathbf{-364.1\text{ kJ}}$$

15-106. **Refer to Section 15-12.** . . . .

The Gibbs-Helmholtz equation states: $\Delta G = \Delta H - T\Delta S$ at constant T and P.

	ΔH	ΔS	ΔG	**Conclusion**
(a)	+	+	+ or -	spontaneous at higher temperatures
(b)	+	-	+	nonspontaneous at all temperatures
(c)	-	+	-	spontaneous at all temperatures
(d)	-	-	- or +	spontaneous at lower temperatures

15-108. **Refer to Section 15-12, Examples 15-13 and 15-14, and Exercise 15-106 Solution.•**

Plan: Evaluate ΔH_{rxn} and ΔS_{rxn}. To assess the temperature range over which the reaction is spontaneous, use the signs of ΔH and ΔS and the Gibbs-Helmholtz equation, $\Delta G = \Delta H - T\Delta S$. Assume that ΔH and ΔS are independent of temperature.

(a) Balanced equation: $2Al(s) + 3Cl_2(g) \rightarrow 2AlCl_3(s)$

$$\Delta H^o_{rxn} = [2\Delta H^o_f\ AlCl_3(s)] - [2\Delta H^o_f\ Al(s) + 3\Delta H^o_f\ Cl_2(g)]$$
$$= [(2\ mol)(-704.2\ kJ/mol)] - [(2\ mol)(0\ kJ/mol) + (3\ mol)(0\ kJ/mol)]$$
$$= -1408.4\ kJ$$

$$\Delta S^o_{rxn} = [2S^o_{AlCl_3(s)}] - [2S^o_{Al(s)} + 3S^o_{Cl_2(g)}]$$
$$= [(2\ mol)(110.7\ J/mol\cdot K)] - [(2\ mol)(28.3\ J/mol\cdot K) + (3\ mol)(223.0\ J/mol\cdot K)]$$
$$= -504.2\ J/K$$

At equilibrium, $\Delta G = 0 = \Delta H_{rxn} - T_{eq}\Delta S_{rxn}$

$$T_{eq} = \frac{\Delta H_{rxn}}{\Delta S_{rxn}} = \frac{-1408.4\ kJ}{-0.5042\ kJ/K} = 2793\ K$$

Since both ΔH_{rxn} and ΔS_{rxn} are negative, the reaction is spontaneous at lower temperatures. Therefore, the temperature range over which the reaction is spontaneous is **T < 2793 K.**

(b) Balanced equation: $2NOCl(g) \rightarrow 2NO(g) + Cl_2(g)$

$$\Delta H^o_{rxn} = [2\Delta H^o_f\ NO(g) + \Delta H^o_f\ Cl_2(g)] - [2\Delta H^o_f\ NOCl(g)]$$
$$= [(2\ mol)(90.25\ kJ/mol) + (1\ mol)(0\ kJ/mol)] - [(2\ mol)(52.59\ kJ/mol)]$$
$$= 75.32\ kJ$$

$$\Delta S^o_{rxn} = [2S^o_{NO(g)} + S^o_{Cl_2(g)}] - [2S^o_{NOCl(g)}]$$
$$= [(2\ mol)(210.7\ J/mol\cdot K) + (1\ mol)(223.0\ J/mol\cdot K)] - [(2\ mol)(264\ J/mol\cdot K)]$$
$$= 116\ J/K$$

$$T_{eq} = \frac{\Delta H_{rxn}}{\Delta S_{rxn}} = \frac{75.32\ kJ}{0.116\ kJ/K} = 647\ K$$

Since both ΔH_{rxn} and ΔS_{rxn} are positive, the temperature range for spontaneity is **T > 649 K.**

(c) Balanced equation: $4NO(g) + 6H_2O(g) \rightarrow 4NH_3(g) + 5O_2(g)$

$$\Delta H^o_{rxn} = [4\Delta H^o_f\ NH_3(g) + 5\Delta H^o_f\ O_2(g)] - [4\Delta H^o_f\ NO(g) + 6\Delta H^o_f\ H_2O(g)]$$
$$= [(4\ mol)(-46.11\ kJ/mol) + (5\ mol)(0\ kJ/mol)] - [(4\ mol)(90.25\ kJ/mol) + (6\ mol)(-241.8\ kJ/mol)]$$
$$= +905.4\ kJ$$

$$\Delta S^o_{rxn} = [4S^o_{NH_3(g)} + 5S^o_{O_2(g)}] - [4S^o_{NO(g)} + 6S^o_{H_2O(g)}]$$
$$= [(4\ mol)(192.3\ J/mol\cdot K) + (5\ mol)(205.0\ J/mol\cdot K)] - [(4\ mol)(210.7\ J/mol\cdot K) + (6\ mol)(188.7\ J/mol\cdot K)]$$
$$= -180.8\ J/K$$

Since ΔH_{rxn} is positive and ΔS_{rxn} is negative, the reaction is <u>not</u> **spontaneous at any temperature.**

(d) Balanced equation: $2PH_3(g) \rightarrow 3H_2(g) + 2P(g)$

$$\Delta H^o_{rxn} = [3\Delta H^o_f\ H_2(g) + 2\Delta H^o_f\ P(g)] - [2\Delta H^o_f\ PH_3(g)]$$
$$= [(3\ mol)(0\ kJ/mol) + (2\ mol)(314.6\ kJ/mol)] - [(2\ mol)(5.4\ kJ/mol)]$$
$$= 618.4\ kJ$$

$$\Delta S^o_{rxn} = [3S^o_{H_2(g)} + 2S^o_{P(g)}] - [2S^o_{PH_3(g)}]$$
$$= [(3\ mol)(130.6\ J/mol\cdot K) + (2\ mol)(153.1\ J/mol\cdot K] - [(2\ mol)(210.1\ J/mol\cdot K)]$$
$$= 277.8\ J/K$$

$$T_{eq} = \frac{\Delta H_{rxn}}{\Delta S_{rxn}} = \frac{618.4\ kJ}{0.2778\ kJ/K} = 2226\ K$$

Since both ΔH_{rxn} and ΔS_{rxn} are positive, the temperature range for spontaneity is T **> 2226 K.**

15-110. **Refer to Section 15-12, Example 15-10 and Exercise 15-108 Solution.** • • • •

(a) $\Delta G^o_{rxn} = \Delta G^o_f\ C(graphite) - \Delta G^o_f\ C(diamond)$

$$= (1\ mol)(0\ kJ/mol) - (1\ mol)(2.900\ kJ/mol) = -2.900\ kJ/mol$$

Since $\Delta G^o_{rxn} < 0$, the reaction is spontaneous at standard conditions.

(b) From common experience, we know that diamonds do not readily change to graphite. Thermodynamics tells us if a reaction will occur but it says nothing about the rate at which a reaction will occur. The rate at which diamonds change to graphite is very, very slow.

(c) $\Delta H^o_{rxn} = \Delta H^o_f\ C(graphite) - \Delta H^o_f\ C(diamond)$

$$= (1\ mol)(0\ kJ/mol) - (1\ mol)(1.897\ kJ/mol) = -1.897\ kJ$$

$$\Delta S^o_{rxn} = S^o_{C(graphite)} - S^o_{C(diamond)}$$
$$= (1\ mol)(5.740\ J/mol\cdot K) - (1\ mol)(2.38\ J/mol\cdot K) = 3.36\ J/K$$

Since ΔH_{rxn} is negative and ΔS_{rxn} is positive, the reaction is spontaneous at all temperatures, i.e., there is no temperature at which diamond and graphite are in equilibrium.

15-112. **Refer to Section 15-12, Appendix K and Example 15-12.** • • • • • • • • • • •

(a) The process is: $H_2O(\ell) \rightarrow H_2O(g)$

$$\Delta H^o_{rxn} = \Delta H^o_f\ H_2O(g) - \Delta H^o_f\ H_2O(\ell)$$
$$= (1\ mol)(-241.8\ kJ/mol) - (1\ mol)(-285.8\ kJ/mol) = 44.0\ kJ$$

$$\Delta S^o_{rxn} = S^o_{H_2O(g)} - S^o_{H_2O(\ell)}$$
$$= (1\ mol)(188.7\ J/mol\cdot K) - (1\ mol)(69.91\ J/mol\cdot K) = 118.8\ J/K$$

$$T_{eq} = \frac{\Delta H_{rxn}}{\Delta S_{rxn}} = \frac{44.0\ kJ}{0.1188\ kJ/K} = 370\ K \text{ or } \mathbf{97^oC}$$

(b) The known boiling point of water is, of course, 100°C. The discrepancy is because we assumed that the standard values of enthalpy of formation and entropy in Appendix K are independent of temperature. However, these tabulated values were determined at 25°C; we are using them to solve a problem at 100°C. Nevertheless, this assumption allows us to estimate the boiling point of water with reasonable accuracy.

15-114. **Refer to Section 15-12, Appendix K and Example 15-12.** . . . .

The sublimation process is: $I_2(s) \rightarrow I_2(g)$.

$$\Delta H^o_{rxn} = \Delta H^o_f\ I_2(g) - \Delta H^o_f\ I_2(s)$$
$$= (1\ mol)(62.44\ kJ/mol) - (1\ mol)(0\ kJ/mol)$$
$$= 62.44\ kJ$$

$$\Delta S^o_{rxn} = S^o_{I_2(g)} - S^o_{I_2(s)}$$
$$= (1\ mol)(260.6\ J/mol\cdot K) - (1\ mol)(116.1\ J/mol\cdot K)$$
$$= 144.5\ J/K$$

$$T_{eq} = \frac{\Delta H_{rxn}}{\Delta S_{rxn}} = \frac{62.44\ kJ}{0.1445\ kJ/K} = 432.1\ K \text{ or } \mathbf{159.1^oC}$$

15-116. **Refer to Section 15-12 and Exercise 15-106.** . . . .

Dissociation reactions, such as $HCl(g) \rightarrow H(g) + Cl(g)$, require energy to break bonds and therefore are endothermic with positive ΔH values. The ΔS values for such reactions are positive since 2 or more particles are being formed from 1 molecule, causing the system to be more disordered. Under the circumstances when ΔH and ΔS are both positive, the spontaneity of the reaction is favored at higher temperatures.

15-118. **Refer to Section 15-12.** . . . .

(a) $H_2O(\ell) \rightarrow H_2O(s)$

Water does not freeze at room temperature and pressure. Therefore, this reaction is nonspontaneous at these conditions.

(b) $2O(g) \rightarrow O_2(g)$

Since $O_2(g)$ is the standard state of oxygen at 25°C and 1 atm, this reaction is expected to be spontaneous under these conditions and quite fast.

(c) $2H_2O(\ell) \rightarrow 2H_2(g) + O_2(g)$

Since water does not break up into its elements at room temperature and pressure, this reaction is nonspontaneous under these conditions. Evidence for this is that our atmosphere consists primarily of N_2 and O_2; in comparison, there is very little NO present, except perhaps in smog. If this had been a rapid, spontaneous reaction, life as we know it would not exist on this planet.

(d) $N_2(g) + O_2(g) \rightarrow 2NO(g)$

Since $O_2(g)$ and $N_2(g)$ are the standard states of oxygen and nitrogen, respectively, at 25°C and 1 atm, we can surmise that this reaction is nonspontaneous under these conditions.

(e) $NaOH(s) + 1/2\ H_2(g) \rightarrow Na(s) + H_2O(\ell)$

We learned in Chapter 9 that sodium metal reacts vigorously with liquid water at room temperature and pressure. Therefore, under these conditions, the reverse reaction as shown above is nonspontaneous.

15-120. **Refer to Section 15-12 and Appendix K.** • • • • • • • • • • • • • • • • • • •

Balanced equation: $4CH_3H_5(NO_3)_3(\ell) \rightarrow 12CO_2(g) + 10H_2O(g) + 6N_2(g) + O_2(g)$
nitroglycerin

(a) We are given that the reaction releases 6.28 kJ/g nitroglycerin. Therefore,

$$\Delta H^o = -6.28 \text{ kJ/g nitroglycerin} \times \frac{227 \text{ g nitroglycerin}}{1 \text{ mol nitroglycerin}}$$

$$= \mathbf{-1430 \text{ kJ/mol nitroglycerin}}$$

(b) ΔH^o_{rxn} = -1430 kJ/mol nitroglycerin × 4 mol nitroglycerin = **-5720 kJ**

(c) $$\Delta H^o_{rxn} = [12\Delta H_f\ CO_2(g) + 10\Delta H_f\ H_2O(g) + 6\Delta H_f\ N_2(g) + \Delta H_f\ O_2(g)] - [4\Delta H_f \text{ nitroglycerin}]$$

Substituting,

$$-5720 \text{ kJ} = [(12 \text{ mol})(-393.5 \text{ kJ/mol}) + (10 \text{ mol})(-241.8 \text{ kJ/mol}) + (6 \text{ mol})(0 \text{ kJ/mol}) + (1 \text{ mol})(0 \text{ kJ/mol})] - [(4 \text{ mol})(\Delta H_f \text{ nitroglycerin})]$$

Solving, ΔH^o_f nitroglycerin = **-355 kJ/mol**

(d) ΔH^o_f is a fairly large negative value, which suggests that nitroglycerin is stable. However, in the above reaction, the products are more stable. The reaction should be spontaneous.

(e) The entropy change for the reaction is expected to be positive and large since the change in the number of moles of gas, Δn = 29. This is a large and positive number, indicating a large increase in the disorder of the system.

(f) Since $\Delta H_{rxn} < 0$ and $\Delta S_{rxn} > 0$, the reaction is spontaneous at all temperatures.

(g) For the reaction as written,

$$\text{work, } w = -\Delta n_{gas}RT = -(29 \text{ moles})(8.314 \text{ J/mol·K})(298 \text{ K}) = -71800 \text{ J}$$

The work done by exploding 50.0 g of nitroglycerin is therefore,

$$w = 50.0 \text{ g nitroglycerin} \times \frac{1 \text{ mol nitroglycerin}}{227 \text{ g nitroglycerin}} \times \frac{-7.15 \times 10^4 \text{ J}}{4 \text{ mol nitroglycerin}}$$

$$= \mathbf{-3.95 \times 10^3 \text{ J}}$$

16 Chemical Kinetics

16-2. **Refer to Sections 16-1 and 16-2.** . . . .

The collision theory of reaction rates states that molecules, atoms or ions must collide in order to react. Moreover, an effective collision must occur in which the reacting species must have (1) at least a minimum amount of energy in order to break old bonds and make new ones, and (2) the proper orientation toward each other.

Transition state theory complements collision theory. When particles collide with enough energy to react, called the activation energy, E_a, the reactants form a short-lived, high energy intermediate, or transition state, before forming the products.

16-4. **Refer to the Introduction to Chapter 16.** . . . .

In Chapter 15, we learned that reactions which are thermodynamically favorable have negative ΔG values and occur spontaneously as written. However, thermodynamics cannot be used to determine the rate of a reaction. Kinetically favorable reactions must be thermodynamically favorable <u>and</u> have a low enough activation energy to occur at a reasonable rate at a certain temperature.

16-6. **Refer to Section 16-6 and Figures 16-6 and 16-7.** . . . .

Catalysts are substances which increase the rate of reaction when added to a system by providing an alternative mechanism with a lower activation energy. Although a catalyst may enter into a reaction, it would not appear in the balanced reaction, i.e., if it is a reactant in one step, it is a product in another. It is not consumed during a reaction.

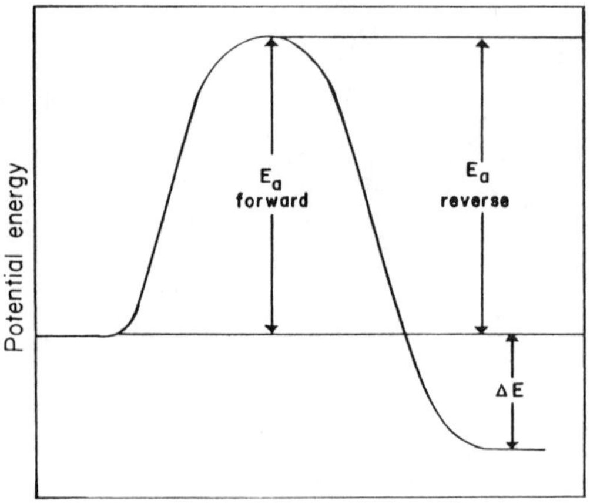

Reaction coordinate for uncatalyzed reaction

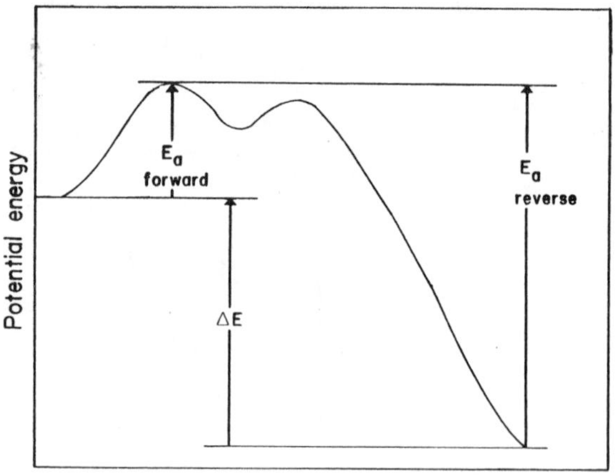

Reaction coordinate for catalyzed reaction

16-8. **Refer to Section 16-7.** . . . .

A reaction mechanism is the pathway or sequence of steps by which reactants are converted into products. Most reaction mechanisms involve only unimolecular isomerizations or decompositions, or bimolecular collisions. Multimolecular collisions are generally not included since simultaneous collisions among 3 or more particles are very improbable.

16-10. **Refer to Section 16-6 and Exercise 16-6 Solution.** . . . .

A homogeneous catalyst is a catalyst that exists in the same phase (solid, liquid or gas) as the reactants. A heterogeneous catalyst, on the other hand, is a catalyst that exists in a different phase from the reactants. An inhibitor is the opposite of a catalyst; it slows down a reaction by providing a pathway with a higher activation energy.

16-12. **Refer to Section 16-4.** . . . .

Plan: Use dimensional analysis and the rate-law expression to determine the units of *k*, the rate constant, in the following general equation:

Rate (M/s) = $k[A]^x$ where x = the overall order of the reaction
[A] = the reactant concentration

	Overall Reaction Order	Example	Units of *k*
(a)	1	Rate = $k[A]$	$(M/s)/M = s^{-1}$
(b)	2	Rate = $k[A]^2$	$(M/s)/M^2 = M^{-1}\cdot s^{-1}$
(c)	3	Rate = $k[A]^3$	$(M/s)/M^3 = M^{-2}\cdot s^{-1}$
(d)	1.5	Rate = $k[A]^{1.5}$	$(M/s)/M^{1.5} = M^{-0.5}\cdot s^{-1}$

16-14. **Refer to Section 16-4 and Examples 16-1 and 16-2.** . . . .

The form of the rate-law expression: Rate = $k[A]^x[B]^y[C]^z$

Step 1: Rate dependence on [A]. Consider Experiments 1 and 3:

Method 1: By observation, [B] and [C] do not change; [A] increases by a factor of 3. However, the reaction rate does not change. Therefore, changing [A] does not affect reaction rate and the reaction is zeroth order with respect to A. In all subsequent determinations, the effect of A can be ignored.

Method 2: A mathematical solution is obtained by substituting the experimental values of Experiments 1 and 3 into rate-law expressions and dividing the latter by the former. Note: the calculations are easier when the experiment with the larger rate is in the numerator.

$$\frac{\text{Expt 3}}{\text{Expt 1}} \quad \frac{4.0 \times 10^{-4}\ M/\text{min}}{4.0 \times 10^{-4}\ M/\text{min}} = \frac{k(0.30\ M)^x(0.20\ M)^y(0.10\ M)^z}{k(0.10\ M)^x(0.20\ M)^y(0.10\ M)^z}$$

$$1 = 3^x$$
$$x = 0$$

Step 2: Rate dependence on [C]. Consider Experiments 1 and 2:

Method 1: [B] does not change; [C] changes by a factor of 3; the reaction rate also changes by a factor of 3 (= $1.2 \times 10^{-3}/4.0 \times 10^{-4}$). The reaction rate is directly proportional to [C] and z must be equal to 1. The reaction is first order with respect to C.

Method 2:

$$\frac{\text{Expt 2}}{\text{Expt 1}} \quad \frac{1.2 \times 10^{-3}\ M/\text{min}}{4.0 \times 10^{-4}\ M/\text{min}} = \frac{k(0.20\ M)^0(0.20\ M)^y(0.30\ M)^z}{k(0.10\ M)^0(0.20\ M)^y(0.10\ M)^z}$$

$$3 = 3^z$$
$$z = 1$$

Step 3: Rate dependence on [B]. Consider Experiments 2 and 4:

Method 1: [C] does not change; [B] changes by a factor of 3; the reaction rate changes by a factor of 3 (= $3.6 \times 10^{-3}/1.2 \times 10^{-3}$). The reaction rate is directly proportional to [B] and y must be equal to 1. The reaction is first order with respect to B.

Method 2:

$$\frac{\text{Expt 4}}{\text{Expt 2}} \qquad \frac{3.6 \times 10^{-3}\ M/\text{min}}{1.2 \times 10^{-3}\ M/\text{min}} = \frac{k(0.40\ M)^0(0.60\ M)^y(0.30\ M)^2}{k(0.20\ M)^0(0.20\ M)^y(0.30\ M)^1}$$

$$3 = 3^y$$
$$y = 1$$

The rate-law expression is: Rate = $k[A]^0[B]^1[C]^1$ = k[B][C]. To calculate the value of k, substitute the values from only one of the experiments into the rate-law expression and solve for k. If we use the data from Experiment 1,

$$4.0 \times 10^{-4}\ M/\text{min} = k(0.20\ M)(0.10\ M)$$
$$k = 2.0 \times 10^{-2}\ M^{-1}\cdot\text{min}^{-1}$$

The rate-law expression is now: **Rate = $(2.0 \times 10^{-2}\ M^{-1}\cdot\text{min}^{-1})$[B][C]**

16-16. **Refer to Section 16-4.** . . . .

The overall reaction order is the sum of the individual orders with respect to the reactants in the rate-law expression. The rate-law expression, Rate = $k[N_2O_5]$, states that the reaction is first order with respect to N_2O_5 and first order overall.

16-18. **Refer to Section 16-4.** . . . .

The simplest approach to this problem is to assume that the initial concentrations of A and B_2 are each 1 M. Then the final concentrations of A and B_2 are each 3 M. Let us substitute these values into the rate-law expression,

Rate = $k[A]^2[B_2]$

Initial: Rate = $k(1\ M)^2(1\ M) = k$

Final: Rate = $k(3\ M)^2(3\ M) = 27k$

Therefore, the rate of reaction would **increase** by a factor of **27**.

16-20. **Refer to Section 16-4, Examples 16-1 and 16-2, and Exercise 16-14 Solution.** . .

The form of the rate-law expression: Rate = $k[A]^x[B]^y$

Step 1: Rate dependence on [A]. Consider Experiments 1 and 2:

Method 1: [B] does not change; [A] changes by a factor of 1.5; reaction rate changes by a factor of 1.5. The reaction rate is directly proportional to [A] and x must be equal to 1. The reaction is first order with respect to A.

Method 2:

$$\frac{\text{Expt 2}}{\text{Expt 1}} \qquad \frac{7.5 \times 10^{-6}\ M/\text{s}}{5.0 \times 10^{-6}\ M/\text{s}} = \frac{k(0.30\ M)^x(0.10\ M)^y}{k(0.20\ M)^x(0.10\ M)^y}$$

$$1.5 = 1.5^x$$
$$x = 1$$

Step 2: Rate dependence on [B].

There is no pair of experiments in which [B] is changing and [A] is constant. Therefore, one may choose any 2 experiments in which [B] is varying and use Method 2. If we choose Experiments 1 and 3:

$$\frac{\text{Expt 3}}{\text{Expt 1}} \quad \frac{4.0 \times 10^{-5}\ M/\text{s}}{5.0 \times 10^{-6}\ M/\text{s}} = \frac{k(0.40\ M)^1(0.20\ M)^y}{k(0.20\ M)^1(0.10\ M)^y}$$

$$8 = (2)^1(2)^y$$
$$4 = 2^y$$
$$y = 2$$

Therefore, the reaction is second order with respect to B.

The rate-law expression is now: Rate = $k[A]^1[B]^2 = k[A][B]^2$
Using the data from Experiment 1 to calculate k, we have

$$5.0 \times 10^{-6}\ M/\text{s} = k(0.20\ M)(0.10\ M)^2$$
$$k = 2.5 \times 10^{-3}\ M^{-2}\cdot\text{s}^{-1}$$

The rate-law expression is: **Rate = $(2.5 \times 10^{-3}\ M^{-2}\cdot\text{s}^{-1})[A][B]^2$**

16-22. **Refer to Section 16-4, Examples 16-1 and 16-2, and Exercises 16-14 and 16-20 Solutions.** •

The form of the rate law expression: Rate = $k[A]^x[B]^y$

Step 1: Rate dependence on [A]. Consider Experiments 1 and 2:

Method 1: [B] is constant; [A] changes by a factor of 2; reaction rate changes by a factor of 4. Therefore, the rate increases as the square of [A] and x must be equal to 2. The reaction is second order with respect to A.

Method 2:

$$\frac{\text{Expt 2}}{\text{Expt 1}} \quad \frac{8.0 \times 10^{-4}\ M/\text{s}}{2.0 \times 10^{-4}\ M/\text{s}} = \frac{k(0.20\ M)^x(0.10\ M)^y}{k(0.10\ M)^x(0.10\ M)^y}$$

$$4 = 2^x$$
$$x = 2$$

Step 2: Rate dependence on [B].

Since there are no two experiments in which [A] stays constant and [B] changes, use Method 2 on any two experiments in which [B] changes, e.g., Experiments 2 and 3.

$$\frac{\text{Expt 3}}{\text{Expt 2}} \quad \frac{2.56 \times 10^{-2}\ M/\text{s}}{8.0 \times 10^{-4}\ M/\text{s}} = \frac{k(0.40\ M)^2(0.20\ M)^y}{k(0.20\ M)^2(0.10\ M)^y}$$

$$32 = (2)^2(2)^y$$
$$8 = 2^y$$
$$y = 3$$

Therefore, the reaction is third order with respect to B.

The rate-law expression is now: Rate = $k[A]^2[B]^3$
Using Experiment 1 to solve for k, we have

$$2.0 \times 10^{-4}\ M/\text{s} = k(0.10\ M)^2(0.10\ M)^3$$
$$k = 20\ M^{-4}\cdot\text{s}^{-1}$$

The rate-law expression is: **Rate = $(20\ M^{-4}\cdot\text{s}^{-1})[A]^2[B]^3$**

16-24. **Refer to Section 16-4.** . . . .

The rate-law expression: Rate = $k[A][B]^2$

Plan: (1) Use the data for Experiment 1 and the rate-law expression to calculate the rate constant, k.
(2) Substitute the given values into the complete rate-law expression to determine the reaction rate.

(1) Substituting, $0.25\ M/s = k(1.0\ M)(0.20\ M)^2$

$$k = 6.25\ M^{-2}\cdot s^{-1}$$

(2) Expt 2: Rate = $(6.25\ M^{-2}\cdot s^{-1})(2.0\ M)(0.20\ M)^2 = \mathbf{0.50\ M/s}$

Expt 3: Rate = $(6.25\ M^{-2}\cdot s^{-1})(2.0\ M)(0.40\ M)^2 = \mathbf{2.0\ M/s}$

16-26. **Refer to Section 16-5 and Figure 16-5.** . . . .

The Arrhenius equation can be presented as follows:

$$\log\frac{k_2}{k_1} = \frac{E_a}{2.303R}\left(\frac{T_2 - T_1}{T_1T_2}\right)$$

where k_2/k_1 = ratio of rate constants
E_a = activation energy (J/mol)
R = 8.314 J/mol·K
T = absolute temperature (K)

Substituting,

$$\log(k_2/k_1) = \frac{100 \times 10^3\ \text{J/mol}}{(2.303)(8.314\ \text{J/mol·K})}\left[\frac{310\ \text{K} - 300\ \text{K}}{(300\ \text{K})(310\ \text{K})}\right] = 0.56$$

$$k_2/k_1 = 3.6 \quad \text{or} \quad k_2 = 3.6k_1$$

Therefore, when a reaction has an activation energy of 100 kJ/mol, the reaction rate increases by a factor of 3.6 when the temperature increases from 300 K to 310 K. Under the same conditions, the reaction rate for a reaction with an activation energy of 50 kJ/mol will increase only by a factor of 1.9. It is therefore concluded that the rates of reactions with higher activation energies will increase by a larger factor as temperature increases. The reason is because the fraction of higher energy molecules is larger at higher temperatures. And so, as the temperature changes from 300 K to 310 K, the fraction of molecules having energies higher than 100 kJ/mol increases more than the fraction of molecules having energies higher than 50 kJ/mol.

16-28. **Refer to Section 16-5.** . . . .

(a) The Arrhenius equation: $k = Ae^{-E_a/RT}$ or $\log k = \log A - E_a/2.303RT$

k = specific rate constant (same units as A)
A = $3.98 \times 10^{13}\ s^{-1}$
E_a = activation energy, 160 kJ/mol
R = 8.314 J/mol·K
T = absolute temperature, 298 K

Substituting and solving for k at 25°C,

$$\log k = \log(3.98 \times 10^{13}\ s^{-1}) - \frac{160 \times 10^3\ \text{J/mol}}{(2.303)(8.314\ \text{J/mol·K})(298\ \text{K})} = -14.4$$

$$k = \mathbf{4 \times 10^{-15}\ s^{-1}}$$

(b) At 227°C, $\log k = \log(3.98 \times 10^{13}\ s^{-1}) - \frac{160 \times 10^3\ \text{J/mol}}{(2.303)(8.314\ \text{J/mol·K})(500\ \text{K})}$

$$= -3.1$$

$$k = \mathbf{8 \times 10^{-4}\ s^{-1}}$$

Note on significant figures and logarithms: A logarithm consists of 2 parts. an integer called the characteristic and a decimal named the mantissa. When working with base 10 logarithms, it is the number of significant digits in the mantissa that give the number of significant figures in the antilogarithm.

16-30. **Refer to Section 16-5 and Exercise 16-28 Solution.** . . . .

(a) $\log(k_2/k_1) = \dfrac{E_a}{2.303R}\left(\dfrac{T_2 - T_1}{T_1T_2}\right)$

Substituting,

$$\log \frac{k_2}{9.16 \times 10^{-3}\ s^{-1}} = \frac{88 \times 10^3\ J/mol}{(2.303)(8.314\ J/mol\cdot K)}\left[\frac{298\ K - 273\ K}{(273\ K)(298\ K)}\right] = 1.4$$

$k_2/(9.16 \times 10^{-3}\ s^{-1}) = 3 \times 10^1$ and $k_2 = \mathbf{0.3\ s^{-1}}$

(b) Substituting into the following equation and solving for T_2,

$$\log \frac{k_2}{k_1} = \frac{E_a}{2.303R}\left(\frac{1}{T_1} - \frac{1}{T_2}\right)$$

$$\log \frac{4.00 \times 10^{-2}\ s^{-1}}{9.16 \times 10^{-3}\ s^{-1}} = \frac{88 \times 10^3\ J/mol}{(2.303)(8.314\ J/mol\cdot K)}\left(\frac{1}{273\ K} - \frac{1}{T_2}\right)$$

$$0.640 = (4.60 \times 10^3\ K)(3.66 \times 10^{-3}\ K^{-1} - 1/T_2)$$

$$T_2 = \mathbf{284\ K}\ \text{or}\ \mathbf{11^oC}$$

16-32. **Refer to Section 16-5.** . . . .

The Arrhenius equation: $\log k = \log A - E_a/2.303RT$

Rearranging,

$$\log A = \log k + E_a/2.303RT$$

$$\log A = \log (1.31 \times 10^3\ M^{-1}\cdot s^{-1}) + \frac{55.6 \times 10^3\ J/mol}{(2.303)(8.314\ J/mol\cdot K)(500°C + 273°)}$$

$$\log A = 6.87$$

$$A = \mathbf{7.5 \times 10^6\ M^{-1}\cdot s^{-1}}$$

16-34. **Refer to Section 16-5.** . . . .

In the equation: $\log \dfrac{k_2}{k_1} = \dfrac{E_a}{2.303R}\left(\dfrac{T_2 - T_1}{T_1T_2}\right)$

we are given: $k_2/k_1 = 3.00$, $T_1 = 300$ K and $T_2 = 310$ K

Substituting,

$$\log 3.00 = \frac{E_a}{(2.303)(8.314\ J/mol\cdot K)}\left[\frac{310\ K - 300\ K}{(300\ K)(310\ K)}\right]$$

$$0.477 = (E_a)(5.6 \times 10^{-6}\ mol/J)$$

$$E_a = \mathbf{8.5 \times 10^4\ J/mol}\ \text{or}\ \mathbf{84.9\ kJ/mol}$$

16-36. **Refer to Section 16-5.** . . . .

(a) let E_a = initial activation energy (J/mol)
E_a - 1000 = final activation energy (J/mol)
k_i = initial rate constant
k_f = final rate constant

When we apply the Arrhenius equations,

$$\frac{k_f}{k_i} = \frac{Ae^{-(E_a - 1000)/RT}}{Ae^{-E_a/RT}} = e^{(-E_a + 1000 + E_a)/RT} = e^{1000\ J/RT}$$

Since e = 2.7183, R = 8.314 J/mol·K, T = 298 K, $k_f/k_i = e^{0.404} = 1.50$

Therefore, the rate constant has increased by a factor of **1.50**, i.e., the rate constant has increased by 50%.

(b) Similarly, let E_a - 10,000 = final activation energy (J/mol)
Then,

$$\frac{k_f}{k_i} = \frac{Ae^{-(E_a - 10{,}000)/RT}}{Ae^{-E_a/RT}} = e^{(-E_a + 10{,}000 + E_a)/RT} = e^{10{,}000\ J/RT}$$

Therefore, $k_f/k_i = e^{4.04}$ = 56.8 and the rate constant has increased by a factor of **56.8**.

16-38. **Refer to Sections 11-8, 16-5, and Exercise 11-28 Solution.** . . . .

The boiling point of water is the temperature at which its vapor pressure equals the atmospheric pressure. At high altitudes, the atmospheric pressure is lower than at sea level and the temperature at which water boils is lower. When we consider the cooking process as a chemical reaction and apply the Arrhenius equation:

$$k = Ae^{-E_a/RT}$$

we see that as the reaction temperature decreases, the rate constant decreases and the reaction slows down. Therefore, reaching the same degree of "doneness" requires longer cooking times.

16-40. **Refer to Sections 11-9 and 16-5.** . . . .

Plan: (1) Use the Clausius-Clapeyron equation to calculate the steam temperature in the pressure cooker. Assume that ΔH_{vap} for H_2O is independent of temperature.
(2) Use the Arrhenius equation to calculate the activation energy for the process of steaming vegetables.

(1) The Clausius-Clapeyron equation: $\log\left(\frac{P_2}{P_1}\right) = \frac{\Delta H_{vap}}{2.303R}\left(\frac{1}{T_1} - \frac{1}{T_2}\right)$

where ΔH_{vap} = molar heat content of vaporization for H_2O, 40.7 kJ/mol
P_1 = atmospheric pressure, 15 psi
P_2 = cooker pressure = P_1 + gauge pressure = (15 + 15) psi
T_1 = boiling point of water at 1 atm, 100.0°C
T_2 = steam temperature in the pressure cooker

Substituting,

$$\log\left(\frac{30\ \text{psi}}{15\ \text{psi}}\right) = \frac{40.7 \times 10^3\ \text{J/mol}}{(2.303)(8.314\ \text{J/mol·K})}\left(\frac{1}{373\ \text{K}} - \frac{1}{T_2}\right)$$

$$1.4 \times 10^{-4}\ \text{K}^{-1} = (2.68 \times 10^{-3}\ \text{K}^{-1} - 1/T_2)$$

$$T_2 = 394\ \text{K}$$

(2) The Arrhenius equation: $\log\left(\frac{k_2}{k_1}\right) = \frac{E_a}{2.303R}\left(\frac{T_2 - T_1}{T_1T_2}\right)$

where T_1 = 373 K
T_2 = 394 K
k_1 = rate constant for cooking vegetables at atmospheric pressure
k_2 = rate constant for cooking vegetables in the pressure cooker
k_2/k_1 = 3 since the cooking process is 3 times faster in the pressure cooker

Substituting,

$$\log 3 = \frac{E_a}{(2.303)(8.314\ \text{J/mol}\cdot\text{K})}\left[\frac{394\ \text{K} - 373\ \text{K}}{(373\ \text{K})(394\ \text{K})}\right]$$

$$E_a = \mathbf{6.4 \times 10^4\ J/mol} \quad \text{or} \quad \mathbf{64\ kJ/mol}$$

16-42. **Refer to Section 16-7.** . . . .

From the rate-law expression for the reaction, Rate = $k[Cl_2][H_2S]$, we can deduce that 1 molecule each of Cl_2 and H_2S is likely to be involved in the slow step.

(a) The rate-law expression consistent with this mechanism is Rate = $k[Cl_2]$. This cannot be a mechanism for the reaction.

(b) The rate-law expression consistent with the mechanism is Rate = $k[Cl_2][H_2S]$ and this is a possible mechanism.

(c) The rate-law expression consistent with the mechanism is likely to be complicated. It is difficut to confirm that this is a possible mechanism.

16-44. **Refer to Section 16-7.** . . . .

The rate-law expression, Rate = $k[NO_2Cl]$, says that only 1 molecule of NO_2Cl is involved in the slow step. The most straightforward mechanism is the following, where Cl is an intermediate. Remember, only bimolecular reactions and unimolecular decompositions are generally included in a possible mechanism.

(1) $NO_2Cl \rightarrow NO_2 + Cl$ slow

(2) $Cl + NO_2Cl \rightarrow NO_2 + Cl_2$ fast

$2NO_2Cl \rightarrow 2NO_2 + Cl_2$

Another possible mechanism is more complex where Cl and N_2O_4 are intermediates.

(1) $NO_2Cl \rightarrow NO_2 + Cl$ slow

(2) $NO_2 + NO_2Cl \rightarrow N_2O_4 + Cl$ fast

(3) $N_2O_4 \rightarrow 2NO_2$ fast

(4) $Cl + Cl \rightarrow Cl_2$ fast

$2NO_2Cl \rightarrow 2NO_2 + Cl_2$

16-46. **Refer to Section 16-7.** . . . .

Trimolecular reactions are rare, and essentially all the known reactions believed to involve the simultaneous collision and reaction of three molecules features NO as one of the reactants. Therefore it is possible that the reaction:

$$2NO + O_2 \rightarrow 2NO_2$$

is a true single step trimolecular reaction. If the reaction occurs in more than one step, a possible mechanism is the following:

(1) $NO + O_2 \rightleftharpoons NO_2 + O$ fast equilibrium

(2) $O + NO \rightarrow NO_2$ slow

$$2NO + O_2 \rightarrow 2NO_2$$

16-48. **Refer to Section 16-8, and Examples 16-3 and 16-4.**

(a) For a first order reaction,

$$t_{1/2} = \frac{0.693}{k} = \frac{0.693}{2.8 \times 10^{-7}\ s^{-1}} = 2.5 \times 10^6\ s \quad \text{or} \quad 29\ \text{days}$$

(b) The integrated first order rate equation: $\log \frac{A_o}{A_t} = \frac{kt}{2.303}$

Substituting,

$$\log \frac{2.00\ g}{0.75\ g} = \frac{(2.8 \times 10^{-7}\ s^{-1})t}{2.303}$$

$$t = \mathbf{3.5 \times 10^6\ s} \quad \text{or} \quad \mathbf{41\ days}$$

(c) If 0.75 g of CS_2 remains, then 1.25 g of CS_2 (= 2.00 g - 0.75 g) had been converted to the products, CS and S.

$$?\ \text{mol } CS_2 \text{ reacted} = \frac{1.25\ g\ CS_2}{76.1\ g/mol} = 0.0164\ \text{mol } CS_2$$

$$?\ \text{mol CS formed} = \text{mol } CS_2 = 0.0164\ \text{mol CS}$$

$$?\ \text{g CS formed} = 0.0164\ \text{mol CS} \times 44.1\ g/mol = \mathbf{0.723\ g\ CS}$$

(d) Substituting into the integrated first order rate equation,

$$\log \left(\frac{2.00\ g}{?\ g\ CS_2}\right) = \frac{(2.8 \times 10^{-7}\ s^{-1})(45\ d \times 24\ hr/d \times 60\ min/hr \times 60\ s/min)}{2.303}$$

$$\log \left(\frac{2.00\ g}{?\ g\ CS_2}\right) = 0.47$$

Solving, ? g CS_2 remaining = **0.68 g**

16-50. **Refer to Section 16-8, and Examples 16-5, 16-6 and 16-7.**

The rate equation, Rate = $(1.4 \times 10^{-10}\ M^{-1}\cdot s^{-1})[NO_2]^2$, tells us that the reaction is second order with respect to NO_2 and second order overall. The specific rate constant, k, is $1.4 \times 10^{-10}\ M^{-1}\cdot s^{-1}$. However, in the following calculations, we must use k' which is equal to $2k$. For an explanation, see the box on p. 489 in your textbook.

(a) For a second order reaction:

$$t_{1/2} = \frac{1}{k'[NO_2]_o} = \frac{1}{(2.8 \times 10^{-10}\ M^{-1}\cdot s^{-1})(3.00\ mol/1.00\ L)}$$

$$= \mathbf{1.2 \times 10^9\ s} \quad \text{or} \quad \mathbf{38\ yrs}$$

(b) The integrated second order rate equation: $\frac{1}{[NO_2]} - \frac{1}{[NO_2]_o} = k't$

Rearranging and substituting,

$$\frac{1}{[NO_2]} = \frac{1}{[NO_2]_o} + k't$$

$$= \frac{1}{3.00\ M} + (2.8 \times 10^{-10}\ M^{-1}\cdot s^{-1})[(150\ yr)(3.15 \times 10^7\ s/yr)$$

$$= 1.65\ M^{-1}$$

Solving, $[NO_2] = 0.61\ M$

? g NO_2 remaining $= 1.00\ L \times \frac{0.61\ mol}{1.00\ L} \times \frac{46.0\ g}{1.0\ mol} =$ **28 g NO_2**

(c) ? $[NO_2]_{reacted} = [NO_2]_o - [NO_2] = 3.00\ M - 0.61\ M = 2.39\ M$

? $[NO]_{produced} = [NO_2]_{reacted} =$ **2.39 M**

16-52. **Refer to Section 16-8 and Exercise 16-28 Solution.** . . . .

From Exercise 16-28, we know that the reaction, $CH_3NC \rightarrow CH_3CN$, is first order.

The calculated specific rate constants are $k_1 = 4 \times 10^{-15}\ s^{-1}$ at 25°C and $k_2 = 8 \times 10^{-4}\ s^{-1}$ at 227°C.

The time required for half of a given amount of CH_3NC to rearrange is the half-life of the reaction.

(a) at 25°C, $t_{1/2} = \frac{0.693}{k_1} = \frac{0.693}{4 \times 10^{-15}\ s^{-1}} =$ **2×10^{14} s** or **6×10^6 yr**

(b) at 227°C, $t_{1/2} = \frac{0.693}{k_2} = \frac{0.693}{8 \times 10^{-4}\ s^{-1}} =$ **9×10^2 s** or **15 min**

16-54. **Refer to Section 16-8 and Exercise 16-30 Solution.** . . . .

From Exercise 16-30, we know that the reaction, $N_2O_5 \rightarrow NO_2 + NO_3$, is first order with specific rate constant, $k = 0.3\ s^{-1}$ at 25°C.

(a) The integrated rate equation: $\log A_o/A = kt/2.303$
where $A_o = 4.50\ mol/2.00\ L$

Substituting, $\log\left(\frac{2.25\ mol}{?\ mol\ N_2O_5}\right) = \frac{(0.3\ s^{-1})(1.00\ min \times 60\ s/min)}{2.303} = 8$

? mol N_2O_5 = **1×10^{-8} mol N_2O_5 left after 1.00 min**

Note: Any small change in the value of k can greatly alter the answer.

(b) If 99.9 % of N_2O_5 have decomposed, then 0.1 % of N_2O_5 remains.
Substituting into the integrated rate equation, we have

$$\log\left(\frac{100.0\ \%}{0.1\ \%}\right) = \frac{(0.3\ s^{-1})t}{2.303}$$ Solving, $t =$ **23 s**

16-56. **Refer to Section 16-8.** . . . .

The first order reaction: $^{223}_{87}Fr \rightarrow ^{223}_{88}Ra + ^{0}_{-1}e$ $\qquad t_{1/2} = 22$ min

Plan: (1) Calculate the specific rate constant, k.
(2) Calculate the mass of francium-223 remaining by using the integrated rate equation: $\log A_0/A = kt/2.303$.

(1) $k = \dfrac{0.693}{t_{1/2}} = \dfrac{0.693}{22 \text{ min}} = 0.032 \text{ min}^{-1}$

(2) Substituting, $\log\left(\dfrac{26\ \mu g}{?\ \mu g\ Fr}\right) = \dfrac{(0.032 \text{ min}^{-1})(2.0 \text{ hr} \times 60 \text{ min/hr})}{2.303} = 1.7$

$? \mu g$ Fr = **0.5 μg**

16-58. **Refer to Section 16-8.** . . . .

The first order reaction: $^{42}_{19}K \rightarrow ^{42}_{20}Ca + ^{0}_{-1}e$ + energy

Plan: (1) Calculate the specific rate constant, k, using the integrated rate equation, $\log A_0/A = kt/2.303$.
(2) Determine the half-life of ^{42}K from the value of k.

(1) Substituting, $\log\left(\dfrac{2.000 \text{ g}}{0.116 \text{ g}}\right) = \dfrac{k[(50 \text{ hr} \times 60 \text{ min/hr}) + 57 \text{ min}]}{2.303}$

$k = 9.316 \times 10^{-4} \text{ min}^{-1}$

(2) $t_{1/2} = \dfrac{0.693}{k} = \dfrac{0.693}{9.314 \times 10^{-4} \text{ min}^{-1}} =$ **744 min** or **12.4 hr**

17 Chemical Equilibrium

17-2. **Refer to Section 17-1.**

A chemical equilibrium exists when 2 opposing reactions, a forward reaction and a reverse reaction, occur at the same rate. Under these conditions, the composition of the system, i.e., the concentration of the products and the reactants, does not change. It is called a dynamic equilibria because on the molecular level, reactant molecules are colliding and producing product molecules at the same rate as product molecules are colliding and producing reactant molecules.

17-4. **Refer to Sections 17-1, 17-2 and 17-5.**

The value of an equilibrium constant tells us nothing about the time required for the reaction to reach equilibrium. It is unrelated to the kinetics of the reaction. Rather, it is a measure of the extent to which a reaction occurs. If the equilibrium constant, *K*, is large, most of the reactants will be converted to products when the system reaches equilibrium. If *K* is small, most of the reactants will remain as reactants when the system reaches equilibrium.

17-6. **Refer to Section 17-2.**

(a) $K_c = \frac{[H_2S]^2[O_2]^3}{[H_2O]^2[SO_2]^2}$ (b) $K_c = \frac{[NO]^4[H_2O]^6}{[NH_3]^4[O_2]^5}$ (c) $K_c = \frac{[SO_2]^2}{[O_2]^3}$

(d) $K_c = \frac{1}{[NH_3][HCl]}$ (e) $K_c = \frac{[PCl_5]}{[PCl_3][Cl_2]}$ (f) $K_c = [HF]$

17-8. **Refer to Section 17-2.**

The equilibrium constant, K_c, is an approximate measure of the extent to which a forward reaction will proceed toward completion. A larger value of K_c indicates that more products will be formed when a system reaches equilibrium. Therefore, in order of the increasing tendency of the forward reaction to go to completion, we have

$$\text{(b) } K_c \simeq 10^{-21} < \text{(e) } K_c \simeq 10^{-3} < \text{(a) } K_c \simeq 10^{3} < \text{(d) } K_c \simeq 10^{19} < \text{(c) } K_c \simeq 10^{69}$$

17-10. **Refer to Section 17-3.**

Many systems are not at equilibrium. The mass action expression, also called the reaction quotient, *Q*, is a measure of how far a system is from equilibrium and in what direction the system must go to get to equilibrium. The reaction quotient has the same form as the equilibrium constant, *K*, but the concentration values put into *Q* are the actual values found in the system at that given moment.

(a) If $Q = K$, the system is at equilibrium.

(b) If $Q < K$, the system has greater concentrations of reactants than it would have if it were at equilibrium. The forward reaction will dominate until equilibrium is established.

(c) If $Q > K$, the system has greater concentrations of products than it would have if it were at equilibrium. The reverse reaction will dominate until equilibrium is reached.

17-12. **Refer to Section 17-3 and Example 17-2.** . . . .

Balanced reaction: $HCHO \rightleftharpoons H_2 + CO$ $K_c = 0.50$

The reaction quotient, Q, for this reaction at the given moment is:

$$Q = \frac{[H_2][CO]}{[HCHO]} = \frac{(1.50)(0.25)}{0.50} = 0.75$$

Under the given conditions, $Q > K_c$.

(a) false - The reaction is not at equilibrium since $Q \neq K_c$.

(b) false - The reaction is not at equilibrium. However, the reaction will continue to proceed until equilibrium is reached.

(c) false - When $Q > K$, the system has more products and less reactants than it would have at equilibrium. Equilibrium will be reached by forming more HCHO.

(d) false - The forward rate of the reaction is less than the reverse reaction since $Q > K_c$. The forward and reverse rates are equal only at equilibrium.

17-14. **Refer to Sections 17-2 and 17-4, and Examples 17-1, 17-3 and 17-4.** . . . . . .

Balanced reaction: $A(g) + B(g) \rightleftharpoons C(g) + 2D(g)$

Plan: (1) Determine the concentrations of the species of interest.
(2) Determine the concentrations of all species after equilibrium is reached.
(3) Calculate K_c.

(1) $[A]_{initial} = [B]_{initial} = \frac{1.00 \text{ mol}}{0.400 \text{ L}} = 2.50\ M$

$[C]_{equil} = \frac{0.20 \text{ mol}}{0.400 \text{ L}} = 0.50\ M$

(2)

	A	+	B	$\rightleftharpoons$	C	+	2D
initial	2.50 M		2.50 M		0 M		0 M
change	- 0.50 M		- 0.50 M		+ 0.50 M		+ 1.00 M
at equilibrium	2.00 M		2.00 M		0.50 M		1.00 M

(3) $K_c = \frac{[C][D]^2}{[A][B]} = \frac{(0.50)(1.00)^2}{(2.00)(2.00)} = 0.12$

17-16. **Refer to Section 17-4, Example 17-3 and Exercise 17-15.** . . . .

From Exercise 17-15, the balanced equation is $CO(g) + H_2O(g) \rightleftharpoons CO_2(g) + H_2(g)$ with $K_c = 1.845$.

Plan: (1) To determine which way the reaction will go to approach equilibrium, calculate the reaction quotient.
(2) Calculate the equilibrium concentrations using the K_c expression.

(1) The reaction quotient, $Q = \frac{[CO_2][H_2]}{[CO][H_2O]} = \frac{(0.500)(0.500)}{(0.500)(0.500)} = 1.00$

Therefore, $Q < K_c$. The forward reaction will predominate: some CO and H_2O will be reacted and some CO_2 and H_2 will be formed as the reaction approaches equilibrium.

(2) Let x = moles per liter of CO that react. Then
x = moles per liter of H_2O that react,
x = moles per liter of CO_2 and H_2 that are formed.

	CO	+	H_2O	⇄	CO_2	+	H_2
initial	0.500 *M*		0.500 *M*		0.500 *M*		0.500 *M*
change	- x *M*		- x *M*		+ x *M*		+ x *M*
at equil	(0.500 - x)*M*		(0.500 - x)*M*		(0.500 + x)*M*		(0.500 + x)*M*

$$K_c = \frac{[CO_2][H_2]}{[CO][H_2]} = \frac{(0.500 + x)^2}{(0.500 - x)^2} = 1.845$$

To solve, take the square root of both sides of the equation, giving

$$\frac{(0.500 + x)}{(0.500 - x)} = 1.358 \qquad \text{Solving, } x = 0.0759$$

At equilibrium,

$[CO] = [H_2O] = (0.500 - x)M = \mathbf{0.424}\ M$

$[CO_2] = [H_2] = (0.500 + x)M = \mathbf{0.576}\ M$

17-18. **Refer to Section 17-4 and Example 17-4.** . . . .

Balanced equation: $COCl_2(g) \rightleftharpoons CO(g) + Cl_2(g)$ $\qquad K_c = 0.083$ at 900°C

Plan: (1) Calculate the initial concentration of $COCl_2$.
(2) Calculate the equilibrium concentrations of $COCl_2$, Cl_2 and CO using the K_c expression.

(1) $[COCl_2]_{initial} = \frac{0.400 \text{ mol}}{2.00 \text{ L}} = 0.200\ M$

(2) Let x = moles per liter of $COCl_2$ that react. Then
x = moles per liter of CO and Cl_2 that are formed.

	$COCl_2$	⇄	CO	+	Cl_2
initial	0.200 *M*		0 *M*		0 *M*
change	- x *M*		+ x *M*		+ x *M*
at equilibrium	(0.200 - x)*M*		x *M*		x *M*

$$K_c = \frac{[CO][Cl_2]}{[COCl_2]} = \frac{x^2}{0.200 - x} = 0.083$$

Rearranging this into a quadratic equation: $x^2 + 0.083x - 0.017 = 0$
Solving,

$$x = \frac{-0.083 \pm \sqrt{(0.083)^2 - (4)(1)(-0.017)}}{(2)(1)} = \frac{-0.083 \pm 0.27}{2} = 0.09 \text{ or } -0.18 \text{ (discard)}$$

Note: There are always 2 solutions when solving quadratic equations, but only 1 is meaningful. The other solution, -0.18 *M*, is discarded because it has no physical meaning in the present problem.

Therefore, at equilibrium, $[COCl_2] = 0.200 - x = \mathbf{0.11}\ M$

$[CO] = [Cl_2] = x = \mathbf{0.09}\ M$

17-20. **Refer to Section 17-4, Example 17-4 and Exercise 17-18.** . . . .

Balanced equation: $COCl_2(g) \rightleftharpoons CO(g) + Cl_2(g)$ $K_c = 0.083$ at 900°C

Plan: (1) Calculate the initial concentrations of CO and Cl_2.
(2) Calculate the equilibrium concentrations of $COCl_2$, Cl_2 and CO, using the K_c expression.

(1) $[CO]_{initial} = [Cl_2]_{initial} = \dfrac{0.400 \text{ mol}}{2.00 \text{ L}} = 0.200\ M$

(2) Since $[COCl_2]_{initial} = 0$, the reverse reaction will predominate: some $COCl_2$ will be produced and some CO and Cl_2 will be reacted.

Let x = moles per liter of $COCl_2$ that are produced. Then
x = moles per liter of CO and Cl_2 that react.

	$COCl_2$	$\rightleftharpoons$	CO	+	Cl_2
initial	0 *M*		0.200 *M*		0.200 *M*
change	+ x *M*		- x *M*		- x *M*
at equilibrium	x *M*		(0.200 - x)*M*		(0.200 - x)*M*

$$K_c = \frac{[CO][Cl_2]}{[COCl_2]} = \frac{(0.200 - x)^2}{x} = 0.083$$

The quadratic equation: $x^2 - 0.483x + 0.0400 = 0$
Solving,

$$x = \frac{0.483 \pm \sqrt{(-0.483)^2 - (4)(1)(0.0400)}}{2(1)} = \frac{0.483 \pm 0.271}{2} = 0.106 \text{ (or } 0.377)$$

Note: The solution, x = 0.377, is meaningless and therefore discarded since it would lead to negative concentrations for CO and Cl_2.

Therefore, at equilibrium, $[COCl_2] = x =$ **0.106** *M*
$[CO] = [Cl_2] = 0.200 - x =$ **0.094** *M*

Note: The answers are the same as for Exercise 17-18. The two systems are identical. In Exercise 17-18, equilibrium was approached from the reactant side, whereas in this exercise, equilibrium was approached from the product side.

17-22. **Refer to Section 17-2 and Example 17-1.** . . . .

Balanced equation: $SbCl_5(g) \rightleftharpoons SbCl_3(g) + Cl_2(g)$

Plan: (1) Determine the equilibrium concentrations of $SbCl_5$, $SbCl_3$ and Cl_2 at 448°C.
(2) Evaluate K_c at 448°C by substituting the equilibrium concentrations into the K_c expression.

(1) $[SbCl_5] = \dfrac{(3.84 \text{ g}/299 \text{ g/mol})}{5.00 \text{ L}} = 2.57 \times 10^{-3}\ M$

$[SbCl_3] = \dfrac{(9.14 \text{ g}/228 \text{ g/mol})}{5.00 \text{ L}} = 8.02 \times 10^{-3}\ M$

$[Cl_2] = \dfrac{(2.84 \text{ g}/70.9 \text{ g/mol})}{5.00 \text{ L}} = 8.01 \times 10^{-3}\ M$

(2) $K_c = \dfrac{[SbCl_3][Cl_2]}{[SbCl_5]} = \dfrac{(8.02 \times 10^{-3})(8.01 \times 10^{-3})}{(2.57 \times 10^{-3})} = \mathbf{2.50 \times 10^{-2}}$

17-24. **Refer to Section 17-4 and Exercise 17-22.** . . . .

Balanced equation: $SbCl_5(g) \rightleftharpoons SbCl_3(g) + Cl_2(g)$ $K_c = 2.51 \times 10^{-2}$

Plan: (1) Determine the initial concentrations of $SbCl_5$ and $SbCl_3$.
(2) Determine the equilibrium concentrations of $SbCl_5$, $SbCl_3$ and Cl_2 using the K_c expression.

(1) $[SbCl_5] = \dfrac{(5.00 \text{ g}/299 \text{ g/mol})}{5.00 \text{ L}} = 3.34 \times 10^{-3}\ M$

$[SbCl_3] = \dfrac{(5.00 \text{ g}/228 \text{ g/mol})}{5.00 \text{ L}} = 4.39 \times 10^{-3}\ M$

(2) Since $[Cl_2] = 0$, the forward reaction will predominate. Some $SbCl_5$ will react and some $SbCl_3$ and Cl_2 will be produced.

Let x = moles per liter of $SbCl_5$ that react. Then
x = moles per liter of $SbCl_3$ and Cl_2 that are produced.

	$SbCl_5$	$\rightleftharpoons$	$SbCl_3$	+	Cl_2
initial	$3.34 \times 10^{-3}\ M$		$4.39 \times 10^{-3}\ M$		$0\ M$
change	$-\ x\ M$		$+\ x\ M$		$+\ x\ M$
at equilibrium	$(3.34 \times 10^{-3} - x)M$		$(4.39 \times 10^{-3}\ M + x)M$		$x\ M$

$$K_c = \frac{[SbCl_3][Cl_2]}{[SbCl_5]} = \frac{(x)(4.39 \times 10^{-3} + x)}{(3.34 \times 10^{-3} - x)} = 0.0251$$

The quadratic equation: $x^2 + 0.0295x - 8.38 \times 10^{-5} = 0$
Solving,

$$x = \frac{-0.0295 \pm \sqrt{(0.0295)^2 - (4)(1)(-8.38 \times 10^{-5})}}{2(1)} = \frac{-0.0295 \pm 0.0347}{2}$$

$= 2.6 \times 10^{-3}$ or -0.0321 (discard)

Therefore, at equilibrium, $[SbCl_5] = 3.34 \times 10^{-3} - x = \mathbf{7 \times 10^{-4}}\ M$
$[SbCl_3] = 4.39 \times 10^{-3} + x = \mathbf{7.0 \times 10^{-3}}\ M$
$[Cl_2] = x = \mathbf{2.6 \times 10^{-3}}\ M$

17-26. **Refer to Section 17-4.** . . . .

Balanced equation: $POCl_3(g) \rightleftharpoons POCl(g) + Cl_2(g)$ $K_c = 0.450$

Plan: (1) Determine the initial concentration of $POCl_3$.
(2) Calculate the concentration of $POCl_3$ that dissociated, using the K_c expression.
(3) Determine the percent dissociation of $POCl_3$ at equilibrium.

(1) $[POCl_3] = \dfrac{0.800 \text{ mol}}{2.00 \text{ L}} = 0.400\ M$

(2) Let x = moles per liter of $POCl_3$ that react (dissociate). Then
x = moles per liter of POCl and Cl_2 that are produced.

	$POCl_3$	$\rightleftharpoons$	$POCl$	+	Cl_2
initial	$0.400\ M$		$0\ M$		$0\ M$
change	$-\ x\ M$		$+\ x\ M$		$+\ x\ M$
at equilibrium	$(0.400 - x)M$		$x\ M$		$x\ M$

$$K_c = \frac{[POCl][Cl_2]}{[POCl_3]} = \frac{x^2}{0.400 - x} = 0.450$$

The quadratic equation: $x^2 + 0.450x - 0.180 = 0$
Solving,

$$x = \frac{-0.450 \pm \sqrt{(0.450)^2 - (4)(1)(-0.180)}}{2(1)} = \frac{-0.450 \pm 0.960}{2} = 0.255 \text{ or } -0.705 \text{ (discard)}$$

Therefore, the concentration of $POCl_3$ that reacted = x = 0.255 *M*

(3) $\% \; PCl_5 \text{ dissociated} = \frac{[POCl_3]_{reacted}}{[POCl_3]_{initial}} \times 100\% = \frac{0.255\ M}{0.400\ M} \times 100\% = \mathbf{63.8\%}$

17-28. **Refer to Section 17-9 and Example 17-8.**

(a) $K_p = \frac{(P_{H_2S})^2(P_{O_2})^3}{(P_{H_2O})^2(P_{SO_2})^2}$ (b) $K_p = \frac{(P_{NO})^4(P_{H_2O})^6}{(P_{NH_3})^4(P_{O_2})^5}$ (c) $K_p = \frac{(P_{SO_2})^2}{(P_{O_2})^3}$

(d) $K_p = \frac{1}{(P_{NH_3})(P_{HCl})}$ (e) $K_p = \frac{(P_{PCl_5})}{(P_{PCl_3})(P_{Cl_2})}$ (f) $K_p = P_{HF}$

17-30. **Refer to Sections 17-7 and 17-8, and Example 17-8.**

Balanced equation: $2Cl_2(g) + 2H_2O(g) \rightleftharpoons 4HCl(g) + O_2(g)$

$$K_p = \frac{(P_{HCl})^4(P_{O_2})}{(P_{Cl_2})^2(P_{H_2O})^2} = 3.2 \times 10^{-14}$$

To determine $K_c = \frac{[HCl]^4[O_2]}{[Cl_2]^2[H_2O]^2}$ (assume T = 298 K):

$$K_c = K_p(RT)^{-\Delta n} = (3.2 \times 10^{-14})[(0.0821\ L{\cdot}atm/mol{\cdot}K)(298\ K)]^{-(5-4)}$$
$$= 1.3 \times 10^{-15}$$

(a) $K_{c(a)} = \frac{[Cl_2]^2[H_2O]^2}{[HCl]^4[O_2]} = 1/K_c = 7.7 \times 10^{14}$

$K_{p(a)} = \frac{(P_{Cl_2})^2(P_{H_2O})^2}{(P_{HCl})^4(P_{O_2})} = 1/K_p = 3.1 \times 10^{13}$

(b) $K_{c(b)} = \frac{[Cl_2][H_2O]}{[HCl]^2[O_2]^{1/2}} = (1/K_c)^{1/2} = 2.8 \times 10^7$

$K_{p(b)} = \frac{(P_{Cl_2})(P_{H_2O})}{(P_{HCl})^2(P_{O_2})^{1/2}} = (1/K_p)^{1/2} = 5.6 \times 10^6$

(c) $K_{c(c)} = \frac{[HCl]^2[O_2]^{1/2}}{[Cl_2][H_2O]} = K_c^{1/2} = 3.6 \times 10^{-8}$

$K_{p(c)} = \frac{(P_{HCl})^2(P_{O_2})^{1/2}}{(P_{Cl_2})(P_{H_2O})} = K_p^{1/2} = 1.8 \times 10^{-7}$

(d) $$K_{c(d)} = \frac{[HCl]^8[O_2]^2}{[Cl_2]^4[H_2O]^4} = K_c^2 = 1.7 \times 10^{-30}$$

$$K_{p(d)} = \frac{(P_{HCl})^8(P_{O_2})^2}{(P_{Cl_2})^4(P_{H_2O})^4} = K_p^2 = 1.0 \times 10^{-27}$$

(e) $$K_{c(e)} = \frac{[HCl][O_2]^{1/4}}{[Cl_2]^{1/2}[H_2O]^{1/2}} = K_c^{1/4} = 1.9 \times 10^{-4}$$

$$K_{p(e)} = \frac{(P_{HCl})(P_{O_2})^{1/4}}{(P_{HCl})^{1/2}(P_{H_2O})^{1/2}} = K_p^{1/4} = 4.2 \times 10^{-4}$$

17-32. **Refer to Section 17-7.** . . . .

Balanced equation: $N_2O_4(g) \rightleftharpoons 2NO_2(g)$ $K_p = 0.143$ at 25°C

Let x = partial pressure (atm) of N_2O_4 that reacts. Then
2x = partial pressure (atm) of NO_2 that is produced.

	N_2O_4	$\rightleftharpoons$	$2NO_2$
initial	0.0500 atm		0 atm
change	- x atm		+ 2x atm
at equilibrium	(0.0500 - x)atm		2x atm

$$K_p = \frac{(P_{NO_2})^2}{(P_{N_2O_4})} = \frac{(2x)^2}{(0.0500 - x)} = 0.143$$

The quadratic equation: $4x^2 + 0.143x - 7.15 \times 10^{-3} = 0$
Solving,

$$x = \frac{-0.143 \pm \sqrt{(0.143)^2 - (4)(4)(-7.15 \times 10^{-3})}}{2(4)} = \frac{-0.143 \pm 0.367}{8}$$

$= 0.0280$ and -0.0638 (discard)

$P_{N_2O_4} = (0.0500 - x) =$ **0.0220 atm** $P_{NO_2} = 2x =$ **0.0560 atm**

$P_{total} = P_{N_2O_4} + P_{NO_2} =$ **0.0780 atm**

17-34. **Refer to Section 17-8.** . . . .

(a) $2SO_3(g) \rightleftharpoons 2SO_2(g) + O_2(g)$

(b) $$K_c = \frac{[SO_2]^2[O_2]}{[SO_3]^2} = \frac{(0.0324)^2(0.0162)}{(0.262)^2} = \mathbf{2.48 \times 10^{-4}}$$

(c) $K_p = K_c(RT)^{\Delta n} = (2.48 \times 10^{-4})(0.0821 \times 900)^{(3-2)}$
$= \mathbf{1.83 \times 10^{-2}}$

17-36. **Refer to Section 17-7.** . . . .

Balanced equation: $2ICl(s) \rightleftharpoons I_2(s) + Cl_2(g)$ $K_p = 0.24$

(a) Since $K_p = P_{Cl_2}$, $P_{Cl_2} = K_p =$ **0.24 atm.**
The partial pressure of Cl_2 at equilibrium must equal K_p no matter how much ICl is present or what the volume is.

(b) If we assume that Cl_2 obeys the ideal gas law, $PV = nRT$, then

$$[Cl_2] = \frac{n}{V} = \frac{P}{RT} = \frac{0.24 \text{ atm}}{(0.0821 \text{ L·atm/mol·K})(298 \text{ K})} = \mathbf{9.8 \times 10^{-3}}\ M$$

Since the concentration of Cl_2 is 9.8×10^{-3} M at equilibrium,

(c) in a 1.00 L container there are 9.8×10^{-3} moles of Cl_2. From the stoichiometry of the balanced equation, we see that for every 1 mole of Cl_2 produced, 2 moles of ICl must have reacted.

? mol ICl reacted = 2 × mol Cl_2 produced = $2 \times 9.8 \times 10^{-3}$ mol
= 2.0×10^{-2} mol

? mol ICl remaining = 2.0 mol - 0.02 mol = **2.0 mol** (2 significant figures)

? g ICl remaining = 2.0 × 162 g/mol = **3.2×10^2 g**

17-38. **Refer to Section 17-8.** . . . .

Balanced equation: $H_2(g) + I_2(g) \rightleftharpoons 2HI(g)$ $K_c = 50$ at 448°C

(a) $K_p = K_c(RT)^{\Delta n} = (50)(0.0821 \times 721)^{(2-2)} = \mathbf{50}$

(b) Plan: (1) Determine the initial pressures of H_2 and I_2 by assuming the gases obey the ideal gas law, $PV = nRT$.
(2) Calculate the partial pressures of H_2, I_2, HI and the total pressure at equilibrium.

(1) $$P_{H_2} = \frac{nRT}{V} = \frac{(1.00 \text{ g}/2.02 \text{ g/mol})(0.0821 \text{ L·torr/mol·K})(448°\text{C} + 273°)}{5.0 \text{ L}}$$
= 5.9 atm

$$P_{I_2} = \frac{nRT}{V} = \frac{(127 \text{ g}/254 \text{ g/mol})(0.0821 \text{ L·torr/mol·K})(448° + 273°)}{5.0 \text{ L}}$$
= 5.9 atm

(2) Let x = partial pressure in atm of H_2 and I_2 that reacts, then
2x = partial pressure in atm of HI that is produced.

	H_2	+	I_2	⇌	2HI
initial	5.9 atm		5.9 atm		0 atm
change	- x atm		- x atm		+ 2x atm
at equilibrium	(5.9 - x) atm		(5.9 - x) atm		2x atm

$$K_p = \frac{(P_{HI})^2}{(P_{H_2})(P_{I_2})} = \frac{(2x)^2}{(5.9 - x)^2} = 50$$

Taking the square root of both sides: $\frac{2x}{5.9 - x} = 7.1$

x = 4.6

$P_{H_2} = P_{I_2} = 5.9 - x = \mathbf{1.3\ atm}$

$P_{HI} = 2x = \mathbf{9.2\ atm}$

$P_{total} = P_{H_2} + P_{I_2} + P_{HI} = \mathbf{11.8\ atm}$

(c) Using the ideal gas law, $PV = nRT$, to calculate n, the moles of unreacted I_2.

$$n = \frac{PV}{RT} = \frac{(1.3\text{ atm})(5.0\text{ L})}{(0.0821\text{ L·atm/mol·K})(448° + 273°)} = 0.11\text{ moles}$$

? g unreacted I_2 = 0.11 mol I_2 × 254 g/mol = **28 g**

17-40. **Refer to Section 17-5.** . . . .

When an equilibrium system involving gases is subjected to an increase in pressure resulting from a decrease in volume, the concentrations of the gases increase and there may or may not be a shift in the equilibrium. If there is the same number of moles of gas on each side of the equation, equilibrium is not affected. If the number of moles of gas on each side of the equation is different, the general rule is that such an increase in pressure shifts a system in the direction that produces the smaller number of moles of gas.

(a) shift to right (b) equilibrium is unaffected (c) shift to left

(d) shift to right (e) shift to right

17-42. **Refer to Sections 17-2, 17-5 and 17-9.** . . . .

If the concentration of a reactant or product which appears in the equilibrium expression is changed, from its equilibrium value, the equilibrium will shift so as to minimize the effect of its change. Due to the applied stress, the following equilibrium will:

(a) shift to right (b) shift to right (c) shift to right

(d) not be affected since $MgSO_4(s)$, because it is a solid, is considered to have a constant concentration (has an activity of unity) and does not appear in the equilibrium expression.

17-44. **Refer to Section 17-9.** . . . .

Balanced equation: $NH_4Cl(s) \rightleftharpoons NH_3(g) + HCl(g)$

$K_c = [NH_3][HCl]$

Let V = volume of the container
  n = moles of NH_3 = moles HCl at equilibrium

Therefore

$$K_c = \left(\frac{n}{V}\right)^2 \qquad \text{Solving for } n, \quad n = V \times K_c^{1/2}$$

If the initial number of moles of $NH_4Cl(s)$ was less than or equal to $V \times K_c^{1/2}$, then all the $NH_4Cl(s)$ would decompose.

17-46. **Refer to Section 17-5.** . . . .

Balanced equation: $PCl_5(g) \rightleftharpoons PCl_3(g) + Cl_2(g)$ $\Delta H^o = 92.5$ kJ

(a) An increase in temperature will shift the equilibrium to the right since the reaction is endothermic ($\Delta H^o > 0$) and the forward reaction will consume the excess heat due to increased temperature.

(b) An increase in pressure caused by a decrease in volume will shift the equilibrium to the left, the side with the lesser number of moles of gas.

(c) An increase in Cl_2 concentration will shift the equilibrium to the left to use up the excess Cl_2.

(d) An increase in PCl_5 concentration will shift the equilibrium to the right to use up the excess PCl_5.

(e) The addition of a catalyst will not affect equilibrium.

17-48. **Refer to Sections 17-3 and 17-4.** . . . .

Balanced equation: $PCl_5(g) \rightleftharpoons PCl_3(g) + Cl_2(g)$ $K_c = 0.063$ at 300°C

Plan: Use the data to calculate the reaction quotient, Q, for the system. Compare Q with K_c to determine whether or not the system is at equilibrium.

$$Q_c = \frac{[PCl_3][Cl_2]}{[PCl_5]} = \frac{(0.30 \text{ mol}/1.00 \text{ L})(0.50 \text{ mol}/1.00 \text{ L})}{(0.080 \text{ mol}/1.00 \text{ L})} = 1.9$$

$Q_c > K_c$. Therefore, the system is not at equilibrium and the reverse reaction will occur to a greater extent than the forward reaction until the system reaches equilibrium, i.e., $Q_c = K_c$.

17-50. **Refer to Section 17-6 and Example 17-5.** . . . .

Balanced equation: $A(g) + B(g) \rightleftharpoons C(g) + D(g)$

(a) $K_c = \frac{[C][D]}{[A][B]} = \frac{(1.60 \text{ mol}/1.00 \text{ L})^2}{(0.40 \text{ mol}/1.00 \text{ L})^2} = \mathbf{16}$

(b) Since we are adding reactant to the system, Le Chatelier's principle says that the equilibrium will then shift to the right.

Let x = moles per liter of A and B that react <u>after</u> the addition of 0.20 moles per liter of A and B. Then
x = moles per liter of C and D that are produced.

	A	+	B	⇌	C	+	D
at equilibrium	0.40 M		0.40 M		1.60 M		1.60 M
mol/L added	+ 0.20 M		+ 0.20 M		0 M		0 M
new system	0.60 M		+ 0.60 M		1.60 M		1.60 M
change	- x M		- x M		+ x M		+ x M
at equilibrium	(0.60 - x)M		(0.60 - x)M		(1.60 + x)M		(1.60 + x)M

$$K_c = \frac{[C][D]}{[A][B]} = \frac{(1.60 + x)^2}{(0.60 - x)^2} = 16$$

Taking the square root of both sides of the equation: $\frac{1.60 + x}{0.60 - x} = 4.0$

$x = 0.16$

Therefore, the new equilibrium concentration of A is (0.6 - x) or **0.44** M.

17-52. **Refer to Sections 17-5 and 17-6, and Example 17-6.** . . . .

Balanced equation: $A(g) \rightleftharpoons B(g) + C(g)$

(a) $K_c = \frac{[B][C]}{[A]} = \frac{(0.20)^2}{0.30\ M} = \mathbf{0.13}$

(b) If the volume is suddenly doubled, the initial concentrations will be halved and the system is no longer at equilibrium. We learned in Section 17-5, that the equilibrium will then shift to the side with the greater number of moles of gas, i.e., the right side.

Let x = number of moles of A that react <u>after</u> the volume is doubled.
x = number of moles of B and C that are produced.

	A	⇄	B	+	C
initial	0.30 *M*		0.20 *M*		0.20 *M*
new system	0.15 *M*		0.10 *M*		0.10 *M*
change	- x *M*		+ x *M*		+ x *M*
at equilibrium	(0.15 - x)*M*		(0.10 + x)*M*		(0.10 + x)*M*

$$K_c = \frac{[B][C]}{[A]} = \frac{(0.10 + x)^2}{(0.15 - x)} = 0.13$$

The quadratic equation: $x^2 + 0.33x - 0.0095 = 0$
Solving,

$$x = \frac{-0.33 \pm \sqrt{(0.33)^2 - (4)(1)(-0.0095)}}{2(1)} = \frac{-0.33 \pm 0.38}{2}$$

$= 0.02$ or -0.36 (discard)

Therefore, [A] = 0.15 - x = **0.13** *M*
[B] = [C] = 0.10 + x = **0.12** *M*

(c) If the volume is suddenly halved, the initial equilibrium concentrations will be doubled and this system is no longer at equilibrium. The equilibrium will shift to the left side, the side with the lesser number of moles of gas.

let x = number of moles of B and C that react <u>after</u> the volume is halved. Then
x = number of moles of A that are produced.

	A	⇄	B	+	C
initial	0.30 *M*		0.20 *M*		0.20 *M*
new system	0.60 *M*		0.40 *M*		0.40 *M*
change	+ x *M*		- x *M*		- x *M*
at equilibrium	(0.60 + x)*M*		(0.40 - x)*M*		(0.40 - x)*M*

$$K_c = \frac{[B][C]}{[A]} = \frac{(0.40 - x)^2}{(0.60 + x)} = 0.13$$

The quadratic equation: $x^2 - 0.93x + 0.082 = 0$
Solving,

$$x = \frac{0.93 \pm \sqrt{(-0.93)^2 - (4)(1)(0.082)}}{2(1)} = \frac{0.93 \pm 0.73}{2} = 0.10 \text{ or } 0.83 \text{ (discard)}$$

Therefore, [A] = 0.60 + x = **0.70** *M*
[B] = [C] = 0.40 - x = **0.30** *M*

17-54. **Refer to Sections 17-6 and 17-8, and Exercises 17-46 and 17-53.** . . . . . .

Balanced equation: $PCl_5(g) \rightleftharpoons PCl_3(g) + Cl_2(g)$

(a) $K_p = K_c(RT)^{\Delta n} = (4.2 \times 10^{-2})(0.0821 \times 523)^{(2-1)} = \mathbf{1.8}$

(b) Plan: (1) Calculate the partial pressure of Cl_2 in the container assuming Cl_2 obeys the ideal gas law, $PV = nRT$.
(2) Since the partial pressures of Cl_2 and PCl_3 must be equal, use the K_p expression to calculate the partial pressure of PCl_5 at equilibrium.

(1) $P_{Cl_2} = (n/V)RT = (0.15\ M)(0.0821\ \text{L·atm/mol·K})(250°\text{C} + 273°) =$ **6.4 atm**

(2) $P_{PCl_3} = P_{Cl_2} =$ **6.4 atm**

$$K_p = \frac{(P_{PCl_3})(P_{Cl_2})}{P_{PCl_5}} = 1.8$$

Rearranging,

$$P_{PCl_5} = \frac{(P_{PCl_3})(P_{Cl_2})}{K_p} = \frac{(6.4)^2}{1.8} = \mathbf{23\ atm}$$

(c) $P_{total} = P_{PCl_5} + P_{PCl_3} + P_{Cl_2} = 23\ \text{atm} + 6.4\ \text{atm} + 6.4\ \text{atm} =$ **36 atm**

(d) At constant temperature and volume, $P \propto n$. For every 1 mole of PCl_5 that reacted, 1 mole of Cl_2 was produced. Therefore

initial $P_{PCl_5} = P_{PCl_5}$ at equilibrium + P_{PCl_5} that reacted
$= P_{PCl_5}$ at equilibrium + P_{Cl_2} produced
$= 23$ atm + 6.4 atm
$=$ **29 atm**

17-56. **Refer to Section 17-6 and Exercise 17-52 Solution.** . . . .

Balanced equation: $N_2O_4(g) \rightleftarrows 2NO_2(g)$ $\qquad K_c = 5.84 \times 10^{-3}$

(a) $$[N_2O_4]_{initial} = \frac{2.50\ g/92.0\ g/mol}{2.00\ L} = 0.0136\ M$$

Let x = moles per liter of N_2O_4 that react. Then
2x = moles per liter of NO_2 that are produced.

	N_2O_4	$\rightleftarrows$	$2NO_2$
initial	0.0136 M		0 M
change	- x M		+ 2x M
at equilibrium	(0.0136 - x)M		2x M

$$K_c = \frac{[NO_2]^2}{[N_2O_4]} = \frac{(2x)^2}{0.0136 - x} = 5.84 \times 10^{-3}$$

The quadratic equation: $0 = 4x^2 + (5.84 \times 10^{-3})x - 7.94 \times 10^{-5}$
Solving,

$$x = \frac{-5.84 \times 10^{-3} \pm \sqrt{(5.84 \times 10^{-3})^2 - (4)(4)(-7.94 \times 10^{-5})}}{2(4)}$$

$$= \frac{-5.84 \times 10^{-3} \pm 3.61 \times 10^{-2}}{8} = 3.78 \times 10^{-3} \text{ or } -5.24 \times 10^{-3}$$
(delete)

Therefore at equilibrium, $[N_2O_4] = 0.0136 - x =$ **9.8×10^{-3}** M
$[NO_2] = 2x =$ **7.56×10^{-3}** M

(b) When the volume is suddenly doubled, (2L → 4L), the concentrations of N_2O_4 and NO_2 are halved and the equilibrium shifts to the right.

Let x = moles per liter of N_2O_4 that react <u>after</u> the volume is doubled. Then
2x = moles per liter of NO_2 that are produced.

	N_2O_4	⇄	$2NO_2$
initial	9.8×10^{-3} *M*		7.56×10^{-3} *M*
new system	4.9×10^{-3} *M*		3.78×10^{-3} *M*
change	- x *M*		+ 2x *M*
at equilibrium	$(4.9 \times 10^{-3} - x)M$		$(3.78 \times 10^{-3} + 2x)M$

$$K_c = \frac{[NO_2]^2}{[N_2O_4]} = \frac{(3.78 \times 10^{-3} + 2x)^2}{(4.9 \times 10^{-3} - x)} = 5.84 \times 10^{-3}$$

The quadratic equation: $4x^2 + 0.0210x - 1.4 \times 10^{-5} = 0$
Solving,

$$x = \frac{-0.0210 \pm \sqrt{(0.0210)^2 - (4)(4)(-1.4 \times 10^{-5})}}{2(4)} = \frac{-0.0210 \pm 0.0258}{8}$$

$$= 6.00 \times 10^{-4} \text{ or } -5.85 \times 10^{-3} \text{ (discard)}$$

Therefore at equilibrium, $[N_2O_4] = 4.9 \times 10^{-3} - x = \mathbf{4.3 \times 10^{-3}}$ *M*

$[NO_2] = 3.78 \times 10^{-3} + 2x = \mathbf{4.98 \times 10^{-3}}$ *M*

(c) When the volume in (a) is suddenly halved (2L → 1L), the concentrations of N_2O_4 and NO_2 are doubled and the equilibrium shifts to the left side.

Let x = moles per liter of N_2O_4 that are produced <u>after</u> the volume is halved. Then
2x = moles per liter of NO_2 that react.

	N_2O_4	⇄	$2NO_2$
initial	9.8×10^{-3} *M*		7.56×10^{-3} *M*
new system	0.020 *M*		0.0151 *M*
change	+ x *M*		- 2x *M*
at equilibrium	(0.020 + x)*M*		(0.0151 - 2x)*M*

$$K_c = \frac{[NO_2]^2}{[N_2O_4]} = \frac{(0.0151 - 2x)^2}{0.020 + x} = 5.84 \times 10^{-3}$$

The quadratic equation: $4x^2 - (0.0662)x + 1.1 \times 10^{-4} = 0$
Solving,

$$x = \frac{0.0662 \pm \sqrt{(-0.0662)^2 - (4)(4)(1.1 \times 10^{-4})}}{2(4)} = \frac{0.0662 \pm 0.051}{8}$$

$$= 1.9 \times 10^{-3} \text{ or } 0.0146 \text{ (discard)}$$

Therefore at equilibrium, $[N_2O_4] = 0.020 + x = \mathbf{0.022}$ *M*

$[NO_2] = 0.0151 - 2x = \mathbf{0.0113}$ *M*

17-58. **Refer to Section 17-10.** . . . .

(a) If *K* >> 1, the forward reaction is likely to be spontaneous, and ΔG^o_{rxn} must be negative.

(b) If *K* = 1, then ΔG^o_{rxn} is 0, and the reaction is at equilibrium when the concentrations of aqueous species are 1 *M* and the partial pressures of gaseous species are 1 atm.

(c) If $K \ll 1$, the forward reaction is likely to be nonspontaneous and ΔG^o_{rxn} must be positive.

17-60. **Refer to Section 17-10, Example 17-9 and Appendix K.** . . . .

Balanced equation: $2SO_2(g) + O_2(g) \rightarrow 2SO_3(g)$

$\Delta G^o_{rxn} = [2\Delta G^o_f\ SO_3(g)] - [2\Delta G^o_f\ SO_2(g) + \Delta G^o_f\ O_2(g)]$

$= [(2\ mol)(-371.1\ kJ/mol)] - [(2\ mol)(-300.2\ kJ/mol) + (1\ mol)(0\ kJ/mol)]$

$= -141.8\ kJ$

We know $\Delta G^o_{rxn} = -2.303RT\log K_p$ for a gas phase reaction

Substituting,

$-141.8 \times 10^3\ J = -2.303(8.314\ J/K)(298.15\ K)\log K_p$

$\log K_p = 24.84$

$K_p = \mathbf{6.9 \times 10^{24}}$

17-62. **Refer to Sections 17-11 and 17-8, and Example 17-12.** . . . .

Balanced equation: $2Cl_2(g) + 2H_2O(g) \rightleftharpoons 4HCl(g) + O_2(g)$

$\Delta H^o = +115\ kJ$

at 25^oC, $K_p = 4.6 \times 10^{-14}$

At 400^oC (673 K): Substituting into the van't Hoff equation, we have

$$\ln \frac{K_{p,T_2}}{K_{p,T_1}} = \frac{\Delta H^o(T_2 - T_1)}{RT_1T_2}$$

$$\ln \frac{K_{p,673\ K}}{4.6 \times 10^{-14}} = \frac{(115 \times 10^3\ J)(673\ K - 298\ K)}{(8.314\ J/K)(298\ K)(673\ K)} = 25.9$$

$$\frac{K_{p,673\ K}}{4.6 \times 10^{-14}} = e^{25.9} = 1.8 \times 10^{11}$$

$$K_{p,673\ K} = \mathbf{8.3 \times 10^{-3}}$$

$$K_c = K_p(RT)^{-\Delta n} = (8.3 \times 10^{-3})(0.0821 \times 673)^{-(5-4)} = \mathbf{1.5 \times 10^{-4}}$$

At 800^oC (1073 K): Substituting into the van't Hoff equation, we have

$$\ln \frac{K_{p,1073\ K}}{4.6 \times 10^{-14}} = \frac{(115 \times 10^3\ J)(1073 - 298\ K)}{(8.314\ J/K)(298\ K)(1073\ K)} = 33.5$$

$$K_{p,873\ K} = \mathbf{16}$$

$$K_c = K_p(RT)^{-\Delta n} = (16)(0.0821 \times 1073)^{-(5-4)} = \mathbf{0.18}$$

17-64. **Refer to Sections 17-10 and 17-11.** . . . .

(a) $\Delta G^o_{rxn} = -2.303RT\log K_p$

$= -2.303(8.314\ J/K)(298.15\ K)\log(4.3 \times 10^6)$

$= \mathbf{-3.79 \times 10^4\ J\ at\ 25^oC}$

(b) Using the van't Hoff equation

$$\ln \frac{K_{T_2}}{K_{T_1}} = \frac{\Delta H^\circ(T_2 - T_1)}{RT_1T_2}$$

$$\ln \frac{K_{1073\ \text{K}}}{4.3 \times 10^6} = \frac{(-78.58 \times 10^3\ \text{J})(1073\ \text{K} - 298\ \text{K})}{(8.314\ \text{J/K})(298\ \text{K})(1073\ \text{K})} = -22.9$$

$$\frac{K_{1073\ \text{K}}}{4.3 \times 10^6} = e^{-22.9} = 1.13 \times 10^{-10}$$

$$K_{1073\ \text{K}} = \mathbf{4.9 \times 10^{-4}}$$

(c) $\Delta G^\circ_{rxn} = -2.303(8.314\ \text{J/K})(1073\ \text{K})\log\ (4.9 \times 10^{-4}) =$ **$+6.80 \times 10^4$ J at 800°C**

(d) The forward reaction at 800°C is nonspontaneous (ΔG is positive) whereas at 25°C the forward reaction is spontaneous (ΔG is negative).

(e) The reaction mixture is heated to speed up the rate at which equilibrium is reached, not to shift the equilibrium toward more product. Heating actually decreases the amount of product present at equilibrium.

(f) Recall from Chapter 16 that the presence of a catalyst increases the rates of both forward and reverse reactions to the same extent without affecting equilibrium.

17-66. **Refer to Sections 17-10 and 17-11 and Appendix K.** . . . .

Balanced equation: $CO(g) + H_2O(g) \rightleftharpoons CO_2(g) + H_2(g)$

(a) Plan: (1) Calculate ΔH°_{rxn} and ΔS°_{rxn} from data in Appendix K and substitute in the Gibbs-Helmholtz equation to determine ΔG°_{rxn}.

(2) Calculate K_p, using $\Delta G^\circ = -2.303RT\log K_p$.

(1) $\Delta H^\circ_{rxn} = [\Delta H^\circ_f\ CO_2(g) + \Delta H^\circ_f\ H_2(g)] - [\Delta H^\circ_f\ CO(g) + \Delta H^\circ_f\ H_2O(g)]$

$= [(1\ \text{mol})(-393.5\ \text{kJ/mol}) + (1\ \text{mol})(0\ \text{kJ/mol})]$
$\quad - [(1\ \text{mol})(-110.5\ \text{kJ/mol}) + (1\ \text{mol})(-241.8\ \text{kJ/mol})]$
$= -41.2\ \text{kJ}$

$\Delta S^\circ_{rxn} = [S^\circ_{CO_2(g)} + S^\circ_{H_2(g)}] - [S^\circ_{CO(g)} + S^\circ_{H_2O(g)}]$

$= [(1\ \text{mol})(213.6\ \text{J/mol}\cdot\text{K}) + (1\ \text{mol})(130.6\ \text{J/mol}\cdot\text{K})]$
$\quad -[(1\ \text{mol})(197.6\ \text{J/mol}\cdot\text{K}) + (1\ \text{mol})(188.7\ \text{J/mol}\cdot\text{K})]$
$= -42.1\ \text{J/K}$

$\Delta G^\circ = \Delta H^\circ - T\Delta S^\circ = -41.2 \times 10^3\ \text{J} - (298\ \text{K})(-42.1\ \text{J/K})$
$= -2.87 \times 10^4\ \text{J}$

(2) Substituting into $\Delta G^\circ = -2.303RT\log K_p$, we have

$-2.87 \times 10^4\ \text{J} = -2.303(8.314\ \text{J/K})(298\ \text{K})\log K_p$

$\log K_p = 5.03$

$K_p = 1.1 \times 10^5$

(b) If we assume that ΔH^o and ΔS^o for the reaction are independent of temperature, we can determine ΔG^o at 200°C (473 K) by substituting into the Gibbs-Helmholtz equation.

$$\Delta G^o = \Delta H^o - T\Delta S^o = -41.2 \times 10^3 \text{ J} - (473 \text{ K})(-42.1 \text{ J/K})$$
$$= -2.13 \times 10^4 \text{ J}$$

Then,

$$\Delta G^o = -2.303RT\log K_p$$

$$-2.13 \times 10^4 \text{ J} = -2.303(8.314 \text{ J/K})(473 \text{ K})\log K_p$$

$$\log K_p = 2.35$$

$$K_p = \mathbf{2.2 \times 10^2}$$

(c) $$\Delta G^o_{rxn} = [\Delta G^o_{CO_2(g)} + \Delta G^o_{H_2(g)}] - [\Delta G^o_{CO(g)} + \Delta G^o_{H_2O(g)}]$$

$$= [(1 \text{ mol})(-394.4 \text{ kJ/mol}) + (1 \text{ mol})(0 \text{ kJ/mol})] - [(1 \text{ mol})(-137.2 \text{ kJ/mol}) + (1 \text{ mol})(-228.6 \text{ kJ/mol})]$$

$$= -28.6 \text{ kJ or } -2.86 \times 10^4 \text{ J}$$

Substituting into $\Delta G^o = -2.303RT\log K_p$, we have

$$-2.86 \times 10^4 \text{ J} = -2.303(8.314 \text{ J/K})(298 \text{ K})\log K_p$$

$$\log K_p = 5.01$$

$$K_p = \mathbf{1.0 \times 10^5}$$

18 Ionic Equilibria I: Acids and Bases

18-2. **Refer to Sections 18-1 and 9-1, and Example 18-1.**

These compounds are strong electrolytes.

(a) 0.10 M HI(aq) → **0.10 M H^+(aq) + 0.10 M I^-(aq)**

(b) 0.040 M RbOH(aq) → **0.040 M Rb^+(aq) + 0.040 M OH^-(aq)**

(c) 0.0020 M Na_2SO_4(aq) → **0.0040 M Na^+(aq) + 0.0020 M SO_4^{2-}(aq)**

18-4. **Refer to Section 18-1 and Example 18-2.**

These compounds are strong electrolytes.

(a) $? \; M \text{ NaOH} = \dfrac{(3.0 \text{ g}/40.0 \text{ g/mol})}{1.50 \text{ L}} = 0.050 \; M$

	NaOH	→	Na^+	+	OH^-
initial	0.050 M		0 M		0 M
change	- 0.050 M		+ 0.050 M		+ 0.050 M
final	0 M		0.050 M		0.050 M

Therefore, $[Na^+]$ = **0.050 M** and $[OH^-]$ = **0.050 M**

(b) $? \; M \text{ Ca(OH)}_2 = \dfrac{(0.61 \text{ g}/74 \text{ g/mol})}{0.250 \text{ L}} = 0.033 \; M$

Similarly, $[Ca^{2+}]$ = **0.033 M** and $[OH^-]$ = **0.066 M**

(c) $? \; M \text{ Ba(NO}_3)_2 = \dfrac{(2.64 \text{ g}/261 \text{ g/mol})}{0.100 \text{ L}} = 0.101 \; M$

Similarly, $[Ba^{2+}]$ = **0.101 M** and $[NO_3^-]$ = **0.202 M**

18-6. **Refer to Section 18-2.**

The ion product for water, K_w, is the equilibrium constant for the auto-ionization of water:

$$H_2O(\ell) + H_2O(\ell) \rightleftharpoons H_3O^+(aq) + OH^-(aq)$$

Since the conventional equilibrium expressions do not include pure liquids or solids, we have

$$K_w = [H_3O^+][OH^-]$$

Careful measurements have found that in pure water **at 25°C**,

$$[H_3O^+] = [OH^-] = 1 \times 10^{-7} \; M.$$

Therefore, $K_w = [H_3O^+][OH^-] = 1 \times 10^{-14}$ **at 25°C.**

18-8. **Refer to Section 18-2.**

(a) A 0.10 M solution of NaOH does have 2 sources of OH^- ion: (1) OH^- from the complete dissociation of NaOH and (2) OH^- from the ionization of water. Since Source 1 dominates Source 2, the concentration of OH^- produced by the ionization of water is therefore neglected.

(b) From Source 1, $NaOH(aq) \rightarrow Na^+(aq) + OH^-(aq)$, $[OH^-] = 0.10\ M$. To calculate $[OH^-]$ from Source 2, consider the ionization of water in a 0.10 M solution of NaOH.

Let x = moles per liter of H_3O^+ and OH^- produced by the ionization of water.

	$2H_2O(\ell)$	$\rightleftharpoons$	$H_3O^+(aq)$	+	$OH^-(aq)$
initial			0 M		0.10 M
change			+ x M		+ x M
at equilibrium			x M		(0.10 + x) M

$K_w = [H_3O^+][OH^-] = x(0.10 + x) = 1.0 \times 10^{-14}$

We know that x << 0.10.
Let us make the approximation that $0.10 + x \simeq 0.10$.

Then, $x \times 0.10 = 1.0 \times 10^{-14}$ and $x = 1.0 \times 10^{-13}$. So, our approximation that x << 0.10 is a good one.

Therefore, $[OH^-]$ from 0.10 M NaOH = 0.10 M
$[OH^-]$ from the ionization of water = 1.0×10^{-13} M, and can be neglected.

18-10. **Refer to Section 18-3 and Exercises 18-2, 18-3 and 18-4 Solutions.** • • • • •

In pure water, $[H_3O^+]$ at 25°C = 1.0×10^{-7} M.

To determine $[H_3O^+]$ in basic solution, use the K_w expression:

$$K_w = [H_3O^+][OH^-] = 1.0 \times 10^{-14}.$$

Solution	$[OH^-]$	$[H_3O^+] = K_w/[OH^-]$
0.040 M RbOH	0.040 M	2.5×10^{-13} M
0.020 M $Ba(OH)_2$	0.040 M	2.5×10^{-13} M
0.033 M $Ca(OH)_2$	0.066 M	1.5×10^{-13} M
pure water	1.0×10^{-7} M	1.0×10^{-7} M

18-12. **Refer to Section 18-3 and Appendix A.2.** • • • • • • • • • • • • • •

(a) log 0.00052 = **-3.28**

(b) log 4.2 = **0.62**

(c) log (5.8×10^{-12}) = **-11.24**

(d) log (4.9×10^{-7}) = **-6.31**

18-14. **Refer to Section 18-3 and Appendix A.2.** • • • • • • • • • • • • • •

(a) antilog 10.73 = **5.4×10^{10}**

(b) antilog (-10.73) = **1.9×10^{-11}**

(c) antilog (-1.84) = **0.014**

(d) antilog 0.60 = **4.0**

18-16. **Refer to Section 18-3, Example 18-4 and Exercise 18-10 Solution.** . . . . . .

Solution	$[H_3O^+]$	$pH = -\log [H_3O^+]$
0.040 *M* RbOH	2.5×10^{-13} *M*	12.60
0.020 *M* $Ba(OH)_2$	2.5×10^{-13} *M*	12.60
0.033 *M* $Ca(OH)_2$	1.5×10^{-13} *M*	12.82
pure water	1.0×10^{-7} *M*	7.00

18-18. **Refer to Section 18-3 and general algebraic principles.** . . . .

(a) $C/D = (1.4 \times 10^{-4})/(4.8 \times 10^{-10}) = \mathbf{2.9 \times 10^5}$

(b) From (a), C is **2.9×10^5** times greater than D.

(c) $D/C = (4.8 \times 10^{-10})/(1.4 \times 10^{-4}) = \mathbf{3.4 \times 10^{-6}}$

18-20. **Refer to Section 18-3 and Example 18-6.** . . . .

	Solution	$[H_3O^+]$	$[OH^-]$	pH	pOH
(a)	0.035 *M* NaOH	2.9×10^{-13} *M*	0.035 *M*	12.54	1.46
(b)	0.035 *M* HCl	0.035 *M*	2.9×10^{-13} *M*	1.46	12.54
(c)	0.035 *M* $Ca(OH)_2$	1.4×10^{-13} *M*	0.070 *M*	12.85	1.15

18-22. **Refer to Section 18-3.** . . . .

	Solution	$[H_3O^+]$	$[OH^-]$	pH	pOH
(a)	A	1.0×10^{-4} *M*	$\underline{1.0 \times 10^{-10}\ M}$	$\underline{4.00}$	$\underline{10.00}$
(b)	B	$\underline{1.00 \times 10^{-2}\ M}$	$\underline{1.00 \times 10^{-12}\ M}$	2.000	$\underline{12.000}$
(c)	C	$\underline{2.1 \times 10^{-9}\ M}$	4.8×10^{-6} *M*	$\underline{8.68}$	$\underline{5.32}$

18-24. **Refer to Section 18-2.** . . . .

(a) $[H_3O^+]_{soln\ E}/[H_3O^+]_{soln\ F} = 0.020\ M/3.0 \times 10^{-4}\ M = \mathbf{67}$

(b) $[H_3O^+]_{soln\ F}/[H_3O^+]_{soln\ E} = 3.0 \times 10^{-4}\ M/0.020\ M = \mathbf{0.015}$

(c) $$[OH^-]_{soln\ E} = \frac{K_w}{[H_3O^+]} = \frac{1.0 \times 10^{-14}}{0.020} = \mathbf{5.0 \times 10^{-13}}\ M$$

$$[OH^-]_{soln\ F} = \frac{K_w}{[H_3O^+]} = \frac{1.0 \times 10^{-14}}{3.0 \times 10^{-4}} = \mathbf{3.3 \times 10^{-11}}\ M$$

(d) $[OH^-]_{soln\ E}/[OH^-]_{soln\ F} = 5.0 \times 10^{-13}\ M/3.3 \times 10^{-11}\ M = \mathbf{0.015}$

(e) $[OH^-]_{soln\ F}/[OH^-]_{soln\ E} = 3.3 \times 10^{-11}\ M/5.0 \times 10^{-13}\ M = \mathbf{66}$

(f) $[H_3O^+]_{soln\ E}/[OH^-]_{soln\ E} = 0.020\ M/5.0 \times 10^{-13}\ M = \mathbf{4.0 \times 10^{10}}$

(g) $[OH^-]_{soln\ E}/[H_3O^+]_{soln\ E} = 5.0 \times 10^{-13}\ M/0.020\ M = \mathbf{2.5 \times 10^{-11}}$

(h) The slight difference between answers (a) and (e) is due to round-off error.

18-26. **Refer to Section 18-3 and Table 18-2.** . . . .

(a) K_w at 37°C = 2.38×10^{-14} = $[H_3O^+][OH^-]$.
In pure water, $[H_3O^+] = [OH^-]$. Therefore, $[H_3O^+] = (2.38 \times 10^{-14})^{1/2}$
$= 1.54 \times 10^{-7}$ at 37°C

pH of pure water = $-\log (1.54 \times 10^{-7})$ = **6.182 at 37°C**

(b) Pure water at 37°C is neutral since $[H_3O^+] = [OH^-]$. The pH of a neutral system is 7.0 only at 25°C.

18-28. **Refer to Section 18-4.** . . . .

Balanced equation: $HA + H_2O \rightleftharpoons H_3O^+ + A^-$

The equilibrium expression is $K_a = \dfrac{[H_3O^+][A^-]}{[HA]}$

For every 1 molecule of HA that dissociates, 1 ion of H_3O^+ and 1 ion of A^- are produced. Therefore, at equilibrium, $[H_3O^+] = [A^-]$.
Substituting,

$$K_a = \frac{(0.0017\ M)^2}{0.0983\ M} = \mathbf{2.9 \times 10^{-5}}$$

18-30. **Refer to Section 18-4 and Example 18-7.** . . . .

Balanced equation: $HX + H_2O \rightleftharpoons H_3O^+ + X^-$

Since HX is 2.7% ionized, $[HX]_{reacted} = 0.10\ M \times 0.027 = 2.7 \times 10^{-3}\ M$

	HX	+	H_2O	$\rightleftharpoons$	H_3O^+	+	X^-
initial	0.10 *M*				~ 0 *M*		0 *M*
change	-2.7×10^{-3} *M*				$+2.7 \times 10^{-3}$ *M*		$+2.7 \times 10^{-3}$ *M*
at equil	0.10 *M*				2.7×10^{-3} *M*		2.7×10^{-3} *M*

$$K_a = \frac{[H_3O^+][X^-]}{[HX]} = \frac{(2.7 \times 10^{-3})^2}{0.10} = \mathbf{7.3 \times 10^{-5}}$$

18-32. **Refer to Section 18-4 and Example 18-9.** . . . .

Balanced equation: $HClO + H_2O \rightleftharpoons H_3O^+ + ClO^-$ $\qquad K_a = 3.5 \times 10^{-8}$

Let x = mol/L of HClO that reacts. Then
x = mol/L of H_3O^+ and ClO^- that are produced.

	HClO	+	H_2O	$\rightleftharpoons$	H_3O^+	+	ClO^-
initial	0.20 *M*				~ 0 *M*		0 *M*
change	- x *M*				+ x *M*		+ x *M*
at equil	(0.20 - x) *M*				x *M*		x *M*

$$K_a = \frac{[H_3O^+][ClO^-]}{[HClO]} = \frac{x^2}{0.20 - x} = 3.5 \times 10^{-8}$$

Assuming that $0.20 - x \approx 0.20$. Then $x^2/0.020 = 3.5 \times 10^{-8}$ and $x = 8.4 \times 10^{-5}$
The simplifying assumption was justified since 8.4×10^{-5} is much less than 5% of 0.20 (= 0.01).

Therefore at equilibrium, [HClO] = **0.20 *M*** $\qquad [OH^-] = K_w/[H_3O^+]$
$[H_3O^+] = [ClO^-]$ = $\mathbf{8.4 \times 10^{-5}}$ $\qquad = \mathbf{1.2 \times 10^{-10}}$ ***M***

18-34. **Refer to Section 18-4 and Examples 18-9 and 18-10.** . . . .

Balanced equation: $HCN + H_2O \rightleftharpoons H_3O^+ + CN^-$ $\qquad K_a = 4.0 \times 10^{-10}$

Let x = $[HCN]_{ionized}$

x = $[H_3O^+]$ and $[CN^-]$ produced

	HCN	+	H_2O	$\rightleftharpoons$	H_3O^+	+	CN^-
initial	0.080 *M*				~ 0 *M*		0 *M*
change	- x *M*				+ x *M*		+ x *M*
at equil	(0.080 - x) *M*				x *M*		x *M*

$$K_a = \frac{[H_3O^+][CN^-]}{[HCN]} = \frac{x^2}{0.080 - x} = 4.0 \times 10^{-10}$$

Assuming (0.080 - x) ≃ 0.080, then $x^2/0.080 = 4.0 \times 10^{-10}$ and $x = 5.7 \times 10^{-6}$.

The simplifying assumption was justified since 5.7×10^{-6} is much less than 5% of 0.080 (= 4×10^{-3}).

Therefore, $[H_3O^+] = x = \mathbf{5.7 \times 10^{-6}}$ *M* $\qquad$ pH = -log $[H_3O^+]$ = **5.24**

$[OH^-] = K_w/[H_3O^+] = \mathbf{1.8 \times 10^{-9}}$ $\qquad$ pOH = -log $[OH^-]$ = **8.74**

$$\text{\% ionization} = \frac{[HCN]_{ionized}}{[HCN]_{initial}} \times 100 = \frac{5.7 \times 10^{-6}\ M}{0.080\ M} \times 100 = \mathbf{7.1 \times 10^{-3}\%}$$

18-36. **Refer to Section 18-4, Exercise 18-34 Solution and Appendix F.** . . . .

The general equation for the ionization of a weak monoprotic acid:

$$HA + H_2O \rightleftharpoons H_3O^+ + A^-$$

Let x = $[HA]_{ionized}$

x = $[H_3O^+]$ and $[A^-]$ that are produced.

	HA	+	H_2O	$\rightleftharpoons$	H_3O^+	+	A^-
initial	0.10 *M*				~ 0 *M*		0 *M*
change	- x *M*				+ x *M*		+ x *M*
at equil	(0.10 - x) *M*				x *M*		x *M*

$$K_a = \frac{[H_3O^+][A^-]}{[HA]} = \frac{x^2}{0.10 - x} \simeq \frac{x^2}{0.10} \text{ if we assume that } 0.10 - x \simeq 0.10$$

(a) for HCN: $K_a = \dfrac{[H_3O^+][CN^-]}{[HCN]} = 4.0 \times 10^{-10} \simeq \dfrac{x^2}{0.10}$. Solving, $x = 6.3 \times 10^{-6}$

$[H_3O^+] = \mathbf{6.3 \times 10^{-6}}$ *M*; pH = -log $[H_3O^+]$ = **5.20**

$$\text{\% ionization} = \frac{[HCN]_{ionized}}{[HCN]_{initial}} \times 100 = \frac{6.3 \times 10^{-6}\ M}{0.10\ M} \times 100 = \mathbf{6.3 \times 10^{-3}\%}$$

(b) for HF: $K_a = \dfrac{[H_3O^+][F^-]}{[HF]} = \dfrac{x^2}{0.10 - x} = 7.2 \times 10^{-4} \simeq \dfrac{x^2}{0.10}$

Solving, $x = 8.5 \times 10^{-3}$

However, in this case, 8.5×10^{-3} is larger than 5% of 0.10. A simplifying assumption cannot be made and the problem must be solved using the original quadratic equation:

$$x^2 + (7.2 \times 10^{-4})x - 7.2 \times 10^{-5} = 0$$

$$x = \frac{-(7.2 \times 10^{-4}) \pm \sqrt{(7.2 \times 10^{-4})^2 - (4)(1)(-7.2 \times 10^{-5})}}{2(1)}$$

$$= \frac{-7.2 \times 10^{-4} \pm 1.7 \times 10^{-2}}{2} = 8.1 \times 10^{-3} \text{ or } -8.9 \times 10^{-3} \text{ (discard)}$$

Note that the 2 answers are about 5% different.

$[H_3O^+]$ = **8.1×10^{-3}** M; pH = -log $[H_3O^+]$ = **2.09**

$$\% \text{ ionization} = \frac{[HF]_{ionized}}{[HF]_{initial}} \times 100 = \frac{8.1 \times 10^{-3}\ M}{0.10\ M} \times 100 = \mathbf{8.1\%}$$

(c) for HBrO: $K_a = \frac{[H_3O^+][BrO^-]}{[HBrO]} = 2.5 \times 10^{-9} \simeq \frac{x^2}{0.10}$. Solving, $x = 1.6 \times 10^{-5}$

$[H_3O^+]$ = **1.6×10^{-5}** M; pH = -log $[H^+]$ = **4.80**

$$\% \text{ ionization} = \frac{[HBrO]_{ionized}}{[HBrO]_{initial}} \times 100 = \frac{1.6 \times 10^{-5}\ M}{0.10\ M} \times 100 = \mathbf{1.6 \times 10^{-2}\%}$$

(d) for HN_3: $K_a = \frac{[H_3O^+][N_3^-]}{[HN_3]} = 1.9 \times 10^{-5} \simeq \frac{x^2}{0.10}$. Solving, $x = 1.4 \times 10^{-3}$

$[H_3O^+]$ = **1.4×10^{-3}** M; pH = -log $[H_3O^+]$ = **2.85**

$$\% \text{ ionization} = \frac{[HN_3]_{ionized}}{[HN_3]_{initial}} \times 100 = \frac{1.4 \times 10^{-3}\ M}{0.10\ M} \times 100 = \mathbf{1.4\%}$$

The observed trend: as K_a increases, $[H_3O^+]$ increases, pH decreases and the % ionization increases.

18-38. **Refer to Section 18-4 and Exercise 18-36 Solution.** . . . .

Balanced equation: $C_2H_4OCOOH + H_2O \rightleftarrows H_3O^+ + C_2H_4OCOO^-$ $K_a = 8.4 \times 10^{-4}$

$$K_a = \frac{[H_3O^+][C_2H_4OCOO^-]}{[C_2H_4OCOOH]} = \frac{x^2}{0.100 - x} = 8.4 \times 10^{-4} \quad \text{where } x = [C_2H_4OCOOH]_{ionized}$$

If $0.100 - x \simeq 0.100$, then $x^2/0.100 \simeq 8.4 \times 10^{-4}$. Solving, $x = 9.2 \times 10^{-3}$

However, 9.2×10^{-3} is more than 5% of 0.100. **A simplifying assumption cannot be made.** Therefore we must use the quadratic equation:

$$x^2 + (8.4 \times 10^{-4})x - 8.4 \times 10^{-5} = 0$$

$$x = \frac{-8.4 \times 10^{-4} \pm \sqrt{(8.4 \times 10^{-4})^2 - (4)(1)(-8.4 \times 10^{-5})}}{2(1)}$$

$$= \frac{-8.4 \times 10^{-4} \pm 1.8 \times 10^{-2}}{2} = 8.6 \times 10^{-3} \text{ or } -9.4 \times 10^{-3} \text{ (discard)}$$

Therefore, $[H_3O^+] = 8.6 \times 10^{-3}$ M; pH = **2.07**

18-40. **Refer to Section 18-4 and Exercise 18-36 Solution.** • • • • • • • • • •

Balanced equation: $C_5H_7O_4COOH + H_2O \rightleftharpoons H_3O^+ + C_5H_7O_4COO^-$ $K_a = 7.9 \times 10^{-5}$

$$K_a = \frac{[H_3O^+][C_5H_7O_4COO^-]}{[C_5H_7O_4COOH]} = \frac{x^2}{0.100 - x} = 7.9 \times 10^{-5} \quad \text{where } x = [C_5H_7O_4COOH]_{ionized}$$

If $0.100 - x \approx 0.100$, then $x^2/0.100 \approx 7.9 \times 10^{-5}$. Solving, $x = 2.8 \times 10^{-3}$.

The simplifying assumption was justified since 2.8×10^{-3} is less than 5% of 0.100 ($= 5 \times 10^{-3}$).

Therefore, $[H_3O^+] = 2.8 \times 10^{-3}$ M; pH = **2.55**

18-42. **Refer to Section 18-4.** • • • • • • • • • • • • • • • • • •

(a) When ascorbic acid is added to amphetamine, which is a weak organic base, a neutralization reaction occurs, yielding a soluble salt. Since the salt is soluble, it is easily excreted in urine.

(b) $C_5H_7O_4COOH \quad + \quad C_9H_{10}NH_2 \quad \rightarrow \quad C_9H_{10}NH_3^+C_5H_7O_4COO^-$
    ascorbic acid   amphetamine   salt

18-44. **Refer to Section 18-4 and Table 18-4.** • • • • • • • • • • • • •

(a) Recall that for a series of weak monoprotic acids, as K_a increases, $[H^+]$ increases, pH decreases and pOH increases.

    1. highest pH - HCN   2. lowest pH - HF
    3. highest pOH - HF   4. lowest pOH - HCN

(b) Consider the dissociation of a weak monoprotic acid: $HA + H_2O \rightleftharpoons H_3O^+ + A^-$. As K_a increases, $[A^-]$ increases.

    1. highest $[A^-]$ - HF   2. lowest $[A^-]$ - HCN

18-46. **Refer to Section 18-4, Example 18-11 and Exercise 18-33.** • • • • • • • •

Balanced equation: $NH_3 + H_2O \rightleftharpoons NH_4^+ + OH^-$ $K_b = 1.8 \times 10^{-5}$

Let $x = [NH_3]_{ionized}$. Then $x = [NH_4^+]$ and $[OH^-]$ that is produced.

	NH_3	$+$	H_2O	$\rightleftharpoons$	NH_4^+	$+$	OH^-
initial	0.080 M				0 M		~ 0 M
change	- x M				+ x M		+ x M
at equil	(0.080 - x) M				x M		x M

$$K_b = \frac{[NH_4^+][OH^-]}{[NH_3]} = 1.8 \times 10^{-5} = \frac{x^2}{0.080 - x} \approx \frac{x^2}{0.080}. \quad \text{Solving, } x = 1.2 \times 10^{-3}$$

Since 1.2×10^{-3} is less than 5% of 0.080, the approximation was justified.

Therefore, $[OH^-] = 1.2 \times 10^{-3}$ M   pOH = **2.92**
              $[H_3O^+] = K_w/[OH^-] = 8.3 \times 10^{-12}$ M   pH = **11.08**

$$\% \text{ ionization} = \frac{[NH_3]_{ionized}}{[NH_3]_{initial}} \times 100 = \frac{1.2 \times 10^{-3} \text{ M}}{0.080 \text{ M}} \times 100 = \mathbf{1.5\%}$$

Exercise 18-33 was a similar problem for a 0.080 M solution of CH_3COOH ($K_a = 1.8 \times 10^{-5}$), giving

$[H_3O^+] = 1.2 \times 10^{-3}\ M$ pH = 2.92

$[OH^-] = 8.3 \times 10^{-12}\ M$ pOH = 11.08

% ionization = 1.5%

It is immediately apparent that $[H_3O^+]$ and pH in one solution have the same value as the $[OH^-]$ and pOH in the other solution, and the percent ionization values are equal. This is because (1) the K_a for CH_3COOH has the same value as the K_b for NH_3, and (2) the solutions are the same concentration. As a result, both weak electrolytes ionize to the same extent, but the former yields H_3O^+ and the latter yields OH^- upon ionization.

18-48. **Refer to Section 18-4 and Table 18-6.** . . . .

Recall that for a series of weak bases, as K_b increases, $[OH^-]$ increases, pOH decreases and pH increases.

(a) highest pH - dimethylamine, $(CH_3)_2NH$

(b) lowest pH - pyridine, C_5H_5N

(c) highest pOH - pyridine

(d) lowest pOH - dimethylamine

18-50. **Refer to Section 3-5, 3-7 and 13-14.** . . . .

Balanced equation: $CH_3COOH(aq) + NH_3(aq) \rightarrow NH_4CH_3COO(aq)$

(a) $$? M\ NH_3 = \frac{5.00\text{ g } NH_3}{100\text{ g soln}} \times \frac{1\text{ mol } NH_3}{17.0\text{ g } NH_3} \times \frac{0.980\text{ g soln}}{1.00\text{ mL soln}} \times \frac{10^3\text{ mL soln}}{1.00\text{ L soln}}$$
$$= \mathbf{2.88\ M\ NH_3}$$

(b) $$? M\ CH_3COOH = \frac{5.00\text{ g } CH_3COOH}{100\text{ g soln}} \times \frac{1\text{ mol } CH_3COOH}{60.1\text{ g } CH_3COOH} \times \frac{1.01\text{ g soln}}{1.00\text{ mL soln}} \times \frac{10^3\text{ mL soln}}{1.00\text{ L soln}}$$
$$= \mathbf{0.840\ M\ CH_3COOH}$$

(c) At equivalence, $N_A \times V_A = N_B \times V_B$

For CH_3COOH and NH_3, normality = molarity, and so,

$$V_A = \frac{N_BV_B}{N_A} = \frac{M_BV_B}{M_A} = \frac{2.88\ M \times 1.00\text{ L}}{0.840\ M} = \mathbf{3.43\ L\ CH_3COOH\ soln}$$

(d) $$V_B = \frac{N_AV_A}{N_B} = \frac{M_AV_A}{M_B} = \frac{0.840\ M \times 1.00\ M}{2.88\ M} = \mathbf{0.292\ L\ NH_3\ soln}$$

18-52. **Refer to Section 18-4.** . . . .

The general balanced equation for monoprotic weak acid: $HA + H_2O \rightleftharpoons H_3O^+ + A^-$

Let x = [nicotinic acid]$_{\text{ionized}}$.

	HA	+	H_2O	$\rightleftharpoons$	H_3O^+	+	A^-
initial	0.050 M				~ 0 M		0 M
change	- x M				+ x M		+ x M
at equil	(0.050 - x) M				x M		x M

$$K_a = \frac{[H_3O^+][A^-]}{[HA]} = \frac{x^2}{0.050 - x} = 1.4 \times 10^{-5} \simeq \frac{x^2}{0.050}.$$ Solving, $x = 8.4 \times 10^{-4}$

The simplifying assumption is justified since 8.4×10^{-4} is less than 5% of 0.050.

Therefore, $[H_3O^+] = 8.4 \times 10^{-4}$; pH = **3.08**.

18-54. **Refer to Section 18-5.** . . . .

(a) Acid-base indicators are organic compounds which behave as weak acids or bases and exhibit different colors in solutions with different acidities.

(b) The essential characteristic of an acid-base indicator is that the conjugate acid-base pair must exhibit different colors. Consider the weak acid indicator, HIn. In solution, HIn dissociates slightly as follows:

$$\underset{\text{acid}}{HIn} + H_2O \rightleftharpoons H_3O^+ + \underset{\text{conjugate base}}{In^-}$$

HIn dominates in more acidic solutions with one characteristic color; In^- dominates in more basic solutions with another color.

(c) The color of an acid-base indicator in an aqueous solution depends upon the ratio, $[In^-]/[HIn]$, which in turn depends upon $[H^+]$ and the K_a value of the indicator. A general rule of thumb: If $[In^-]/[HIn] \le 0.1$, then the indicator will show its true acid color. If $[In^-]/[HIn] \ge 10$, then the indicator will show its true base color.

18-56. **Refer to Section 18-5 and Table 18-7.** . . . .

	pH = 2.00	pH = 11.00
methyl orange	red	yellow
bromthymol blue	yellow	blue
phenolphthalein	colorless	red

18-58. **Refer to Section 18-5.** . . . .

Balanced equation: $HIn + H_2O \rightleftharpoons H_3O^+ + In^-$

At pH 8.2, $[HIn] = [In^-]$ and $[H_3O^+]$ = antilog (-pH) = 6×10^{-9}

Substituting into the K_a expression, we have

$$K_a = \frac{[H_3O^+][In^-]}{[HIn]} = \mathbf{6 \times 10^{-9}}$$ since $[In^-]$ and $[HIn]$ cancel.

18-60. **Refer to Section 18-6, Table 18-8 and Figure 18-3.** . . . .

	mL of 0.100 *M* HNO_3 added	mmol HNO_3 added	mmol of excess acid or base	mL of total volume	*M* of OH^- or H_3O^+	pH
initial	0.0	0.00	2.50 OH^-	25.0	0.100 OH^-	13.000
(a)	5.0	0.50	2.00	30.0	0.0667	12.824
(b)	10.0	1.00	1.50	35.0	0.0429	12.632
(c)	12.5	1.25	1.25	37.5	0.0333	12.522
(d)	20.0	2.00	0.50	45.0	0.011	12.04
(e)	24.0	2.40	0.10	49.0	0.0020	11.30
(f)	24.9	2.49	0.01	49.5	0.0002	10.3
(g)	25.0	2.50	0.00 (eq. pt.)	50.0	1.00×10^{-7}	7.000
(h)	25.1	2.51	0.01 H_3O^+	50.1	0.0002 H_3O^+	3.7
(i)	27.0	2.70	0.20	52.0	0.0038	2.42
(j)	30.0	3.00	0.50	55.0	0.0091	2.04

(1) Titration of 25.0 mL of 0.100 *M* NaOH solution with 0.100 *M* HNO_3 solution:

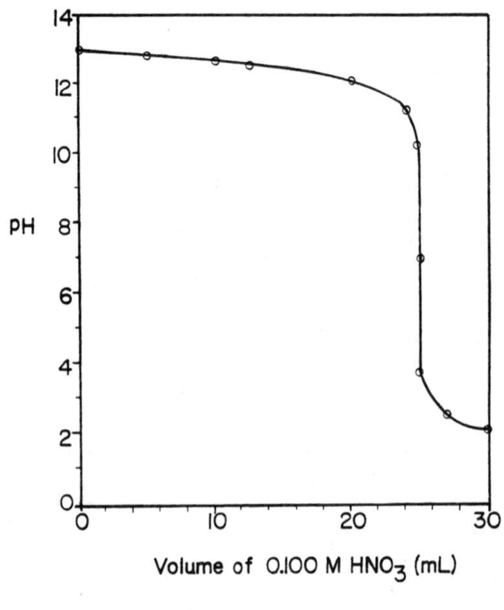

(2) Yes, it is similar to the titration curve in Figure 18-3b, but inverted to the one in Figure 18-3a. It is similar to Figure 18-3b because both involve the addition of a strong acid to a strong base.

(3) Indicators that change color in the pH range 4 to 10 may be used, although the ideal indicator should change color at pH 7, the equivalence point.

18-62. **Refer to Sections 18-7 and 18-8.** . . . .

A buffer solution resists changes in pH. Additions of small amounts of acid or base to a buffer cause only a small shift in an equilibrium involving H^+ or OH^- ions, thus minimizing pH changes. A buffer solution is generally a weak acid and a weak conjugate base mixture. The weak acid will neutralize the added strong base, and the weak base will neutralize the added strong acid.

A solution of a strong acid and a salt of the acid is not a good buffer solution. A typical example of such a solution would be a HCl/NaCl mixture. The main drawback to this system is the lack of a species to react with any added strong acid such as HNO_3.

A solution of a weak base and a salt of a strong soluble base is also not a buffer solution. A typical example of such a solution would be a NH_3/NaCl mixture. The main drawback to this system is the lack of a species to react with any added strong base such as NaOH.

18-64. **Refer to Sections 18-7 and 18-8.** . . . .

(a) 0.15 *M* CH_2COOH and 0.30 *M* NaCl - not a buffer since NaCl will not neutralize any strong acid added to the solution.

(b) 0.10 *M* NH_3 and 0.15 *M* NaCl - not a buffer since NaCl will not neutralize any strong base added to the solution.

(c) 0.10 *M* NH_3 and 0.10 *M* NH_4NO_3 - is a buffer solution containing a weak base and a salt which will provide NH_4^+ ion, a weak acid.

18-66. **Refer to Section 18-7.** . . . .

(a) $NaHCO_3 \xrightarrow{H_2O} Na^+ + HCO_3^-$ (to completion)

$H_2CO_3 + H_2O \rightleftharpoons H_3O^+ + HCO_3^-$ (reversible)

When a small amount of base is added to the buffer:

$H_2CO_3 + OH^- \rightarrow HCO_3^- + H_2O$

When a small amount of acid is added to the buffer:

$HCO_3^- + H_3O^+ \rightarrow H_2CO_3 + H_2O$

(b) $NaH_2PO_4 \xrightarrow{H_2O} Na^+ + H_2PO_4^-$ (to completion)

$Na_2HPO_4 \xrightarrow{H_2O} 2Na^+ + HPO_4^{2-}$ (to completion)

$H_2PO_4^- + H_2O \rightleftharpoons H_3O^+ + HPO_4^{2-}$ (reversible)

When a small amount of base is added to the buffer:

$H_2PO_4^- + OH^- \rightarrow HPO_4^{2-} + H_2O$

When a small amount of acid is added to the buffer:

$HPO_4^{2-} + H_3O^+ \rightarrow H_2PO_4^- + H_2O$

18-68. **Refer to Section 18-7, Example 18-13, Appendix F and Exercise 18-36 Solution.** •

(a) Balanced equations: $NaCN \rightarrow Na^+ + CN^-$ (to completion)
$HCN + H_2O \rightleftharpoons H_3O^+ + CN^-$ (reversible)

Since NaCN dissociates completely,
$[CN^-]$ from the salt = $[NaCN]_{initial}$ = 0.10 M

Let x = $[HCN]_{ionized}$. Then
x = $[H_3O^+]$ and $[CN^-]$ produced from the ionization of HCN

	HCN	+	H_2O	$\rightleftharpoons$	H_3O^+	+	CN^-
initial	0.10 M				~ 0 M		0.10 M
change	- x M				x M		+ x M
at equil	(0.10 - x) M				x M		(0.10 + x) M

$$K_a = \frac{[H_3O^+][CN^-]}{[HCN]} = \frac{(x)(0.10 + x)}{(0.10 - x)} = 4.0 \times 10^{-10} \simeq \frac{x(0.10)}{(0.10)} = x$$

Therefore, $[H_3O^+]$ = **4.0×10^{-10}** M; pH = **9.40**

(b) Balanced equations: $NaF \rightarrow Na^+ + F^-$ (to completion)
$HF + H_2O \rightleftharpoons H_3O^+ + F^-$ (reversible)

Since NaF dissociates completely,
$[F^-]$ from the salt = $[NaF]_{initial}$ = 0.10 M

Let x = $[HCN]_{ionized}$. Then
x = $[H_3O^+]$ and $[CN^-]$ produced from the ionization of HCN.

	HF + H_2O ⇌	H_3O^+ +	F^-
initial	0.10 *M*	~ 0 *M*	0.10 *M*
change	- x *M*	+ x *M*	+ x *M*
at equil	(0.10 - x) *M*	x *M*	(0.10 + x) *M*

$$K_a = \frac{[H_3O^+][F^-]}{[HF]} = \frac{(x)(0.10 + x)}{(0.10 - x)} = 7.2 \times 10^{-4} \simeq \frac{x(0.10)}{(0.10)} = x$$

Therefore, $[H_3O^+]$ = **7.2×10^{-4} *M***; pH = **3.14**

(c) Balanced equations: $KBrO \rightarrow K^+ + BrO^-$ (to completion)
$HBrO + H_2O \rightleftharpoons H_3O^+ + BrO^-$ (reversible)

Since KBrO dissociates completely, $[BrO^-]$ from the salt = [KBrO] = 0.10 *M*

Let x = $[HBrO]_{ionized}$. Then
x = $[H_3O^+]$ and $[BrO^-]$ produced from the ionization of HBrO.

	HBrO + H_2O ⇌	H_3O^+ +	BrO^-
initial	0.10 *M*	~ 0 *M*	0.10 *M*
change	- x *M*	+ x *M*	+ x *M*
at equil	(0.10 - x) *M*	x *M*	(0.10 + x) *M*

$$K_a = \frac{[H_3O^+][BrO^-]}{[HBrO]} = \frac{(x)(0.10 + x)}{(0.10 - x)} = 2.5 \times 10^{-9} \simeq \frac{(x)(0.10)}{(0.10)} = x$$

Therefore, $[H_3O^+]$ = **2.5×10^{-9} *M***; pH = **8.60**

If we compare the values to those we obtained for 0.10 *M* solutions of the weak acids alone in Exercise 18-36, we conclude that buffer solutions prepared from weak acids and their salts are more basic than solutions of weak acids alone.

	$[H_3O^+]$	pH
0.10 *M* HCN	6.3×10^{-6} *M*	5.20
0.10 *M* HCN/0.10 *M* NaCN	4.0×10^{-10} *M*	9.40
0.10 *M* HF	8.1×10^{-3} *M*	2.09
0.10 *M* HF/0.10 *M* NaF	7.2×10^{-4} *M*	3.14
0.10 *M* HBrO	1.6×10^{-5} *M*	4.80
0.10 *M* HBrO/0.10 *M* NaBrO	2.5×10^{-9} *M*	8.60

18-70. **Refer to Section 18-7, Example 18-13 and Appendix F.** . . . .

(a) Balanced equations: $KF \rightarrow K^+ + F^-$ (to completion)
$HF + H_2O \rightleftharpoons H_3O^+ + F^-$ (reversible)

Since KF dissociates completely, $[F^-]$ from the salt = $[KF]_{ionized}$ = 0.20 *M*

Let x = $[HF]_{ionized}$. Then
x = $[H_3O^+]$ and $[F^-]$ produced from the ionization of HF.

	HF + H_2O ⇌	H_3O^+ +	F^-
initial	0.10 *M*	~ 0 *M*	0.20 *M*
change	- x *M*	+ x *M*	+ x *M*
at equil	(0.10 - x) *M*	x *M*	(0.20 + x) *M*

$$K_a = \frac{[H_3O^+][F^-]}{[HF]} = \frac{(x)(0.20 + x)}{(0.10 - x)} = 7.2 \times 10^{-4} \simeq \frac{x(0.20)}{(0.10)}$$

Solving, $x = 3.6 \times 10^{-4}$

Therefore, $[H_3O^+] = 3.6 \times 10^{-4}\ M$; pH = **3.44**

Alternatively, this problem can be solved using the Henderson-Hasselbalch equation,

$$pH = pK_a + \log \frac{[salt]}{[acid]}. \quad \text{Substituting, } pH = 3.14 + \log \frac{(0.20)}{(0.10)} = 3.44$$

(b) Note than the Henderson-Hasselbalch equation cannot be used in this case since it is valid only for solutions containing a weak monoprotic acid and a soluble, ionic salt of the weak acid with a <u>univalent</u> cation. The cation, Ba^{2+}, in $Ba(CH_3COO)_2$, is divalent.

Since $Ba(CH_3COO)_2$ dissociates totally,

$[CH_3COO^-]$ from the salt = $2 \times [Ba(CH_3COO^-)_2]$.

Let $x = [CH_3COOH]_{ionized}$. Then

$x = [H_3O^+]$ and $[CH_3COO^-]$ produced from the ionization of CH_3COOH.

	CH_3COOH	+	H_2O	⇌	H_3O^+	+	CH_3COO^-
initial	0.050 M				~ 0 M		0.050 M
change	- x M				+ x M		+ x M
at equil	(0.050 - x) M				x M		(0.050 + x) M

$$K_a = 1.8 \times 10^{-5} = \frac{[H_3O^+][CH_3COO^-]}{[CH_3COOH]} = \frac{(x)(0.050 + x)}{(0.050 - x)} \simeq \frac{x(0.050)}{(0.050)} = x$$

Therefore, $[H_3O^+] = 1.8 \times 10^{-5}\ M$; pH = **4.74**

18-72. **Refer to Section 18-7, Example 18-14 and Appendix G.** . . . .

Balanced equation: $NH_4NO_3 \rightarrow NH_4^+ + NO_3^-$ (to completion)
$NH_3 + H_2O \rightleftharpoons NH_4^+ + OH^-$ (reversible)

(a) Since NH_4NO_3 is a soluble salt,

$[NH_4^+]$ from the salt = $[NH_4NO_3]_{initial}$ = 0.30 M

Let $x = [NH_3]_{ionized}$. Then

$x = [NH_4^+]$ and $[OH^-]$ produced by the ionization of NH_3.

	NH_3	+	H_2O	⇌	NH_4^+	+	OH^-
initial	0.20 M				0.30 M		~ 0 M
change	- x M				+ x M		+ x M
at equil	(0.20 - x) M				(0.30 + x) M		x M

$$K_b = \frac{[NH_4^+][OH^-]}{[NH_3]} = \frac{(0.30 + x)(x)}{(0.20 - x)} = 1.8 \times 10^{-5} \simeq \frac{(0.3)x}{(0.2)}$$

Solving, $x = 1.2 \times 10^{-5}$

Therefore, $[OH^-] = \mathbf{1.2 \times 10^{-5}}\ M$; pOH = 4.92; pH = **9.08**

Alternatively, using the Henderson-Hasselbalch equation,

$$\text{pOH} = \text{p}K_b + \log \frac{[\text{salt}]}{[\text{base}]} = -\log (1.8 \times 10^{-5}) + \log \frac{(0.30)}{(0.20)} = 4.92$$

$\text{pH} = \mathbf{9.08}$

(b) Since $(NH_4)_2SO_4$ is a soluble salt,

$[NH_4^+]$ from the salt $= 2 \times [(NH_4)_2SO_4]_{\text{initial}} = 0.30\ M$

Note that we now have the same system as (a) with $[OH^-] = \mathbf{1.2 \times 10^{-5}}\ M$ and $\text{pH} = \mathbf{9.08}$.

18-74. **Refer to Section 18-8.** . . . .

Balanced equation: $C_8H_7O_2COOH + H_2O \rightleftharpoons H_3O^+ + C_8H_7O_2COO^-$ $\quad K_a = 3.3 \times 10^{-4}$

(a) Plan: (1) Use the K_a expression for aspirin and the given data to calculate $[C_8H_7O_2COOH]$ that ionizes.
(2) Determine the percent ionization of aspirin at these conditions.

(1) $[H_3O^+]_{\text{initial}} = 0.010\ M$ since pH = 2.00.

Let $x = [C_8H_7O_2COOH]_{\text{ionized}}$. Then
$x = [H_3O^+]$ and $[C_8H_7O_2COO^-]$ produced by the ionization of aspirin.

	$C_8H_7O_2COOH$	+	H_2O	$\rightleftharpoons$	H_3O^+	+	$C_8H_7O_2COO^-$
initial	$1.0 \times 10^{-4}\ M$				$0.010\ M$		$0\ M$
change	$-\ x\ M$				$+\ x\ M$		$+\ x\ M$
at equil	$(1.0 \times 10^{-4} - x)\ M$				$(0.010 + x)\ M$		$x\ M$

$$K_a = \frac{[H_3O^+][C_8H_7O_2COO^-]}{[C_8H_7O_2COOH]} = \frac{(0.010 + x)(x)}{(1.0 \times 10^{-4} - x)} = 3.3 \times 10^{-4} \approx \frac{(0.010)(x)}{(1.0 \times 10^{-4})}$$

Solving, $x = 3.3 \times 10^{-6}$

Therefore, $[C_8H_7O_2COOH]_{\text{ionized}} = \mathbf{3.3 \times 10^{-6}}\ M$

(2) $$\%\ \text{ionization} = \frac{[C_8H_7O_2COOH]_{\text{ionized}}}{[C_8H_7O_2COOH]_{\text{initial}}} \times 100 = \frac{3.3 \times 10^{-6}\ M}{1.0 \times 10^{-4}\ M} \times 100 = \mathbf{3.3\%}$$

(b) Let $x = [C_8H_7O_2COOH]_{\text{ionized}}$. Then $x = [H_3O^+]$ and $[C_8H_7O_2COO^-]$ produced.

	$C_8H_7O_2COOH$	+	H_2O	$\rightleftharpoons$	H_3O^+	+	$C_8H_7O_2COO^-$
initial	$1.0 \times 10^{-4}\ M$				$\sim 0\ M$		$0\ M$
change	$-\ x\ M$				$+\ x\ M$		$+\ x\ M$
at equil	$(1.0 \times 10^{-4} - x)\ M$				$x\ M$		$x\ M$

$$K_a = \frac{[H_3O^+][C_8H_7O_2COO^-]}{[C_8H_7O_2COOH]} = \frac{x^2}{(1.0 \times 10^{-4} - x)} = 3.3 \times 10^{-4} \approx \frac{x^2}{1.0 \times 10^{-4}}$$

Solving, $x = 1.8 \times 10^{-4}\ M$

The simplifying assumption gives a meaningless result and we must use the quadratic equation:

$$x^2 + (3.3 \times 10^{-4})x - 3.3 \times 10^{-8} = 0$$

Solving,

$$x = \frac{-3.3 \times 10^{-4} \pm \sqrt{(3.3 \times 10^{-4})^2 - (4)(1)(-3.3 \times 10^{-8})}}{2(1)}$$

$$= \frac{-3.3 \times 10^{-4} \pm 4.9 \times 10^{-4}}{2} = 8.0 \times 10^{-5} \text{ or } -4.1 \times 10^{-4} \text{ (discard)}$$

Therefore, $[C_8H_7O_2COOH]_{ionized}$ = **8.0 × 10^{-5}** M

Then,

$$\% \text{ ionization} = \frac{[C_8H_7O_2COOH]_{ionized}}{[C_8H_7O_2COOH]_{initial}} \times 100 = \frac{8.0 \times 10^{-5}\ M}{1.0 \times 10^{-4}\ M} \times 100 = \mathbf{80\%}$$

Therefore, aspirin, a weak acid, ionizes to a much greater extent in distilled water than it does in stomach acid having pH = 2.00.

18-76. **Refer to Section 18-7 and 18-8, Example 18-15 and Appendices F and G. . . . .**

From Section 18-7, we learn that if the concentrations of the weak acid or base and its salt are ~0.05 M or greater, and the salt contains a univalent cation, then

for an acidic buffer: $$[H_3O^+] = \frac{[\text{Acid}]}{[\text{Salt}]} \times K_a = \frac{\text{mol acid}}{\text{mol salt}} \times K_a$$

for a basic buffer: $$[OH^-] = \frac{[\text{Base}]}{[\text{Salt}]} \times K_b = \frac{\text{mol base}}{\text{mol salt}} \times K_b$$

(a) Original Buffer:

$$[H_3O^+] = \frac{[\text{Acid}]}{[\text{Salt}]} \times K_a = \frac{0.10\ M}{0.10\ M} \times (6.3 \times 10^{-5}) = 6.3 \times 10^{-5}\ M;\ \text{pH} = 4.20$$

New Buffer: When 0.010 moles of HCl is added to the original buffer, it reacts with the base component of the buffer, $C_6H_5COO^-$, to form more of the acid component, C_6H_5COOH, thereby creating a new buffer solution with a slightly more acidic pH.

	HCl	+	$C_6H_5COO^-$	→	C_6H_5COOH
initial	0.010 mol		0.10 mol		0.10 mol
change	- 0.010 mol		- 0.010 mol		+ 0.010 mol
after rxn	0 mol		0.09 mol		0.11 mol

$$[H_3O^+] = \frac{\text{mol acid}}{\text{mol salt}} \times K_a = \frac{0.11\ \text{mol}}{0.09\ \text{mol}} (6.3 \times 10^{-5}) = 8 \times 10^{-5}\ M;\ \text{pH} = 4.1$$

Change in $[H_3O^+]$ = $8 \times 10^{-5} - 6.3 \times 10^{-5}$ = **2 × 10^{-5}** M

Change in pH = 4.1 - 4.2 = **-0.1**

(b) Original Buffer:

$$[OH^-] = \frac{[\text{Base}]}{[\text{Salt}]} \times K_b = \frac{(0.10\ M)}{(0.20\ M)} (7.4 \times 10^{-5}) = 3.7 \times 10^{-5}\ M$$

$[H_3O^+] = 2.7 \times 10^{-10}\ M$; pH = 9.57

New Buffer: When 0.010 mole of HNO_3 is added to Buffer (1), it reacts with the base component of the buffer, $(CH_3)_3N$, to form more of the acid component, $(CH_3)_3NH^+$, creating a slightly more acidic buffer solution.

? mol $(CH_3)_3N$ = 0.10 M × 0.500 L = 0.050 mol

? mol $(CH_3)_3NHCl$ = 0.20 M × 0.500 L = 0.10 mol

	HNO_3	+	$(CH_3)_3N$	→	$(CH_3)_3NHNO_3$
initial	0.010 mol		0.050 mol		0.10 mol
change	- 0.010 mol		- 0.010 mol		+ 0.010 mol
after rxn	0 mol		0.040 mol		0.11 mol

$$[OH^-] = \frac{\text{mol base}}{\text{mol salt}} \times K_b = \frac{0.040 \text{ mol}}{0.11 \text{ mol}} (7.4 \times 10^{-5}) = 2.7 \times 10^{-5} \; M$$

$[H_3O^+] = 3.7 \times 10^{-10}$ M; pH = 9.43

Change in $[H_3O^+] = 3.7 \times 10^{-10} - 2.7 \times 10^{-10} = \mathbf{1.0 \times 10^{-10}}$ M
Change in pH = 9.43 - 9.57 = **-0.14**

(c) Original Buffer: Same as (b)
New Buffer: When 0.010 mole of NaOH is added to Buffer (1), it reacts with the acid component of the buffer, $(CH_3)_3NH^+$, to form more of the base component, $(CH_3)_3N$, creating a slightly more basic buffer solution.

	NaOH	+	$(CH_3)_3NH^+$	→	$(CH_3)_3N$
initial	0.010 mol		0.10 mol		0.050 mol
change	- 0.010 mol		- 0.010 mol		+ 0.010 mol
after rxn	0 mol		0.09 mol		0.060 mol

$$[OH^-] = \frac{\text{mol base}}{\text{mol acid}} \times K_b = \frac{0.060 \text{ mol}}{0.09 \text{ mol}} (7.4 \times 10^{-5}) = 5 \times 10^{-5} \; M$$

$[H_3O^+] = 2 \times 10^{-10}$ M; pH = 9.7

Change in $[H_3O^+] = 2 \times 10^{-10} - 2.7 \times 10^{-10} = \mathbf{-1 \times 10^{-10}}$ M
Change in pH = 9.7 - 9.43 = **+0.3**

18-78. **Refer to Section 18-9 and Appendix F.** . . . .

Balanced equation: HCl + NaCN → HCN + NaCl

Plan: (1) Calculate the final concentrations of the products and reactants by first determining the number of moles of HCl and NaCN.
(2) Calculate the pH of the resulting solution.

(1) ? mol HCl = 0.20 M × 0.100 L = 0.020 mol.
? mol NaCN = 0.20 M × 0.100 L = 0.020 mol.

Therefore, we have stoichiometric amounts of both HCl and NaCN, reacting to give 0.020 moles of HCN and NaCl.

$$[HCN] = \frac{0.020 \text{ mol}}{0.200 \text{ L}} = 0.10 \; M$$

(2) After the reaction, the pH is determined by the ionization of the weak acid, HCN.

Let x = $[HCN]_{ionized} = [H_3O^+]$ and $[CN^-]$ produced.

	HCN	+	H_2O	⇄	H_3O^+	+	CN^-
initial	0.10 M				~ 0 M		0 M
change	- x M				+ x M		+ x M
at equil	(0.10 - x) M				x M		x M

$$K_a = \frac{[H_3O^+][CN^-]}{[HCN]} = \frac{x^2}{0.10 - x} = 4.0 \times 10^{-10} \approx \frac{x^2}{0.10}$$ Solving, x = 6.3×10^{-6}

Therefore, $[H_3O^+] = 6.3 \times 10^{-6}$ *M*; pH = **5.2**

This reaction should not be performed in an open area since HCN is a volatile and very toxic substance.

18-80. **Refer to Sections 18-4, 18-7 and 18-8.** . . . .

Balanced equation: $NH_3 + H_2O \rightleftharpoons NH_4^+ + OH^-$

(a) Let x = $[NH_3]_{ionized}$ = $[NH_4^+]$ and $[OH^-]$ produced.

	NH_3	+	H_2O	$\rightleftharpoons$	NH_4^+	+	OH^-
initial	2.0 *M*				0 *M*		~ 0 *M*
change	- x *M*				+ x *M*		+ x *M*
at equil	(2.0 - x) *M*				x *M*		x *M*

$$K_b = \frac{[NH_4^+][OH^-]}{[NH_3]} = \frac{x^2}{2.0 - x} = 1.8 \times 10^{-5} \simeq \frac{x^2}{2.0}.\quad \text{Solving, } x = 6.0 \times 10^{-3}$$

Therefore,
$[OH^-]$ = **6.0×10^{-3}** *M*; $[H_3O^+]$ = **1.7×10^{-12}**; pOH = **2.22**; pH = **11.78**

(b) $$[OH^-] = \frac{[Base]}{[Salt]} \times K_b = \frac{2.0\ M}{1.0\ M}(1.8 \times 10^{-5}) = \mathbf{3.6 \times 10^{-5}}\ M;$$

$[H_3O^+]$ = **2.8×10^{-10}** *M*; pOH = **4.44**; pH = **9.56**

(c) $$[OH^-] = \frac{[Base]}{2[Salt]} \times K_b = \frac{2.0\ M}{2(1.0\ M)}(1.8 \times 10^{-5}) = \mathbf{1.8 \times 10^{-5}}\ M;$$

$[H_3O^+]$ = **5.6×10^{-10}** *M*, pOH = **4.74**; pH = **9.26**

(d) ? mol HNO_3 = 1.0 *M* × 0.010 L = 0.010 mol
? mol NH_3 = 2.0 *M* × 0.050 L = 0.10 mol

	HNO_3	+	NH_3	→	NH_4NO_3
initial	0.010 mol		0.10 mol		0 mol
change	- 0.010 mol		- 0.010 mol		+ 0.010 mol
after rxn	0 mol		0.09 mol		+ 0.010 mol

Therefore, the reaction produces a buffer solution.

$$[OH^-] = \frac{\text{mol base}}{\text{mol salt}} \times K_b = \frac{0.09\ \text{mol}}{0.010\ \text{mol}}(1.8 \times 10^{-5}) = \mathbf{2.0 \times 10^{-4}}\ M;$$

$[H_3O^+]$ = **5×10^{-11}** *M*; pOH = **3.7**; pH = **10.3**

Therefore, in adding 0.010 moles of HNO_3 to the NH_3 solution in (a), the pH changes from 11.78 to 10.3, a decrease of 1.5 pH units.

(e) The added HNO_3 reacts with the base component of the buffer to produce more of the acid component of the buffer, creating a new, slightly more acidic buffer.

? mol HNO_3 = 1.0 *M* × 0.010 L = 0.010 mol
? mol NH_3 = 2.0 *M* × 0.050 L = 0.10 mol
? mol NH_4Cl = 1.0 *M* × 0.050 L = 0.050 mol

	HNO_3	+	NH_3	→	NH_4^+
initial	0.010 mol		0.10 mol		0.050 mol
change	- 0.010 mol		- 0.010 mol		+ 0.010 mol
after rxn	0 mol		0.09 mol		0.060 mol

$$[OH^-] = \frac{\text{mol base}}{\text{mol salt}} \times K_b = \frac{0.09 \text{ mol}}{0.060 \text{ mol}} (1.8 \times 10^{-5}) = 3 \times 10^{-5}\ M;$$

$[H_3O^+]$ = **3 × 10^{-10}** M; pOH = **4.5**; pH = **9.5**

Therefore, in adding 0.010 moles of HNO_3 to the buffer solution in (b), the pH only decreases from 9.56 to 9.5.

(f) The added KOH reacts with the acidic component of the buffer to produce more of the basic component of the buffer, creating a new, slightly more basic buffer.

? mol KOH = 1.0 M × 0.010 L = 0.010 mol
? mol NH_3 = 2.0 M × 0.050 L = 0.10 mol
? mol NH_4^+ = 2 × mol$(NH_4)_2SO_4$ = 2(1.0 M × 0.050 L) = 0.10 mol

	KOH	+	NH_4^+	→	NH_3	+	H_2O	+	K^+
initial	0.010 mol		0.10 mol		0.10 mol				
change	- 0.010 mol		- 0.010 mol		+ 0.010 mol				
after rxn	0 mol		0.09 mol		0.11 mol				

$$[OH^-] = \frac{\text{mol base}}{\text{mol salt}} \times K_b = \frac{0.11 \text{ mol}}{0.09 \text{ mol}}(1.8 \times 10^{-5}) = 2 \times 10^{-5}\ M;$$

$[H_3O^+]$ = **5 × 10^{-10}**; pOH = **4.7**; pH = **9.3**

Similarly, in adding 0.010 moles of KOH to the buffer solution in (c), the pH only increases from 9.26 to 9.3.

18-82. **Refer to Section 18-9 and Example 18-17.** . . . .

Balanced equation: $CH_3CH_2COOH + H_2O \rightleftharpoons H_3O^+ + CH_3CH_2COO^-$

$$K_a = \frac{[H_3O^+][CH_3CH_2COO^-]}{[CH_3CH_2COOH]} = 1.3 \times 10^{-5}$$

In this buffer system, $[H_3O^+] = 3.2 \times 10^{-4}\ M$ since pH = 3.50
$[CH_3CH_2COOH] \simeq [CH_3CH_2COOH]_{\text{initial}} = 2.0\ M$

$$\text{Solving, } [CH_3CH_2COO^-] = \frac{K_a[CH_3CH_2COOH]}{[H_3O^+]} \simeq \frac{(1.3 \times 10^{-5})(2.0)}{(3.2 \times 10^{-4})} = 0.081\ M$$

$$?\text{ g } NaCH_3CH_2COO = 0.300 \text{ L} \times \frac{0.081 \text{ mol}}{1.00 \text{ L}} \times \frac{96 \text{ g}}{1 \text{ mol}} = \mathbf{2.3\ g}$$

18-84. **Refer to Section 18-9 and Appendix F.** . . . .

Balanced equation: $HF + H_2O \rightleftharpoons H_3O^+ + F^-$

Recall for a weak acid buffer: $[H_3O^+] = \frac{[\text{acid}]}{[\text{salt}]} \times K_a = \frac{\text{mol acid}}{\text{mol salt}} \times K_a$

In this buffer system, $[H_3O^+] = 6.0 \times 10^{-4}\ M$ since pH = 3.22
mol HF = 0.15 M × 0.100 L = 0.015 mol HF
$K_a = 7.2 \times 10^{-4}$

Rearranging and solving,

$$? \text{ mol NaF} = \frac{\text{mol HF} \times K_a}{[H_3O^+]} = \frac{(0.015)(7.2 \times 10^{-4})}{(6.0 \times 10^{-4})} = 0.018 \text{ mol NaF}$$

$$V_{NaF} = \frac{0.018 \text{ mol}}{0.35\ M} = 5.1 \times 10^{-2} \text{ L or } \mathbf{51\ mL}$$

Note that by working in moles of acid and salt, it was unnecessary to worry about the effect of increasing solution volume on molarity.

18-86. **Refer to Section 18-10, Example 18-20 and Appendix F.** . . . .

Balanced equations:

$H_2CO_3 + H_2O \rightleftharpoons H_3O^+ + HCO_3^-$ $K_1 = 4.2 \times 10^{-7}$

$HCO_3^- + H_2O \rightleftharpoons H_3O^+ + CO_3^{2-}$ $K_2 = 4.8 \times 10^{-11}$

First Step:

Let x = $[H_2CO_3]_{ionized}$. Then $[H_2CO_3] = (0.070 - x)M$; $[H_3O^+] = [HCO_3^-] = x\ M$

$$K_1 = \frac{[H_3O^+][HCO_3^-]}{[H_2CO_3]} = \frac{x^2}{(0.070 - x)} = 4.2 \times 10^{-7} \simeq \frac{x^2}{0.070}$$

Solving, $x = 1.7 \times 10^{-4}$

Therefore, $[H_2CO_3] = 0.070\ M$ since the simplifying assumption is valid.

$[H_3O^+] = [HCO_3^-] = 1.7 \times 10^{-4}\ M$

Second Step:

Let y = $[HCO_3^-]_{ionized}$. Then $[HCO_3^-] = (1.7 \times 10^{-4} - y)M$

$[H_3O^+] = (1.7 \times 10^{-4} + y)M$

$[CO_3^{2-}] = y\ M$

$$K_2 = \frac{[H_3O^+][CO_3^{2-}]}{[HCO_3^-]} = \frac{(1.7 \times 10^{-4} + y)(y)}{(1.7 \times 10^{-4} - y)} = 4.8 \times 10^{-11} \simeq \frac{(1.7 \times 10^{-4})y}{(1.7 \times 10^{-4})}$$

Solving, $y = 4.8 \times 10^{-11}$

Therefore, since the simplifying assumptions were valid,

$[H_3O^+] = \mathbf{1.7 \times 10^{-4}}\ M$ $[HCO_3^-] = \mathbf{1.7 \times 10^{-4}}\ M$

$[OH^-] = K_w/[H_3O^+] = \mathbf{5.9 \times 10^{-11}}\ M$ $[CO_3^{-2}] = \mathbf{4.8 \times 10^{-11}}\ M$

18-88. **Refer to Section 18-10, Example 18-18 and Appendix F.** . . . .

Balanced equations:

$H_3AsO_4 + H_2O \rightleftharpoons H_3O^+ + H_2AsO_4^-$ $K_1 = 2.5 \times 10^{-4}$

$H_2AsO_4^- + H_2O \rightleftharpoons H_3O^+ + HAsO_4^{2-}$ $K_2 = 5.6 \times 10^{-8}$

$HAsO_4^{2-} + H_2O \rightleftharpoons H_3O^+ + AsO_4^{3-}$ $K_3 = 3.0 \times 10^{-13}$

First Step:

Let x = $[H_3AsO_4]_{ionized}$. Then $[H_3AsO_4] = (0.100 - x)M$

$[H_3O^+] = [H_2AsO_4^-] = x.$

$$K_1 = \frac{[H_3O^+][H_2AsO_4^-]}{[H_3AsO_4]} = \frac{x^2}{(0.100 - x)} = 2.5 \times 10^{-4} \simeq \frac{x^2}{0.100}$$

Solving, $x = 5.0 \times 10^{-3}$

Since 5.0×10^{-3} is 5% of 0.100, we should solve the quadratic equation,

$$x^2 + (2.5 \times 10^{-4})x - 2.5 \times 10^{-5} = 0$$

$$x = \frac{-2.5 \times 10^{-4} \pm \sqrt{(2.5 \times 10^{-4})^2 - (4)(1)(-2.5 \times 10^{-5})}}{2(1)}$$

$$= \frac{-2.5 \times 10^{-4} \pm 1.0 \times 10^{-2}}{2} = 4.9 \times 10^{-3} \text{ or } -5.1 \times 10^{-3} \text{ (discard)}$$

Therefore, $[H_3AsO_4] = 0.100 - x = 0.095\ M$
$[H_3O^+] = [H_2AsO_4^-] = 4.9 \times 10^{-3}\ M$

Second Step:

Let $y = [H_2AsO_4^-]_{ionized}$. Then $[H_2AsO_4^-] = (4.9 \times 10^{-3} - y)M$
$[H_3O^+] = (4.9 \times 10^{-3} + y)M$
$[HAsO_4^{2-}] = y\ M$

$$K_2 = \frac{[H_3O^+][HAsO_4^{2-}]}{[H_2AsO_4^-]} = \frac{(4.9 \times 10^{-3} + y)(y)}{(4.9 \times 10^{-3} - y)} = 5.6 \times 10^{-8} \simeq \frac{(4.9 \times 10^{-3})y}{(4.9 \times 10^{-3})}$$

Solving, $y = 5.6 \times 10^{-8}$

Therefore, since the simplifying assumptions are valid,
$[H_3O^+] = [H_2AsO_4^-] = 4.9 \times 10^{-3}\ M$
$[HAsO_4^{2-}] = 5.6 \times 10^{-8}\ M$

Third Step:

Let $z = [HAsO_4^{2-}]_{ionized}$. Then $[HAsO_4^{2-}] = (5.6 \times 10^{-8} - z)M$
$[H_3O^+] = (4.9 \times 10^{-3} + z)M$
$[AsO_4^{3-}] = z\ M$

$$K_2 = \frac{[H_3O^+][AsO_4^{3-}]}{[HAsO_4^{2-}]} = \frac{(4.9 \times 10^{-3} + z)(z)}{(5.6 \times 10^{-8} - z)} = 3.0 \times 10^{-13} \simeq \frac{(4.9 \times 10^{-3})(z)}{(5.6 \times 10^{-8})}$$

Solving, $z = 3.4 \times 10^{-18}$

Therefore, $[AsO_4^{3-}] = 3.4 \times 10^{-18}\ M$

0.100 *M* H_3AsO_4 Solution		0.100 *M* H_3PO_4 Solution	
Species	Concentration (*M*)	Species	Concentration (*M*)
H_3AsO_4	0.095	H_3PO_4	0.076
H_3O+	0.0049	H_3O+	0.024
$H_2AsO_4^-$	0.0049	$H_2PO_4^-$	0.024
$HAsO_4^{2-}$	5.6×10^{-8}	HPO_4^{2-}	6.2×10^{-8}
OH^-	2.0×10^{-12}	OH^-	4.2×10^{-13}
AsO_4^{3-}	3.4×10^{-18}	PO_4^{3-}	9.3×10^{-19}

19 Ionic Equilibria II: Hydrolysis

19-2. **Refer to the Introduction to Chapter 19.** . . . .

Dilute aqueous solutions of weak acids, such as HF and HNO_2, contain relatively few ions because they are weak electrolytes and dissociate only slightly to form ions.

19-4. **Refer to the Introduction to Chapter 19.** . . . .

(a) Solvolysis is the reaction of a substance with the solvent in which it is dissolved. Common solvents used include $H_2O(\ell)$, $NH_3(\ell)$, $H_2SO_4(\ell)$ and $CH_3COOH(\ell)$. There are many others. For example, glacial acetic acid, $CH_3COOH(\ell)$, is commonly used in non-aqueous titrations with weak acids:

$$\underset{\text{solute}}{C_6H_5NH_3^+} + \underset{\text{solvent}}{CH_3COOH} \rightleftharpoons CH_3COOH_2^+ + \underset{\text{aniline}}{C_6H_5NH_2}$$

(b) Hydrolysis is the reaction of a substance with the solvent, water, or its ions, OH^- and H_3O^+, e.g. the hydrolysis of the weak acid, CH_3COOH:

$$\underset{\text{solute}}{CH_3COOH} + \underset{\text{solvent}}{H_2O} \rightleftharpoons H_3O^+ + CH_3COO^-$$

19-6. **Refer to Section 19-2.** . . . .

The solution of a salt derived from a strong soluble base and weak acid is basic because the anion of a weak acid reacts with water (hydrolysis) to form hydroxide ions. Consider the soluble salt NaClO prepared by reacting NaOH, a strong soluble base, and HClO, a weak acid. The salt dissociates completely in water and the conjugate base of the weak acid, ClO^-, hydrolyzes, producing OH^- ions.

$$NaClO(s) \xrightarrow[100\%]{H_2O} Na^+(aq) + ClO^-(aq)$$

$$ClO^-(aq) + H_2O(\ell) \rightleftharpoons HClO(aq) + OH^-(aq)$$

19-8. **Refer to Section 19-2 and Appendix F.** . . . .

(a) for NO_2^-, $K_b = \dfrac{K_w}{K_{a(HNO_2)}} = \dfrac{1.0 \times 10^{-14}}{4.5 \times 10^{-4}} = 2.2 \times 10^{-11}$

(b) for ClO^-, $K_b = \dfrac{K_w}{K_{a(HClO)}} = \dfrac{1.0 \times 10^{-14}}{3.5 \times 10^{-8}} = 2.9 \times 10^{-7}$

(c) for $C_6H_5COO^-$, $K_b = \dfrac{K_w}{K_{a(C_6H_5COOH)}} = \dfrac{1.0 \times 10^{-14}}{6.3 \times 10^{-5}} = 1.6 \times 10^{-10}$

The mathematical relationship between K_a, the ionization constant for a weak acid, and K_b, the base hydrolysis constant for the anion of the weak acid is $K_w = K_a \times K_b$. The weaker the acid, the smaller is its K_a, the more its anion will hydrolyze, and the larger is K_b for the anion.

19-10. **Refer to Section 19-2, Example 19-1 and Exercise 19-8 Solution.** . . . . . .

(a) Balanced equations: $NaNO_2 \rightarrow Na^+ + NO_2^-$ (to completion)
$NO_2^- + H_2O \rightleftharpoons HNO_2 + OH^-$ (reversible)

Let $x = [NO_2^-]_{hydrolyzed}$. Then $0.15 - x = [NO_2^-]$ and $x = [HNO_2] = [OH^-]$

$$K_b = \frac{[HNO_2][OH^-]}{[NO_2^-]} = \frac{x^2}{0.15 - x} = 2.2 \times 10^{-11} \simeq \frac{x^2}{0.15}.$$ Solving, $x = 1.8 \times 10^{-6}$

$$\%\ hydrolysis = \frac{[NO_2^-]_{hydrolyzed}}{[NO_2^-]_{initial}} \times 100 = \frac{1.8 \times 10^{-6}\ M}{0.15\ M} \times 100 = \mathbf{1.2 \times 10^{-3}\ \%}$$

(b) Balanced equations: $NaClO \rightarrow Na^+ + ClO^-$ (to completion)
$ClO^- + H_2O \rightleftharpoons HClO + OH^-$ (reversible)

Let $x = [ClO^-]_{hydrolyzed}$. Then $0.15 - x = [ClO^-]$ and $x = [HClO] = [OH^-]$

$$K_b = \frac{[HClO][OH^-]}{[ClO^-]} = \frac{x^2}{0.15 - x} = 2.9 \times 10^{-7} \simeq \frac{x^2}{0.15}.$$ Solving, $x = 2.1 \times 10^{-4}$

$$\%\ hydrolysis = \frac{[ClO^-]_{hydrolyzed}}{[ClO^-]_{initial}} \times 100 = \frac{2.1 \times 10^{-4}\ M}{0.15\ M} \times 100 = \mathbf{0.14\%}$$

(c) Balanced equations: $NaC_6H_5COO \rightarrow Na^+ + C_6H_5COO^-$ (to completion)
$C_6H_5COO^- + H_2O \rightleftharpoons C_6H_5COOH + OH^-$ (reversible)

Let $x = [C_6H_5COO^-]_{hydrolyzed}$. Then $0.15 - x = [C_6H_5COO^-]$ and
$x = [C_6H_5COOH] = [OH^-]$

$$K_b = \frac{[C_6H_5COOH][OH^-]}{[C_6H_5COO^-]} = \frac{x^2}{0.15 - x} = 1.6 \times 10^{-10} \simeq \frac{x^2}{0.15}$$

Solving, $x = 4.9 \times 10^{-6}$

$$\%\ hydrolysis = \frac{[C_6H_5COO^-]_{hydrolyzed}}{[C_6H_5COO^-]_{initial}} \times 100 = \frac{4.9 \times 10^{-6}\ M}{0.15\ M} \times 100$$

$$= \mathbf{3.3 \times 10^{-3}\ \%}$$

19-12. **Refer to Section 19-3.** . . . .

NH_4Cl	ammonium chloride
$CH_3NH_3NO_3$	methylammonium nitrate
$C_5H_5NHClO_4$	pyridinium perchlorate
$[(CH_3)_3NH]_2SO_4$	trimethylammonium sulfate

19-14. **Refer to Section 19-3, Example 19-2 and Appendix G.** . . . .

(a) Balanced equations: $NH_4NO_3 \rightarrow NH_4^+ + NO_3^-$ (to completion)
$NH_4^+ + H_2O \rightleftharpoons NH_3 + H_3O^+$ (reversible)

Let $x = [NH_4^+]_{hydrolyzed}$. Then $0.12 - x = [NH_4^+]$ and $x = [NH_3] = [H_3O^+]$

$$K_a = \frac{K_w}{K_{b(NH_3)}} = \frac{1.0 \times 10^{-14}}{1.8 \times 10^{-5}} = 5.6 \times 10^{-10} = \frac{[NH_3][H_3O^+]}{[NH_4^+]} = \frac{x^2}{0.12 - x} \simeq \frac{x^2}{0.12}$$

Solving, x = 8.2×10^{-6}. Therefore, $[H_3O^+] = 8.2 \times 10^{-6}$ *M*; pH = **5.1**

(b) Balanced equations: $CH_3NH_3NO_3 \rightarrow CH_3NH_3^+ + NO_3^-$ (to completion)
$CH_3NH_3^+ + H_2O \rightleftharpoons CH_3NH_2 + H_3O^+$ (reversible)

Let $x = [CH_3NH_3^+]_{hydrolyzed}$. Then $0.12 - x = [CH_3NH_3^+]$ and $x = [CH_3NH_2] = [H_3O^+]$

$$K_a = \frac{K_w}{K_{b(CH_3NH_2)}} = \frac{1.0 \times 10^{-14}}{5.0 \times 10^{-4}} = 2.0 \times 10^{-11} = \frac{[CH_3NH_2][H_3O^+]}{[CH_3NH_3^+]} = \frac{x^2}{0.12 - x} \simeq \frac{x^2}{0.12}$$

Solving, x = 1.5×10^{-6}. Therefore, $[H_3O^+] = 1.5 \times 10^{-6}$ *M*; pH = **5.82**

(c) Balanced equations: $C_5H_5NHNO_3 \rightarrow C_5H_5NH^+ + NO_3^-$ (to completion)
$C_5H_5NH^+ + H_2O \rightleftharpoons C_5H_5N + H_3O^+$ (reversible)

Let $x = [C_5H_5NH^+]_{hydrolyzed}$. Then $0.12 - x = [C_5H_5NH^+]$ and $x = [C_5H_5N] = [H_3O^+]$

$$K_a = \frac{K_w}{K_{b(C_5H_5N)}} = \frac{1.0 \times 10^{-14}}{1.5 \times 10^{-9}} = 6.7 \times 10^{-6} = \frac{[C_5H_5N][H_3O^+]}{[C_5H_5NH^+]} = \frac{x^2}{0.12 - x} \simeq \frac{x^2}{0.12}$$

Solving, x = 9.0×10^{-4}. Therefore, $[H_3O^+] = 9.0 \times 10^{-4}$ *M*; pH = **3.05**

19-16. **Refer to Section 19-3 and Exercise 19-14a.** . . . .

(a) NH_4Cl and NH_4NO_3 are both salts derived from monoprotic strong acids and the weak base, NH_3. Since the concentration of NH_4^+ is the same in each solution, the pH values will be identical.

(b) NH_4ClO_4 is a salt derived from the strong monoprotic acid, $HClO_4$, and the weak base, NH_3. $(NH_4)_2SO_4$ is a salt derived from the strong diprotic acid, H_2SO_4, and the weak base, NH_3. The initial concentration of NH_4^+ in 0.010 *M* NH_4ClO_4 is 0.010 *M*, whereas the initial concentration of NH_4^+ in 0.010 *M* $(NH_4)_2SO_4$ is 0.020 *M*. In the $(NH_4)_2SO_4$ solution, the 0.02*M* NH_4^+ ions will hydrolyze to give more H_3O^+ ions. As a result, the pH of 0.010 *M* $(NH_4)_2SO_4$ will be lower than for 0.010 *M* NH_4ClO_4.

Note: The pH of the 0.010 *M* $(NH_4)_2SO_4$ solution will not be quite as low as predicted due to the very slight hydrolysis of the sulfate ion:
$SO_4^{2-} + H_2O \rightleftharpoons HSO_4^- + OH^-$

19-18. **Refer to Section 19-4 and Appendices F and G.** . . . .

(a) A salt of a weak acid and a weak base for which $K_a = K_b$ gives a neutral solution, e.g., ammonium acetate, NH_4CH_3COO.

$(K_{a(CH_3COOH)} = K_{b(NH_3)} = 1.8 \times 10^{-5})$

(b) A salt of a weak acid and a weak base for which $K_a > K_b$ gives an acidic solution, e.g., pyridinium fluoride, C_5H_5NHF.

$(K_{a(HF)} = 7.2 \times 10^{-4}; K_{b(C_5H_5N)} = 1.5 \times 10^{-9})$

(c) A salt of a weak acid and a weak base for which $K_b > K_a$ gives a basic solution, e.g., methylammonium cyanide, CH_3NH_3CN.

$(K_{a(HCN)} = 4.0 \times 10^{-10}; K_{b(CH_3NH_2)} = 5.0 \times 10^{-4})$

19-20. **Refer to Section 19-5, Tables 18-7 and 19-2, and Figure 19-1.** . . . .

The calculations for determining the pH at every point in the titration of 1 liter of 0.0100 *M* CH_3COOH with solid NaOH, assuming no volume change, can be divided into 4 types:

(1) Initially, the pH is determined by the concentration of the weak acid, CH_3COOH.

(2) Before the equivalence point, the pH is determined by the buffer solution consisting of the unreacted CH_3COOH and $NaCH_3COO$ produced by the reaction. Each calculation is a limiting reagent problem using the original concentration of CH_3COOH. (Refer to Exercise 18-76 Solution). For example, at point (c) in the following table:

	NaOH	+	CH_3COOH	→	$NaCH_3COO$	+	H_2O
initial	4.00 mmol		10.0 mmol		0 mmol		
change	- 4.00 mmol		- 4.00 mmol		+ 4.00 mmol		
after rxn	0 mmol		6.0 mmol		4.00 mmol		

After the reaction, we have a 1 liter buffer solution consisting of 6.0 mmol CH_3COOH and 4.00 mmol $NaCH_3COOH$.

$$[H_3O^+] = \frac{\text{mol } CH_3COOH}{\text{mol } NaCH_3COO} \times K_a = \frac{6.0 \text{ mmol}}{4.00 \text{ mmol}} (1.8 \times 10^{-5}) = 2.7 \times 10^{-5}\ M;$$

Therefore, pH = 4.57

Halfway to the equivalence point (i.e., when half of the required amount of base needed to reach the equivalence point is added.), pH = pK_a. At point (d):

	NaOH	+	CH_3COOH	→	$NaCH_3COO$	+	H_2O
initial	5.00 mmol		10.0 mmol		0 mmol		
change	- 5.00 mmol		- 5.00 mmol		+ 5.00 mmol		
after rxn	0 mmol		5.0 mmol		5.00 mmol		

$$[H_3O^+] = \frac{\text{mol } CH_3COOH}{\text{mol } NaCH_3COO} \times K_a = \frac{5.0 \text{ mmol}}{5.00 \text{ mmol}} (1.8 \times 10^{-5}) = 1.8 \times 10^{-5};$$

Therefore, pH = pK_a = 4.74

(3) At the equivalence point, there is no excess acid or base. The concentration of $NaCH_3COO$ determines the pH of the system (Refer to Exercises 19-9 and 19-10 Solutions). At point (h),

	NaOH	+	CH_3COOH	→	$NaCH_3COO$	+	H_2O
initial	10.0 mmol		10.0 mmol		0 mmol		
change	- 10.0 mmol		- 10.0 mmol		+ 10.0 mmol		
after rxn	0 mmol		0 mmol		10.0 mmol		

and $[NaCH_3COO] = \frac{0.0100 \text{ mol}}{1.00 \text{ L}} = 0.0100\ M.$

Note: if aqueous NaOH had been added, $[NaCH_3COO]$ = 0.0100 mol/total volume.

Then the salt hydrolyzes to produce a basic solution:

$$CH_3COO^- + H_2O \rightleftharpoons CH_3COOH + OH^- \qquad K_b = 5.6 \times 10^{-10}$$

Let x = $[CH_3COO^-]_{hydrolyzed}$. Then 0.0100 - x = $[CH_3COO^-]$ and x = $[CH_3COOH]$ = $[OH^-]$

$$K_b = \frac{[CH_3COOH][OH^-]}{[CH_3COO^-]} = \frac{x^2}{0.0100 - x} = 5.6 \times 10^{-10} \approx \frac{x^2}{0.0100}$$

Solving, $x = 2.4 \times 10^{-6}$; $[OH^-] = 2.4 \times 10^{-6}$ *M*; pOH = 5.62; pH = 8.38.

(4) After the equivalence point, the pH is determined directly from the concentration of excess NaOH since CH_3COOH is now the limiting reagent. In the presence of the strong base, the effect of the weak base, CH_3COO^-, derived from the salt is negligible. For example, at point (j):

	NaOH	+	CH_3COOH	→	$NaCH_3COO$	+	H_2O
initial	12.0 mmol		10.0 mmol		0 mmol		
change	- 10.0 mmol		- 10.0 mmol		+ 10.0 mmol		
after rxn	2.0 mmol		0 mmol		10.0 mmol		

$$[NaOH]_{excess} = \frac{0.0020 \text{ moles}}{1.00 \text{ L}} = 2.0 \times 10^{-3}\ M$$

Therefore, $[OH^-] = 2.0 \times 10^{-3}$ *M*; pOH = 2.70; pH = 11.30.

Note: if aqueous NaOH had been added,

$$[NaOH]_{excess} = \frac{0.0020 \text{ mol}}{\text{total volume}}.$$

Data Table:

	Mol NaOH Added	Type of Solution	$[H_3O^+]$ (*M*)	$[OH^-]$ (*M*)	pH	pOH
(a)	none	weak acid	4.2×10^{-4}	2.4×10^{-11}	3.38	10.62
(b)	0.00200	buffer	7.2×10^{-5}	1.4×10^{-10}	4.14	9.86
(c)	0.00400	buffer	2.7×10^{-5}	3.7×10^{-10}	4.57	9.43
(d)	0.00500 (halfway to equivalence point)	buffer	1.8×10^{-5}	5.5×10^{-10}	4.74 (= pK_a)	9.26
(e)	0.00700	buffer	7.7×10^{-6}	1.3×10^{-9}	5.11	8.89
(f)	0.00900	buffer	2.0×10^{-6}	4.9×10^{-9}	5.70	8.31
(g)	0.00950	buffer	9×10^{-7}	1.1×10^{-8}	6.0	8.0
(h)	0.0100 (at equivalence)	salt	4.2×10^{-9}	2.4×10^{-6}	8.38	5.62
(i)	0.0105	strong base	2×10^{-11}	5×10^{-4}	10.7	3.3
(j)	0.0120	strong base	5.0×10^{-12}	2.0×10^{-3}	11.30	2.70
(k)	0.0150	strong base	2.0×10^{-12}	5.0×10^{-3}	11.70	2.30

Titration Curve II: CH_3COOH vs. NaOH

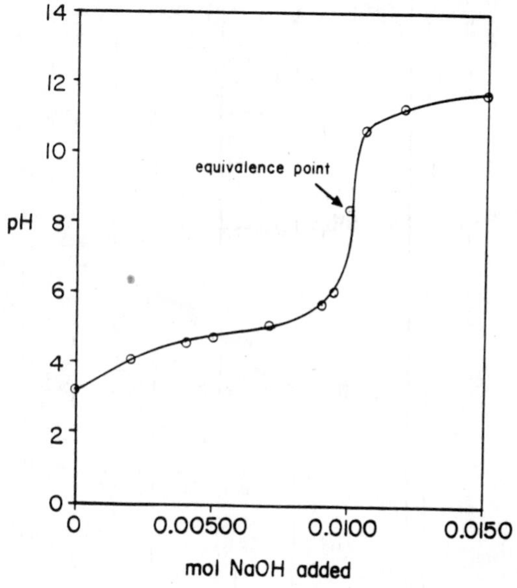

An appropriate indicator would change color in the pH range, 7 - 10. From Table 18-7, phenolphthalein is an acceptable indicator for this titration.

19-22. **Refer to Sections 19-4 and 19-5, Figure 19-3, Table 18-7 and Exercise 19-20 Solution.** . . . .

The calculations for determining the pH in the titration of 1 liter of 0.0100 *M* CH_3COOH with gaseous NH_3 can be divided into 4 types:

(1) Initially, the pH is determined by the concentration of the weak acid, CH_3COOH.

(2) Before the equivalence point, the pH is essentially determined by the buffer solution consisting of the unreacted CH_3COOH and CH_3COO^- produced by the neutralization reaction. For example, consider the limiting reagent problem for point (c) in the following table:

	NH_3	+	CH_3COOH	→	NH_4CH_3COOH
initial	4.00 mmol		10.0 mmol		0 mmol
change	- 4.00 mmol		- 4.00 mmol		+ 4.00 mmol
after rxn	0 mmol		6.00 mmol		4.00 mmol

After the reaction, we have a 1 liter solution of CH_3COOH and the soluble salt, NH_4CH_3COOH, produced by the two equilibria:

$$CH_3COOH + H_2O \rightleftharpoons CH_3COO^- + H_3O^+ \qquad K_a = 1.8 \times 10^{-5}$$
$$NH_4^+ + H_2O \rightleftharpoons NH_3 + H_3O^+ \qquad K_a = 5.6 \times 10^{-10}$$

However, we can ignore the hydrolysis of NH_4^+ in our pH calculation since NH_4^+ is a much weaker acid than CH_3COOH (K_a for NH_4^+ << K_a for CH_3COOH).

$$[H_3O^+] = \frac{\text{mol } CH_3COOH}{\text{mol } NH_4CH_3COO} \times K_a = \frac{6.00 \text{ mmol}}{4.00 \text{ mmol}} (1.8 \times 10^{-5}) = 2.7 \times 10^{-5}\ M;$$

Therefore, pH = 4.57

Hence, the titration curve before the equivalence point is identical to that calculated in Exercise 19-20 for 0.01 *M* CH_3COOH titrated with a strong soluble base.

(3) At the equivalence point, there is no excess acid or base. The pH of the salt solution, $NH_4CH_3COO(aq)$, is 7 since $K_{a(CH_3COOH)} = K_{b(NH_3)}$.

(4) After the equivalence point, the pH is essentially determined by the buffer solution consisting of excess NH_3 and NH_4^+ produced by the neutralization reaction. For example, at point (i):

	NH_3	+	CH_3COOH	→	NH_4CH_3COOH
initial	13.0 mmol		10.0 mmol		0 mmol
change	- 10.0 mmol		- 10.0 mmol		+ 10.00 mmol
after rxn	3.0 mmol		0 mmol		10.00 mmol

After the reaction, we have a 1 liter solution of NH_3 and the soluble salt, NH_4CH_3COOH. OH^- ion is theoretically produced by two equilibria.

$$NH_3 + H_2O \rightleftharpoons NH_4^+ + OH^- \qquad K_b = 1.8 \times 10^{-5}$$
$$CH_3COO^- + H_2O \rightleftharpoons CH_3COOH + OH^- \qquad K_b = 5.6 \times 10^{-10}$$

However, we can ignore the hydrolysis of CH_3COO^- in our calculations since CH_3COO^- is a much weaker base than NH_3 (K_b for CH_3COOH << K_b for NH_3).

$$[OH^-] = \frac{\text{mol } NH_3}{\text{mol } NH_4CH_3COO} \times K_b = \frac{3.0 \text{ mmol}}{10.0 \text{ mmol}} (1.8 \times 10^{-5}) = 5.4 \times 10^{-6}\ M;$$

Therefore, pH = 8.73

Data Table:

	Mol NH_3 Added	Type of Solution	$[H_3O^+]$ (M)	$[OH^-]$ (M)	pH	pOH
(a)	none	weak acid	4.2×10^{-4}	2.4×10^{-11}	3.38	10.62
(b)	0.00100	buffer	1.6×10^{-4}	6.2×10^{-11}	3.80	10.21
(c)	0.00400	buffer	2.7×10^{-5}	3.7×10^{-10}	4.57	9.43
(d)	0.00500 (halfway to equivalence point)	buffer	1.8×10^{-5}	5.6×10^{-10}	4.74 (= pK_a)	9.26
(e)	0.00900	buffer	2.0×10^{-6}	5.0×10^{-9}	5.70	8.30
(f)	0.00950	buffer	9×10^{-7}	1.1×10^{-8}	6.0	8.0
(g)	0.0100 (at equivalence point)	salt	1.0×10^{-7}	1.0×10^{-7}	7.00	7.00
(h)	0.0105	buffer	1.1×10^{-8}	9×10^{-7}	8.0	6.0
(i)	0.0130	buffer	1.8×10^{-9}	5.4×10^{-6}	8.73	5.27

Titration Curve IV: CH_3COOH vs. NH_3

The major difference between this titration curve of the CH_3COOH/NH_3 reaction and the other titrations is that the system is buffered both before and after the equivalence point. There is no satisfactory indicator for this titration since the vertical portion of the curve is too short.

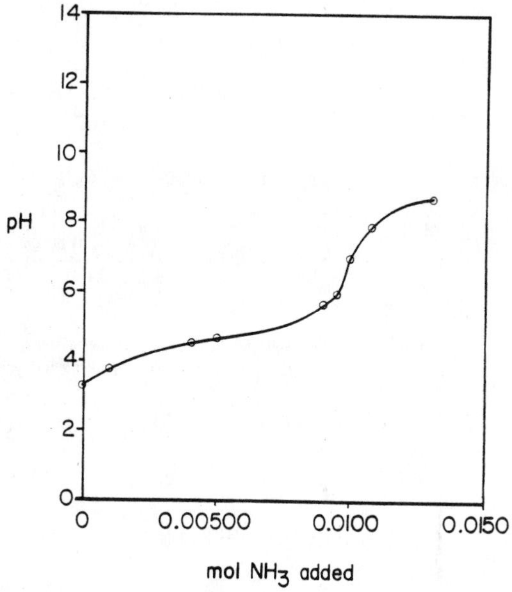

19-24. **Refer to Section 19-6, Example 19-3 and Table 19-3.** . . . .

(a) Balanced equations: $Al(NO_3) \rightarrow Al^{3+} + 3NO_3^-$ (to completion)

$Al^{3+} + 2H_2O \rightleftharpoons Al(OH)^{2+} + H_3O^+$ (reversible)

Let $x = [Al^{3+}]_{hydrolyzed}$. Then $0.15 - x = [Al^{3+}]$; $x = [Al(OH)^{2+}] = [H_3O^+]$.

$$K_a = \frac{[Al(OH)^{2+}][H_3O^+]}{[Al^{3+}]} = \frac{x^2}{0.15 - x} = 1.2 \times 10^{-5} \simeq \frac{x^2}{0.15}$$

Solving, $x = 1.3 \times 10^{-3}$

Therefore, $[H_3O^+] = 1.3 \times 10^{-3}$ M; pH = **2.89**

$$\% \text{ hydrolysis} = \frac{[Al^{3+}]_{hydrolyzed}}{[Al^{3+}]_{initial}} \times 100 = \frac{1.3 \times 10^{-3}\ M}{0.15\ M} \times 100 = \mathbf{0.87\%}$$

(b) Balanced equations: $Co(ClO_4)_2 \rightarrow Co^{2+} + 2ClO_4^-$ (to completion)

$Co^{2+} + 2H_2O \rightleftharpoons Co(OH)^+ + H_3O^+$ (reversible)

Let x = $[Co^{2+}]_{hydrolyzed}$. Then 0.13 - x = $[Co^{2+}]$; x = $[Co(OH)^{2+}]$ = $[H_3O^+]$

$$K_a = \frac{[Co(OH)^{2+}][H_3O^+]}{[Co^{2+}]} = \frac{x^2}{0.13 - x} = 5.0 \times 10^{-10} \simeq \frac{x^2}{0.13}$$

Solving, x = 8.1×10^{-6}

Therefore, $[H_3O^+] = 8.1 \times 10^{-6}$ *M*; pH = **5.09**

$$\% \text{ hydrolysis} = \frac{[Co^{2+}]_{hydrolyzed}}{[Co^{2+}]_{initial}} \times 100 = \frac{8.1 \times 10^{-6}\ M}{0.13\ M} \times 100 = \mathbf{6.2 \times 10^{-3}\ \%}$$

(c) Balanced equations: $MgCl_2 \rightarrow Mg^{2+} + 2Cl^-$ (to completion)

$Mg^{2+} + 2H_2O \rightleftharpoons Mg(OH)^+ + H_3O^+$ (reversible)

Let x = $[Mg^{2+}]_{hydrolyzed}$. Then 0.12 - x = $[Mg^{2+}]$; x = $[Mg(OH)^+]$ = $[H_3O^+]$

$$K_a = \frac{[Mg(OH)^+][H_3O^+]}{[Mg^{2+}]} = \frac{x^2}{0.12 - x} = 3.0 \times 10^{-12} \simeq \frac{x^2}{0.12}$$

Solving, x = 6.0×10^{-7}

Therefore, $[H_3O^+] = 6.0 \times 10^{-7}$ *M*; pH = **6.22** (ignoring the H_3O^+ produced by the ionization of water).

$$\% \text{ hydrolysis} = \frac{[Mg^{2+}]_{hydrolyzed}}{[Mg^{2+}]_{initial}} \times 100 = \frac{6.0 \times 10^{-7}\ M}{0.12\ M} \times 100 = \mathbf{5.0 \times 10^{-4}\ \%}$$

Note: To calculate the actual $[H_3O^+]$ in this problem, let x = $[OH^-]$ = $[H_3O^+]$ produced by the ionization of water in this system.

Therefore $[H_3O^+]_{total}$ = $[H_3O^+]$ produced by hydrolysis + $[H_3O^+]$ produced by the ionization of water
$= 6.0 \times 10^{-7} + x$

We know that $K_w = 1.0 \times 10^{-14} = [H_3O^+][OH^-] = (6.0 \times 10^{-7} + x)(x)$
$= (6.0 \times 10^{-7})x + x^2$

Solving the quadratic equation, $x^2 + (6.0 \times 10^{-7})x - 1.0 \times 10^{-14} = 0$
we have x = 1.5×10^{-8}

Therefore, $[H_3O^+]_{total} = 6.0 \times 10^{-7} + x = 6.2 \times 10^{-7}$ *M*; pH = **6.21**

19-26. **Refer to Sections 18-2 and 19-2, and Appendices F and G.** . . . .

(a) Base strength increases with increasing K_b value. Consider the following weak bases:

Bases	K_b
NH_3	1.8×10^{-5}
$C_6H_3NH_2$	4.2×10^{-10}
CN^-	$K_w/K_{a(HCN)} = 2.5 \times 10^{-5}$
OCN^-	$K_w/K_{a(HOCN)} = 2.9 \times 10^{-11}$

The order of increasing base strength is : $OCN^- < C_6H_5NH_2 < NH_3 < CN^-$

(b) If we had equal concentrations of NH_3, $C_6H_5NH_2$, NaCN and NaOCN, the NaCN solution would be the most basic and the NaOCN solution would be the least basic.

19-28. **Refer to Sections 19-1, 19-2 and 19-3.** . . . .

Balanced equations:

(a) $Na_2CO_3 \rightarrow 2Na^+ + CO_3^{2-}$

$CO_3^{2-} + H_2O \rightleftharpoons HCO_3^- + OH^-$ $K_{b(1)} = 2.1 \times 10^{-4}$

$HCO_3^- + H_2O \rightleftharpoons H_2CO_3 + OH^-$ $K_{b(2)} = 2.4 \times 10^{-8}$

(b) $Na_2SO_4 \rightarrow 2Na^+ + SO_4^{2-}$

$SO_4^{2-} + H_2O \rightleftharpoons HSO_4^- + OH^-$ $K_{b(1)} = 8.3 \times 10^{-13}$

$HSO_4^- + H_2O \rightleftharpoons H_2SO_4 + OH^-$ $K_{b(2)}$ = very small

(c) $(NH_4)_2SO_4 \rightarrow 2NH_4^+ + SO_4^{2-}$

$NH_4^+ + H_2O \rightleftharpoons NH_3 + H_3O^+$ $K_a = 5.6 \times 10^{-10}$

$SO_4^{2-} + H_2O \rightleftharpoons HSO_4^- + OH^-$ $K_{b(1)} = 8.3 \times 10^{-13}$

$HSO_4^- + H_2O \rightleftharpoons H_2SO_4 + OH^-$ $K_{b(2)}$ = very small

(d) $Na_3PO_4 \rightarrow 3Na^+ + PO_4^{3-}$

$PO_4^{3-} + H_2O \rightleftharpoons HPO_4^{2-} + OH^-$ $K_{b(1)} = 2.8 \times 10^{-2}$

$HPO_4^{2-} + H_2O \rightleftharpoons H_2PO_4^- + OH^-$ $K_{b(2)} = 1.6 \times 10^{-7}$

$H_2PO_4^- + H_2O \rightleftharpoons H_3PO_4 + OH^-$ $K_{b(3)} = 1.3 \times 10^{-12}$

$(NH_4)_2SO_4$ definitely could not be used in cleaning materials since it produces an acidic solution, not a basic solution. Also, Na_2SO_4 would not be preferred either since SO_4^{2-} is a very weak base (has a very small K_b).

20 Ionic Equilibria III: The Solubility Product Principle

20-2. **Refer to Section 20-1.**

The solubility product principle states that the solubility product expression for a slightly soluble compound is the product of the concentrations of its constituent ions, each raised to the power that corresponds to the number of ions in one formula unit of the compound. The quantity, K_{sp}, is constant at constant temperature for a saturated solution of the compound, when the system is at equilibrium. The significance of the solubility product is that it can be used to calculate the concentrations of the ions in solutions for such slightly soluble compounds.

20-4. **Refer to Section 20-1.**

(a) $PbSO_4(s) \rightleftharpoons Pb^{2+}(aq) + SO_4^{2-}(aq)$ $\quad K_{sp} = [Pb^{2+}][SO_4^{2-}]$

(b) $AgBr(s) \rightleftharpoons Ag^+(aq) + Br^-(aq)$ $\quad K_{sp} = [Ag^+][Br^-]$

(c) $Mg(OH)_2(s) \rightleftharpoons Mg^{2+}(aq) + 2OH^-(aq)$ $\quad K_{sp} = [Mg^{2+}][OH^-]^2$

(d) $Ag_2CO_3(s) \rightleftharpoons 2Ag^+(aq) + CO_3^{2-}(aq)$ $\quad K_{sp} = [Ag^+]^2[CO_3^{2-}]$

(e) $As_2S_3(s) \rightleftharpoons 2As^{3+}(aq) + 3S^{2-}(aq)$ $\quad K_{sp} = [As^{3+}]^2[S^{2-}]^3$

20-6. **Refer to Section 20-2, Examples 20-1 and 20-2, and Appendix H.**

Plan: (1) Calculate the molar solubility of the slightly soluble salt, which is the number of moles of the salt that will dissolve in 1 liter of solution.
(2) Determine the concentrations of the ions in solution.
(3) Substitute the ion concentrations into the K_{sp} expression to calculate K_{sp}.

(a) Balanced equation: $AgI(s) \rightleftharpoons Ag^+(aq) + I^-(aq)$ $\quad K_{sp} = [Ag^+][I^-]$

(1) $$\text{molar solubility (mol AgI/L)} = \frac{2.8 \times 10^{-6}\text{ g AgI}}{1.0\text{ L}} \times \frac{1\text{ mol AgI}}{235\text{ g AgI}}$$
$$= 1.2 \times 10^{-8}\text{ mol AgI/L (dissolved)}$$

(2) $[Ag^+] = [I^-] = \text{molar solubility} = 1.2 \times 10^{-8}\ M$

(3) $K_{sp} = [Ag^+][I^-] = (1.2 \times 10^{-8})^2 = \mathbf{1.4 \times 10^{-16}}$
$(1.5 \times 10^{-16}$ from Appendix H)

(b) Balanced equations: $BaF_2(s) \rightleftharpoons Ba^{2+}(aq) + 2F^-(aq)$ $\quad K_{sp} = [Ba^{2+}][F^-]^2$

(1) $$\text{molar solubility} = \frac{0.012\text{ g BaF}_2}{10\text{ mL}} \times \frac{1000\text{ mL}}{1\text{ L}} \times \frac{1\text{ mol BaF}_2}{175\text{ g BaF}_2}$$
$$= 6.9 \times 10^{-3}\text{ mol BaF}_2\text{/L (dissolved)}$$

(2) $[Ba^{2+}] = \text{molar solubility} = 6.9 \times 10^{-3}\ M$
$[F^-] = 2 \times \text{molar solubility} = 1.4 \times 10^{-2}\ M$

(3) $K_{sp} = [Ba^{2+}][F^-]^2 = (6.9 \times 10^{-3})(1.4 \times 10^{-2})^2 = \mathbf{1.4 \times 10^{-6}}$
$(1.7 \times 10^{-6}$ from Appendix H)

(c) Balanced equation: $Ag_2SO_4(s) \rightleftharpoons 2Ag^+(aq) + SO_4^{2-}(aq)$ $K_{sp} = [Ag^+]^2[SO_4^{2-}]$

(1) molar solubility $= \dfrac{5.7 \text{ g } Ag_2SO_4}{1 \text{ L}} \times \dfrac{1 \text{ mol}}{312 \text{ g } Ag_2SO_4}$

$= 1.8 \times 10^{-2}$ mol Ag_2SO_4/L (dissolved)

(2) $[Ag^+] = 2 \times$ molar solubility $= 3.6 \times 10^{-2}$ M

$[SO_4^{2-}] =$ molar solubility $= 1.8 \times 10^{-2}$ M

(3) $K_{sp} = [Ag^+]^2[SO_4^{2-}] = (3.6 \times 10^{-2})^2(1.8 \times 10^{-2}) = \mathbf{2.3 \times 10^{-5}}$
(1.7×10^{-5} from Appendix H)

(d) Balanced equation: $Ag_3[Fe(CN)_6](s) \rightleftharpoons 3Ag^+(aq) + Fe(CN)_6^{3-}(aq)$

$K_{sp} = [Ag^+]^3[Fe(CN)_6^{3-}]$

(1) molar solubility $= \dfrac{6.6 \times 10^{-5} \text{ g } Ag_3[Fe(CN)_6]}{100 \text{ mL}} \times \dfrac{1000 \text{ mL}}{1 \text{ L}}$

$\times \dfrac{1 \text{ mol } Ag_3[Fe(CN)_6]}{536 \text{ g } Ag_3[Fe(CN)_6]} = 1.2 \times 10^{-6}$ mol $Ag_3[Fe(CN)_6]$ (dissolved)

(2) $[Ag^+] = 3 \times$ molar solubility $= 3.6 \times 10^{-6}$ M
$[Fe(CN)_6]^{3-} =$ molar solubility $= 1.2 \times 10^{-6}$ M

(3) $K_{sp} = [Ag^+]^3[Fe(CN)_6^{3-}] = \mathbf{5.6 \times 10^{-23}}$ (not in Appendix H)

20-8. **Refer to Section 20-2, Table 20-1, Exercise 20-7 and Appendix H.** . . . . . .

Compound	Molar Solubility	Solubility	K_{sp}
$SrCrO_4$	6.0×10^{-3} mol/L	1.2 g/L	$[Sr^{2+}][CrO_4^{2-}] = 3.6 \times 10^{-5}$
$Fe(OH)_2$	1.3×10^{-5} mol/L	1.2×10^{-3} g/L	$[Fe^{2+}][OH^-]^2 = 7.9 \times 10^{-15}$
BiI_3	1.3×10^{-5} mol/L	7.7×10^{-3} g/L	$[Bi^{3+}][I^-]^3 = 8.1 \times 10^{-19}$
$Pb_3(PO_4)_2$	7.7×10^{-10} mol/L	6.3×10^{-7} g/L	$[Pb^{2+}]^3[PO_4^{3-}]^2 = 3.0 \times 10^{-44}$

Sample calculation for $Pb_3(PO_4)_2$:

let x = molar solubility of $Pb_3(PO_4)_2$. Then $3x = [Pb^{2+}]$ and $2x = [PO_4^{3-}]$

$K_{sp} = [Pb^{2+}]^3[PO_4^{3-}]^2 = (3x)^3(2x)^2 = 108x^5 = 3.0 \times 10^{-44}$
$x = 7.7 \times 10^{-10}$

Therefore, molar solubility $= 7.7 \times 10^{-10}$ mol $Pb_3(PO_4)_2$/L (dissolved)

solubility (g/L) $= \dfrac{7.7 \times 10^{-10} \text{ mol } Pb_3(PO_4)_2}{1.00 \text{ L}} \times \dfrac{812 \text{ g } Pb_3(PO_4)_2}{1 \text{ mol } Pb_3(PO_4)_2}$

$= 6.3 \times 10^{-7}$ g/L

(a) $SrCrO_4$ has the highest molar solubility.

(b) $Pb_3(PO_4)_2$ has the lowest molar solubility.

(c) $SrCrO_4$ has the high solubility in g/L.

(d) $Pb_3(PO_4)_2$ has the lowest solubility in g/L.

20-10. **Refer to Section 20-3, Example 20-5 and Appendix H.** • • • • • • • • • • •

A precipitate will form if $Q_{sp} > K_{sp}$. A precipitate can be seen by the human eye if $Q_{sp} > K_{sp}$ by a factor of 1000.

Plan: (1) Identify the ions that may be involved in a precipitation reaction and calculate their concentrations immediately after mixing but before any precipitation occurs.
(2) Substitute these values into the Q_{sp} expression and compare Q_{sp} with K_{sp}.

(a) Balanced molecular equation: $KCl(aq) + K_2S(aq) \rightarrow$ no reaction
No precipitation will occur since both potential products are soluble.

(b) Balanced molecular equation: $AgNO_3(aq) + KCl(aq) \rightarrow AgCl(s) + KNO_3(aq)$
Balanced net ionic equation: $Ag^+(aq) + Cl^-(aq) \rightarrow AgCl(s)$

(1) AgCl will precipitate if $Q_{sp} > K_{sp}$ for the equilibrium:

$AgCl(s) \rightleftharpoons Ag^+(aq) + Cl^-(aq)$ $\qquad K_{sp} = 1.8 \times 10^{-10}$

$$\text{After mixing, } [Ag^+] = \frac{\text{mol } Ag^+}{\text{total volume}} = \frac{0.00050\ M \times 0.100\ L}{0.200\ L} = 2.5 \times 10^{-4}\ M$$

$$[Cl^-] = \frac{\text{mol } Cl^-}{\text{total volume}} = \frac{0.0019\ M \times 0.100\ L}{0.200\ L} = 9.5 \times 10^{-4}\ M$$

(2) $Q_{sp} = [Ag^+][Cl^-] = (2.5 \times 10^{-4})(9.5 \times 10^{-4}) = 2.4 \times 10^{-7}$

Since $Q_{sp} > K_{sp}$ by a factor of 1300, **a visible precipitate will form.**

(c) Balanced molecular equation: $Pb(NO_3)_2(aq) + 2NaI(aq) \rightarrow PbI_2(s) + 2NaNO_3(aq)$
Balanced net ionic equation: $Pb^{2+}(aq) + 2I^-(aq) \rightarrow PbI_2(s)$

(1) PbI_2 will precipitate if $Q_{sp} > K_{sp}$ for the equilibrium:
$PbI_2(s) \rightleftharpoons Pb^{2+}(aq) + 2I^-(aq)$ $\qquad K_{sp} = 8.7 \times 10^{-9}$

$$\text{After mixing, } [Pb^{2+}] = \frac{\text{mol } Pb^{2+}}{\text{total volume}} = \frac{0.015\ M \times 0.200\ L}{0.300\ L} = 1.0 \times 10^{-2}\ M$$

$$[I^-] = \frac{\text{mol } I^-}{\text{total volume}} = \frac{0.033\ M \times 0.100\ L}{0.300\ L} = 1.1 \times 10^{-2}\ M$$

(2) $Q_{sp} = [Pb^{2+}][I^-]^2 = (1.0 \times 10^{-2})(1.1 \times 10^{-2})^2 = 1.2 \times 10^{-6}$

A precipitate will form since $Q_{sp} > K_{sp}$ by a factor of 140, but it **will not be visible** to the human eye.

(d) Balanced molecular equation: $3AgNO_3(aq) + K_3PO_4(aq) \rightarrow Ag_3PO_4(s) + 3KNO_3(aq)$

Balanced net ionic equation: $3Ag^+(aq) + PO_4^{3-}(aq) \rightarrow Ag_3PO_4(s)$

(1) Ag_3PO_4 will precipitate if $Q_{sp} > K_{sp}$ for the equilibrium:
$Ag_3PO_4(s) \rightleftharpoons 3Ag^+(aq) + PO_4^{3-}(aq)$ $\qquad K_{sp} = 1.3 \times 10^{-20}$

$$\text{After mixing, } [Ag^+] = \frac{\text{mol } Ag^+}{\text{total volume}} = \frac{0.0015\ M \times 0.020\ L}{0.030\ L} = 1.0 \times 10^{-3}\ M$$

$$[PO_4^{3-}] = \frac{\text{mol } PO_4^{3-}}{\text{total volume}} = \frac{0.0033\ M \times 0.010\ L}{0.030\ L} = 1.1 \times 10^{-3}\ M$$

(2) $Q_{sp} = [Ag^+]^3[PO_4^{3-}] = (1.0 \times 10^{-3})^3(1.1 \times 10^{-3}) = 1.1 \times 10^{-12}$

Since $Q_{sp} > K_{sp}$ by a factor of ~10^8, **a visible precipitate will occur.**

20-12. **Refer to Section 20-3, Example 20-7 and Appendix H.** . . . .

(a) Plan: Use the K_{sp} expression for the dissolution of AgCl(s) to calculate the Cl^- concentration required to reduce $[Ag^+]$ to 1.0×10^{-8} *M*.

Balanced equation: $AgCl(s) \rightleftharpoons Ag^+(aq) + Cl^-(aq)$

$K_{sp} = [Ag^+][Cl^-] = 1.8 \times 10^{-10}$

$$[Cl^-]_{soln} = \frac{K_{sp}}{[Ag^+]} = \frac{1.8 \times 10^{-10}}{1.0 \times 10^{-8}} = 1.8 \times 10^{-2}\ M$$

This is the concentration of excess chloride ion remaining in solution after the precipitation of AgCl has occurred.

Therefore,

$[NaCl]_{added} = [Cl^-]_{added} = [Cl^-]_{soln}$ + moles of Cl^- that precipitated as AgCl per liter of solution

$= 1.8 \times 10^{-2}\ M$ + moles of Cl^- that precipitated as AgCl per liter of solution

(b) $$?\ g\ Ag^+ = \frac{1.0 \times 10^{-8}\ mol\ Ag^+}{1.0\ L} \times \frac{108\ g\ Ag^+}{1\ mol\ Ag^+} = 1.1 \times 10^{-6}\ g\ Ag^+$$

(c) Use the ratio method: $$\frac{1.1 \times 10^{-6}\ g\ Ag^+}{1.00\ L} = \frac{1.0\ g\ Ag^+}{?\ L\ soln}$$

Solving, ? L soln = $\mathbf{9.1 \times 10^5\ L}$

20-14. **Refer to Section 20-3, Example 20-7 and Appendix H.** . . . .

(1) Plan: Use the K_{sp} expression for the dissolution of $CaCO_3(s)$ to calculate the CO_3^{2-} concentration required to reduce $[Ca^{2+}]$ to 8.0×10^{-5} *M*.

Balanced equation: $CaCO_3(s) \rightleftharpoons Ca^{2+}(aq) + CO_3^{2-}(aq)$ $\qquad K_{sp} = [Ca^{2+}][CO_3^{2-}] = 4.8 \times 10^{-9}$

$$[CO_3^{2-}] = \frac{K_{sp}}{[Ca^{2+}]} = \frac{4.8 \times 10^{-9}}{0.000080} = \mathbf{6.0 \times 10^{-5}}\ M$$

(2) $$?\ V\ soln\ containing\ 1\ g\ Ca^{2+} = \frac{1.0\ L\ soln}{0.000080\ mol\ Ca^{2+}} \times \frac{1\ mol\ Ca^{2+}}{40\ g\ Ca^{2+}}$$

$= \mathbf{3.1 \times 10^2\ L\ soln}$

20-16. **Refer to Section 20-4 and the Key Terms for Chapter 20.** . . . .

Fractional precipitation refers to a separation process whereby some ions are removed from solution by precipitation, leaving other ions with similar properties in solution.

20-18. **Refer to Section 20-4, Examples 20-8 and 20-9, and Appendix H.** . . . . . .

(a) The compound with the smallest K_{sp} is the least soluble, in this case, and will precipitate first. Hence, AuCl ($K_{sp} = 2.0 \times 10^{-13}$) will precipitate first, then AgCl ($K_{sp} = 1.8 \times 10^{-10}$), and finally CuCl ($K_{sp} = 1.9 \times 10^{-7}$).

(b) AgCl will begin to precipitate when $Q_{sp(AgCl)} = K_{sp(AgCl)} = [Ag^+][Cl^-]$. At this point,

$$[Cl^-] = \frac{K_{sp(AgCl)}}{[Ag^+]} = \frac{1.8 \times 10^{-10}}{0.10} = 1.8 \times 10^{-9}\ M$$

At this concentration of $[Cl^-]$, the $[Au^+]$ still in solution is governed by the K_{sp} expression for AuCl: $K_{sp(AuCl)} = [Au^+][Cl^-]$

$$[Au^+] = \frac{K_{sp(AuCl)}}{[Cl^-]} = \frac{2.0 \times 10^{-13}}{1.8 \times 10^{-9}} = 1.1 \times 10^{-4}\ M$$

Therefore, the percentage of Au^+ that is still in solution when $[Cl^-] = 1.8 \times 10^{-9}\ M$ is

$$\%\ Au^+ \text{ in solution} = \frac{[Ag^+]}{[Ag^+]_{initial}} \times 100 = \frac{1.1 \times 10^{-4}\ M}{0.10\ M} \times 100 = 0.11\%$$

$$\%\ Au^+ \text{ precipitated out} = 100.00\% - 0.11\% = \mathbf{99.89\% \simeq 100\%}$$

(c) CuCl will begin to precipitate when $Q_{sp(CuCl)} = K_{sp(CuCl)} = [Cu^+][Cl^-]$

$$[Cl^-] = \frac{K_{sp(CuCl)}}{[Cu^+]} = \frac{1.9 \times 10^{-7}}{0.10} = 1.9 \times 10^{-6}\ M$$

At this concentration of Cl^-, $[Au^+]$ and $[Ag^+]$ in solution are governed by their K_{sp} expressions:

$$[Au^+] = \frac{K_{sp(AuCl)}}{[Cl^-]} = \frac{2.0 \times 10^{-13}}{1.9 \times 10^{-6}} = \mathbf{1.1 \times 10^{-7}\ M}$$

$$[Ag^+] = \frac{K_{sp(AgCl)}}{[Cl^-]} = \frac{1.8 \times 10^{-10}}{1.9 \times 10^{-6}} = \mathbf{9.5 \times 10^{-5}\ M}$$

20-20. **Refer to Section 20-5, Example 20-10 and Appendices G and H.** . . . .

Plan: Calculate the concentrations of Mg^{2+} and OH^- ions and determine Q_{sp}. If $Q_{sp} > K_{sp}$ for $Mg(OH)_2$, then precipitation will occur.

Two equilibria must be considered:

$$Mg(OH)_2(s) \rightleftarrows Mg^{2+}(aq) + 2OH^-(aq) \qquad K_{sp} = 1.5 \times 10^{-11}$$

$$NH_3(aq) + H_2O(\ell) \rightleftarrows NH_4^+(aq) + OH^-(aq) \qquad K_b = 1.8 \times 10^{-5}$$

We recognize that the given solution is a buffer. The NH_3/NH_4^+ equilibrium determines $[OH^-]$. Recall from Chapter 18

$$[OH^-] = \frac{[base]}{[salt]} \times K_b = \frac{0.080\ M}{3.0\ M}(1.8 \times 10^{-5}) = 4.8 \times 10^{-7}\ M$$

At this low $[OH^-]$, we must include the effect of the ionization of water. Let x = $[OH^-]$ and $[H^+]$ produced by the ionization of water.

$$K_w = [H^+][OH^-] = (x)(4.8 \times 10^{-7} + x) = 1.0 \times 10^{-14}$$

Solving the quadratic equation, $x^2 + (4.8 \times 10^{-7})x - 1.0 \times 10^{-14} = 0$,

$$x = \frac{-4.8 \times 10^{-7} \pm \sqrt{(4.8 \times 10^{-7})^2 - (4)(1)(-1.0 \times 10^{-14})}}{2(1)}$$

$$= \frac{-4.8 \times 10^{-7} \pm 5.2 \times 10^{-7}}{2} = 2 \times 10^{-8} \text{ or } -5.0 \times 10^{-7} \text{ (discard)}$$

Therefore, $[OH^-] = 4.8 \times 10^{-7}\ M + 0.2 \times 10^{-7}\ M = 5.0 \times 10^{-7}\ M$

Since $Mg(NO_3)_2$ is a soluble salt, $[Mg^{2+}] = [Mg(NO_3)_2] = 0.085\ M$

For $Mg(OH)_2$, $Q_{sp} = [Mg^{2+}][OH^-]^2 = (0.085)(5.0 \times 10^{-7})^2 = 2.1 \times 10^{-14}$

We see that $Q_{sp} < K_{sp}$ and **$Mg(OH)_2$ will not precipitate.**

Also, since $[OH^-] = 5.0 \times 10^{-7}\ M$; pOH = 6.30; pH = **7.70**

20-22. **Refer to Sections 20-3 and 20-5, and Appendices G and H.** . . . .

(a) Plan: (1) Calculate the concentration of OH^- required to initiate precipitation of $Mg(OH)_2$.
(2) Determine the concentration of NH_4NO_3 needed for an ammonia buffer with the $[OH^-]$ calculated in Step (1).

(1) Balanced equation: $Mg(OH)_2(s) \rightleftharpoons Mg^{2+}(aq) + 2OH^-(aq)$

$$K_{sp} = [Mg^{2+}][OH^-]^2 = 1.5 \times 10^{-11}$$

$$[OH^-] = \left(\frac{K_{sp}}{[Mg^{2+}]}\right)^{1/2} = \left(\frac{1.5 \times 10^{-11}}{0.1}\right)^{1/2} = 1.2 \times 10^{-5}\ M$$

(2) For this buffer, $NH_3 + H_2O \rightleftharpoons NH_4^+ + OH^-$

$$K_b = \frac{[NH_4^+][OH^-]}{[NH_3]} = 1.8 \times 10^{-5}$$

$$\text{Therefore, } [NH_4NO_3] = [NH_4^+] = \frac{K_b[NH_3]}{[OH^-]} = \frac{(1.8 \times 10^{-5})(0.085)}{(1.2 \times 10^{-5})} = \mathbf{0.13}\ M$$

(b) $$?\ g\ NH_4NO_3/L\ soln = \frac{0.13\ mol\ NH_4NO_3}{1.00\ L} \times \frac{80.0\ g\ NH_4NO_3}{1\ mol\ NH_4NO_3} = \mathbf{10\ g\ NH_4NO_3/L\ soln}$$

(c) $[OH^-] = 1.2 \times 10^{-5}\ M$; pOH = 4.92; pH = **9.08**

20-24. **Refer to Section 20-5, Example 20-10 and Appendices G and H.** . . . .

Plan: Calculate the concentrations of Mn^{2+} and OH^- ions and determine Q_{sp}. If $Q_{sp} > K_{sp}$ for $Mn(OH)_2$, then precipitation will occur.

Balanced equations:

$Mn(OH)_2(s) \rightleftharpoons Mn^{2+}(aq) + 2OH^-(aq)$ $\quad K_{sp} = [Mn^{2+}][OH^-]^2 = 4.6 \times 10^{-14}$

$[Mn^{2+}] = [Mn(NO_3)_2] = 1.0 \times 10^{-5}\ M$

$[OH^-]$ is determined by the ionization of NH_3.

$$NH_3(aq) + H_2O(\ell) \rightleftharpoons NH_4^+(aq) + OH^-(aq) \qquad K_b = \frac{[NH_4^+][OH^-]}{[NH_3]} = 1.8 \times 10^{-5}$$

Let x = $[NH_3]_{ionized}$. Then $1.5 \times 10^{-3} - x = [NH_3]$ and $x = [OH^-] = [NH_4^+]$

$$K_b = \frac{[NH_4^+][OH^-]}{[NH_3]} = \frac{x^2}{(1.5 \times 10^{-3} - x)} = 1.8 \times 10^{-5} \simeq \frac{x^2}{1.5 \times 10^{-3}}$$

Solving, $x = 1.6 \times 10^{-4}$

Since the value for x is greater than 5% of 1.5×10^{-3}, the simplifying assumption may not hold.

Solving the quadratic equation, $x^2 + (1.8 \times 10^{-5})x - 2.7 \times 10^{-8} = 0$,

$$x = 1.6 \times 10^{-4}$$

Therefore, $[OH^-] = 1.6 \times 10^{-4}$ *M* (Note we see that the simplifying assumption was adequate to 2 significant figures).

And so, $Q_{sp} = [Mn^{2+}][OH^-]^2 = (1.0 \times 10^{-5})(1.6 \times 10^{-4})^2 = 2.6 \times 10^{-13}$

Thus $Q_{sp} > K_{sp}$ by a factor of ~6. Therefore a precipitate will form but will not be seen.

20-26. Refer to Section 20-6. . . . .

A slightly soluble compound will dissolve when the concentration of its ions in solution are reduced to such a level that $Q_{sp} < K_{sp}$. The following hydroxides and carbonates dissolve in strong acid, such as nitric acid.

(a)

$$\begin{array}{l} Cu(OH)_2(s) \rightleftharpoons Cu^{2+}(aq) + 2OH^-(aq) \\ 2H^+(aq) + 2OH^-(aq) \rightarrow 2H_2O(\ell) \\ \hline Cu(OH)_2(s) + 2H^+(aq) \rightarrow Cu^{2+}(aq) + 2H_2O(\ell) \end{array}$$

The H^+ from the acid reacts with OH^- and lowers the concentration of OH^- by forming H_2O, a weak electrolyte in an acid/base neutralization reaction. Whenever $[OH^-]$ is low enough such that $[Cu^{2+}][OH^-] < K_{sp}$, $Cu(OH)_2(s)$ will dissolve.

(b)

$$\begin{array}{l} Al(OH)_3(s) \rightleftharpoons Al^{3+}(aq) + 3OH^-(aq) \\ 3H^+(aq) + 3OH^-(aq) \rightarrow 3H_2O(\ell) \\ \hline Al(OH)_3(s) + 3H^+(aq) \rightarrow Al^{3+}(aq) + 3H_2O(\ell) \end{array}$$

The H^+ from the acid reacts with OH^- and thus lowers $[OH^-]$ in solution in an acid/base neutralization reaction. Whenever $[OH^-]$ is low enough such that $[Al^{3+}][OH^-]^3 < K_{sp}$, $Al(OH)_3(s)$ will dissolve.

(c)

$$\begin{array}{l} MnCO_3(s) \rightleftharpoons Mn^{2+}(aq) + CO_3^{2-}(aq) \\ 2H^+(aq) + CO_3^{2-}(aq) \rightarrow CO_2(g) + H_2O(\ell) \\ \hline MnCO_3(s) + 2H^+(aq) \rightarrow Mn^{2+}(aq) + CO_2(g) + H_2O(\ell) \end{array}$$

The H^+ from the acid removes CO_3^{2-} from solution in a reaction which forms $CO_2(g)$ and $H_2O(\ell)$. Whenever $[CO_3^{2-}]$ is low enough such that $[Mn^{2+}][CO_3^{2-}] < K_{sp}$, $MnCO_3(s)$ will dissolve.

(d) $$(PbOH)_2CO_3(s) \rightleftharpoons 2Pb^{2+}(aq) + 2OH^-(aq) + CO_3^{2-}(aq)$$

$$4H^+(aq) + 2OH^-(aq) + CO_3^{2-}(aq) \rightarrow 3H_2O(\ell) + CO_2(g)$$

$$(PbOH)_2CO_3(s) + 4H^+(aq) \rightarrow 2Pb^{2+}(aq) + 3H_2O(\ell) + CO_2(g)$$

The H^+ from the acid removes both OH^- and CO_3^{2-} ions from solution by forming $H_2O(\ell)$ and $CO_2(g)$. Whenever $[OH^-]$ and $[CO_3^{2-}]$ are low enough such that

$$[Pb^{2+}]^2[OH^-]^2[CO_3^{2-}] < K_{sp},$$

$(PbOH)_2CO_3(s)$ will dissolve.

20-28. **Refer to Section 20-6.** . . . .

Nonoxidizing acids dissolve some insoluble sulfides, including MnS and FeS. The H^+ ions react with S^{2-} ions to form gaseous H_2S, which bubbles out of the solutions. Then, Q_{sp} of the sulfide becomes less than the corresponding K_{sp} value and the metal sulfide dissolves.

(a) $$MnS(s) \rightleftharpoons Mn^{2+}(aq) + S^{2-}(aq)$$

$$2H^+(aq) + S^{2-}(aq) \rightarrow H_2S(g)$$

$$MnS(s) + 2H^+(aq) \rightarrow Mn^{2+}(aq) + H_2S(g)$$

(b) $$FeS(s) \rightleftharpoons Fe^{2+}(aq) + S^{2-}(aq)$$

$$2H^+(aq) + S^{2-}(aq) \rightarrow H_2S(g)$$

$$FeS(s) + 2H^+(aq) \rightarrow Fe^{2+}(aq) + H_2S(g)$$

20-30. **Refer to Section 20-6 and Table 13-3.** . . . .

Generally, elements of intermediate electronegativities form insoluble, amphoteric hydroxides which will dissolve in basic solutions containing high concentrations of hydroxide ion. We predict:

(a) $Mn(OH)_2$ - soluble (b) $Be(OH)_2$ - soluble (c) $Fe(OH)_2$ - soluble

(d) $Cr(OH)_3$ - soluble (e) $Co(OH)_2$ - soluble

However, Table 13-3 states that only (b) $Be(OH)_2$ and (d) $Cr(OH)_2$ are amphoteric and will dissolve in excess base. $Co(OH)_2$ is only slightly amphoteric and requires a large excess (> 6 *M*) before a small amount will dissolve.

20-32. **Refer to Section 20-6 and Example 20-14.** . . . .

The important equilibria are:

$$Be(OH)_2(s) \rightleftharpoons Be^{2+}(aq) + 2OH^-(aq) \qquad K_{sp} = 4.9 \times 10^{-22} \text{ (for } \alpha \text{ form)}$$

$$Be(OH)_4^{2-} \rightleftharpoons Be^{2+}(aq) + 4OH^-(aq) \qquad K_d = 2.5 \times 10^{-19}$$

(Data obtained from **Critical Stability Constants**, 1976, R. M. Smith and A. E. Martell)

The basis for this type of calculation is that the concentrations of ions present must obey all the equilibrium expressions simultaneously.

(a) Plan: I. Determine [total Be] and the total number of moles of Be in the original saturated solution by

(1) calculating the molar solubility and $[Be^{2+}]$ in the saturated solution using K_{sp}, and

(2) calculating $[Be(OH)_4{}^{2-}]$ in the same saturated solution using K_d.

II. Determine [total Be] and the total number of moles of Be in a saturated solution of $Be(OH)_2$ at pH = 13.85, by

(1) calculating $[Be^{2+}]$ in the solution using K_{sp}, and

(2) calculating $[Be(OH)_4{}^{2-}]$ using K_d.

III. The difference between the total moles of Be in the solution with pH = 13.85 and the original saturated solution is equal to the number of moles of $Be(OH)_2$ that will dissolve (Remember: there is 1 mole of Be in the solution for every 1 mole of $Be(OH)_2$ that dissolves). If the difference is less than 0.010 mol, then not all of the $Be(OH)_2$ suspended in the solution will dissolve.

I. We probably could assume that [total Be] and the total number of moles of Be is very small in the original saturated solution, in comparison to that in the solution at pH = 13.85, by looking at the small value of K_{sp}. However, let's proceed through the very lengthy calculation for practice.

(1) Let x = molar solubility of $Be(OH)_2$. Then x = $[Be^{2+}]$
2x = $[OH^-]$

$$K_{sp} = [Be^{2+}][OH^-]^2 = (x)(2x)^2 = 4x^3 = 4.9 \times 10^{-22}$$
$$x = 5.0 \times 10^{-8}$$

Therefore, $[Be^{2+}] = x = 5.0 \times 10^{-8}$ *M*; $[OH^-] = 2x = 1.0 \times 10^{-7}$ *M*

However, these results are not accurate. We must include the effect of the ionization of water in our calculations because $[OH^-]$ produced by the dissolution of $Be(OH)_2$ to create the original saturated solution is so close to the $[OH^-]$ produced by the ionization of H_2O in pure water.

Let x' = $[Be^{2+}]$.
Then 2x' = $[OH^-]$ produced by the dissolution of $Be(OH)_2$.

Let y = $[H_3O^+]$ = $[OH^-]$ produced by the ionization of water.

Therefore, $[OH^-]_{total} = 2x' + y$

Now, we can solve 2 equations in 2 unknowns:

$$K_{sp} = [Be^{2+}][OH^-]^2 = (x')(2x' + y)^2$$
$$K_w = [H_3O^+][OH^-] = (y)(2x' + y) = 2x'y + y^2$$

An easy way to solve for x' and y is by the method of successive approximation. Choose a value for y; substitute it into the K_w expression and solve for x'. Then, substitute your x' and y values into the K_{sp} expression and see how close the calculated K_{sp} value comes to the actual value. Continue to change the value of y until the calculated value of K_{sp} agrees with the actual value within acceptable error limits. For this series of calculations, you will obtain:

$$x' = 2.8 \times 10^{-8} \quad \text{and} \quad y = 7.6 \times 10^{-8}$$

Therefore, in a saturated solution of $Be(OH)_2$,

$$[Be^{2+}] = x' = 2.8 \times 10^{-8}\ M$$
$$[OH^-] = 2x' + y = 1.3 \times 10^{-7}\ M$$

(2) $$K_d = \frac{[Be^{2+}][OH^-]^4}{[Be(OH)_4{}^{2-}]} = 2.5 \times 10^{-19}$$

Substituting and rearranging, we have

$$[Be(OH)_4{}^{2-}] = \frac{[Be^{2+}][OH^-]^4}{K_d} = \frac{(2.8 \times 10^{-8})(1.3 \times 10^{-7})^4}{(2.5 \times 10^{-19})}$$
$$= 3.2 \times 10^{-7}\ M$$

Therefore, $[\text{total Be}] = [Be^{2+}] + [Be(OH)_4{}^{2-}]$
$$= 2.8 \times 10^{-8}\ M + 3.2 \times 10^{-17}\ M = 2.8 \times 10^{-8}\ M$$

$$\text{total moles of Be} = (2.8 \times 10^{-8}\ M)(0.500\ L) = 1.4 \times 10^{-8}\ \text{mol Be}$$

II. After the addition of NaOH, pH = 13.85, $[OH^-] = 0.71\ M$

(1) Rearranging the K_{sp} expression,

$$[Be^{2+}] = \frac{K_{sp}}{[OH^-]^2} = \frac{4.9 \times 10^{-22}}{(0.71)^2} = 9.7 \times 10^{-22}\ M$$

(2) Rearranging the K_d expression,

$$[Be(OH)_4{}^{2-}] = \frac{[Be^{2+}][OH^-]^4}{K_d} = \frac{(9.7 \times 10^{-22})(0.71)^4}{2.5 \times 10^{-19}} = 9.9 \times 10^{-4}\ M$$

Therefore, $[\text{total Be}] = [Be^{2+}] + [Be(OH)_4{}^{2-}]$
$$= 9.7 \times 10^{-22}\ M + 9.9 \times 10^{-4}\ M = 9.9 \times 10^{-4}\ M$$

$$\text{total moles of Be} = (9.9 \times 10^{-4}\ M)(0.500\ L) = 5.0 \times 10^{-4}\ \text{mol Be}$$

Note: In distilled water, the soluble Be exists primarily as Be^{2+}; in a more basic solution, the soluble Be exists primarily as $Be(OH)_4{}^{2-}$.

III. The amount of $Be(OH)_2$ that could be dissolved in this solution

= mol Be in basic solution saturated with $Be(OH)_2$
 - mole Be in pure water saturated with $Be(OH)_2$

$= 5.0 \times 10^{-4}$ mol $- 1.4 \times 10^{-8}$ mol

$= 5.0 \times 10^{-4}$ mol

Therefore, not all the 0.010 mol of $Be(OH)_2$ would dissolve at the higher pH; only 5.0×10^{-4} moles of it would dissolve.

(b) The minimum $[OH^-]$ that will dissolve 0.010 moles of $Be(OH)_2$ in 500 mL of solution can be calculated easily by assuming that all the Be in solution will occur as the complex ion, $Be(OH)_4{}^{2-}$, at the high pH. Justification for this assumption is the calculations done in (a).

Therefore $$[Be(OH)_4{}^{2-}] = \frac{0.010\ \text{mol}}{0.500\ L} = 0.020\ M$$

We know: $K_{sp} = [Be^{2+}][OH^-]^2$

Rearranging and solving for $[Be^{2+}]$ yields

$$[Be^{2+}] = \frac{K_{sp}}{[OH^-]^2}$$

Substituting the expression for $[Be^{2+}]$ into the K_d expression,

$$K_d = \frac{[Be^{2+}][OH^-]^4}{[Be(OH)_4{}^{2-}]}$$

gives:

$$K_d = \frac{(K_{sp}/[OH^-]^2)[OH^-]^4}{[Be(OH)_4{}^{2-}]} = \frac{K_{sp}[OH^-]^2}{[Be(OH)_4{}^{2-}]}$$

Solving for $[OH^-]$,

$$[OH^-]^2 = \frac{K_d[Be(OH)_4{}^{2-}]}{K_{sp}} = \frac{(2.5 \times 10^{-19})(0.020)}{4.9 \times 10^{-22}} = 10.2$$

$$[OH^-] = 3.2\ M$$

21 Electrochemistry

21-2. **Refer to the Introduction to Chapter 21.** . . . .

All electrochemical reactions occurring in electrochemical cells involve the transfer of electrons and are therefore oxidation-reduction reactions.

21-4. **Refer to Section 21-2 and the Key Terms for Chapter 21.** . . . .

Electrodes are surfaces upon which oxidation or reduction reactions occur. They may or may not participate in the electrochemical reactions. Inert electrodes do not take part in the reactions.

21-6. **Refer to Sections 21-2, 21-3 and 21-10.** . . . .

The cathode is defined as the electrode at which reduction occurs, i.e., where electrons are consumed, regardless of whether the electrochemical cell is an electrolytic or voltaic cell. However, the statement, "The cathode in any electrochemical cell is the negative electrode," is false. In both electrolytic and voltaic cells, the electrons flow through the wire from the anode, where electrons are produced, to the cathode, where electrons are consumed. In an electrolytic cell, the dc source forces the electrons to travel nonspontaneously through the wire. Thus, the electrons flow from the positive electrode (the anode) to the negative electrode (the cathode). And so, the statement is true for electrolytic cells. However, in a voltaic cell, the electrons flow spontaneously, **away** from the negative electrode (the anode) and toward the positive electrode (the cathode). The statement is false for a voltaic cell.

21-8. **Refer to Section 21-4.** . . . .

Magnesium metal is too reactive in water to be obtained by the electrolysis of $MgCl_2(aq)$. In other words, $H_2O(\ell)$ is more easily reduced to $OH^-(aq)$ and $H_2(g)$ than is $Mg^{2+}(aq)$ to $Mg(s)$. In electrochemical reactions, the species that is most easily reduced (or oxidized) will be reduced (or oxidized) first.

21-10. **Refer to Sections 21-8 and 22-3, and Table 21-2.** . . . .

Electrolysis should be a useful method for obtaining Ca from $CaCO_3$, Al from Al_2O_3, K from KCl and Mg from $MgCO_3$. It is not used to obtain Ni from NiS, Pb from PbS, Fe from Fe_2O_3, Cr from Cr_2O_3 or Cu from Cu_2S. Metals forming cations with more negative standard reduction potentials than that of $H^+(aq)$ (such as Ca, Al, K and Mg) are usually produced by electrolysis, while the others can be obtained by chemical reduction methods which are less costly. Electrolytic refining may be used to purify some of the latter group of metals.

21-12. **Refer to Section 21-4.** . . . .

Sodium ions do not appear in the overall cell reaction for the electrolysis of NaCl(aq) because Na^+ ions are spectator ions and do not react. Since H_2O is more easily reduced than Na^+ ions, the reduction reaction involves H_2O:

reduction at cathode: $2e^- + 2H_2O(\ell) \rightarrow H_2(g) + 2OH^-(aq)$

oxidation at anode: $2Cl^-(aq) \rightarrow Cl_2(g) + 2e^-$

overall cell reaction: $2H_2O(\ell) + 2Cl^-(aq) \rightarrow H_2(g) + Cl_2(g) + 2OH^-(aq)$

21-14. **Refer to Section 21-3 and Figure 21-2.** . . . .

Cell diagram:

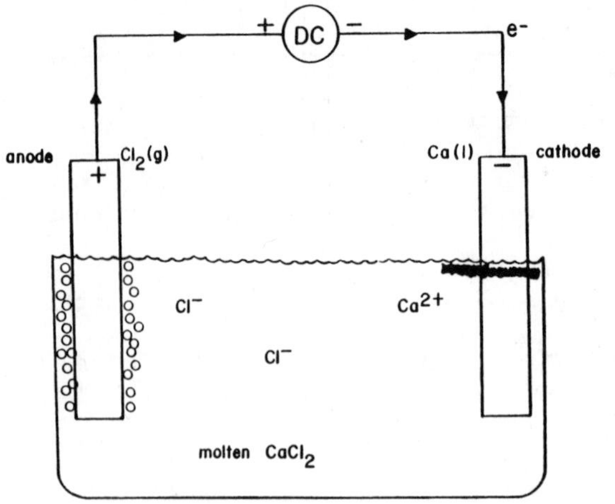

oxidation at anode	$2Cl^-(molten) \rightarrow Cl_2(g) + 2e^-$
reduction at cathode	$Ca^{2+}(molten) + 2e^- \rightarrow Ca(\ell)$
overall cell reaction	$Ca^{2+}(molten) + 2Cl^-(molten) \rightarrow Ca(\ell) + Cl_2(g)$

21-16. **Refer to Section 21-4 and Figure 21-3.** . . . .

Cell diagram:

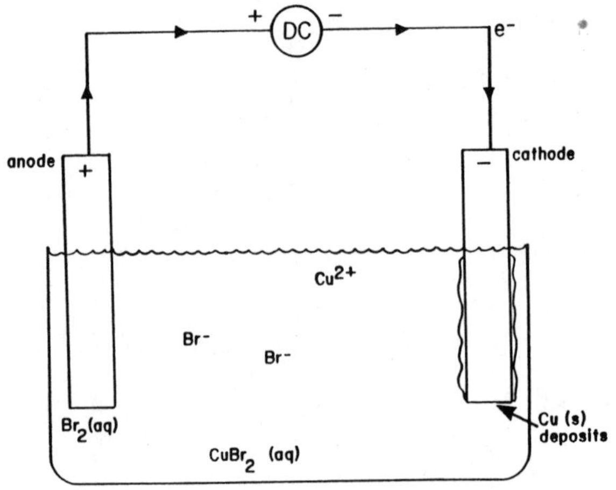

oxidation at anode	$2Br^-(aq) \rightarrow Br_2(aq) + 2e^-$
reduction at cathode	$Cu^{2+}(aq) + 2e^- \rightarrow Cu(s)$
overall cell reaction	$Cu^{2+}(aq) + 2Br^-(aq) \rightarrow Cu(s) + Br_2(aq)$

21-18. **Refer to Section 21-6 and the Key Terms for Chapter 21.** . . . .

(a) A coulomb (C) is the amount of charge that passes a given point when one ampere of current flows for one second.

(b) Electrical current (I) is the rate of transfer of electricity.

(c) An ampere (A) is the practical unit of current equal to the transfer of 1 coulomb per second. So, 1 A = 1 C/s.

(d) A faraday of electricity corresponds to the charge on 6.022×10^{23} electrons, or 96,487 coulombs. It is the amount of electricity that reduces 1 equivalent weight of a substance at the cathode and oxidizes 1 equivalent weight of a substance at the anode.

21-20. **Refer to Section 21-6 and Table 21-1.** . . . .

If the same current is used in electrolysis of different solutions for the same period of time, the same number of electrons is being passed through the solutions. Let us assume that 1 mole of electrons or 1 Faraday is being passed through the given solutions, and therefore, 1 equivalent weight of each metal is plated out. The metal produced in the greatest mass is the metal with the largest equivalent weight.

Half-Reaction	Equivalent Weight	Ratio of Moles Plated
$Ag^+(aq) + e^- \rightarrow Ag(s)$	107.9 g/eq = 1 × atomic weight	1
$Cu^+(aq) + e^- \rightarrow Cu(s)$	63.55 g/eq = 1 × atomic weight	1
$Cu^{2+}(aq) + 2e^- \rightarrow Cu(s)$	31.78 g/eq = 1/2 × atomic weight	1/2
$Cr^{3+}(aq) + 3e^- \rightarrow Cr(s)$	17.33 g/eq = 1/3 × atomic weight	1/3
$Hg^{2+}(aq) + 2e^- \rightarrow Hg(\ell)$	100.3 g/eq = 1/2 × atomic weight	1/2
$Hg_2^{2+}(aq) + 2e^- \rightarrow 2Hg(\ell)$	200.6 g/eq = 1 × atomic weight	1

Therefore, Hg from $Hg_2(NO_3)_2$ would be produced in the greatest mass, whereas Cr from $CrCl_3$ would be produced in the least mass. The ratio of the numbers of moles of metal plated out is given in the table above.

21-22. **Refer to Section 21-6.** . . . .

$$? \text{ coulombs/electron} = \frac{1 \text{ faraday}}{6.022 \times 10^{23}\ e^-} \times \frac{96487 \text{ coulombs}}{1 \text{ faraday}} = \mathbf{1.602 \times 10^{-19}\ C/e^-}$$

21-24. **Refer to Section 21-6.** . . . .

Recall that 1 faraday of electricity is equivalent to 1 mole of electrons passing through a system. Consider the general balanced half-reaction:

$$M^{n+} + ne^- \rightarrow M$$

In accordance with stoichiometry, to form 1 mole of M requires n moles of electrons, hence n faradays of electricity.

	Balanced Half-Reaction	No. of Faradays
(a)	$Hg^{2+}(aq) + 2e^- \rightarrow Hg(\ell)$	2
(b)	$Hg_2^{2+}(aq) + 2e^- \rightarrow 2Hg(\ell)$	1
(c)	$K^+(\text{molten}) + e^- \rightarrow K(s)$	1
(d)	$Al^{3+}(aq) + 3e^- \rightarrow Al(s)$	3

21-26. Refer to Section 21-6 and Exercise 21-24 Solution.

The amount of charge required to deposit 1.00 g of each of the metals according to the reactions in Exercise 21-24:

(a) $? \text{ coulombs} = 1.00 \text{ g Hg} \times \frac{1 \text{ mol Hg}}{200.6 \text{ g Hg}} \times \frac{2 \text{ mol } e^-}{1 \text{ mol Hg}} \times \frac{96500 \text{ C}}{1 \text{ mol } e^-} = \mathbf{962 \text{ C}}$

(b) $? \text{ coulombs} = 1.00 \text{ g Hg} \times \frac{1 \text{ mol Hg}}{200.6 \text{ g Hg}} \times \frac{2 \text{ mol } e^-}{2 \text{ mol Hg}} \times \frac{96500 \text{ C}}{1 \text{ mol } e^-} = \mathbf{481 \text{ C}}$

(c) $? \text{ coulombs} = 1.00 \text{ g K} \times \frac{1 \text{ mol K}}{39.10 \text{ g K}} \times \frac{1 \text{ mol } e^-}{1 \text{ mol K}} \times \frac{96500 \text{ C}}{1 \text{ mol } e^-} = \mathbf{2470 \text{ C}}$

(d) $? \text{ coulombs} = 1.00 \text{ g Al} \times \frac{1 \text{ mol Al}}{26.98 \text{ g Al}} \times \frac{3 \text{ mol } e^-}{1 \text{ mol Al}} \times \frac{96500 \text{ C}}{1 \text{ mol } e^-} = \mathbf{1.07 \times 10^4 \text{ C}}$

21-28. Refer to Section 21-6, Example 21-2, Exercise 21-27 and Exercise 21-26 Solution.

The balanced half-reaction: $2H_2O \rightarrow O_2 + 4H^+ + 4e^-$

Plan: (1) Calculate the number of coulombs passing through the cell.
(2) Calculate the amount of O_2 produced.

(1) $? \text{ coulombs} = \text{current (A)} \times \text{time (s)}$

$= 0.250 \text{ A} \times 48.0 \text{ hr} \times 60 \text{ min/hr} \times 60 \text{ s/min} = 4.32 \times 10^4 \text{ s}$

(2) $? \text{ g } O_2 = 4.32 \times 10^4 \text{ coul} \times \frac{1 \text{ mol } e^-}{96500 \text{ C}} \times \frac{1 \text{ mol } O_2}{4 \text{ mol } e^-} \times \frac{32.0 \text{ g } O_2}{1 \text{ mol } O_2} = \mathbf{3.58 \text{ g}}$

$? \text{ mL}_{STP} O_2 = 4.32 \times 10^4 \text{ coul} \times \frac{1 \text{ mol } e^-}{96500 \text{ C}} \times \frac{1 \text{ mol } O_2}{4 \text{ mol } e^-} \times \frac{22400 \text{ mL}_{STP} O_2}{1 \text{ mol } O_2}$

$= \mathbf{2.51 \times 10^3 \text{ mL}_{STP}}$

21-30. Refer to Section 21-6 and Exercises 21-26, 21-27 and 21-28 Solutions.

The balanced half-reaction: $2H_2O \rightarrow O_2 + 4H^+ + 4e^-$

$? \text{ time (s)} = 5.00 \text{ L}_{STP} O_2 \times \frac{1 \text{ mol } O_2}{22.4 \text{ L}_{STP} O_2} \times \frac{4 \text{ mol } e^-}{1 \text{ mol } O_2} \times \frac{96500 \text{ C}}{1 \text{ mol } e^-} \times \frac{1 \text{ s}}{0.250 \text{ C}}$

$= \mathbf{3.44 \times 10^5 \text{ s}}$ or **3.99 days**

21-32. Refer to Section 21-6 and Exercise 21-26 Solution.

(a) $Fe^{3+} + 3e^- \rightarrow Fe$

$? \text{ coulombs} = 1.00 \text{ g Fe} \times \frac{1 \text{ mol Fe}}{55.85 \text{ g Fe}} \times \frac{3 \text{ mol } e^-}{1 \text{ mol Fe}} \times \frac{96500 \text{ C}}{1 \text{ mol } e^-} = 5.18 \times 10^3 \text{ C}$

$? \text{ amperes} = (5.18 \times 10^3 \text{ C})/(12.0 \text{ hr} \times 3600 \text{ s/hr}) = \mathbf{0.120 \text{ A}}$

(b) $Cr^{2+} + 2e^- \rightarrow Cr$

$$? \text{ coulombs} = 1.00 \text{ g Cr} \times \frac{1 \text{ mol Cr}}{52.00 \text{ g Cr}} \times \frac{2 \text{ mol } e^-}{1 \text{ mol Cr}} \times \frac{96500 \text{ C}}{1 \text{ mol } e^-} = 3.71 \times 10^3 \text{ C}$$

$$? \text{ amperes} = (3.71 \times 10^3 \text{ C})/(12.0 \text{ hr} \times 3600 \text{ s/hr}) = \mathbf{0.0859 \text{ A}}$$

(c) $Ag^+ + e^- \rightarrow Ag$

$$? \text{ coulombs} = 1.00 \text{ g Ag} \times \frac{1 \text{ mol Ag}}{107.9 \text{ g Ag}} \times \frac{1 \text{ mol } e^-}{1 \text{ mol Ag}} \times \frac{96500 \text{ C}}{1 \text{ mol } e^-} = 8.94 \times 10^2 \text{ C}$$

$$? \text{ amperes} = (8.94 \times 10^2 \text{ C})/(12.0 \text{ hr} \times 3600 \text{ s/hr}) = \mathbf{0.0207 \text{ A}}$$

(d) $2I^- \rightarrow I_2 + 2e^-$

$$? \text{ coulombs} = 1.00 \text{ g } I_2 \times \frac{1 \text{ mol } I_2}{253.8 \text{ g } I_2} \times \frac{2 \text{ mol } e^-}{1 \text{ mol } I_2} \times \frac{96500 \text{ C}}{1 \text{ mol } e^-} = 7.60 \times 10^2 \text{ C}$$

$$? \text{ amperes} = (7.60 \times 10^2 \text{ C})/(12.0 \text{ hr} \times 3600 \text{ s/hr}) = \mathbf{0.0176 \text{ A}}$$

21-34. **Refer to Section 21-6 and Exercise 21-26 Solution.** . . . .

The balanced half-reaction is $M^{2+} + 2e^- \rightarrow M$

$$? \text{ atomic weight (g/mol)} = \frac{8.43 \text{ g M}}{14{,}475 \text{ C}} \times \frac{96500 \text{ C}}{1 \text{ mol } e^-} \times \frac{2 \text{ mol } e^-}{1 \text{ mol M}} = 112 \text{ g/mol}$$

The metal must be **Cd**.

21-36. **Refer to Section 21-6 and Exercise 21-26 Solution.** . . . .

The two half-reactions involved in the electrolysis of water are:

(oxidation, anode) $2H_2O \rightarrow O_2 + 4H^+ + 4e^-$

(reduction, cathode) $2e^- + 2H_2O \rightarrow H_2 + 2OH^-$

(1) $$? \text{ g } H_2 = 4.00 \text{ hr} \times \frac{3600 \text{ s}}{1 \text{ hr}} \times \frac{1.80 \text{ C}}{1 \text{ s}} \times \frac{1 \text{ mol } e^-}{96500 \text{ C}} \times \frac{1 \text{ mol } H_2}{2 \text{ mol } e^-} \times \frac{2.016 \text{ g } H_2}{1 \text{ mol } H_2}$$

$$= \mathbf{0.271 \text{ g}}$$

$$? \text{ L}_{STP} \text{ } H_2 = 0.271 \text{ g} \times \frac{1 \text{ mol } H_2}{2.016 \text{ g } H_2} \times \frac{22.4 \text{ L}_{STP} \text{ } H_2}{1 \text{ mol } H_2} = \mathbf{3.01 \text{ L}_{STP}}$$

(2) $$? \text{ g } O_2 = 4.00 \text{ hr} \times \frac{3600 \text{ s}}{1 \text{ hr}} \times \frac{1.80 \text{ C}}{1 \text{ s}} \times \frac{1 \text{ mol } e^-}{96500 \text{ C}} \times \frac{1 \text{ mol } O_2}{4 \text{ mol } e^-} \times \frac{32.00 \text{ g } O_2}{1 \text{ mol } O_2}$$

$$= \mathbf{2.15 \text{ g}}$$

$$? \text{ L}_{STP} \text{ } O_2 = 2.15 \text{ g} \times \frac{1 \text{ mol } O_2}{32.00 \text{ g } O_2} \times \frac{22.4 \text{ L}_{STP} \text{ } O_2}{1 \text{ mol } O_2} = \mathbf{1.51 \text{ L}_{STP}}$$

Note: The volumes of H_2 and O_2 obtained above are in the ratio 3.01:1.51 or 2:1 in accordance with the 2:1 mole ratio in the overall reaction:

$$2H_2O(\ell) \rightarrow 2H_2(g) + O_2(g)$$

21-38. **Refer to Section 21-6 and Exercise 21-26 Solution.** . . . .

Balanced half-reaction: $Ag^+ + e^- \rightarrow Ag$

$$? \text{ g Ag} = 20.0 \text{ min} \times \frac{60 \text{ s}}{1 \text{ min}} \times \frac{1.00 \text{ C}}{1 \text{ s}} \times \frac{1 \text{ mol } e^-}{96500 \text{ C}} \times \frac{1 \text{ mol Ag}}{1 \text{ mol } e^-} \times \frac{107.9 \text{ g Ag}}{1 \text{ mol Ag}} = 1.34 \text{ g}$$

21-40. **Refer to Section 21-6 and Exercise 21-26 Solution.** . . . .

Balanced net ionic half-reaction: $Fe(CN)_6^{4-} + 2e^- \rightarrow Fe + 6CN^-$

$$? \text{ coulombs} = 36.0 \text{ g } K_4[Fe(CN)_6] \times \frac{1 \text{ mol } K_4[Fe(CN)_6]}{368 \text{ g } K_4[Fe(CN)_6]} \times \frac{1 \text{ mol Fe}}{1 \text{ mol } K_4[Fe(CN)_6]}$$

$$\times \frac{2 \text{ mol } e^-}{1 \text{ mol Fe}} \times \frac{96500 \text{ C}}{1 \text{ mol } e^-}$$

$$= 1.89 \times 10^4 \text{ C}$$

21-42. **Refer to Section 21-8.** . . . .

Rapid plating of a metal results in rough, grainy, black surfaces because the metal atoms are deposited so rapidly that they are not able to form an extended lattice using metal bonds. Slower plating produces smooth surfaces.

21-44. **Refer to Section 21-8 and Figures 21-5 and 21-6.** . . . .

(a) Electroplating is a process that plates metal onto a cathodic surface by electrolysis.

(b) A simple silver electroplating apparatus for a jeweler consists of a dc generator (a battery) with the negative lead attached to the piece of jewelry (cathode) and the positive lead attached to a piece of silver metal (anode). The jewelry and the silver metal are both immersed in a beaker containing aqueous silver ions. The apparatus is similar to that shown in Figure 21-6 for electroplating with Cu.

(c) Highly purified silver as the anode is not necessary in an electroplating operation. As the electrolytic cell operates, Ag and other metals from an impure Ag anode oxidize to form metal cations in solution. However, only Ag^+ ions are reduced to Ag metal at the cathode because of its ease of reduction and higher concentration.

21-46. **Refer to Section 21-9 and Figure 21-7.** . . . .

A salt bridge in a voltaic or galvanic cell has three functions: it allows electrical contact between the two solutions; it prevents mixing of the electrode solutions; and it maintains electrical neutrality in each half-cell.

21-48. **Refer to the Introduction to Voltaic or Galvanic Cells.** . . . .

In a voltaic cell, the solutions in the two half-cells must be kept separate in order to produce usable electrical energy since electricity is produced when electron transfer is forced to occur through the wiring of an external circuit. If the two half-cells were mixed, electron transfer would happen directly in the solution. It could not be exploited to give electricity.

21-50. **Refer to Sections 21-9, 21-10 and 21-11.** . . . .

Cell diagram:

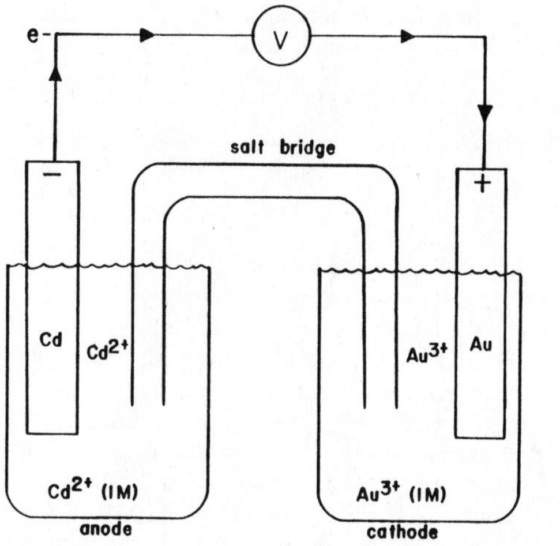

oxidation at anode		$3(Cd \rightarrow Cd^{2+} + 2e^-)$
reduction at cathode		$2(Au^{3+} + 3e^- \rightarrow Au)$
overall cell reaction		$3Cd^{2+} + 2Au^{3+} \rightarrow 3Cd^{2+} + 2Au$

21-52. **Refer to Sections 21-9, 21-10 and 21-11.** . . . .

Balanced equation: $2Al(s) + 3Pd^{2+}(aq) \rightarrow 2Al^{3+}(aq) + 3Pd(s)$

(a) reduction half-reaction: $Pd^{2+}(aq) + 2e^- \rightarrow Pd(s)$

(b) oxidation half-reaction: $Al(s) \rightarrow Al^{3+}(aq) + 3e^-$

(c),(d) Al is the anode and Pd is the cathode.

(e) cell diagram:

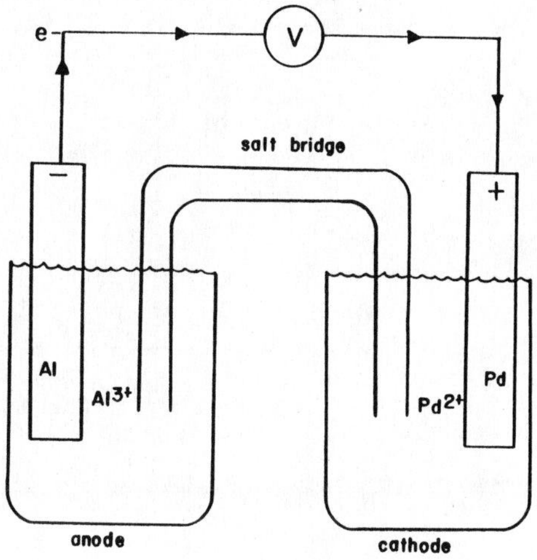

21-54. **Refer to Section 21-9.** . . . .

A voltaic cell is operating under standard electrochemical conditions and is called a standard cell when all the reactants are in their thermodynamic standard states: dissolved ions are at 1 *M* and gases are present at 1 atmosphere partial pressure. The temperature for the standard condition is 25°C unless otherwise specified.

21-56. **Refer to Exercise 21-48 Solution.** • • • • • • • • • • • • • • • • •

No electricity is produced when Cu(s) is placed into $AgNO_3(aq)$ even though a spontaneous redox reaction occurs:

$$Cu(s) + 2AgNO_3(aq) \rightarrow Cu(NO_3)_2(aq) + 2Ag(s)$$

The electron transfer occurs within the solution; it is not forced to occur through an external circuit where it produces useful electrical energy.

21-58. **Refer to Sections 21-22, 21-23 and 21-25.** • • • • • • • • • • • • • • •

(a) The dry cell (Leclanche cell) is shown in Figure 21-14. The container is made of zinc, which also acts as one of the electrodes. The other electrode is a carbon rod in the center of the cell. The cell is filled with a moist mixture of NH_4Cl, MnO_2 and $ZnCl_2$ and a porous inert filler and is separated from the zinc container by a porous paper. Dry cells are sealed to keep moisture from evaporating. As the cell operates, the Zn electrode is the anode and is oxidized to Zn^{2+} ions. The ammonium ion is reduced to give NH_3 and H_2 at the carbon cathode. The ammonia produced combines with Zn^{2+} ion and forms a soluble compound containing the complex ion, $[Zn(NH_3)_4]^{2+}$; H_2 is removed by oxidizing with MnO_2. This type of battery cannot be recharged.

(b) The lead storage battery is shown in Figure 21-15. It consists of a group of lead plates bearing compressed spongy lead alternating with a group of lead plates bearing lead(IV) oxide, PbO_2. The electrodes are immersed in a solution of about 40% sulfuric acid. When the cell discharges, the spongy lead is oxidized to give Pb^{2+} ions which then combine with sulfate ions to form insoluble $PbSO_4$, coating the anode. Electrons produced at the anode by oxidation of spongy lead travel through the external circuit to the cathode and reduce lead(IV) to lead(II) in the presence of H^+. The cathode also becomes coated with insoluble lead sulfate. The lead storage battery can be recharged by reversal of all reactions.

(c) The hydrogen-oxygen fuel cell is shown in Figure 21-16. Hydrogen (the fuel) is supplied to the anode compartment. Oxygen is fed into the cathode compartment. Oxygen is reduced at the cathode to OH^- ions. The OH^- ions migrate through the electrolyte, an aqueous solution of a base, to the anode, where H_2 is oxidized to H_2O. The net reaction of the cell is the same as the burning of hydrogen in oxygen to form water, but combustion does no occur. Rather, most of the chemical energy from the formation of O-H bonds is converted directly into electrical energy.

21-60. **Refer to Section 21-15.** •

If the sign of the standard reduction potential, E^o, of a half-reaction is positive, the half-reaction is the cathodic (reduction) reaction when connected to the standard hydrogen electrode (SHE). Half-reactions with more positive E^o values have greater tendencies to occur in the forward direction. If the E^o of a half-reaction is negative, the half-reaction is the anodic (oxidation) reaction when connected to the SHE. Half-reactions with more negative E^o values have greater tendencies to occur in the reverse direction.

21-62. **Refer to Section 21-15 and Table 21-2.** • • • • • • • • • • • • • • •

The electromotive series was constructed by experimentally measuring the potentials of each electrode under the standard conditions versus the SHE. By international convention, the standard potentials are tabulated for reduction half-reactions:

$$\text{oxidized form} + ne^- \rightarrow \text{reduced form},$$

and are listed in order of increasing (more positive) E^{o} values. As E^{o} becomes more positive, the strength of the oxidized form as an oxidizing agent increases, whereas the strength of the reduced form as a reducing agent decreases.

The elements with high ionization energies and highly negative electron affinities have the greatest tendencies to exist as anions. They usually have very positive standard reduction potentials. On the other hand, the elements with low ionization energies and positive or slightly negative electron affinities have the greatest tendencies to exist as cations. They usually have very negative standard reduction potentials.

21-64. **Refer to Section 21-16 and Appendix J.** . . . .

(a) cell diagram:

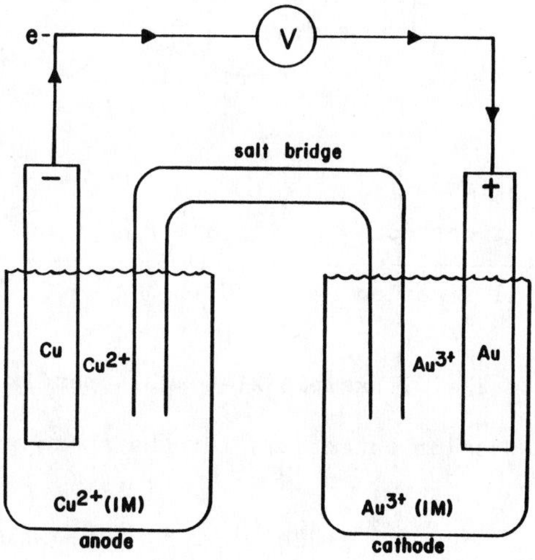

		E^{o}
oxidation at anode	$3(Cu \rightarrow Cu^{2+} + 2e^{-})$	-0.337 V
reduction at cathode	$2(Au^{3+} + 3e^{-} \rightarrow Au)$	1.50 V
overall cell reaction	$3Cu + 2Au^{3+} \rightarrow 3Cu^{2+} + 2Au$	**1.16** V = E^{o}_{cell}

(b) cell diagram:

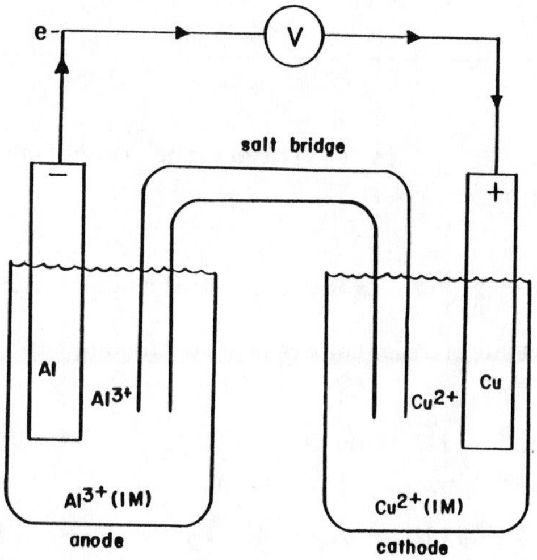

		E^{o}
oxidation at anode	$2(Al \rightarrow Al^{3+} + 3e^{-})$	1.66 V
reduction at cathode	$3(Cu^{2+} + 2e^{-} \rightarrow Cu)$	0.337 V
overall cell reaction	$3Cu^{2+} + 2Al \rightarrow 3Cu + 2Al^{3+}$	**2.00** V = E^{o}_{cell}

(c) cell diagram:

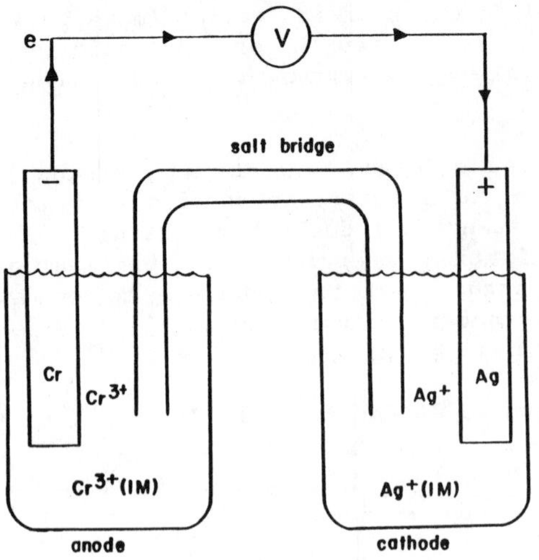

		E^o
oxidation at anode	$Cr \rightarrow Cr^{3+} + 3e^-$	0.74 V
reduction at cathode	$3(Ag^+ + e^- \rightarrow Ag)$	0.7994 V
overall cell reaction	$Cr + 3Ag^+ \rightarrow Cr^{3+} + 3Ag$	**1.54** V = E^o_{cell}

21-66. **Refer to Section 21-16, Example 21-4 and Appendix J.**

The standard reduction potentials for the appropriate half-reactions are:

	E^o
$Pb^{2+} + 2e^- \rightarrow Pb$	-0.126 V
$Fe^{3+} + e^- \rightarrow Fe^{2+}$	0.771 V

The E^o for the reduction of Fe^{3+} is more positive than E^o for the reduction of Pb^{2+}. Therefore, Fe^{3+} is a better oxidizing agent than Pb^{2+} and will oxidize Pb to Pb^{2+}. The spontaneous reaction is formulated and E^o for the cell is calculated as follows:

		E^o
reduction	$2(Fe^{3+} + e^- \rightarrow Fe^{2+})$	0.771 V
oxidation	$Pb \rightarrow Pb^{2+} + 2e^-$	0.126 V
cell reaction	$2Fe^{3+} + Pb \rightarrow Pb^{2+} + 2Fe^{2+}$	**0.897 V** = E^o_{cell}

And also, since E^o_{cell} is positive, the reaction is spontaneous as written and Fe^{3+} will oxidize Pb to Pb^{2+}.

21-68. **Refer to Section 21-17, Example 21-5 and Appendix J.**

The standard reduction potentials for the appropriate half-reactions are:

	E^o
$SO_4^{2-} + 4H^+ + 2e^- \rightarrow H_2SO_3 + H_2O$	0.17 V
$SO_4^{2-} + 4H^+ + 2e- \rightarrow SO_2 + 2H_2O$	0.20 V
$H_3AsO_4 + 2H^+ + 2e^- \rightarrow H_3AsO_3 + H_2O$	0.58 V

The E^o for the reduction of H_3AsO_4 is more positive than either E^o for sulfate reduction. Therefore, H_3AsO_4 is a better oxidizing agent than sulfate ions and sulfate ions will not oxidize H_3AsO_3 to H_3AsO_4.

21-70. **Refer to Section 21-17, Example 21-5 and Appendix J.**

The standard reduction potentials for the appropriate half-reactions are:

	E^o
$MnO_4^- + 8H^+ + 5e^- \rightarrow Mn^{2+} + 4H_2O$	1.51 V
$Cr_2O_7^{2-} + 14H^+ + 6e^- \rightarrow 2Cr^{3+} + 7H_2O$	1.33 V

The E^o for the reduction of MnO_4^- is more positive than E^o for the reduction of $Cr_2O_7^{2-}$. Therefore, MnO_4^- is a better oxidizing agent than $Cr_2O_7^{2-}$ and will oxidize Cr^{3+} to $Cr_2O_7^{2-}$. The spontaneous reaction is formulated and E^o for the cell is calculated as follows:

		E^o
reduction	$6(MnO_4^- + 8H^+ + 5e^- \rightarrow Mn^{2+} + 4H_2O)$	1.51 V
oxidation	$5(2Cr^{3+} + 7H_2O \rightarrow Cr_2O_7^{2-} + 14H^+ + 6e^-)$	-1.33 V
	$6MnO_4^- + 48H^+ + 10Cr^{3+} + 35H_2O \rightarrow 6Mn^{2+} + 24H_2O + 5Cr_2O_7^{2-} + 70H^+$	**0.18 V**
OR	$6MnO_4^- + 10Cr^{3+} + 11H_2O \rightarrow 6Mn^{2+} + 5Cr_2O_7^{2-} + 22H^+$	$= E^o_{cell}$

21-72. **Refer to Section 21-16, Example 21-4 and Appendix J.**

		E^o
reduction	$2(Au^{3+} + 3e^- \rightarrow Au)$	+1.50 V
oxidation	$3(Cd \rightarrow Cd^{2+} + 2e^-)$	+0.403 V
cell reaction	$2Au^{3+} + 3Cd \rightarrow 2Au + 3Cd^{2+}$	**+1.90 V** $= E^o_{cell}$

21-74. **Refer to Sections 21-16 and 21-17, and Appendix J.**

Plan: Calculate E_{cell} for each reaction as written. If E_{cell} is positive, the reaction is spontaneous.

			E^o
(a)	reduction	$H_2 + 2e^- \rightarrow 2H^-$	-2.25 V
	oxidation	$H_2 \rightarrow 2H^+ + 2e^-$	0.00 V
		$2H_2 \rightarrow 2H^- + 2H^+$	-2.25 V $= E^o_{cell}$
	or	$H_2 \rightarrow H^- + H^+$	

Since $E_{cell} < 0$, the reaction is nonspontaneous.

			E^o
(b)	reduction	$MnO_2 + 4H^+ + 2e^- \rightarrow Mn^{2+} + 2H_2O$	1.23 V
	oxidation	$Pd \rightarrow Pd^{2+} + 2e^-$	-0.987 V
		$MnO_2 + 4H^+ + Pd \rightarrow Mn^{2+} + 2H_2O + Pd^{2+}$	0.24 V $= E^o_{cell}$

Since $E_{cell} > 0$, the reaction is spontaneous.

			E^o
(c)	reduction	$ZnS + 2e^- \rightarrow Zn + S^{2-}$	-1.44 V
	oxidation	$Cl_2 + 2H_2O \rightarrow 2HClO + 2H^+ + 2e^-$	-1.63 V
		$Cl_2 + 2H_2O + ZnS \rightarrow 2HClO + H_2S + Zn$	-3.07 V $= E^o_{cell}$

Since $E_{cell} < 0$, the reaction is nonspontaneous. Note that we write H_2S, not $2H^+ + S^{2-}$ in the net ionic equation since H_2S is a weak acid.

			E^o
(d)	reduction	$Ag_2CrO_4 + 2e^- \rightarrow 2Ag + CrO_4^{2-}$	0.446 V
	oxidation	$Zn + 4CN^- \rightarrow [Zn(CN)_4]^{2-} + 2e^-$	1.26 V
		$Zn + 4CN^- + Ag_2CrO_4 \rightarrow [Zn(CN)_4]^{2-} + 2Ag + CrO_4^{2-}$	1.71 V = E^o_{cell}

Since $E^o_{cell} > 0$, the reaction is spontaneous.

21-76. **Refer to Sections 21-16 and 21-17, and Appendix J.** . . . .

The substance that is the stronger oxidizing agent is the more easily reduced and has the more positive reduction potential. Therefore, the stronger oxidizing agents are listed below.

(a) Tl^+ (-0.34 V) > Cd^{2+} (-0.403 V)

(b) Sn^{4+} (0.15 V) > Sn^{2+} (-0.14 V)

(c) Cu^{2+} (0.153 or 0.337 V) > H^+ (0.00 V)

(d) Br_2 (1.08 V) > I_2 (0.535 V)

(e) MnO_4^- in acidic soln (1.51 V) > MnO_4^- in basic soln (0.564 V or 0.588 V)

(f) Cl_2 (1.360 V) > Pt^{2+} (1.2 V)

21-78. **Refer to Sections 21-16 and 21-17, and Appendix J.** . . . .

The substance that is the stronger reducing agent is the more easily oxidized. The reduced form of a species is a stronger reducing agent when the half-reaction has a more negative standard reduction potential. The stronger reducing agents are given below.

(a) K (-2.925 V) > H^- (-2.25 V)

(b) V (-1.18 V) > Fe^{2+} (0.771 V)

(c) Mg (-2.37 V) > H^- (-2.25 V)

(d) $Mn(OH)_2$ in basic solution (-0.05 V) > Mn^{2+} in acidic solution (1.23 V)

(e) Zr (-1.53 V) > H_2S (0.14 V)

(f) Sr (-2.89 V) > Zr (-1.53 V)

21-80. **Refer to Section 21-20.** . . . .

The Nernst equation is used to calculate electrode potentials or cell potentials when the concentrations and partial pressures are other than standard state values. The Nernst equation in base-10 log is given by:

$$E = E^o - \frac{2.303RT}{nF} \log Q$$

where E = potential at nonstandard conditions (V)
E^o = standard potential (V)
R = gas constant, 8.314 J/mol·K
T = absolute temperature, 298.15 K
F = Faraday's constant, 96487 J/V·mol e^-
n = number of moles of e^- transferred
Q = reaction quotient

If we wished to use natural logarithms, we have

$$E = E^o - \frac{RT}{nF} \ln Q$$

Therefore, at 25° C, we have

$$E = E^o - \frac{0.0257}{n} \ln Q$$

21-82. **Refer to Section 21-20 and Appendix J.** . . . .

The reduction half-reaction is: $Zn^{2+} + 2e^- \rightarrow Zn$ $E^o = -0.763$ V

For the standard half-cell, $[Zn^{2+}] = 1$ *M*. Substituting this data into the Nernst equation, we have

$$E = E^o - \frac{0.0592}{n} \log \frac{1}{[Zn^{2+}]} = -0.763 \text{ V} - \frac{0.0592}{2} \log \frac{1}{1} = -0.763 \text{ V}$$

Therefore, the Nernst equation predicts that the voltage of a standard half-cell equals E^o.

21-84. **Refer to Section 21-20, Example 21-8 and Appendix J.** . . . .

(a) Balanced half-reaction: $Cl_2 + 2e^- \rightarrow 2Cl^-$ $E^o = 1.360$ V

$$E = E^o - \frac{0.0592}{2} \log \frac{[Cl^-]^2}{P_{Cl_2}} = 1.360 - \frac{0.0592}{2} \log \frac{(1.00)^2}{2.50} = \mathbf{1.372\ V}$$

(b) Balanced half-reaction: $Cl_2 + 2e^- \rightarrow 2Cl^-$ $E^o = 1.360$ V

$$E = E^o - \frac{0.0592}{2} \log \frac{[Cl^-]^2}{P_{Cl_2}} = 1.360 - \frac{0.0592}{2} \log \frac{(1.00)^2}{(1140/760)} = \mathbf{1.365\ V}$$

(c) Balanced half-reaction: $TeO_2 + 4H^+ + 4e^- \rightarrow Te + 2H_2O$ $E^o = 0.529$ V

$$E = E^o - \frac{0.0592}{4} \log \frac{1}{[H^+]^4} = 0.529 - \frac{0.0592}{4} \log \frac{1}{(4.40 \times 10^{-3})^4} = \mathbf{0.389\ V}$$

(d) Balanced half-reaction: $MnO_4^- + 8H^+ + 5e^- \rightarrow Mn^{2+} + 4H_2O$ $E^o = 1.51$ V

$$E = E^o - \frac{0.0592}{5} \log \frac{[Mn^{2+}]}{[MnO_4^-][H^+]^8} = 1.51 - \frac{0.0592}{5} \log \frac{0.662}{(0.0500)(0.0100)^8}$$

$$= \mathbf{1.31\ V}$$

21-86. **Refer to Section 21-20, Example 21-9, Exercise 21-84 Solution and Appendix J.** .

(a) The two electrodes described in (a) and (b) of Exercise 21-84 are exactly identical except for the pressure of Cl_2. The cell potential, E_{cell}, for the combination of the two electrodes can be calculated from the difference in the *E* values of the two electrodes. This is called a concentration cell.

Cl_2 (2.50 atm) + $2Cl^-$ (1.00 *M*) → Cl_2 (1140 torr or 1.50 atm) + $2Cl^-$ (1.00 *M*)

$$E_{cell} = 1.372 \text{ V} - 1.365 \text{ V} = 0.007 \text{ V}$$

(b) E_{cell} can be calculated in two ways:

(1) by adding the non-standard potentials for the reduction and oxidation half-reactions; or
(2) by substituting E^o_{cell} and species concentrations into the Nernst equation for the cell reaction.

Therefore,

			E
(1)	reduction	$4(MnO_4^- + 8H^+ + 5e^- \rightarrow Mn^{2+} + 4H_2O)$	1.31 V
	oxidation	$5(Te + 2H_2O \rightarrow TeO_2 + 4H^+ + 4e^-)$	-0.389 V
		$4MnO_4^- + 32H^+ + 5Te + 10H_2O \rightarrow 4Mn^{2+} + 16H_2O + 5TeO_2 + 20H^+$	**0.92 V** = E_{cell}

Note: Because the half-cells are physically separated by a salt bridge, the H^+ ions (and H_2O) in the permanganate half-cell do <u>not</u> have the same concentration as the H^+ (and H_2O) in the tellurium half-cell. For this reason, the total cell reaction is not simplified.

			E^o
(2)	reduction	$4(MnO_4^- + 8H^+ + 5e^- \rightarrow Mn^{2+} + 4H_2O)$	1.51 V
	oxidation	$5(Te + 2H_2O \rightarrow TeO_2 + 4H^+ + 4e^-)$	-0.529 V
		$4\,MnO_4^- + 32H^+_{(MnO_4^-)} + 5Te + 10H_2O \rightarrow 4Mn^{2+} + 16H_2O + 5TeO_2 + 20H^+_{(Te)}$	0.98 V $= E^o_{cell}$

$$E = E^o - \frac{0.0592}{20}\log\frac{[H^+_{(Te)}]^{20}[Mn^{2+}]^4}{[MnO_4^-]^4[H^+_{(MnO_4^-)}]^{32}}$$

$$= 0.98 - \frac{0.0592}{20}\log\frac{(4.40\times10^{-3})^{20}(0.662)^4}{(0.0500)^4(0.0100)^{32}}$$

$= \mathbf{0.92\ V}$

21-88. **Refer to Section 21-20 and Appendix J.** . . . .

			E^o
(a)	reduction	$PbO_2 + SO_4^{2-} + 4H^+ + 2e^- \rightarrow PbSO_4 + 2H_2O$	1.685 V
	oxidation	$Zn \rightarrow Zn^{2+} + 2e^-$	0.763 V
		$PbO_2 + SO_4^{2-} + 4H^+ + Zn \rightarrow PbSO_4 + 2H_2O + Zn^{2+}$	2.448 V $= E^o_{cell}$

$$E = E^o - \frac{0.0592}{2}\log\frac{[Zn^{2+}]}{[SO_4^{2-}][H^+]^4} = 2.448 - \frac{0.0592}{2}\log\frac{2.0\times10^{-5}}{(0.010)(0.20)^4}$$

$= \mathbf{2.445\ V}$

			E^o
(b)	reduction	$2(MnO_4^- + 8H^+ + 5e^- \rightarrow Mn^{2+} + 4H_2O)$	1.51 V
	oxidation	$5(2Cl^- \rightarrow Cl_2 + 2e^-)$	-1.360 V
		$2MnO_4^- + 16H^+ + 10Cl^- \rightarrow 2Mn^{2+} + 5Cl_2 + 8H_2O$	0.15 V $= E^o_{cell}$

$$E = E^o - \frac{0.0592}{10}\log\frac{[Mn^{2+}]^2(P_{Cl_2})^5}{[MnO_4^-]^2[H^+]^{16}[Cl^-]^{10}}$$

$$= 0.15 - \frac{0.0592}{10}\log\frac{(0.010)^2(0.80)^5}{(0.10)^2(4.0)^{16}(4.0)^{10}} = \mathbf{0.26\ V}$$

			E^o
(c)	reduction	$Sn^{2+} + 2e^- \rightarrow Sn$	-0.14 V
	oxidation	$Ni \rightarrow Ni^{2+} + 2e^-$	0.25 V
		$Sn^{2+} + Ni \rightarrow Sn + Ni^{2+}$	0.11 V $= E^o_{cell}$

$$E = E^o - \frac{0.0592}{2}\log\frac{[Ni^{2+}]}{[Sn^{2+}]} = 0.11 - \frac{0.0592}{2}\log\frac{6.0}{0.010} = \mathbf{0.03\ V}$$

21-90. **Refer to Section 21-21.**

Because $\Delta G^o = nFE^o_{cell}$ and $\Delta G^o = -RT\ln K$, the signs and magnitudes of E^o_{cell}, ΔG^o and K are related as shown in the following table for different types of reactions under standard state conditions.

Forward Reaction	E^o_{cell}	ΔG^o	K
spontaneous	+	-	>1
at equilibrium	0	0	1
nonspontaneous	-	+	<1

From the above equations, it is seen that the value of K is related to the value of ΔG^o and E^o of the cell, but not ΔG and E of the cell. E^o, ΔG^o and K are indicators of the thermodynamic tendency of an oxidation-reduction reaction to occur under standard conditions.

On the other hand, E and ΔG are related to the value of Q and are indicators of the spontaneity of a reaction under any given conditions. The reaction proceeds until $Q = K$ at which point $\Delta G = 0$ and $E_{cell} = 0$. Then

$$\log K = \frac{nFE^o_{cell}}{2.303RT}$$

21-92. **Refer to Section 21-21 and Examples 21-10, 21-11 and 21-12.**

(a)

		E^o
reduction	$Sn^{4+} + 2e^- \rightarrow Sn^{2+}$	0.15 V
oxidation	$2(Fe^{2+} \rightarrow Fe^{3+} + e^-)$	-0.771 V
	$Sn^{4+} + 2Fe^{2+} \rightarrow Sn^{2+} + 2Fe^{3+}$	-0.62 V = E^o_{cell}

The reaction is not spontaneous as written under standard conditions since E^o for the cell < 0.

$\Delta G^o = -nFE^o_{cell} = -(2\text{ mol }e^-)(96500\text{ J/V·mol }e^-)(-0.62\text{ V}) = \mathbf{1.2 \times 10^5\ J}$ or **120 kJ**

$$\text{at } 25^oC,\ E^o_{cell} = \frac{2.303RT\log K}{nF} = \frac{(2.303)(8.314\text{ J/mol·K})(298.15\text{ K})}{n(96500\text{ J/V·mol }e)}\log K$$

$$= \frac{0.0592}{n}\log K$$

Substituting and rearranging,

$$\log K = \frac{nE^o_{cell}}{0.0592} = \frac{(2)(-0.62)}{0.0592} = -21$$ Solving, $K = \mathbf{10^{-21}}$

(b)

		E^o
reduction	$Cu^+ + e^- \rightarrow Cu$	0.521 V
oxidation	$Cu^+ \rightarrow Cu^{2+} + e^-$	-0.153 V
	$2Cu^+ \rightarrow Cu + Cu^{2+}$	0.368 V = E^o_{cell}

The reaction is spontaneous as written since $E^o_{cell} > 0$.

$\Delta G^o = -nFE^o_{cell} = -(1\text{ mol }e^-)(96500\text{ J/V·mol }e^-)(0.368\text{ V}) = \mathbf{-3.55 \times 10^4\ J}$ or **-35.5 kJ**

$$\text{at } 25^oC,\ \log K = \frac{nE^o_{cell}}{0.0592} = \frac{(1)(0.368)}{0.0592} = 6.22$$ Solving, $K = 10^{6.22} = \mathbf{1.6 \times 10^6}$

E^o

(c) reduction $2(MnO_4^- + 2H_2O + 3e^- \rightarrow MnO_2 + 4OH^-)$ 0.588 V

oxidation $3(Zn + 2OH^- \rightarrow Zn(OH)_2 + 2e^-)$ 1.245 V

$3Zn + 2MnO_4^- + 4H_2O \rightarrow 2MnO_2 + 3Zn(OH)_2 + 2OH^-$ 1.833 V = E^o_{cell}

The reaction is spontaneous as written, since $E^o_{cell} > 0$.

$\Delta G^o = -nFE^o_{cell} = -(6 \text{ mol } e^-)(96500 \text{ J/V·mol } e^-)(1.833 \text{ V}) =$ **-1.061×10^6 V** or **-1061 kJ**

at 25°C, $\log K = \frac{nE^o_{cell}}{0.0592} = \frac{(6)(1.833)}{0.0592} = 186$ Solving, K = **10^{186}**

21-94. **Refer to Section 21-21, Example 21-10 and Appendix J.** . . . .

E^o

(a) reduction $Cr_2O_7^{2-} + 14H^+ + 6e^- \rightarrow 2Cr^{3+} + 7H_2O$ 1.33 V

oxidation $3(2Br^- \rightarrow Br_2 + 2e^-)$ -1.08 V

$Cr_2O_7^{2-} + 14H^+ + 6Br^- \rightarrow 2Cr^{3+} + 3Br_2 + 7H_2O$ 0.25 V = E^o_{cell}

$\Delta G^o_{overall} = -nFE^o_{cell} = -(6 \text{ mol } e^-)(96500 \text{ J/V·mol } e^-)(0.25 \text{ V}) =$ **-1.4×10^5 J** or **-140 kJ**

$\Delta G^o/\text{mol } K_2Cr_2O_7 = \frac{-140 \text{ kJ}}{1 \text{ mol } K_2Cr_2O_7} =$ **-140 kJ/mol $K_2Cr_2O_7$**

$\Delta G^o/\text{mol NaBr} = \frac{-140 \text{ kJ}}{6 \text{ mol NaBr}} =$ **-23 kJ/mol NaBr**

E^o

(b) reduction $2(MnO_4^- + 8H^+ + 5e^- \rightarrow Mn^{2+} + 4H_2O)$ 1.51 V

oxidation $5(H_2SO_3 + H_2O \rightarrow SO_4^{2-} + 4H^+ + 2e^-)$ -0.17 V

$2MnO_4^- + 5H_2SO_3 \rightarrow 2Mn^{2+} + 5SO_4^{2-} + 4H^+ + 3H_2O$ 1.34 V = E^o_{cell}

$\Delta G^o_{overall} = -nFE^o_{cell} = -(10 \text{ mol } e^-)(96500 \text{ J/V·mol } e^-)(1.34 \text{ V})$
= **-1.29×10^6 J** or **-1290 kJ**

$\Delta G^o/\text{mol } KMnO_4 = -1290 \text{ kJ/2 mol } KMnO_4 =$ **-645 kJ/mol $KMnO_4$**

$\Delta G^o/\text{mol } H_2SO_3 = -1290 \text{ kJ/5 mol } H_2SO_3 =$ **-258 kJ/mol H_2SO_3**

E^o

(c) reduction $Cr_2O_7^{2-} + 14H^+ + 6e^- \rightarrow 2Cr^{3+} + 7H_2O$ 1.33 V

oxidation $3[(COOH)_2 \rightarrow 2CO_2 + 2H^+ + 2e^-]$ 0.49 V

$Cr_2O_7^{2-} + 3(COOH)_2 + 8H^+ \rightarrow 2Cr^{3+} + 6CO_2 + 7H_2O$ 1.82 V = E^o_{cell}

$\Delta G^o_{overall} = -nFE^o_{cell} = -(6 \text{ mol } e^-)(96500 \text{ J/V·mol } e^-)(1.82 \text{ V})$
= **-1.05×10^6 J**

$\Delta G^o/\text{mol } K_2Cr_2O_7 = -1.05 \times 10^6 \text{ J/1 mol } K_2Cr_2O_7 =$ **-1.05×10^6 J/mol $K_2Cr_2O_7$**

$\Delta G^o/\text{mol } (COOH)_2 = -1.05 \times 10^6 \text{ J/3 mol } (COOH)_2 =$ **-3.50×10^5 J/mol $(COOH)_2$**

21-96. Refer to Section 21-21, Exercise 21-89, Exercise 21-92 Solution and Appendix J.

Recall: The thermodynamic equilibrium constant, K, is related to E^o_{cell}, not E_{cell}.

			E^o
(a)	reduction	$2H^+ + 2e^- \rightarrow H_2$	0.000 V
	oxidation	$Zn \rightarrow Zn^{2+} + 2e^-$	0.763 V
		$Zn + 2H^+ \rightarrow Zn^{2+} + H_2$	0.763 V = E^o_{cell}

at 25°C, $\log K = \frac{nE^o_{cell}}{0.0592} = \frac{(2)(0.763)}{0.0592} = 25.8$ Solving, K = **6 × 10^25^**

			E^o
(b)	reduction	$2(Ag^+ + e^- \rightarrow Ag)$	0.7994 V
	oxidation	$Cu \rightarrow Cu^{2+} + 2e^-$	-0.337 V
		$Cu + 2Ag^+ \rightarrow Cu^{2+} + 2Ag$	0.462 V = E^o_{cell}

at 25°C, $\log K = \frac{nE^o_{cell}}{0.0592} = \frac{(2)(0.462)}{0.0592} = 15.6$ Solving, K = **4×10^{15}**

			E^o
(c)	reduction	$Sn^{4+} + 2e^- \rightarrow Sn^{2+}$	0.15 V
	oxidation	$2(Fe^{2+} \rightarrow Fe^{3+} + e^-)$	-0.771 V
		$Sn^{4+} + 2Fe^{2+} \rightarrow Sn^{2+} + 2Fe^{3+}$	-0.62 V = E^o_{cell}

at 25°C, $\log K = \frac{nE^o_{cell}}{0.0592} = \frac{(2)(-0.62)}{0.0592} = -21$ Solving, K = **10^{-21}**

21-98. Refer to Section 21-21, Exercise 21-92 Solution and Appendix J. • • • • • •

Recall: At 25°C, $\log K = (nE^o_{cell})/0.0592$. Therefore, $E^o_{cell} = (0.0592/n)\log K$

(a) This net reaction is the result of

reduction $Cl_2 + 2e^- \rightarrow 2Cl^-$

oxidation $2Br^- \rightarrow Br_2 + 2e^-$

Therefore, $n = 2$

At 25°C, $E^o_{cell} = (0.0592/2)\log(2.9 \times 10^9) =$ **0.280 V**

(b) This net reaction is the result of

reduction $ClO_3^- + 3H_2O + 6e^- \rightarrow Cl^- + 6OH^-$

oxidation $3(ClO_3^- + 2OH^- \rightarrow ClO_4^- + H_2O + 2e^-)$

Therefore, $n = 6$

At 25°C, $E^o_{cell} = (0.0592/6)\log(2.4 \times 10^{26}) =$ **0.260 V**

(c) This net reaction is the result of

reduction $Cl_2 + 2e^- \rightarrow 2Cl^-$

oxidation $Cl^- + 2OH^- \rightarrow ClO^- + H_2O + 2e^-$

Therefore, $n = 2$

At 25°C, $E_{cell} = (0.0592/2)\log(2.9 \times 10^{32}) = \mathbf{0.961\ V}$

21-100. **Refer to Section 21-21 and Appendix J.** . . . .

Recall: $\Delta G^o = -nFE_{cell}$ and $\Delta G^o = -2.303RT\log K$

Therefore, $E^o_{cell} = -\Delta G^o/nF$ and $\log K = -\Delta G^o/2.303RT$

Each of the following systems are at 25°C for the following calculations.

(a) This net reaction is the result of

reduction $MnO_4^- + 8H^+ + 5e^- \rightarrow Mn^{2+} + 4H_2O$

oxidation $5(Fe^{2+} \rightarrow Fe^{3+} + e^-)$

Therefore, $n = 5$

$$E^o_{cell} = \frac{-(-3.57 \times 10^5\ J)}{(5\ mol\ e^-)(96500\ J/V \cdot mol\ e^-)} = \mathbf{0.740\ V}$$

$$\log K = \frac{-(-3.57 \times 10^5)}{(2.303)(8.314)(298)} = 62.6$$ Solving, $K = \mathbf{4 \times 10^{62}}$

(b) The net reaction is the result of

reduction $Cr_2O_7^{2-} + 14H^+ + 6e^- \rightarrow 2Cr^{3+} + 7H_2O$

oxidation $3(2I^- \rightarrow I_2 + 2e^-)$

Therefore, $n = 6$

$$E^o_{cell} = \frac{-(-4.60 \times 10^5\ J)}{(6\ mol\ e^-)(96500\ J/V \cdot mol\ e^-)} = \mathbf{0.794\ V}$$

$$\log K = \frac{-(-4.60 \times 10^5)}{(2.303)(8.314)(298)} = 80.6$$ Solving, $K = \mathbf{4 \times 10^{80}}$

(c) The net reaction is the result of

reduction $2(NO_3^- + 4H^+ + 3e^- \rightarrow NO + 2H_2O)$

oxidation $3(Zn \rightarrow Zn^{2+} + 2e^-)$

Therefore, $n = 6$

$$E^o_{cell} = \frac{-(-9.97 \times 10^5\ J)}{(6\ mol\ e^-)(96500\ J/V \cdot mol)} = \mathbf{1.72\ V}$$

$$\log K = \frac{-(-9.97 \times 10^5)}{(2.303)(8.314)(298)} = 175$$ Solving, $K = \mathbf{10^{175}}$

22 Metals and Metallurgy

22-2. **Refer to Section 22-1, Table 22-2 and Figure 22-1.** . . . .

Anion Name	Formula	Example	Mineral Name
oxide	O^{2-}	Fe_2O_3	hematite
sulfide	S^{2-}	Cu_2S	chalcocite
chloride	Cl^-	$NaCl$	halite (rock salt)
carbonate	CO_3^{2-}	$CaCO_3$	limestone
sulfate	SO_4^{2-}	$BaSO_4$	barite
silicate	Si_xO_y	$Al_2(Si_2O_8)(OH)_4$	kaolinite

The silicates are the most widespread minerals. However, extraction of metals from silicates is very difficult.

22-4. **Refer to Section 22-2.** . . . .

The flotation method of separating a crushed ore from the gangue is a physical separation method used with ores, e.g., sulfides, carbonates or silicates, which either are not "wet" by water or can be made water repellent by treatment. Their surfaces are covered by layers of oil or other flotation agents. A stream of air is blown through a swirled suspension of such an ore in a mixture of water and oil; bubbles form on the oil surfaces of the mineral particles, causing them to rise to the surface of the suspension. The bubbles are prevented from breaking and escaping by a layer of oil and emulsifying agent. A frothy ore concentrate forms on the surface.

22-6. **Refer to Section 22-3 and Table 22-3.** . . . .

Aluminum and the metals of Groups IA and IIA are metals which are easily oxidized to ions that are difficult to reduce. So, we predict that electrolysis would be required to obtain the free metals from the molten, anhydrous salts, KCl, Al_2O_3 and $MgSO_4$.

22-8. **Refer to Section 22-3.** . . . .

In order to obtain free metals from ores, it is necessary to break bonds between metal ions and anions. The stronger the bonds, the more energy is required to break them. Since the more active metals usually form stronger bonds, more energy is required to reduce active metal ions and thus, the reduction process for these metals is more expensive. Chemical reduction is less costly than electrolytic reduction, since production of electricity is a very inefficient process from an energy standpoint.

22-10. **Refer to Section 22-6.** . . . .

The basic oxygen furnace is used to purify pig iron, which is the iron obtained from the blast furnace process. It is impure and contains carbon, among other substances, but it can be converted to steel by burning out most of the carbon with oxygen in a basic oxygen furnace. The method involves blowing oxygen through the molten iron at high temperatures. The carbon is converted to carbon monoxide and finally to carbon dioxide.

22-12. **Refer to Section 22-5, 22-6 and 22-8.** . . . .

(a) The procedure for obtaining Al from bauxite, Al_2O_3 or $Al_2O_3 \cdot H_2O$, is as follows. The first three steps are necessary since Al^{3+} ions are reduced to Al by electrolysis only in the absence of H_2O.

(1) The crushed ore is purified by dissolving it in a concentrated solution of NaOH to form soluble $Na[Al(OH)_4]$.

(2) Then, $Al(OH)_3 \cdot xH_2O$ is precipitated from the solution by blowing carbon dioxide into the solution to neutralize the unreacted NaOH and remove one OH^- unit from $Na[Al(OH)_4]$.

(3) $Al(OH)_3 \cdot xH_2O$ is converted to Al_2O_3 by heating.

(4) A mixture of molten Al_2O_3, NaF and AlF_3 is subjected to electrolysis to produce aluminum metal and oxygen gas in the Hall process (Figure 22-5). The fluorides are added to lower the electrolysis temperature.

(b) The procedure for obtaining Fe from Fe_2O_3 or Fe_3O_4 is as follows:

(1) The oxides are reduced in blast furnaces by CO. First, coke (C), limestone ($CaCO_3$) and the crushed ore (Fe_2O_3 or Fe_3O_4 in very hard SiO_2 rock) are loaded into the top of the blast furnace.

(2) Most of the oxides are reduced to molten iron by CO, although some are reduced by coke directly. Carbon dioxide, a reaction product, reacts with excess coke to provide more CO to reduce the next charge of iron ore.

(3) The obtained product contains C as an impurity and is called pig iron. After all the C is removed, it is remelted and cooled into cast iron. Alternatively, if some C is removed and other metals, such as Mn, Cr, Ni, W, Mo and V, are added to increase the tensile strength, the mixture is known as steel.

(c) The procedure for obtaining Au from very low grade ores by the cyanide process is as follows:

(1) The ore containing native Au is mixed with a solution of NaCN and converted to an aqueous slurry.

(2) Air is bubbled through the agitated slurry to oxidize the gold metal to a water soluble complex ion, $[Au(CN)_2]^-$.

(3) Free gold can then be regenerated by reduction of $[Au(CN)_2]^-$ with zinc or by electrolytic reduction.

22-14. **Refer to Section 21-8 and Figure 21-5.** . . . .

Impure metallic Cu obtained from the chemical reduction of Cu_2S and CuS can be refined with the following arrangement.

(1) Thin sheets of very pure Cu are made cathodes by connecting them to the negative terminal of a d.c. generator. Impure chunks of copper connected to the positive terminal function as anodes. The electrodes are immersed in a solution of $CuSO_4$ and H_2SO_4.

(2) When the cell operates, Cu from impure anodes is oxidized and goes into solution as Cu^{2+} ions; Cu^{2+} ions from the solution are reduced and plate out as metallic Cu on the pure Cu cathode.

22-16. **Refer to Sections 22-5 and 21-6.** . . . .

Plan: (1) Calculate the number of coulombs of charge and the time necessary to convert all of the aluminum in 45 pounds of Al_2O_3 to aluminum metal.

(2) Calculate the number of moles of oxygen in 45 pounds of Al_2O_3 and use the ideal gas law, $PV = nRT$, to determine the volume of O_2 gas produced.

(1) Balanced reduction reaction: $Al^{3+} + 3e^- \rightarrow Al$

$$? \text{ C} = 45.0 \text{ lb } Al_2O_3 \times \frac{454 \text{ g } Al_2O_3}{1.00 \text{ lb } Al_2O_3} \times \frac{1 \text{ mol } Al_2O_3}{102 \text{ g } Al_2O_3} \times \frac{2 \text{ mol Al}}{1 \text{ mol } Al_2O_3} \times \frac{3 \text{ mol } e^-}{1 \text{ mol Al}}$$

$$\times \frac{96500 \text{ C}}{1 \text{ mol } e^-} = 1.16 \times 10^8 \text{ C}$$

$? \text{ time (s)} = 1.16 \times 10^8 \text{ C}/0.600 \text{ A} = \mathbf{1.93 \times 10^8}$ **s** or **6.12 years**

(2) Let us assume that all the oxygen in Al_2O_3 is converted to $O_2(g)$.

$$? \text{ mol } O_2 = 45.0 \text{ lb } Al_2O_3 \times \frac{454 \text{ g } Al_2O_3}{1.00 \text{ lb } Al_2O_3} \times \frac{1 \text{ mol } Al_2O_3}{102 \text{ g } Al_2O_3} \times \frac{3 \text{ mol O}}{1 \text{ mol } Al_2O_3}$$

$$\times \frac{1 \text{ mol } O_2}{2 \text{ mol O}} = 3 \times 10^2 \text{ mol } O_2$$

$$V = \frac{nRT}{P} = \frac{(3.00 \times 10^2 \text{ mol})(0.0821 \text{ L}\cdot\text{atm/mol}\cdot\text{K})(145^\circ\text{C} + 273^\circ)}{(840/760) \text{ atm}} = \mathbf{9.31 \times 10^3 \text{ L}}$$

22-18. **Refer to Sections 22-3, 22-5 and 22-6, and Appendix K.** • • • • • • • • • •

The data from Appendix K gives:

Compound	ΔH_f^o	Extractive Metallurgy Method
HgS(s)	-58.2 kJ/mol	roasting HgS
Fe_2O_3(s)	-824.2 kJ/mol	chemical reduction by CO
Al_2O_3(s)	-1676 kJ/mol	electrolysis of molten Al_2O_3

As the heats of formation of minerals become more exothermic, i.e., more negative, we surmise that they have increasing thermodynamic stability. And so the difficulty by which free metals can be separated from the minerals also increases. In other words, the more active is the metal, the easier it is to form compounds and the more difficult it is to retrieve the metal from its compounds. The parallel is seen in the methods by which the metals are removed from their mineral matrix as shown in the third column of the above table.

23 The Representative Metals

23-2. **Refer to the Introduction to Chapter 23 and Section 5-2.** . . . .

(a) The representative elements have valence electrons in *s* and/or *p* orbitals in the outermost occupied energy level.

(b) The *d*-transition metals must have a partially filled set of *d* orbitals.

(c) The *f*-transition metals have a partially filled set of *f* orbitals.

23-4. **Refer to Table 5-8 and Exercise 5-36 Solution.** . . . .

Metals are located at the left side of the periodic table and therefore, in comparison with nonmetals, have (a) fewer outer shell electrons, (b) lower electronegativities, (c) more negative standard reduction potentials and (d) less endothermic ionization energies.

23-6. **Refer to Sections 4-16, 6-1 and 28-2, and Appendix B.** . . . .

(a) Mg [Ne] $\underline{\uparrow\downarrow}$ (3*s*) (b) Mg^{2+} [Ne] (c) Na [Ne] $\underline{\uparrow}$ (3*s*) (d) Na^{+} [Ne]

(e) Sn [Kr] $\underline{\uparrow\downarrow}$ $\underline{\uparrow\downarrow}$ $\underline{\uparrow\downarrow}$ $\underline{\uparrow\downarrow}$ $\underline{\uparrow\downarrow}$ (4*d*) $\underline{\uparrow\downarrow}$ (5*s*) $\underline{\uparrow}$ $\underline{\uparrow}$ __ (5*p*)

(f) Sn^{2+} [Kr] $\underline{\uparrow\downarrow}$ $\underline{\uparrow\downarrow}$ $\underline{\uparrow\downarrow}$ $\underline{\uparrow\downarrow}$ $\underline{\uparrow\downarrow}$ (4*d*) $\underline{\uparrow\downarrow}$ (5*s*)

(g) Sn^{4+} [Kr] $\underline{\uparrow\downarrow}$ $\underline{\uparrow\downarrow}$ $\underline{\uparrow\downarrow}$ $\underline{\uparrow\downarrow}$ $\underline{\uparrow\downarrow}$ (4*d*)

23-8. **Refer to Sections 23-1 and 23-4.** . . . .

(a) Alkali metals are larger than alkaline earth metals in the same period due to increasing effective nuclear charge.

(b) Alkaline earth metals have higher densities since they are both heavier and smaller than alkali metals of the same period.

(c) Alkali metals have lower first ionization energies than alkaline earth metals of the same period due to both the increasing effective nuclear charge and decreasing size.

(d) Alkali metals have much higher second ionization energies than alkaline earth metals of the same period. This is because removal of a second electron from an alkali metal ion involves destroying the very stable noble gas electronic configuration of the ion whereas removal of a second electron from an alkaline earth metal ion involves creating a stable noble gas configuration.

23-10. **Refer to Section 23-4.** . . . .

physical properties: Alkaline earth metals are silvery-white, malleable, ductile metals, somewhat harder than alkali metals, and are excellent electrical and thermal conductors.

chemical properties: Alkaline earth metals are easily oxidized and thus are strong reducing agents. They are not as reactive as IA metals, but are too reactive to occur as free elements in nature. Alkaline earths are characterized by the loss of 2 electrons per metal atom and form basic metal oxides (except BeO) which react with water to produce hydroxides.

23-12. **Refer to Section 23-6.** . . . .

(a) Calcium metal is used (1) as a reducing agent in the metallurgy of U, Th and other metals, (2) as a scavenger to remove dissolved impurities in molten metals and residual gases in vacuum tubes, and (3) as a component in many alloys. Slaked lime, $Ca(OH)_2$, is a cheap base used in industry and is also a major component of mortar and lime plaster. Careful heating of gypsum, $CaSO_4 \cdot 2H_2O$, produces plaster of Paris, $2CaSO_4 \cdot H_2O$.

(b) Magnesium metal is used (1) in photographic flash accessories, fireworks and incendiary bombs, (2) as a structural component in alloys, and (3) as a reagent in organic syntheses. Magnesia, MgO, is an excellent heat insulator used in furnaces, ovens and crucibles. Milk of magnesia, an aqueous suspension of $Mg(OH)_2$, is a stomach antacid and laxative. Anhydrous $MgSO_4$ and $Mg(ClO_4)_2$ are used as drying agents.

23-14. **Refer to Section 23-2 and Table 23-2.** . . . .

let M = alkali metal, X = halogen

(a) $2M + 2H_2O \rightarrow 2MOH + H_2$ (b) $12M + P_4 \rightarrow 4M_3P$ (c) $2M + X_2 \rightarrow 2MX$

23-16. **Refer to Section 23-5 and Table 23-4.** . . . .

let M = alkaline earth metal

(a) $M + 2H_2O \rightarrow M(OH)_2 + H_2$ (b) $6M + P_4 \rightarrow 2M_3P_2$ (c) $M + Cl_2 \rightarrow MCl_2$

23-18. **Refer to Section 23-2 and the Key Terms for Chapter 23.** . . . .

Diagonal similarities refer to chemical similarities of Period 2 elements of a certain group to Period 3 elements, one group to the right. This effect is particularly evident toward the left side of the periodic table. For example, due to the fact that the ionic charge densities and electronegativities of Li and Mg are similar, the compounds of these two elements are similar in many ways:

(1) Li is the only IA metal that forms a nitride, Li_3N. Mg readily forms the nitride, Mg_3N_2.
(2) Li and Mg both form carbides.
(3) The solubilities of Li compounds are similar to those of Mg compounds.
(4) Li and Mg form normal oxides, Li_2O and MgO, when oxidized in air at 1 atm pressure, while the other members of Group IA form peroxides and superoxides.

23-20. **Refer to Section 23-4 and Table 23-3.** . . . .

Hydration energy is the energy released when a mole of ions in the gaseous phase forms a mole of ions in the aqueous phase. The higher the charge to size ratio of a cation, the stronger is its interaction with polar water molecules and the more exothermic is its hydration energy. Therefore, hydration energies of the alkaline earth metals become less exothermic from top to bottom within a group because the size of the ions increases, whereas the charge of the ions remains +2.

23-22. **Refer to Section 23-4 and Table 23-3.** . . . .

Standard reduction potentials of the alkaline earth metals are, in general, very negative, indicating that alkaline earth metals are easily oxidized and hence are good reducing agents. Progressing down Group IIA, metallic character increases, the metals become better reducing agents and standard reduction potentials become more negative.

23-24. **Refer to Sections 23-2 and 23-5, and Appendix K.** . . . .

(a) Balanced equation: $Li(s) + H_2O(\ell) \rightarrow LiOH(aq) + 1/2\ H_2(g)$

$$\Delta H^o_{rxn} = [\Delta H^o_{f,LiOH(aq)} + 1/2\ \Delta H^o_{f,H_2(g)}] - [\Delta H^o_{f,Li(s)} + \Delta H^o_{f,H_2O(\ell)}]$$

$$= [(1\ mol)(-508.4\ kJ/mol) + (1/2\ mol)(0\ kJ/mol)] - [(1\ mol)(0\ kJ/mol) + (1\ mol)(-285.8\ kJ/mol)]$$

$$= \mathbf{-222.6\ kJ}$$

(b) Balanced equation: $K(s) + H_2O(\ell) \rightarrow KOH(aq) + 1/2\ H_2(g)$

$$\Delta H^o_{rxn} = [\Delta H^o_{f,KOH(aq)} + 1/2\ \Delta H^o_{f,H_2(g)}] - [\Delta H^o_{f,K(s)} + \Delta H^o_{f,H_2O(\ell)}]$$

$$= [(1\ mol)(-481.2\ kJ/mol) + (1/2\ mol)(0\ kJ/mol)] - [(1\ mol)(0\ kJ/mol) + (1\ mol)(-285.8\ kJ/mol)]$$

$$= \mathbf{-195.4\ kJ}$$

(c) Balanced equation: $Ca(s) + 2H_2O(\ell) \rightarrow Ca(OH)_2(aq) + H_2(g)$

$$\Delta H^o_{rxn} = [\Delta H^o_{f,Ca(OH)_2(aq)} + \Delta H^o_{f,H_2(g)}] - [\Delta H^o_{f,Ca(s)} + 2\Delta H^o_{f,H_2O(\ell)}]$$

$$= [(1\ mol)(-1002.8\ kJ/mol) + (1\ mol)(0\ kJ/mol)] - [(1\ mol)(0\ kJ/mol) + (2\ mol)(-285.8\ kJ/mol)]$$

$$= \mathbf{-431.2\ kJ}$$

All three hydroxides produced in the reactions above are strong electrolytes. The resultant ions will undergo hydration in aqueous solutions. The differences in the standard enthalpy changes for these reactions are due mainly to the different hydration energies for the cations of the elements.

	$\Delta H_{hydration}$ (kJ/mol)	ΔH^o_{rxn} from above (kJ)
K^+	-351	-195.4
Li^+	-544	-222.6
Ca^{2+}	-1650	-431.2

It is obvious from the above table that as the hydration energy for the metal cation becomes more negative, so does the standard enthalpy change for the reaction of the metal with water.

23-26. **Refer to Section 23-2, Appendix K and Exercise 23-25 Solution.** . . . . . .

Balanced equation: $Rb(s) + H_2O(\ell) \rightarrow RbOH(aq) + 1/2\ H_2(g)$

The thermodynamic data for Rb(s) and RbOH(aq) are found in *The Handbook of Chemistry and Physics* (54th Ed.).

$$\Delta H^o_{rxn} = [\Delta H^o_{f,RbOH(aq)} + 1/2\ \Delta H^o_{f,H_2(g)}] - [\Delta H^o_{f,Rb(s)} + \Delta H^o_{f,H_2O(\ell)}]$$

$$= [(1\ mol)(-476.6\ kJ/mol) + (1/2\ mol)(0\ kJ/mol)] - [(1\ mol)(0\ kJ/mol) + (1\ mol)(-285.8\ kJ/mol)]$$

$$= \mathbf{-190.8\ kJ/mol\ Rb(s)}$$

$\Delta S^{\circ}_{rxn} = [S^{\circ}_{RbOH(aq)} + 1/2\ S^{\circ}_{H_2(g)}] - [S^{\circ}_{Rb(s)} + S^{\circ}_{H_2O(\ell)}]$

$= [(1\ mol)(113.8\ J/molJK) + (1/2\ mol)(130.60\ J/molJK)]$
$\quad - [(1\ mol)(69.5\ J/molJK) + (1\ mol)(69.91\ J/molJK)]$

= 39.7 J/K per 1 mol Rb(s)

$\Delta G^{\circ}_{rxn} = [\Delta G^{\circ}_{f,RbOH(aq)} + 1/2\ \Delta G^{\circ}_{f,H_2(g)}] - [\Delta G^{\circ}_{f,Rb(s)} + \Delta G^{\circ}_{f,H_2O(M)}]$

$= [(1\ mol)(-439.53\ kJ/mol) + (1/2\ mol)(0\ kJ/mol)]$
$\quad - [(1\ mol)(0\ kJ/mol) + (1\ mol)(-237.2\ kJ/mol)]$

= -202.3 kJ/mol Rb(s)

In Exercise 23-25, the ΔG°_{rxn} was calculated for the following reaction:

$$Na(s) + H_2O(\ell) \rightarrow NaOH(aq) + 1/2\ H_2(g)$$

$\Delta G^{\circ}_{rxn} = [\Delta G^{\circ}_{f,NaOH(aq)} + 1/2\ \Delta G^{\circ}_{f,H_2(g)}] - [\Delta G^{\circ}_{f,Na(s)} + \Delta G^{\circ}_{f,H_2O(\ell)}]$

$= [(1\ mol)(-419.2\ kJ/mol) + (1/2\ mol)(0\ kJ/mol)]$
$\quad - [(1\ mol)(0\ kJ/mol) + (1\ mol)(-237.2\ kJ/mol)]$

$= -182.0\ kJ/mol\ Na(s)$

Therefore, the reaction between Rb(s) and water is more spontaneous than the reaction between Na(s) and water, since the ΔG° for the reaction between Rb(s) and water is more negative.

23-28. **Refer to Sections 20-3 and 20-4, and Appendix H. . . . .**

(a) From Appendix H, K_{sp} for $BaCO_3 = [Ba^{2+}][CO_3^{2-}] = 8.1 \times 10^{-9}$
K_{sp} for $SrCO_3 = [Sr^{2+}][CO_3^{2-}] = 9.4 \times 10^{-10}$

$BaCO_3$ begins to precipitate when $[CO_3^{2-}] = K_{sp}/[Ba^{2+}] = 8.1 \times 10^{-9}/0.10\ M$
$= 8.1 \times 10^{-8}\ M$

$SrCO_3$ begins to precipitate when $[CO_3^{2-}] = K_{sp}/[Sr^{2+}] = 9.4 \times 10^{-10}/0.10\ M$
$= 9.4 \times 10^{-9}\ M$

Therefore, $SrCO_3$ will precipitate first.

(b) When $BaCO_3$ begins to precipitate, $[CO_3^{2-}] = 8.1 \times 10^{-8}\ M$

At this point, $[Sr^{2+}] = K_{sp}/[CO_3^{2-}] = (9.4 \times 10^{-10})/(8.1 \times 10^{-8}) = 0.012\ M$

The fraction of Sr^{2+} remaining in solution is

$$\frac{[Sr^{2+}]_{remaining}}{[Sr^{2+}]_{initial}} = \frac{0.012\ M}{0.10\ M} = \mathbf{0.12}$$

This is not a practical method for separating Ba^{2+} and Sr^{2+} ions since 12% of Sr^{2+} ions are still in solution when $BaCO_3$ begins to precipitate.

23-30. **Refer to Sections 23-7, 23-8 and 23-9. . . . .**

The post-transition metals are found below the stepwise division in the periodic table in Groups IIIA, IVA and VA. The likely oxidation numbers for Group IIIA are +1 and +3, for Group IVA are +2 and +4, and for Group VA is +3.

23-32. **Refer to Section 23-8.** . . . .

Aluminum has many uses: (1) as a lightweight structural material, (2) as protection against corrosion, (3) in electrical transmission lines, and (4) as a strong reducing agent.

23-34. **Refer to Section 23-7.** . . . .

(1) The distortion of the electron cloud of an anion by a small highly charged cation is called polarization. In general, the greater the charge of the ion, the larger is the polarization effect. This is why M^{3+} ions are highly polarizing.

(2) Ions with greater charges and higher polarizing power generally interact so strongly with the electron clouds of other ions in compounds that the electron clouds are severely distorted and considerable covalent character results in the bonds.

(3) The lighter, smaller members of a family of ions of the same oxidation state are more polarizing than the heavier, larger members. This ability of a cation to polarize an anion increases with increasing charge density, i.e., with increasing charge and decreasing size of the cation.

23-36. **Refer to Sections 23-7 and 23-9, Table 23-6 and Appendix J.** . . . .

		E^o
From Appendix J:	$Sn^{2+} + 2e^- \rightarrow Sn$	-0.14 V
	$Pb^{2+} + 2e^- \rightarrow Pb$	-0.126 V
	$Sn^{4+} + 2e^- \rightarrow Sn^{2+}$	0.15 V
	$Pb^{4+} + 2e^- \rightarrow Pb^{2+}$	1.8 V

From the E^o values for the standard reduction potentials, we can deduce that

(a) Sn is a stronger reducing agent than Pb.

(b) Pb^{2+} is a stronger oxidizing agent than Sn^{2+}.

(c) Sn^{2+} is a stronger reducing agent than Pb^{2+}.

(d) Pb^{4+} is a stronger oxidizing agent than Sn^{4+}.

23-38. **Refer to Section 19-6 and Example 19-3.** . . . .

Balanced equation: $Sn^{2+} + 2H_2O \rightleftharpoons Sn(OH)^+ + H_3O^+$

$$K_a = \frac{[Sn(OH)^+][H_3O^+]}{[Sn^{2+}]} = 1.0 \times 10^{-2}$$

Let $x = [Sn^{2+}]_{hydrolyzed}$. Then, $x = [Sn(OH)^+] = [H_3O^+]$

$0.020 - x = [Sn^{2+}]$

Substituting into the equilibrium expression, we have

$$1.0 \times 10^{-2} = \frac{x^2}{0.020 - x}$$

Since the value of K_a is fairly large, it is unlikely that the x can be omitted in comparison with 0.020. We must solve the quadratic equation:

$$x^2 + (1.0 \times 10^{-2})x - 2.0 \times 10^{-4} = 0$$

$$x = 1.0 \times 10^{-2}$$

Therefore, $[H_3O^+] = 1.0 \times 10^{-2}$ M; pH = 2.00

23-40. **Refer to Sections 20-3 and 20-4, and Appendix H.** . . . .

(a) From Appendix H, K_{sp} for $PbI_2 = [Pb^{2+}][I^-]^2 = 8.7 \times 10^{-9}$

K_{sp} for $SnI_2 = [Sn^{2+}][I^-]^2 = 1.0 \times 10^{-4}$

PbI_2 will begin to precipitate when $[I^-] = (K_{sp}/[Pb^{2+}])^{1/2}$

$$= (8.7 \times 10^{-9}/0.010)^{1/2}$$

$$= 9.3 \times 10^{-4}\ M$$

SnI_2 will begin to precipitate when $[I^-] = (K_{sp}/[Sn^{2+}])^{1/2}$

$$= (1.0 \times 10^{-4}/0.010)^{1/2} = 0.10\ M$$

Therefore, PbI_2 will precipitate first.

(b) When SnI_2 begins to precipitate, $[I^-] = 0.10\ M$

At this point, $[Pb^{2+}] = K_{sp}/[I^-]^2 = 8.7 \times 10^{-9}/(0.10)^2 = 8.7 \times 10^{-7}\ M$

The fraction of Pb^{2+} remaining in solution is

$$\frac{[Pb^{2+}]_{remaining}}{[Pb^{2+}]_{initial}} = \frac{8.7 \times 10^{-7}\ M}{0.010\ M} = 8.7 \times 10^{-5}$$

This is an adequate method for separating Pb^{2+} and Sn^{2+} ions since only $8.7 \times 10^{-3}\%$ of Pb^{2+} ions remain in solution when SnI_2 begins to precipitate.

24 Nonmetallic Elements, Part I: Noble Gases and Group VIIA

24-2. **Refer to Sections 24-1 and 24-3.**

The inert nature of the noble gases and their low abundances in the atmosphere were two factors causing their late discovery.

24-4. **Refer to Section 24-2 and Table 24-3.**

(a) In order of increasing size: He < Ne < Ar < Kr < Xe < Rn.

(b) In order of increasing melting points and boiling points: He < Ne < Ar < Kr < Xe < Rn. This order is because the only forces of attraction between the monatomic noble gas atoms are London dispersion forces, which increase with increasing atomic size.

(c) See (b).

(d) In order of increasing density: He < Ne < Ar < Kr < Xe < Rn. This order is because the masses of noble gases increase going down the group, but the molar volumes of all these gases are the same at constant temperature and pressure.

(e) In order of increasing first ionization energy: Rn < Xe < Kr < Ar < Ne < He. From top to bottom within a group, the outer electrons are further from the nucleus, and thus the first ionization energy, the energy required to remove the outermost electron from the attractive force of the nucleus, decreases.

24-6. **Refer to Section 24-3.**

The accidental discovery of $O_2^+PtF_6^-$ by the reaction of O_2 with PtF_6 led Bartlett to reason that xenon should also be oxidized by PtF_6, since the first ionization energy of molecular oxygen is actually slightly larger than that of xenon. He obtained a red crystalline solid initially believed to be $Xe^+PtF_6^-$, but now known to be more complex.

24-8. **Refer to Section 24-4.**

All the fluorides of xenon are reasonably stable having Xe-F bond energies of about 125 kJ/mol of bonds. To compare, stable covalent and ionic bonds generally have an energy range of about 170 - 500 kJ/mol, whereas hydrogen bonds and other weak intermolecular forces generally have energies less than 40 kJ/mol.

24-10. **Refer to Section 24-4.**

Balanced equation: $6XeF_4(s) + 8H_2O(\ell) \rightarrow 2XeOF_4(\ell) + 4Xe(g) + 16HF(g) + 3O_2(g)$

(1) Plan: $\text{g XeF}_4 \xrightarrow{(1)} \text{mol XeF}_4 \xrightarrow{(2)} \text{mol XeOF}_4 \xrightarrow{(3)} \text{g XeOF}_4$

$$? \text{ g XeOF}_4 = 6.50 \text{ g XeF}_4 \times \underbrace{\frac{1 \text{ mol XeF}_4}{207 \text{ g XeF}_4}}_{\text{Step 1}} \times \underbrace{\frac{2 \text{ mol XeOF}_4}{6 \text{ mol XeF}_4}}_{\text{Step 2}} \times \underbrace{\frac{223 \text{ g XeOF}_4}{1 \text{ mol XeOF}_4}}_{\text{Step 3}} = 2.33 \text{ g}$$

(2) Plan: g XeF_4 $\xrightarrow{(1)}$ mole XeF_4 $\xrightarrow{(2)}$ mol HF $\xrightarrow{(3)}$ V_{STP} HF

Step 1 **Step 2** **Step 3**

$$? \ V_{STP}\ HF = 6.50\ g\ XeF_4 \times \frac{1\ mol\ XeF_4}{207\ g\ XeF_4} \times \frac{16\ mol\ HF}{6\ mol\ XeF_4} \times \frac{22.4\ L_{STP}\ HF}{1\ mol\ HF}$$

= 1.88 L at STP

24-12. **Refer to Section 24-5 and Table 24-5.** . . . .

Halogen	Physical State (25°C, 1 atm)	Color
F_2	gas	pale yellow
Cl_2	gas	yellow-green
Br_2	liquid	red-brown
I_2	solid	black (s), violet (g)
At_2	solid	-

24-14. **Refer to Section 24-5 and Table 24-5.** . . . .

(a) In order of increasing atomic radii: F < Cl < Br < I < At

(b) In order of increasing ionic radii: $F^- < Cl^- < Br^- < I^-$

(c) In order of increasing electronegativity: At < I < Br < Cl < F

(d) In order of increasing melting points and boiling points: $F_2 < Cl_2 < Br_2 < I_2$. In nature, the halogens exist as nonpolar diatomic molecules. London dispersion forces are the only forces of attraction acting between the molecules. These forces increase with increasing molecular size.

(e) See (d).

(f) In order of increasing standard reduction potentials: $I_2 < Br_2 < Cl_2 < F_2$. F_2 has the most positive standard reduction potential and therefore is the strongest of all common oxidizing agents. Oxidizing strengths of the diatomic halogen molecules decrease down Group VIIA.

24-16. **Refer to Section 24-5.** . . . .

Anion polarizability increases as the size of the anion increases because a larger electron cloud is more easily distorted when it interacts with a positively-charged ion. Therefore, in order of increasing polarizability:

$$F^- < Cl^- < Br^- < I^-$$

24-18. **Refer to Section 24-6 and Table 21-2.** . . . .

F_2 is an extremely strong oxidizing agent which reduces to F^- ions with great ease. Therefore, it cannot be produced by electrolysis of aqueous solutions of fluoride salts through the oxidation of F^- ions.

24-20. **Refer to Section 24-6.** . . . .

Christie's preparation of F_2 is not a direct chemical oxidation, but rather it involves the formation of unstable MnF_4, which spontaneously decomposes into MnF_3 and F_2.

24-22. **Refer to Section 24-6.** . . . .

Balanced equation: $2K_2MnF_6 + 4SbF_5 \rightarrow 4KSbF_6 + 2MnF_3 + F_2$

Plan: (1) Determine the limiting reagent.
(2) Calculate the theoretical yield of F_2 based on the amount of limiting reagent.
(3) Calculate % yield.

(1) $$? \text{ mol } K_2MnF_6 = \frac{120 \text{ g } K_2MnF_6}{247 \text{ g/mol}} = 0.486 \text{ mol } K_2MnF_6$$

$$? \text{ mol } SbF_5 = \frac{217 \text{ g } SbF_5}{217 \text{ g/mol}} = 1.00 \text{ mol } SbF_5$$

If K_2MnF_6 is the limiting reagent, then

$$? \text{ mol } SbF_5 \text{ reacted} = 0.486 \text{ mol } K_2MnF_6 \times \frac{4 \text{ mol } SbF_5}{2 \text{ mol } K_2MnF_6} = 0.972 \text{ mol } SbF_5$$

Since we have 1.00 mol SbF_5, then K_2MnF_6 is the limiting reagent and all stoichiometric calculations will be based on the given amount of K_2MnF_6.

(2) $$? \text{ g } F_2 = 120 \text{ g } K_2MnF_6 \times \frac{1 \text{ mol } K_2MnF_6}{247 \text{ g } K_2MnF_6} \times \frac{1 \text{ mol } F_2}{2 \text{ mol } K_2MnF_6} \times \frac{38.0 \text{ g } F_2}{1 \text{ mol } F_2} = 9.23 \text{ g } F_2$$

(3) $$\% \text{ yield} = \frac{\text{actual yield}}{\text{theoretical yield}} \times 100 = \frac{3.80 \text{ g}}{9.23 \text{ g}} \times 100 = \mathbf{41.2\%}$$

24-24. **Refer to Section 24-6.** . . . .

Laboratory preparation of Cl_2: $4HCl(aq) + MnO_2(s) \rightarrow MnCl_2(aq) + Cl_2(g) + 2H_2O(\ell)$

24-26. **Refer to Section 24-6.** . . . .

F_2 used in the production of (1) fluorocarbons for refrigerants, lubricants, plastics, insecticides, and coolants, and (2) UF_6 which aids in separating fissionable and nonfissionable uranium isotopes for nuclear reactors.

Cl_2 used in the production of (1) many chlorinated hydrocarbons which are starting materials for some plastics, and (2) many chlorine compounds which are household bleaches and bleaches for wood pulp and textiles.

Br_2 used in the production of (1) $C_2H_4Br_2$, a leaded gasoline additive that reduces Pb buildup in engines, and (2) AgBr used in light-sensitive eyeglasses and photographic film.

I_2 (1) used to prepare KI, an additive to table salt which prevents goiter and (2) used as an antiseptic and germicide in the form of tincture of iodine, a solution in alcohol.

24-28. **Refer to Section 24-7.** . . . .

F_2 is a strong oxidizing agent, whereas F^- ions are difficult to oxidize. Other highly oxidized cations will not be easily reduced in the presence of F^- ions.

$2Fe + 3F_2 \rightarrow 2FeF_3$ (only) $\qquad$ $Cu + F_2 \rightarrow CuF_2$

I_2 is only a mild oxidizing agent, but I^- ions are fairly easily oxidized. I^- ions are able to reduce cations in high oxidation states and stabilize low oxidation states.

$Fe + I_2 \rightarrow FeI_2$ (only) $\qquad$ $2Cu + I_2 \rightarrow 2CuI$ (only)

24-30. **Refer to Sections 24-8 and Chapter 7.** . . . .

(1) BrF_3

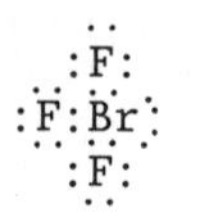

The Lewis dot formula predicts 5 regions of high electron density around the central atom, a trigonal bipyramidal electronic geometry and sp^3d hybridization for Br. The molecular geometry is T-shaped due to the presence of 2 lone pairs of electrons in the equatorial positions.

(2) ClF_5

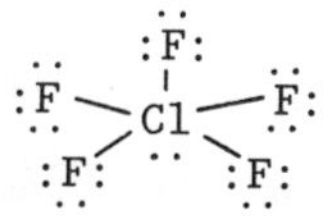

The Lewis dot formula predicts 6 regions of high electron density around the central atom, an octahedral electronic geometry and sp^3d^2 hybridization for Cl. The molecular geometry is square pyramidal due to the presence of 1 lone pair of electrons in the axial position.

(3) IF_7

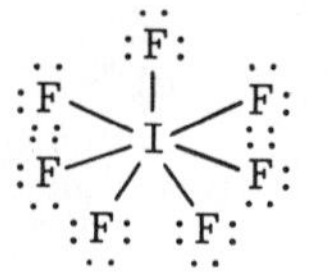

The Lewis dot formula predicts 7 regions of high electron density around the central atom, a pentagonal bipyramidal electronic geometry and sp^3d^3 hybridization for I. The molecular geometry is the same as the electronic geometry since there are no lone pairs of electrons.

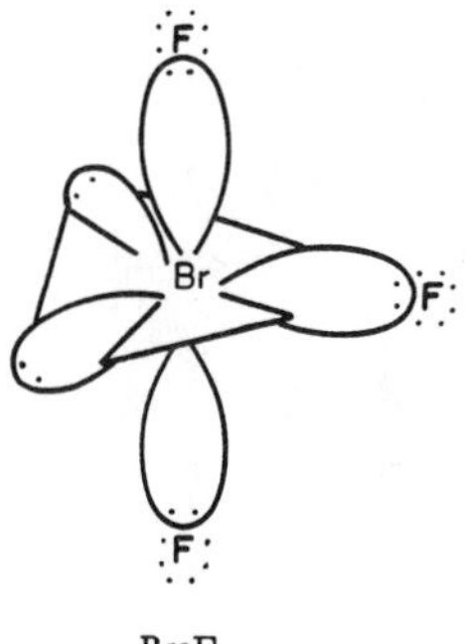

BrF_3

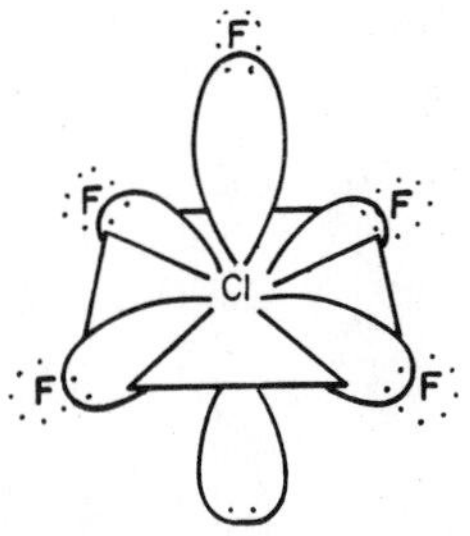

ClF_5

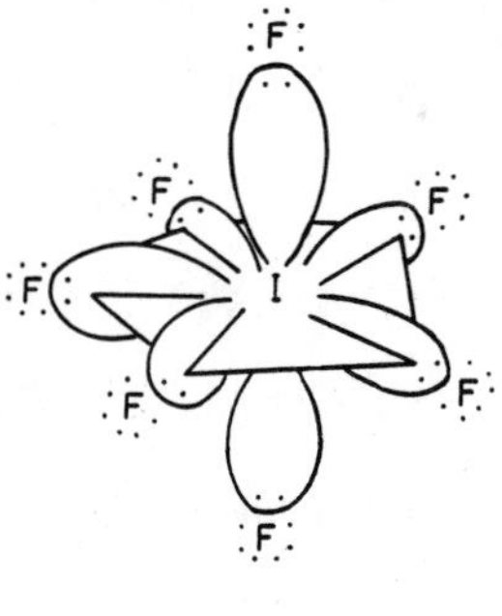

IF_7

24-32. **Refer to Section 24-9.** . . . .

Hydrogen bromide, HBr(g), is a colorless gas which dissolves in water to give hydrobromic acid, HBr(aq). The latter is a strong acid which completely dissociates in aqueous solutions giving H_3O^+(aq) and Br^-(aq).

24-34. **Refer to Section 24-10.** . . . .

Preparations of hydrogen halides:

(1) Direct combination of the elements: $H_2 + Br_2 \rightarrow 2HBr$

(2) Reaction of metal halide with nonvolatile acid: $NaI + H_3PO_4 \rightarrow NaH_2PO_4 + HI$

(3) Hydrolysis of nonmetal halide: $BCl_3 + 3H_2O \rightarrow H_3BO_3 + 3HCl$

(4) Halogenation of saturated hydrocarbons to give HCl as a byproduct:

$$C_3H_8 + Cl_2 \rightarrow C_3H_7Cl + HCl$$

24-36. **Refer to Section 24-11.** . . . .

Hydrofluoric acid is used to etch glass by reacting with the silicates in glass to produce a very volatile and thermodynamically stable compound, silicon tetrafluoride, SiF_4. For example,

$$CaSiO_3(s) + 6HF(aq) \rightarrow CaF_2(s) + SiF_4(g) + 3H_2O(\ell)$$

24-38. **Refer to Sections 9-9 and 24-13.** . . . .

The acid anhydride of perchloric acid, $HClO_4$, is explosive dichlorine heptoxide, Cl_2O_7. Recall that an acid anhydride reactions with water to produce the corresponding acid with no change in oxidation state. The reaction is:

$$Cl_2O_7 + H_2O \rightarrow 2HClO_4.$$

24-40. **Refer to Section 24-13 and Table 24-9.** . . . .

(a)	$KBrO_3$	potassium bromate	(b)	KBrO	potassium hypobromite
(c)	$NaClO_4$	sodium perchlorate	(d)	$NaClO_2$	sodium chlorite
(e)	HBrO	hypobromous acid	(f)	$HBrO_3$	bromic acid
(g)	HIO_3	iodic acid	(h)	$HClO_4$	perchloric acid

24-42. **Refer to Section 24-13.** . . . .

(a) $X_2 + H_2O \rightarrow HX + HOX$ (X = Cl, Br, I)

(b) $X_2 + 2NaOH \rightarrow NaX + NaOX + H_2O$ (X = Cl, Br, I)

(c) $Ba(ClO_2)_2 + H_2SO_4 \rightarrow BaSO_4 + 2HClO_2$

(d) $3NaOX \xrightarrow{\Delta} NaXO_3 + 2NaX$ (X = Cl, Br, I)

(e) $KOH + HClO_4 \rightarrow KClO_4 + H_2O$

(f) $NaClO_4 + H_2SO_4 \rightarrow NaHSO_4 + HClO_4$

24-44. **Refer to Section 24-7 and Table 24-6.** . . . .

Cyanogen, $(CN)_2$, is expected to add across the double bond of ethene in a manner similar to the diatomic halogens:

$$CH_2{=}CH_2 + (CN)_2 \rightarrow CH_2(CN){-}CH_2CN \quad \text{or} \quad N{\equiv}C{-}CH_2{-}CH_2{-}C{\equiv}N$$

24-46. **Refer to Section 10-13 and Example 10-90 Solution.** . . . .

From Graham's Law, we have: $\dfrac{\text{diffusion rate of } X_2}{\text{diffusion rate of } F_2} = [(MW_{F_2})/(MW_{X_2})]^{1/2}$

For Cl_2: diffusion rate of Cl_2/diffusion rate of F_2 = $(38.0/70.9)^{1/2}$ = **0.732**

For Br_2: diffusion rate of Br_2/diffusion rate of F_2 = $(38.0/159.8)^{1/2}$ = **0.488**

For I_2: diffusion rate of I_2/diffusion rate of F_2 = $(38.0/253.8)^{1/2}$ = **0.387**

For At_2: diffusion rate of At_2/diffusion rate of F_2 = $(38.0/420)^{1/2}$ = **0.301**

24-48. **Refer to Section 21-3 and Example 21-26.** . . . .

Plan: (1) Determine the moles of Cl_2 produced using the ideal gas equation, $PV = nRT$.
(2) Calculate the amount of NaCl that was electrolyzed and the amount of Na produced.
(3) Use Faraday's Law to calculate the time of cell operation.

(1) $$n = \frac{PV}{RT} = \frac{(774/760\text{ atm})(32700\text{ L})}{(0.0821\text{ L·atm/mol·K})(26°\text{C} + 273°)} = 1.36 \times 10^3 \text{ mol } Cl_2$$

(2) Balanced equation: $2NaCl \rightarrow 2Na + Cl_2$

$$? \text{ mol NaCl} = 1.36 \times 10^3 \text{ mol } Cl_2 \times \frac{2 \text{ mol NaCl}}{1 \text{ mol } Cl_2} = \mathbf{2.72 \times 10^3 \text{ mol NaCl}}$$

$$? \text{ g NaCl} = 2.72 \times 10^3 \text{ mol NaCl} \times 58.4 \text{ g/mol} = \mathbf{1.59 \times 10^5 \text{ g NaCl}}$$

$$? \text{ lb NaCl} = 1.59 \times 10^5 \text{ g NaCl} \times \frac{1 \text{ lb}}{454 \text{ g}} = \mathbf{3.50 \times 10^2 \text{ lb NaCl}}$$

$$? \text{ lb Na} = 1.36 \times 10^3 \text{ mol } Cl_2 \times \frac{2 \text{ mol Na}}{1 \text{ mol } Cl_2} \times \frac{23.0 \text{ g Na}}{1 \text{ mol Na}} \times \frac{1 \text{ lb Na}}{454 \text{ g Na}} = \mathbf{138 \text{ lb Na}}$$

(3) Balanced oxidation reaction: $2Cl^- \rightarrow Cl_2 + 2e^-$

Substituting into the Faraday's Law ratio:

$$\frac{1.36 \times 10^3 \text{ mol } Cl_2}{0.250 \text{ A} \times ? \text{ time (s)}} = \frac{1 \text{ mol } Cl_2}{2 \text{ mol } e^- \times 96500 \text{ C/mol } e^-}$$

$$? \text{ time (s)} = \mathbf{1.05 \times 10^9 \text{ s}} \quad \text{or} \quad \mathbf{33.3 \text{ years}}$$

24-50. **Refer to Section 24-6.** . . . .

Balanced equation: $2IO_3^- + 5HSO_3^- \rightarrow 3HSO_4^- + 2SO_4^{2-} + H_2O + I_2$

(1) Balanced reduction reaction: $12H^+ + 2IO_3^- + 10e^- \rightarrow I_2 + 6H_2O$

Therefore, 1 mole of I_2 is equal to 10 equivalents of I_2 for this reaction.

At the equivalence point,

$$\text{no. equivalents } HSO_3^- = \text{no. equivalents } I_2$$

$$0.150\ N \times V_{HSO_3^-} = 500 \text{ g } I_2 \times \frac{1 \text{ mol } I_2}{254 \text{ g } I_2} \times \frac{10 \text{ eq } I_2}{1 \text{ mol } I_2}$$

$$V_{HSO_3^-} = \mathbf{131 \text{ L}}$$

(2) $$? \text{ g } NaIO_3 = 500 \text{ g } I_2 \times \frac{1 \text{ mol } I_2}{254 \text{ g } I_2} \times \frac{2 \text{ mol } NaIO_3}{1 \text{ mol } I_2} \times \frac{198 \text{ g } NaIO_3}{1 \text{ mol } NaIO_3} = \mathbf{780 \text{ g } NaIO_3}$$

24-52. **Refer to Section 24-11.** . . . .

Balanced equation: $CaSiO_3 + 6HF \rightarrow CaF_2 + SiF_4 + 3H_2O$

Plan: (1) Determine the limiting reagent.
(2) Calculate the amount of SiF_4 produced.

(1) ? mol HF = 3.00 *M* × 0.100 L = 0.300 mol HF

? mol $CaSiO_3$ = (65.4 g)/(116 g/mol) = 0.564 mol $CaSiO_3$

If HF is the limiting reagent, then

$$? \text{ mol } CaSiO_3 = 0.300 \text{ mol HF} \times \frac{1 \text{ mol } CaSiO_3}{6 \text{ mol}} = 0.0500 \text{ mol } CaSiO_3$$

Since 0.564 mol $CaSiO_3$ are present (an excess), HF is the limiting reagent and all stoichiometric calculations must be based on the amount of HF.

(2) $$? \text{ g } SiF_4 = 0.300 \text{ mol HF} \times \frac{1 \text{ mol } SiF_4}{6 \text{ mol HF}} \times \frac{104 \text{ g } SiF_4}{1 \text{ mol } SiF_4} = \mathbf{5.20 \text{ g } SiF_4}$$

$$? V_{STP} \text{ } SiF_4 = 0.300 \text{ mol HF} \times \frac{1 \text{ mol } SiF_4}{6 \text{ mol HF}} \times \frac{22400 \text{ mL}_{STP} \text{ } SiF_4}{1 \text{ mol } SiF_4}$$

$$= \mathbf{1120 \text{ mL } SiF_4 \text{ at STP}}$$

24-54. **Refer to Section 24-13.** . . . .

Balanced disproportionation equation: $4KClO_3 \rightarrow 3KClO_4 + KCl$

$$? \text{ g } KClO_4 = 5.271 \text{ g } KClO_3 \times \frac{1 \text{ mol } KClO_3}{122.55 \text{ g } KClO_3} \times \frac{3 \text{ mol } KClO_4}{4 \text{ mol } KClO_3} \times \frac{138.55 \text{ g } KClO_4}{1 \text{ mol } KClO_4}$$

$$= \mathbf{4.469 \text{ g } KClO_4}$$

25 Nonmetallic Elements, Part II: Group VIA Elements

25-2. **Refer to the Introduction to Chapter 25.** . . . .

oxide, O^{2-} [Ne] or $1s^2\ 2s^2\ 2p^6$

sulfide, S^{2-} [Ar] or $1s^2\ 2s^2\ 2p^6\ 3s^2\ 3p^6$

selenide, Se^{2-} [Kr] or $1s^2\ 2s^2\ 2p^6\ 3s^2\ 3p^6\ 3d^{10}\ 4s^2\ 4p^6$

25-4. **Refer to the Introduction to Chapter 25.** . . . .

Every Group VIA element has six valence electrons in the highest energy level, and is therefore two electrons away from achieving an octet of electrons. This is why all of them exhibit an oxidation state of -2. In addition, Group VIA elements below oxygen can share their six valence electrons to various degrees with other elements to give different positive oxidation states up to +6. An oxidation state of -3 is impossible because it would require placing an electron into the next higher energy *d* or *s* orbitals. An oxidation state of +7 is impossible because the Group VIA elements only possess six valence electrons to share or to lose.

25-6. **Refer to the Introduction to Chapter 25 and Table 25-1.** . . . .

The fact that first ionization energies decrease from oxygen to tellurium is consistent with the order of increasing metallic character in the Group VIA elements. Metallic elements, such as polonium, have lower ionization energies and greater tendencies to form cations than do nonmetallic elements, such as oxygen and sulfur.

25-8. **Refer to Chapter 27 and the Exercises as stated.** . . . .

(a) H_2S (Exercise 7-46c)

(b) SF_6 (Exercise 7-46d, replacing Se with S)

(c) SF_4

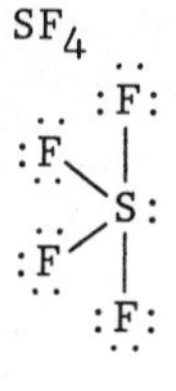

The Lewis dot formula predicts 5 regions of high electron density around the central sulfur atom, a trigonal bipyramidal electronic geometry and a see-saw molecular geometry. The S atom has sp^3d hybridization. The valence electrons available for bonding from F are in 2*p* orbitals and the three-dimensional structure is shown below.

(d) SO_2

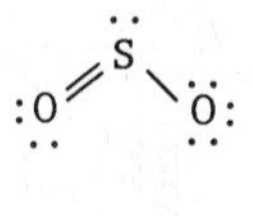

The Lewis dot formulas for the two resonance structures (one is shown) predict 3 regions of high electron density around the central S atom, a trigonal planar electronic geometry and an angular molecular geometry. The S atom has sp^2 hybridization. The valence electrons available for bonding from the O atoms are in 2*p* orbitals. A delocalized π bond is formed between the remaining *p* orbitals of S and O, as shown in Exercise 8-30a. The three-dimensional structure is shown below.

(e) SO_3

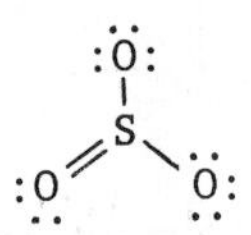

The Lewis dot formulas for the three resonance structures (one is shown) predict 3 regions of high electron density around the central S atom and trigonal planar electronic and molecular geometries. The S atom has sp^2 hybridization. The valence electrons available from the O atoms are in $2p$ orbitals. A delocalized π bond is formed between the remaining p orbitals of S and O, as shown in Exercise 8-30c. The three-dimensional structure is shown below.

SO_2

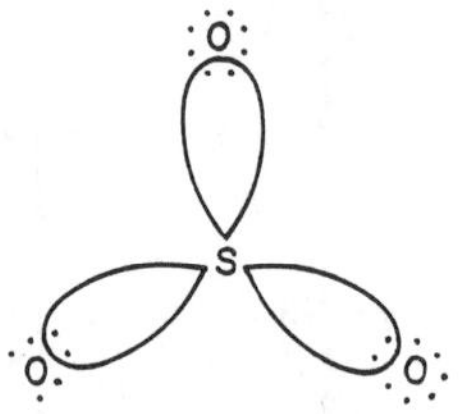

SO_3

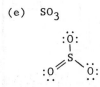

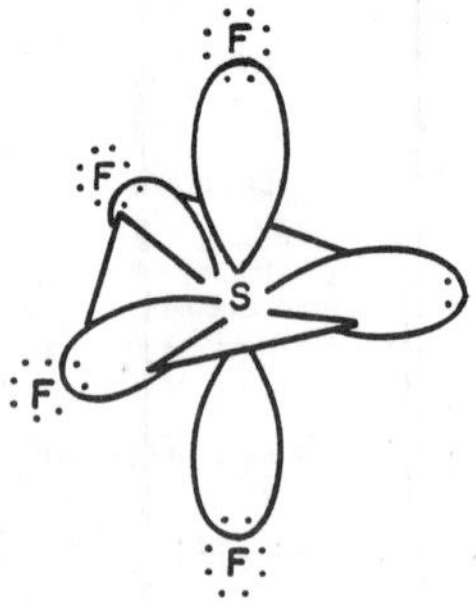

SF_4

Note: Pi bonding is not shown.

25-10. **Refer to Sections 25-2 and 25-8, and Table 25-2.** . . . .

(a) $E + 3F_2 \text{ (excess)} \rightarrow EF_6$ (E = S, Se, Te)

(b) $O_2 + 2H_2 \rightarrow 2H_2O$

$E + H_2 \rightarrow H_2E$ (E = S, Se, Te)

(c) $E + O_2 \rightarrow EO_2$ (E = S, Se, Te)

25-12. **Refer to Sections 11-9 and 25-3, Table 25-3 and Figure 11-6.** . . . .

The melting points, boiling points and heats of vaporization of the Group VIA hydrides all increase with the following order: $H_2S < H_2Se < H_2Te < H_2O$. Under normal conditions, H_2S, H_2Se and H_2Te are gases whereas H_2O is a liquid. The order observed with H_2S, H_2Se and H_2Te follows an order of increasing London dispersion forces. H_2O is out of line because of its extremely strong hydrogen bonding.

25-14. **Refer to Section 25-4.** . . . .

Group VIA hydrides dissociate in two stages. Their acid ionization constants, K_1 and K_2, are shown below:

		H_2S	H_2Se	H_2Te
$H_2E \rightleftharpoons H^+ + HE^-$	K_1:	1.0×10^{-7}	1.9×10^{-4}	2.3×10^{-3}
$HE^- \rightleftharpoons H^+ + E^{2-}$	K_2:	1.3×10^{-13}	$\sim 10^{-11}$	$\sim 1.6 \times 10^{-11}$

Acid strength increases upon descending the group: $H_2O < H_2S < H_2Se < H_2Te$. This results from the corresponding decrease in the average E-H bond energy.

25-16. **Refer to Section 25-7.** . . . .

Sulfur tetrafluoride, SF_4, reacts with water readily to produce HF which destroys lung tissue and SO_2, an irritant which causes respiratory problems.

$$SF_4(g) + 2H_2O(\ell) \rightarrow 4HF(g) + SO_2(g)$$

However, SF_6 is a stable compound that is inert towards H_2O. Although the reaction between SF_6 and H_2O is thermodynamically favorable, it does not occur within a reasonable time. This is an example of <u>kinetic</u> stability. SF_6 can actually be inhaled with oxygen for x-ray examinations of the lungs. This difference in reactivity is because, unlike SF_6, SF_4 has a lone pair of electrons on the sulfur atom that can be attacked by electron-seeking species, such as water. On the other hand, the relative chemical stability of SF_6 is due in part to the absence of nonbonding valence electrons around the central atom.

25-18. **Refer to Sections 25-8, 25-9 and 25-10.** . . . .

(a) $Cu(s) + 2H_2SO_4(\ell) \xrightarrow{\Delta} CuSO_4(aq) + SO_2(g) + 2H_2O(\ell)$

(b) $2SO_2(g) + O_2(g) \rightleftarrows 2SO_3(g)$ (in the presence of a catalyst)

(c) $SO_2(g) + H_2O(\ell) \rightleftarrows H_2SO_3(aq)$

25-20. **Refer to Section 25-9.** . . . .

Preparation of pyrosulfuric acid:

(1) $SO_3 + H_2SO_4 \rightarrow H_2S_2O_7$

(2) $2H_2SO_4 \rightarrow H_2S_2O_7 + H_2O$

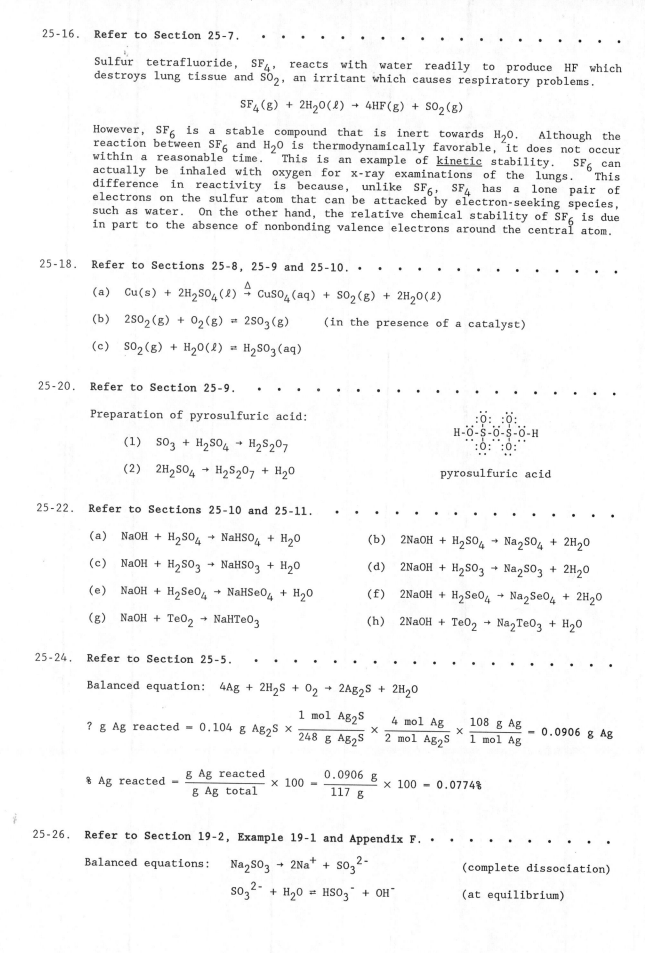

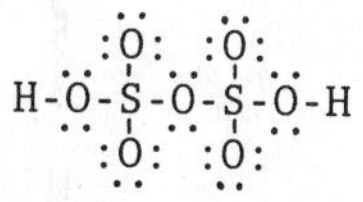

pyrosulfuric acid

25-22. **Refer to Sections 25-10 and 25-11.** . . . .

(a) $NaOH + H_2SO_4 \rightarrow NaHSO_4 + H_2O$

(b) $2NaOH + H_2SO_4 \rightarrow Na_2SO_4 + 2H_2O$

(c) $NaOH + H_2SO_3 \rightarrow NaHSO_3 + H_2O$

(d) $2NaOH + H_2SO_3 \rightarrow Na_2SO_3 + 2H_2O$

(e) $NaOH + H_2SeO_4 \rightarrow NaHSeO_4 + H_2O$

(f) $2NaOH + H_2SeO_4 \rightarrow Na_2SeO_4 + 2H_2O$

(g) $NaOH + TeO_2 \rightarrow NaHTeO_3$

(h) $2NaOH + TeO_2 \rightarrow Na_2TeO_3 + H_2O$

25-24. **Refer to Section 25-5.** . . . .

Balanced equation: $4Ag + 2H_2S + O_2 \rightarrow 2Ag_2S + 2H_2O$

$$? \text{ g Ag reacted} = 0.104 \text{ g } Ag_2S \times \frac{1 \text{ mol } Ag_2S}{248 \text{ g } Ag_2S} \times \frac{4 \text{ mol Ag}}{2 \text{ mol } Ag_2S} \times \frac{108 \text{ g Ag}}{1 \text{ mol Ag}} = \mathbf{0.0906 \text{ g Ag}}$$

$$\% \text{ Ag reacted} = \frac{\text{g Ag reacted}}{\text{g Ag total}} \times 100 = \frac{0.0906 \text{ g}}{117 \text{ g}} \times 100 = \mathbf{0.0774\%}$$

25-26. **Refer to Section 19-2, Example 19-1 and Appendix F.** . . . .

Balanced equations: $Na_2SO_3 \rightarrow 2Na^+ + SO_3^{2-}$ (complete dissociation)

$SO_3^{2-} + H_2O \rightleftarrows HSO_3^- + OH^-$ (at equilibrium)

Let x = $[SO_3^{2-}]_{hydrolyzed}$. Then, 0.10 - x = $[SO_3^{2-}]$ and x = $[HSO_3^-]$ = $[OH^-]$

	SO_3^{2-}	+	H_2O	$\rightleftharpoons$	HSO_3^-	+	OH^-
initial	0.10 *M*				0 *M*		~0 *M*
change	- x *M*				+ x *M*		+ x *M*
at equil	(0.10 - x) *M*				x *M*		x *M*

$$K_b = \frac{K_w}{K_{a_2}} = \frac{1.00 \times 10^{-14}}{6.2 \times 10^{-8}} = 1.6 \times 10^{-7} = \frac{[HSO_3^-][OH^-]}{[SO_3^{2-}]} = \frac{(x)(x)}{0.10 - x} \simeq \frac{x^2}{0.10}$$

Solving, x = 1.3×10^{-4}

Therefore, $[OH^-] = 1.3 \times 10^{-4}$ *M*; pOH = 3.89; pH = 10.11

26 Nonmetallic Elements, Part III: Group VA Elements

26-2. **Refer to the Introduction to Chapter 26 and Appendix B.** . . . .

N [He] $2s^2\ 2p^3$ P [Ne] $3s^2\ 3p^3$ As [Ar] $3d^{10}\ 4s^2\ 4p^3$

Sb [Kr] $4d^{10}\ 5s^2\ 5p^3$ Bi [Xe] $4f^{14}\ 5d^{10}\ 6s^2\ 6p^3$

N^{3-} [Ne] or $1s^2\ 2s^2\ 2p^6$ P^{3-} [Ar] or $1s^2\ 2s^2\ 2p^6\ 3s^2\ 3p^6$

26-4. **Refer to Section 26-20.** . . . .

PCl_3 and PCl_5 are covalent molecules because the P-Cl bond is a polar covalent bond. This is predicted since the difference in electronegativities of the nonmetals, P and Cl, is only 0.9 units (EN of P = 2.1 and EN of Cl = 3.0).

PCl_3

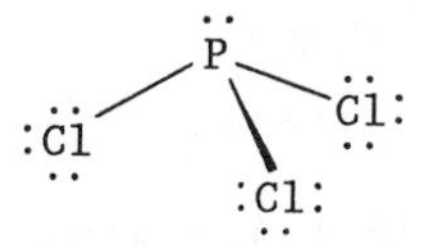

pyramidal

PCl_5

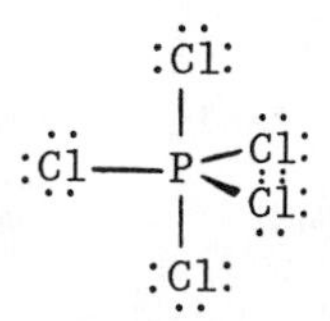

trigonal bipyramidal

26-6. **Refer to Section 8-4 and Exercise 8-28 Solution.** . . . .

Compounds of phosphorus and arsenic are different from nitrogen compounds for two reasons:

(1) Atoms of phosphorus and arsenic can use empty 3*d* and 4*d* orbitals, respectively, to form sp^3d or sp^3d^2 hybrid orbitals. They can therefore form compounds with trigonal bipyramidal or octahedral electronic geometries. Nitrogen atoms, on the other hand, do not have access to *d* orbitals and its compounds almost always obey the octet rule.

(2) Phosphorus and arsenic atoms are much larger than nitrogen atoms. Therefore, in a compound, their *p* orbitals are too far away from the *p* orbitals on adjacent atoms to overlap side-on to form pi bonds. Thus, phosphorus and arsenic only bond with single, sigma bonds to other atoms. Nitrogen atoms readily form pi bonds in its compounds.

26-8. **Refer to Section 26-2.** . . . .

The nitrogen cycle is the complex series of reactions by which nitrogen is slowly but continually recycled in the atmosphere, lithosphere (earth) and hydrosphere (water). Atmospheric nitrogen is made accessible to us and other life-forms in mainly two ways. (1) A class of plants, called legumes, have bacteria which extract N_2 directly, converting it to NH_3. This nitrogen fixation process, catalyzed by an enzyme produced by the bacteria, is highly efficient at usual temperatures and pressures. (2) N_2 and O_2 react in the atmosphere near lightning, forming NO and NO_2, which dissolve in rainwater and fall to earth. These nitrogen compounds are adsorbed and incorporated into plants forming amino acids and proteins. The plants are eaten by animals or die and decay, releasing their nitrogen to the environment. The animals, in turn, excrete waste and/or die, releasing their nitrogen to the environment.

26-10. **Refer to Section 26-3 and Table 26-2.** . . . .

	Oxidation No. of N		Oxidation No. of N
(a) N_2	0	(b) NO	+2
(c) N_2O_4	+4	(d) HNO_3	+5
(e) HNO_2	+3		

26-12. **Refer to Section 26-1 and Appendix K.** . . . .

The bond energy of N≡N is equal to ΔH^o_{rxn} for: $N_2(g) \rightarrow 2N(g)$

$$\Delta H^o_{rxn} = 2\Delta H^o_{f,N(g)} - \Delta H^o_{f,N_2(g)} = (2\text{ mol})(472.704\text{ kJ/mol}) - (1\text{ mol})(0\text{ kJ/mol})$$

$$= \mathbf{945.408\ kJ}$$

The 945 kJ value indicates that the nitrogen-nitrogen triple bond is very strong. Since the N≡N bond cannot easily be broken, the reactions of N_2 have high activation energies and N_2 is not very reactive. Because of the chemical inertness of N_2, the concentration of N_2 in the atmosphere is very high and the concentrations of N-containing compounds in the environment are relatively low.

26-14. **Refer to Tables 7-2 and 7-3, and the Sections as stated.** . . . .

(a) NH_2Br

H:N:Br:
  H

The Lewis dot formula predicts 4 regions of high electron density around the central N atom, a tetrahedral electronic geometry and a pyramidal molecular geometry. The N atom has sp^3 hybridization **(Section 26-5)**. The three-dimensional structure is shown below.

(b) HN_3

H:N::N::N:

Around the outer two N atoms, the Lewis dot formula predicts 3 regions of high electron density, a trigonal planar electronic geometry and a bent molecular geometry; each of the two N atoms has sp^2 hybridization. The Lewis dot formula also predicts 2 regions of high electron density around the central N atom and a linear electronic and molecular geometry; the N atom has *sp* hybridization **(Section 26-3)**. The three-dimensional structure is shown below.

(c) N_2O_2

:O::N:N::O:

The Lewis dot formula predicts 3 regions of high electron density, trigonal planar electronic geometry and angular molecular geometry around each N atom. The N atoms have sp^2 hybridization **(Section 26-9)**. The three-dimensional structure is shown below.

(d) NO_2^+

:O::N::O:$^+$

The Lewis dot formula predicts 2 regions of high electron density, a linear electronic and ionic geometry around the N atom and *sp* hybridization for the N atom **(Section 26-11)**.

NO_3^-

:O::N:O:$^-$
  :O:

The Lewis dot formulas for the three resonance structures (one is shown) predicts 3 regions of high electron density around the central N atom and a trigonal planar electronic and ionic geometry. The N atom has sp^2 hybridization **(Section 26-11)**.

The three dimensional structures for the NO_2^+ and NO_3^- ions are shown below.

(e) HNO_3

H:O:N::O:
  :O:

The Lewis dot formulas for the two resonance structures (one is shown) predicts 3 regions of high electron density for the N atom, and a trigonal planar electronic and molecular geometry about the N atom. The N atom has sp^2 hybridization **(Section 26-14)**. The three-dimensional structure is shown.

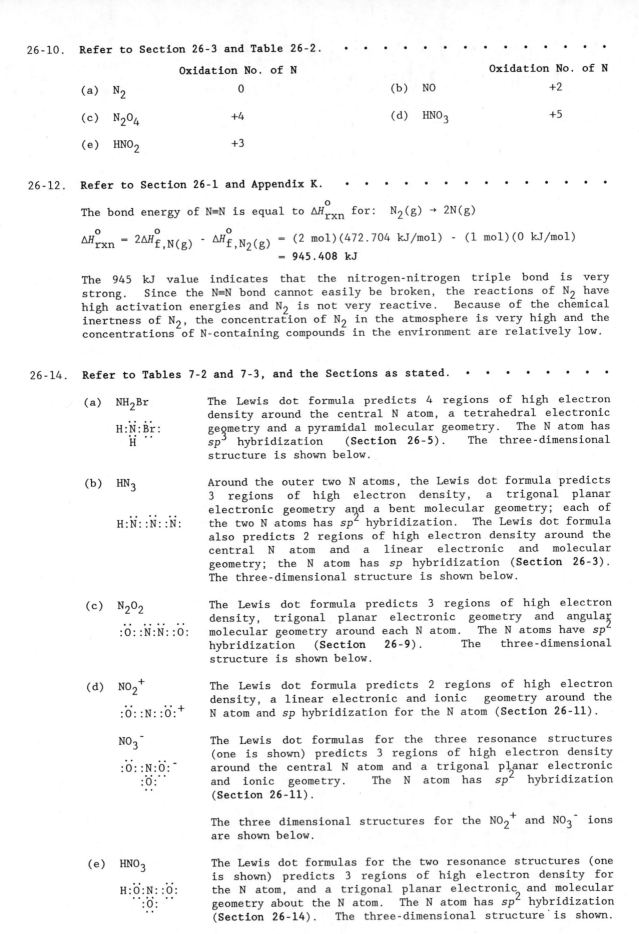

(f) NO_2^-

:O::N:O: $^-$

The Lewis dot formulas for the two resonance structures (one is shown) predicts 3 regions of high electron density for the central N atom, and a trigonal planar electronic geometry and a bent ionic geometry. The N atom has sp^2 hybridization. The three-dimensional structure is shown below.

(a) NH_2Br

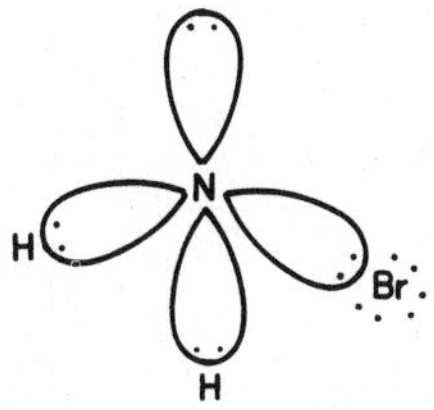

(b) HN_3

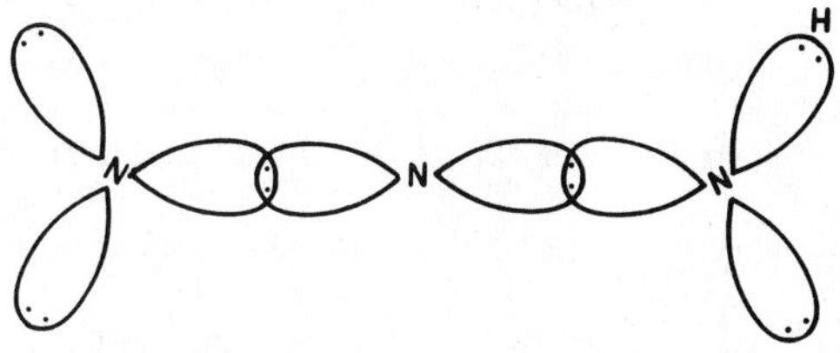

(c) N_2O_2

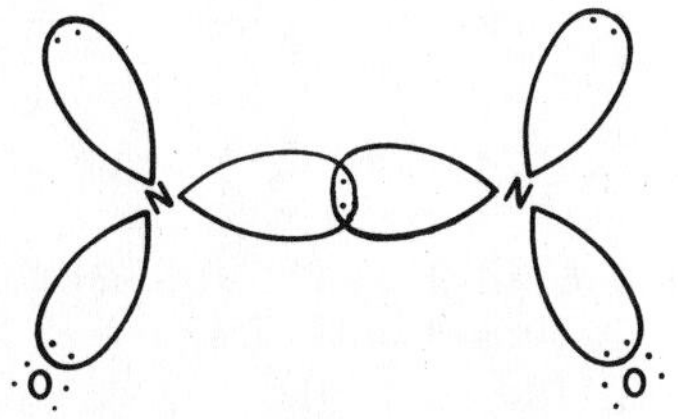

(d) NO_2^+

NO_3^-

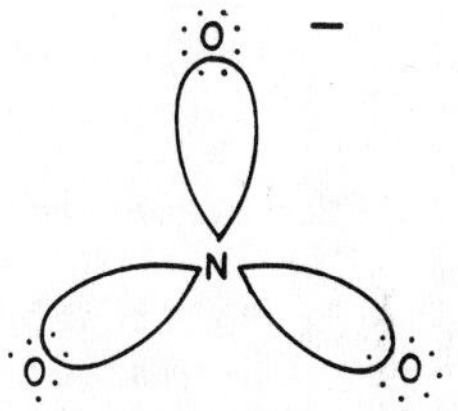

(e) HNO_3

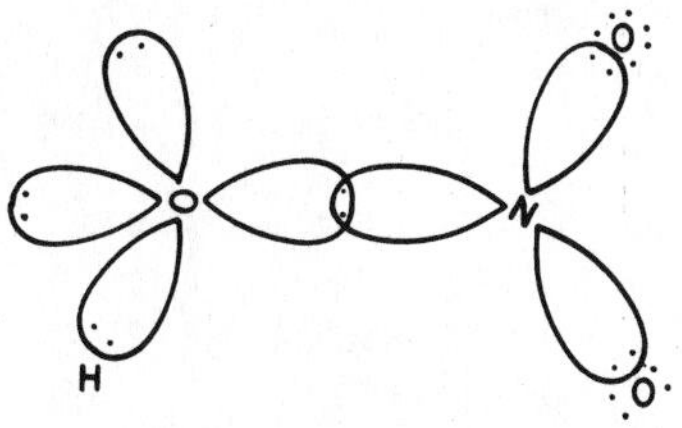

(f) NO_2^-

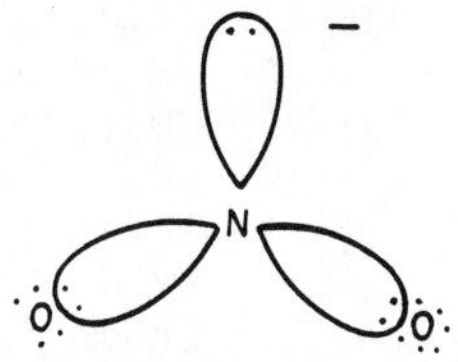

Note: Pi bonding is not shown.

26-16. **Refer to the Sections as stated.** . . . .

(a) $2KN_3(s) \xrightarrow{\Delta} 2K(\ell) + 3N_2(g)$ **(Section 26-3)**

(b) $NH_3(g) + HCl(g) \rightarrow NH_4Cl(s)$ **(Section 26-4)**

(c) $NH_3(g) + HCl(aq) \rightarrow NH_4Cl(aq)$ **(Section 26-4)**

(d) $2NH_4NO_3(s) \xrightarrow{\Delta} 2N_2(g) + 4H_2O(g) + O_2(g)$ **(Section 26-4)**

(e) $4NH_3(g) + 5O_2(g) \xrightarrow[Pt]{\Delta} 4NO(g) + 6H_2O(g)$ **(Section 26-4)**

(f) $2N_2O(g) \xrightarrow{\Delta} 2N_2(g) + O_2(g)$ **(Section 26-7)**

(g) $3NO_2(g) + H_2O(\ell) \rightarrow 2HNO_3(aq) + NO(g)$ **(Section 26-14)**

26-18. **Refer to Section 26-4.** . . . .

$Co^{2+}(aq) + 6NH_3(aq) \rightarrow [Co(NH_3)_6]^{2+}(aq)$

$BF_3(g) + :NH_3(g) \rightarrow F_3B:NH_3(s)$

26-20. **Refer to Exercise 7-32 Solution and the Sections as stated.** . . . .

	Molecule	Structure	Polarity of Molecule	
(a)	NH_3	pyramidal	polar	(Section 7-6)
(b)	NH_2Cl	distorted pyramidal	polar	(Section 7-6)
(c)	NO	heteronuclear diatomic	polar	(Section 6-3)
(d)	N_2H_4	unsymmetric	polar	(Section 26-6)
(e)	HNO_3	unsymmetric	polar	(Section 26-14)
(f)	PH_3	pyramidal	polar	(Section 26-19)
(g)	As_4O_6	unsymmetric	polar	(Section 26-21)

26-22. **Refer to Section 26-8.** . . . .

Nitrogen oxide, NO, is very reactive because each NO molecule contains an unpaired electron. Note: Molecules with unpaired electrons are called radicals.

26-24. **Refer to the Sections as stated.** . . . .

When an acid anhydride reacts with water, the corresponding acid is formed. There is no change in oxidation number of the elements.

(a) $N_2O_5(s) + H_2O(\ell) \rightarrow 2HNO_3(aq)$ (Section 26-11)

(b) $N_2O_3(s) + H_2O(\ell) \rightarrow 2HNO_2(aq)$ (Section 26-9)

(c) $P_4O_{10}(s) + 6H_2O(\ell) \rightarrow 4H_3PO_4(aq)$ (Section 26-21)

(d) $P_4O_6(s) + 6H_2O(\ell) \rightarrow 4H_3PO_3(aq)$ (Section 26-21)

26-26. **Refer to Section 26-14.** . . . .

$Cu(s) + 4HNO_3(aq) \rightarrow Cu(NO_3)_2(aq) + 2NO_2(g) + 2H_2O(\ell)$

$3Zn(s) + 8HNO_3(aq) \rightarrow 3Zn(NO_3)_2(aq) + 2NO(g) + 4H_2O(\ell)$

$P_4(s) + 20\ HNO_3(aq) \rightarrow 4H_3PO_4(aq) + 20\ NO_2(g) + 4H_2O(\ell)$

$S(s) + 6HNO_3(aq) \rightarrow H_2SO_4(aq) + 6NO_2(g) + 2H_2O(\ell)$

26-28. **Refer to Section 26-15.** . . . .

The function of sodium nitrite, $NaNO_2$, as a food additive is two-fold: (1) it inhibits the oxidation of blood, preventing the discoloring of red meat, and (2) it prevents the growth of botulism bacteria. There is now some controversy regarding this food additive because nitrites are suspected of combining with amines under the acidic conditions of the stomach to produce carcinogenic nitrosoamines.

26-30. **Refer to Sections 2-9 and 19-3.** . . . .

(a) $\% \text{ N in } NH_4NO_3 = \dfrac{28.0 \text{ g N}}{80.0 \text{ g } NH_4NO_3} \times 100 = \mathbf{35.0\% \text{ N}}$

$\% \text{ N in } NaNO_3 = \dfrac{14.0 \text{ g N}}{85.0 \text{ g } NaNO_3} \times 100 = \mathbf{16.5\% \text{ N}}$

Therefore, NH_4NO_3 contains a greater percentage of nitrogen by mass.

(b) Balanced equations: $NH_4NO_3 \rightarrow NH_4^+ + NO_3^-$ (to completion)

$NH_4^+ + H_2O \rightleftharpoons NH_3 + H_3O^+$ (in equilibrium)

$$K_a = \frac{[NH_3][H_3O^+]}{[NH_4^+]} = \frac{K_w}{K_b} = \frac{1.00 \times 10^{-14}}{1.8 \times 10^{-5}} = 5.6 \times 10^{-10}$$

Let x = $[NH_4^+]_{hydrolyzed}$. Then, $0.10 - x = [NH_4^+]$ and $x = [NH_3] = [H_3O^+]$

Substituting,

$$5.6 \times 10^{-10} = \frac{x^2}{0.10 - x} \approx \frac{x^2}{0.10}$$ Solving, $x = 7.5 \times 10^{-6}$

Therefore, $[H_3O^+] = 7.5 \times 10^{-6}$ *M*; **pH = 5.12**

26-32. **Refer to Section 26-12 and Appendix K.** . . . .

(1) Bond energy of N≡N is equal to ΔH^o for the reaction: $N_2(g) \rightarrow 2N(g)$

$$\Delta H^o_{rxn} = 2\Delta H^o_{f,N(g)} - \Delta H^o_{f,N_2(g)} = (2 \text{ mol})(472.7 \text{ kJ/mol}) - (1 \text{ mol})(0 \text{ kJ/mol}) = \mathbf{945.4 \text{ kJ}}$$

Bond energy of O=O is equal to ΔH^o for the reaction: $O_2(g) \rightarrow 2O(g)$

$$\Delta H^o_{rxn} = 2\Delta H^o_{f,O(g)} - \Delta H^o_{f,O_2(g)} = (2 \text{ mol})(249.2 \text{ kJ/mol}) - (1 \text{ mol})(0 \text{ kJ/mol}) = \mathbf{498.4 \text{ kJ}}$$

Bond energy of N=O is equal to ΔH^o for the reaction: $NO(g) \rightarrow N(g) + O(g)$

$$\Delta H^o_{rxn} = [\Delta H^o_{f,N(g)} + \Delta H^o_{f,O(g)}] - [\Delta H^o_{f,NO(g)}]$$
$$= [(1 \text{ mol})(472.7 \text{ kJ/mol}) + (1 \text{ mol})(249.2 \text{ kJ/mol})] - [(1 \text{ mol})(90.25 \text{ kJ/mol})]$$
$$= \mathbf{631.6 \text{ kJ}}$$

It is seen that the bond energy of N=O is somewhere between those of N≡N and O=O.

(2) For the reaction, $N_2(g) + O_2(g) \rightleftharpoons 2NO(g)$, the heat of reaction can be calculated as follows:

$$\Delta H_{rxn} = \Sigma \text{ B.E.}_{reactants} - \Sigma \text{ B.E.}_{products}$$
$$= (945.4 \text{ kJ} + 498.4 \text{ kJ}) - 2(631.6 \text{ kJ}) = 180.6 \text{ kJ}$$

This indicates that the energy required to break one mole of N≡N bonds and one mole of O=O bonds is greater than the energy that is released when two moles of N=O bonds are formed. Since the reaction is endothermic, it is expected to have a high activation energy and it should not go appreciably at room temperature.

26-34. **Refer to Section 3-4 and Exercise 3-46 Solution.** . . . .

Balanced equation: $3NaNH_2 + NaNO_3 \rightarrow NaN_3 + 3NaOH + NH_3$

Plan: (1) Solve for the theoretical yield of NaN_3.
(2) Calculate the mass of $NaNH_2$ required to produce the theoretical yield of NaN_3.

(1) $$\% \text{ yield} = \frac{\text{actual yield}}{\text{theoretical yield}} \times 100$$

Substituting,

$$75\% = \frac{22.0 \text{ g}}{\text{theoretical yield}} \times 100$$ And, theoretical yield = 29 g NaN_3

(2) $$? \text{ g } NaNH_2 = 29 \text{ g } NaN_3 \times \frac{1 \text{ mol } NaN_3}{65 \text{ g } NaN_3} \times \frac{3 \text{ mol } NaNH_2}{1 \text{ mol } NaN_3} \times \frac{39 \text{ g } NaNH_2}{1 \text{ mol } NaNH_2}$$

$$= \mathbf{52 \text{ g } NaNH_2}$$

26-36. **Refer to Section 26-16.** . . . .

Balanced equation: $Ca_3(PO_4)_2 + 2H_2SO_4 + 4H_2O \rightarrow [Ca(H_2PO_4)_2 + 2(CaSO_4 \cdot 2H_2O)]$
superphosphate of lime

$$? \text{ g superphosphate of lime} = 32.7 \text{ g rock} \times \frac{66.3 \text{ g } Ca_3(PO_4)_2}{100 \text{ g rock}} \times \frac{1 \text{ mol } Ca_3(PO_4)_2}{310 \text{ g } Ca_3(PO_4)_2}$$

$$\times \frac{1 \text{ mol superphosphate}}{1 \text{ mol } Ca_3(PO_4)_2} \times \frac{578 \text{ g superphosphate}}{1 \text{ mol superphosphate}}$$

$$= \mathbf{40.4 \text{ g}}$$

26-38. **Refer to Section 2-9.** . . . .

Let us assume we have 1 mole of each compound. Then

$$\% \text{ P in } P_4O_{10} = \frac{124 \text{ g P}}{284 \text{ g } P_4O_{10}} \times 100 = \mathbf{43.7\% \text{ P}}$$

$$\% \text{ P in } H_3PO_4 = \frac{31.0 \text{ g P}}{98.0 \text{ g } H_3PO_4} \times 100 = \mathbf{31.6\% \text{ P}}$$

27 Nonmetallic Elements, Part IV: Carbon, Silicon, and Boron

27-2. **Refer to the Introduction to Chapter 27, Sections 27-3, 27-9, 27-13 and 23-2.** •

Carbon dioxide, CO_2, and silicon dioxide, SiO_2, are both Group IVA oxides. However, at room temperature and pressure, carbon dioxide is a colorless, odorless gas, while silicon dioxide exists in either of two solid forms: as quartz, a mineral which occurs in sand, and as flint, an uncrystallized amorphous type of silica. Carbon dioxide consists of discrete nonpolar molecules, and so has only weak London dispersion forces between its molecules. The carbon atom is able to fulfill its octet of electrons by forming double bonds with oxygen, thereby creating a molecule. Due to its larger size, a silicon atom cannot form double bonds with oxygen; the *p* orbitals on the Si are simply too far from the *p* orbitals on adjacent oxygen atoms to effectively overlap side-on. Therefore, silicon obtains its octet of electrons by forming four single bonds with neighboring oxygen atoms. Silicon dioxide is properly represented as $(SiO_2)_n$, because it is actually a polymeric solid of SiO_4 tetrahedra sharing all oxygen atoms among surrounding tetrahedra.

Using the concept of diagonal similarity, the structures and properties of SiO_2 and B_2O_3 should be similar. Both oxides are solids, with B_2O_3 being a colorless, transparent, glass-like solid. Boron, however, is a IIIA element and often does not have a complete octet of electrons in its compounds. The structure and bonding for B_2O_3 is not well defined.

27-4. **Refer to Section 27-1 and Figure 11-25.** • • • • • • • • • • • • • • •

The structures of graphite and diamond, given in the figure below, can explain the properties of these two allotropes of carbon. At ordinary temperatures and pressures, graphite is more stable than diamond. Graphite is soft, black, slippery, is less dense than diamond and a reasonably good conductor of electricity. Diamond is hard, colorless, a poor conductor of electricity, and is more dense than graphite.

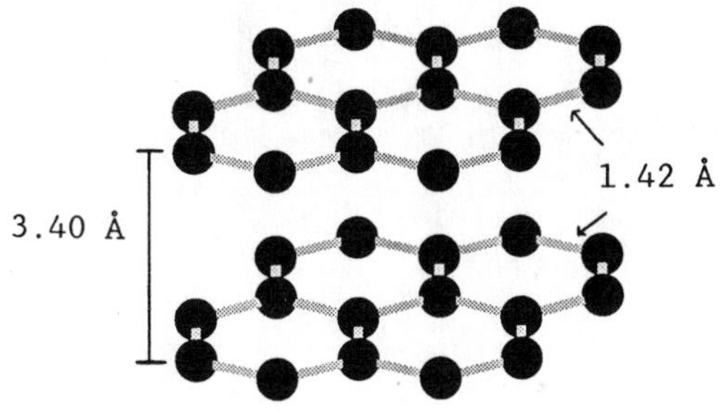

graphite

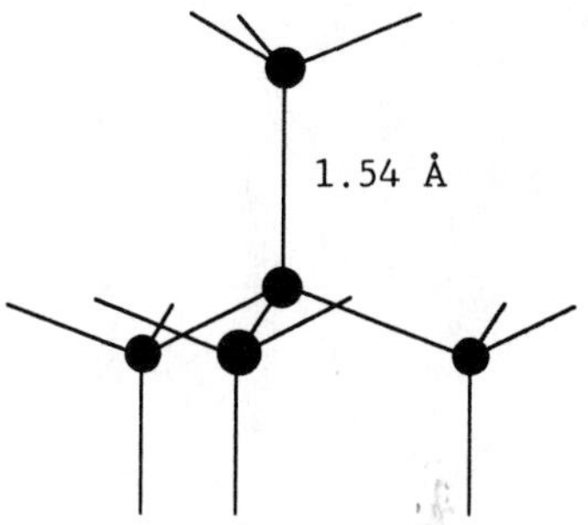

diamond

In graphite, the sp^2 hybridized carbon atoms form planar layers composed of 6-membered rings. Each carbon atom is at the center of a trigonal planar arrangement. Layers easily slide past one another, since the only forces holding the layers together are weak London dispersion forces, making graphite a good lubricant. Only 3 of the 4 valence electrons of carbon are tied up in sigma bonds; the fourth electron is in an extended pi bonding system, and is quite mobile, which accounts for graphite's ability to conduct electricity. Diamond, on the other hand, consists of a network of tetrahedral sp^3 hybridized carbon atoms. All of the valence electrons are in sigma bonds; there are no mobile electrons. This accounts for diamond's hardness, poor conductivity, and very high melting point.

27-6.   **Refer to Sections 22-3 and 27-1.** • • • • • • • • • • • • • • • •

Coke is the impure carbon residue left when coal or petroleum is destructively distilled to release valuable volatile hydrocarbons. It is important industrially because it can be used to reduce metal oxides in metallurgy. It can also be converted to graphite for use as electrodes by the Acheson process, in which a large electric current is passed through a pressed rod of coke for several days to heat it sufficiently to recrystallize the carbon in the form of graphite.

27-8.   **Refer to Sections 27-3 and 27-4, and Tables 7-2 and 7-3.** • • • • • • • • •

(a) CO_2

:Ö::C::Ö:

The Lewis dot formula predicts 2 regions of high electron density around the central C atom and a linear electronic and molecular geometry. The C atom has *sp* hybridization. The three-dimensional structure is shown below.

(b) H_2CO_3

H:Ö:C:Ö:H

:Ö:

The Lewis dot formula predicts 3 regions of high electron density around the central C atom, a trigonal planar electronic geometry and a trigonal planar arrangement of atoms about the C atom. The C atom has sp^2 hybridization. The three-dimensional structure is shown below.

(c) CO_3^{2-}

:Ö::C:Ö:$^{2-}$

:Ö:

The Lewis dot formulas for the three resonance structures (only one is shown) predict 3 regions of high electron density around the central C atom and a trigonal planar electronic and ionic geometry. The C atom has sp^2 hybridization. The three-dimensional structure is shown below.

(a) CO_2

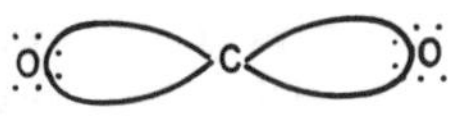

(b) H_2CO_3

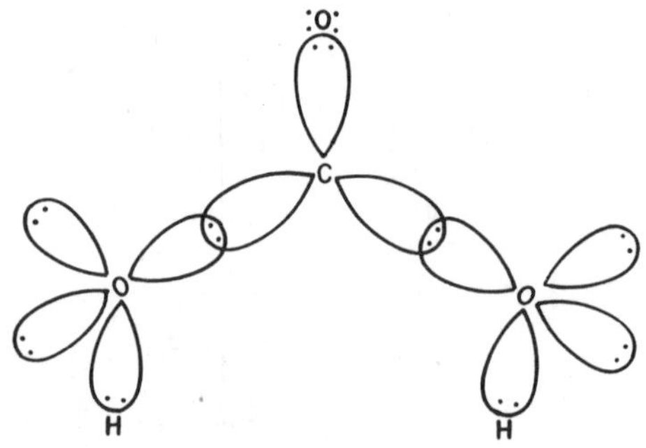

(c) CO_3^{2-}

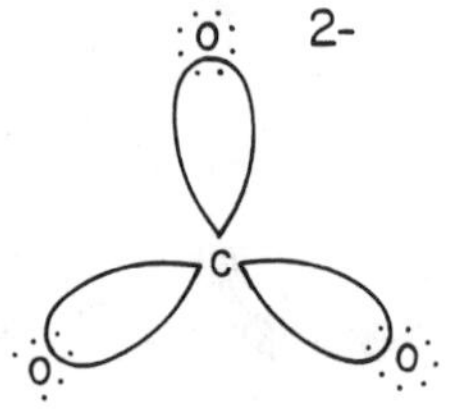

Note: No pi bonding is shown.

27-10.   **Refer to Sections 5-4 and 5-5.** • • • • • • • • • • • • • • • • • •

Simple, single-element ions with a charge of 4+ or 4- are rare because it is energetically unfeasible to either remove 4 electrons successively from or add 4 electrons successively to any element. The ionization energy required to remove an electron from an ion with a 3+ charge is so endothermic that the likelihood of the process occurring is very small. In the same manner, the electron affinity required to add an electron to an ion with a 3- charge is so endothermic that the probability of forming a simple 4- charged ion is very small.

27-12. **Refer to Section 27-3.** . . . .

(a) $CO(g) + 2H_2(g) \xrightarrow[\Delta,\ pressure]{catalyst} CH_3OH(\ell)$

(b) $CO(g) + 3H_2(g) \xrightarrow[250°C]{catalyst} H_2O(g) + CH_4(g)$

(c) $2Co(s) + 8CO(g) \rightarrow Co_2(CO)_8(s)$

27-14. **Refer to Section 27-4.** . . . .

Limestone caves are formed when ground water containing dissolved CO_2, in the form of H_2CO_3, selectively dissolves $CaCO_3$ (limestone), forming soluble $Ca(HCO_3)_2$, as shown in Reaction (1). Stalactites and stalagmites are formed very slowly on the ceilings and floors, respectively, of these caves when Reaction (1) reverses itself, giving Reaction (2). The driving force for (2) is the degassing of dissolved CO_2 and the evaporation of water.

(1) $CaCO_3(s) + H_2CO_3(aq) \rightarrow Ca(HCO_3)_2(aq)$

(2) $Ca(HCO_3)_2(aq) \rightarrow CaCO_3(s) + CO_2(g) + H_2O(g)$

27-16. **Refer to Section 27-5.** . . . .

Carbon tetrachloride, CCl_4, is effective in removing greasy stains from clothes and has been used as a dry cleaning agent because it is an excellent nonpolar solvent. However, it is toxic to the liver, a known carcinogen and has a high vapor pressure, and its use is now illegal.

27-18. **Refer to Section 27-4.** . . . .

Balanced equation: $Li_2CO_3(s)\ [+ \underset{impurity}{Li_2O(s)}] \xrightarrow{\Delta} Li_2O(s) + CO_2(g)$

Plan: (1) Determine the number of moles of CO_2 using the ideal gas law, $PV = nRT$.

(2) Calculate the mass and % by mass of Li_2CO_3 directly from the moles of CO_2 formed.

(1) $$n = \frac{PV}{RT} = \frac{[(747/760)\text{ atm}](1.644\text{ L})}{(0.0821\text{ L·atm/mol·K})(32°C + 273°)} = 0.0645\text{ mol }CO_2$$

(2) $$?\text{ g }Li_2CO_3 = 0.0645\text{ mol }CO_2 \times \frac{1\text{ mol }Li_2CO_3}{1\text{ mol }CO_2} \times \frac{73.9\text{ g }Li_2CO_3}{1\text{ mol }Li_2CO_3} = \mathbf{4.77\text{ g }Li_2CO_3}$$

$$\%\ Li_2CO_3 = \frac{4.77\text{ g }Li_2CO_3}{6.617\text{ g mixture}} \times 100 = \mathbf{72.1\%}$$

$$\%\ Li_2O = 100\% - 72.1\% = \mathbf{27.9\%}$$

27-20. **Refer to Sections 27-6 and 22-4.** . . . .

(1) Elemental silicon, Si, is usually prepared by the high-temperature reduction of silica (sand) with coke (impure carbon). An excess of SiO_2 prevents the formation of SiC.

$$SiO_2(s,excess) + 2C(s) \xrightarrow{\Delta} Si(s) + 2CO(g)$$

(2) Ultrapure silicon is prepared by reducing a tetrahalide of silicon in the vapor phase with an active metal such as Na. The tetrahalide first must be distilled carefully to remove impurities of boron, aluminum and arsenic halides.

$$SiCl_4(g) + 4Na(\ell) \rightarrow 4NaCl(s) + Si(s)$$

The sodium chloride is dissolved out of the product mixture with hot water, leaving behind relatively pure silicon, which is melted and cast into bars.

(3) The silicon can be further purified by a process called zone refining. An induction heater surrounds a bar of the impure solid and passes slowly from one end to the other (Figure 22-4a). As it passes, it melts a portion of the bar, which slowly recrystallizes into a more perfect, hence purer crystal, as the heating element moves away. The impurities are concentrated in the melt and most are carried away. The impurities do not fit into the lattice as easily as the silicon and most of them are carried along in the molten portion until the induction heater reaches the end. By repeating the process several times, silicon with less than one part per billion impurities can be obtained.

27-22. **Refer to Sections 27-7 and 27-8.** . . . .

(a) $Si(s) + 2NaOH(aq) + H_2O(\ell) \rightarrow Na_2SiO_3(aq) + 2H_2(g)$

(b) $Si(s) + 2Br_2(\ell) \rightarrow SiBr_4(\ell)$

(c) $SiCl_4(\ell) + 4H_2O(\ell) \rightarrow H_4SiO_4(s) + 4HCl(aq)$

(d) $CCl_4(\ell) + 2H_2O(\ell) \rightarrow CO_2(aq) + 4HCl(aq)$ (Note: too slow to be observed)

27-24. **Refer to Section 27-11 and Table 27-2.** . . . .

Fused sodium silicate, Na_2SiO_3, and calcium silicate, $CaSiO_3$, are the basic ingredients of glass. They are produced by heating a mixture of Na_2CO_3 and $CaCO_3$ with sand (silica) until it melts at about 700°C, forming the corresponding silicates and carbon dioxide.

$$[CaCO_3 + SiO_2](\ell) \xrightarrow{\Delta} CaSiO_3(\ell) + CO_2(g)$$

$$[Na_2CO_3 + SiO_2](\ell) \xrightarrow{\Delta} Na_2SiO_3(\ell) + CO_2(g)$$

Under carefully controlled conditions, clear and colorless "soda lime" glass results. Compounds can be added in small amounts to add color to glass. Examples include CoO (blue), MnO_2 (violet) and colloidal Se (red).

27-26. **Refer to Section 27-8 and Appendix K.** . . . .

$Si(s) + 2F_2(g) \rightarrow SiF_4(g)$ $\Delta G^o_{rxn} = \Delta G^o_{f,SiF_4(g)} = -1573$ kJ

$Si(s) + 2Cl_2(g) \rightarrow SiCl_4(g)$ $\Delta G^o_{rxn} = \Delta G^o_{f,SiCl_4(g)} = -617.0$ kJ

Therefore, the reaction between Si and F_2 to form SiF_4 has the more negative ΔG^o and is the more spontaneous reaction at 25°C; in fact, it is explosive.

27-28. **Refer to Section 27-13.** . . . .

$$2H_3BO_3(\ell) \xrightarrow{\Delta} B_2O_3(\ell) + 3H_2O(g)$$

$$\underset{\text{orthoboric acid}}{4H_3BO_3(s)} \xrightarrow[-4H_2O]{\Delta} \underset{\text{metaboric acid}}{4HBO_2(s)} \xrightarrow[-H_2O]{\Delta} \underset{\text{tetraboric acid (pyroboric acid)}}{H_2B_4O_7(s)} \xrightarrow[-H_2O]{\Delta} \underset{\text{boric oxide}}{2B_2O_3(\ell)}$$

27-30. Refer to Sections 13-12 and 27-17. . . . .

(1) $BCl_3 + NH_3 \rightarrow H_3NBCl_3$
Lewis acid, Lewis base

(2) $BCl_3 + Cl^- \rightarrow BCl_4^-$
Lewis acid, Lewis base

27-32. Refer to Sections 27-15, 27-16 and 27-17. . . . .

(1) Boron forms B-B-B three-center two-electron bonds. The two electrons are delocalized over a molecular orbital resulting from the overlap of three pseudo-*sp* hybrid orbitals, one orbital from each B atom, as illustrated below.

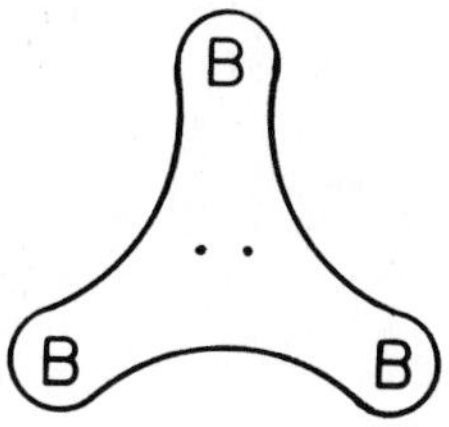

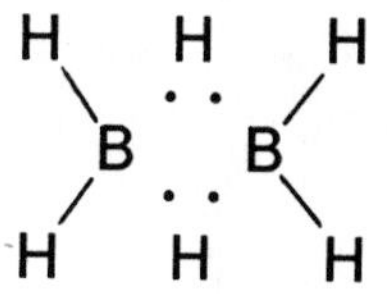

(2) Boron also forms B-H-B three-center two-electron bonds in B_2H_6. In this compound, only four electrons can be allocated to form the two B-H and the one B-B bonds. It is thought that each B-H-B bridge is held together by a three-center two electron bond as illustrated above.

27-34. Refer to Sections 18-10 and 27-13, Example 18-18, and Exercise 18-88 Solution. •

Note: $B(OH)_3$ can be written as H_3BO_3.

Balanced equations:

$H_3BO_3 + H_2O \rightleftharpoons H_2BO_3^- + H_3O^+$ $\quad K_1 = 7.3 \times 10^{-10}$

$H_2BO_3^- + H_2O \rightleftharpoons HBO_3^{2-} + H_3O^+$ $\quad K_2 = 1.8 \times 10^{-13}$

$HBO_3^{2-} + H_2O \rightleftharpoons BO_3^{3-} + H_3O^+$ $\quad K_3 = 1.6 \times 10^{-14}$

First Step:

Let x = $[H_3BO_3]_{ionized}$. Then $0.10 - x = [H_3BO_3]$ and $x = [H_3O^+] = [H_2BO_3^-]$

$$K_1 = \frac{[H_2BO_3^-][H_3O^+]}{[H_3BO_3]} = \frac{x^2}{0.10 - x} = 7.3 \times 10^{-10} \simeq \frac{x^2}{0.10}$$

Solving, $x = 8.5 \times 10^{-6}$

Therefore, $[H_3O^+] = [H_2BO_3^-] = 8.5 \times 10^{-6}$ *M*; $[H_3BO_3] = 0.10$ *M*

Second Step:

Let y = $[H_2BO_3^-]_{ionized}$.

Then $[H_2BO_3^-] = 8.5 \times 10^{-6} - y$; $[H_3O^+] = 8.5 \times 10^{-6} + y$; $[HBO_3^{2-}] = y$

$$K_2 = \frac{[H_3O^+][HBO_3^{2-}]}{[H_2BO_3^-]} = \frac{(8.5 \times 10^{-6} + y)(y)}{(8.5 \times 10^{-6} - y)} = 1.8 \times 10^{-13} \simeq \frac{(8.5 \times 10^{-6})y}{(8.5 \times 10^{-6})} = y$$

Therefore, $[HBO_3^{2-}] = 1.8 \times 10^{-13}$ *M*

Third Step:

Let z = $[HBO_3^{2-}]_{ionized}$.

Then $[HBO_3^{2-}] = 1.8 \times 10^{-13} - z$; $[H_3O^+] = 8.5 \times 10^{-6} + z$; $[BO_3^{3-}] = z$

$$K_3 = \frac{[H_3O^+][BO_3^{3-}]}{[HBO_3^{2-}]} = \frac{(8.5 \times 10^{-6} + z)(z)}{(1.8 \times 10^{-13} - z)} = 1.6 \times 10^{-14} \simeq \frac{(8.5 \times 10^{-6})z}{(1.8 \times 10^{-13})}$$

Solving, $z = 3.4 \times 10^{-22}$; $[BO_3^{3-}] = 3.4 \times 10^{-22}$ *M*

Therefore, $[H_3BO_3]$ = **0.10** *M*

$[H_3O^+] = [H_2BO_3^-]$ = **8.5 × 10⁻⁶** *M*; pH = **5.07**

$[OH^-] = K_w/[H_3O^+] = \mathbf{1.2 \times 10^{-9}}$ *M*

$[HBO_3^{2-}] = \mathbf{1.8 \times 10^{-13}}$ *M*

$[BO_3^{3-}] = \mathbf{3.4 \times 10^{-22}}$ *M*

27-36. **Refer to Section 2-9 and Exercise 2-64 Solution.** • • • • • • • • • • • •

Plan: (1) Determine the simplest formula for a boron hydride, B_xH_y.
(2) Determine the molecular weight and the molecular formula for the boron hydride.

(1) Assume there is 100 g of boron hydride.

$$? \text{ mol B} = \frac{78.4 \text{ g B}}{10.81 \text{ g/mol}} = 7.25 \text{ mol B} \qquad \text{Ratio} = \frac{7.25}{7.25} = 1.0$$

$$? \text{ mol H} = \frac{21.6 \text{ g H}}{1.008 \text{ g/mol}} = 21.4 \text{ mol H} \qquad \text{Ratio} = \frac{21.4}{7.25} = 3.0$$

Therefore the simplest formula is BH_3 with a formula weight of 13.83 g/mol.

(2) $$? \text{ mol boron hydride} = 131.7 \text{ mL}_{STP} \times \frac{1 \text{ mol gas}}{22400 \text{ mL}_{STP}} = 5.88 \times 10^{-3} \text{ mol}$$

$$MW = \frac{0.2438 \text{ g boron hydride}}{5.88 \times 10^{-3} \text{ mol}} = 41.5 \text{ g/mol}$$

$$n = \frac{\text{molecular weight}}{\text{simplest formula weight}} = \frac{41.5}{13.83} = 3$$

Therefore, the true molecular formula is $(BH_3)_3$ or $\mathbf{B_3H_9}$.

28 The Transition Metals

28-2. **Refer to Section 28-1.** . . . .

The following are properties of most transition elements:

(1) All are metals.
(2) Most are harder, more brittle and have higher melting points and boiling points and higher heats of vaporization than nontransition metals.
(3) Their ions and compounds are usually colored.
(4) They form many complex ions.
(5) With few exceptions, they exhibit multiple oxidation states.
(6) Many of the metals and their compounds are paramagnetic.
(7) Many of the metals and their compounds are effective catalysts.

28-4. **Refer to Section 28-2, Table 28-2 and Appendix B.** . . . .

(a) V [Ar] $3d^3\ 4s^2$ (b) Fe [Ar] $3d^6\ 4s^2$

(c) Cu [Ar] $3d^{10}\ 4s^1$ (d) Zn [Ar] $3d^{10}\ 4s^2$

(e) Fe^{3+} [Ar] $3d^5$ (f) Ni^{2+} [Ar] $3d^8$

(g) Ag [Kr] $4d^{10}\ 5s^1$ (h) Ag^+ [Kr] $4d^{10}$

28-6. **Refer to Section 28-4.** . . . .

Unlike the compounds of representative elements, transition metal compounds are usually colored. The reason for this is that the *d* orbitals of the metal <u>in compounds</u> do not have the same energy as they do in isolated atoms, but are often split into sets of orbitals separated by energies that correspond to wavelengths of light in the visible region. The absorption of visible light causes electronic transitions between orbitals in these sets. The color of the compounds corresponds to the wavelengths of light that are <u>not</u> absorbed, but reflected.

28-8. **Refer to Section 28-3.** . . . .

Group VIIIB consists of three columns of three metals each, which have no counterparts among the representative elements. Each horizontal row is called a triad and is named after the best-known metal of the row; they are called the iron (Fe, Co, Ni), palladium (Ru, Rh, Pd) and platinum (Os, Ir, Pt) triads.

28-10. **Refer to Section 28-5, Table 28-6 and the Key Terms for Chapter 28.** . . . .

The lanthanide contraction refers to the decrease in atomic radii of the elements following the lanthanides compared to what would be expected if there were no *f*-transition metals. The lanthanide contraction affects the transition metals of the sixth period which have nearly the same radii as the metals of the fifth period above them. This observation goes against the trend of having larger radii upon descending a group of representative metals. The observed lanthanide contraction is attributed to the insertion of fourteen lanthanide elements, with fourteen poorly shielding *f* electrons two shells inside the outermost shell. The higher effective nuclear charges felt by the outermost electrons pull the outer electrons closer to the nucleus, which results in smaller radii and greater densities than expected.

28-12. **Refer to Sections 28-6 and 28-7, and Tables 28-7 and 28-9.** • • • • • • • •

The acidity and the covalent nature of transition metal oxides generally increases with increasing oxidation state of the metals. This is shown by the oxides of manganese and chromium.

Mn Oxide	Ox. No. of Mn	Character	Cr Oxide	Ox. No. of Cr	Character
MnO	+2	basic	CrO	+2	basic
Mn_2O_3	+3	weakly basic	Cr_2O_3	+3	amphoteric
MnO_2	+4	amphoteric	CrO_3	+6	weakly acidic or acidic
MnO_3	+6	acidic			
Mn_2O_7	+7	strongly acidic			

28-14. **Refer to Section 28-7 and Table 28-9.** • • • • • • • • • • • • • •

Chromium(VI) oxide, CrO_3, is the acid anhydride of chromic acid, H_2CrO_4, and dichromic acid, $H_2Cr_2O_7$. Recall that there is no change in oxidation state when an acid anhydride is converted to a corresponding acid and so the oxidation state of Cr is +6 in both acids.

$CrO_3 + H_2O \rightarrow H_2CrO_4$ $\qquad$ $2CrO_3 + H_2O \rightarrow H_2Cr_2O_7$

28-16. **Refer to Section 28-5 and the Key Terms for Chapter 28.** • • • • • • • •

Ferromagnetism is the ability of a substance to become permanently magnetized by exposure to an external magnetic field. This happens when unpaired electrons of adjacent metal atoms which are at certain distances from each other interact cooperatively to form regions called domains. These are clusters of atoms, all of which are aligned in the same direction. The only free elements to exhibit ferromagnetism are the metals of the iron triad: iron, cobalt and nickel.

28-18. **Refer to Sections 28-6 and 28-7.** • • • • • • • • • • • • • • • •

Mn_2O_7, $Cr_2O_7^{2-}$ and Cl_2O_7 are isoelectronic and have the same geometry. All three species involve two tetrahedra sharing a common corner, as depicted below:

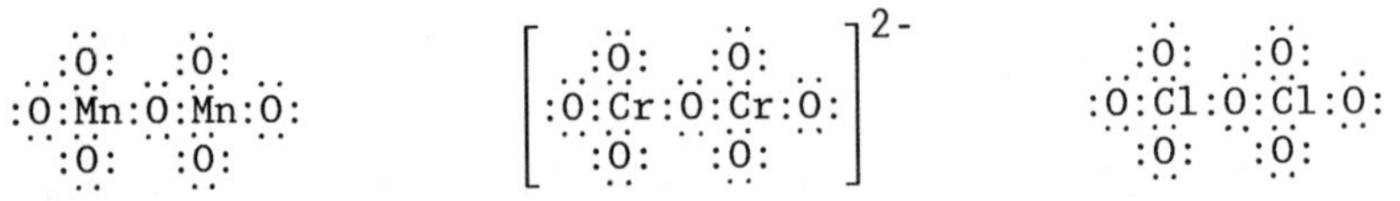

Manganese (VII) oxide, Mn_2O_7, a dark brown, explosive liquid, is the acid anhydride of $HMnO_4$. Dichlorine heptoxide, Cl_2O_7, a colorless, explosive liquid, is the acid anhydride of $HClO_4$. The dichromate anion, $Cr_2O_7^{2-}$, from dichromic acid, $H_2Cr_2O_7$, is a powerful strong oxidizing agent in acidic conditions.

28-20. **Refer to Section 3-4 and Exercise 3-46 Solution.** • • • • • • • • • • •

Balanced equation: $3Co_3O_4 + 8Al \xrightarrow{\Delta} 9Co + 4Al_2O_3$

Plan: (1) Calculate the theoretical yield of Co metal.
(2) Calculate the mass of Co_3O_4 required to produce the theoretical yield of Co.

(1) $\%\ \text{yield} = \dfrac{\text{actual yield}}{\text{theoretical yield}} \times 100$

Substituting,

$$68\% = \frac{270\ \text{g}}{\text{theoretical yield}} \times 100$$

And, theoretical yield = 400 g Co

(2) $? \text{ g } Co_3O_4 = 400 \text{ g Co} \times \frac{1 \text{ mol Co}}{59 \text{ g Co}} \times \frac{3 \text{ mol } Co_3O_4}{9 \text{ mol Co}} \times \frac{241 \text{ g } Co_3O_4}{1 \text{ mol } Co_3O_4}$ = **540 g Co_3O_4**

28-22. **Refer to Section 28-7.** . . . .

Balanced equation: $2CrO_4^{2-} + 2H^+ \rightleftharpoons Cr_2O_7^{2-} + H_2O$

$$K_c = \frac{[Cr_2O_7^{2-}]}{[CrO_4^{2-}]^2[H^+]^2} = 4.2 \times 10^{14}$$

$[Cr_2O_7^{2-}]_{initial} = [Na_2Cr_2O_7] = (1.0 \times 10^{-3} \text{ mol})/0.200 \text{ L} = 5.0 \times 10^{-3}$ M

The solution is buffered at pH = 12.00. Therefore, $[H^+] = 1.0 \times 10^{-12}$ *M* at all times in the calculation.

Let x = $[Cr_2O_7^{2-}]$ that reacts. Then, $5.0 \times 10^{-3} - x = [Cr_2O_7^{2-}]_{equil}$

$2x = [CrO_4^{2-}]_{equil}$

	$2CrO_4^{2-}$	+	$2H^+$	$\rightleftharpoons$	$Cr_2O_7^{2-}$	+ H_2O
initial	0 *M*		1.0×10^{-12} *M*		5.0×10^{-3} *M*	
change	+ 2x *M*				- x *M*	
at equilibrium	2x *M*		1.0×10^{-12} *M*		$(5.0 \times 10^{-3} - x)$*M*	

If we blindly continue this calculation and make the assumption that x is small relative to 5.0×10^{-3} *M*, we obtain:

$$K_c = \frac{[Cr_2O_7^{2-}]}{[CrO_4^{2-}]^2[H^+]^2} = \frac{(5.0 \times 10^{-3} - x)}{(2x)^2(1.0 \times 10^{-12})^2} = 4.2 \times 10^{14} \approx \frac{5.0 \times 10^{-3}}{(2x)^2(1.0 \times 10^{-12})^2}$$

Solving, $(2x)^2 = 1.2 \times 10^7$ and $x = 1.7 \times 10^3$

This is a totally ridiculous answer because in order to have a positive value for $[Cr_2O_7^{2-}]$, x cannot be any larger than 5.0×10^{-3}. The reason for this error is due to the conditions of the system. The pH is so high and $[H^+]$ is so low, that the equilibrium is shifted almost entirely to the left. Without using the quadratic formula, let us examine the quadratic equation,

$$(1.7 \times 10^{-9})x^2 + x - 5.0 \times 10^{-3} = 0.$$

We can readily see by examination that one of the solutions is $x = 5.0 \times 10^{-3}$. The x^2 term can be omitted because it is so small: 4.2×10^{-13}. Therefore,

$$x \approx 5.0 \times 10^{-3}$$
$$[Cr_2O_7^{2-}]_{equil} = 5.0 \times 10^{-3} - x \approx 0 \; M$$
$$[CrO_4^{2-}]_{equil} = 2x \approx 0.010 \; M$$

We can now substitute $[CrO_4^{2-}]_{equil}$ and $[H^+]$ into the equilibrium expression and solve for $[Cr_2O_7^{2-}]$.

$$[Cr_2O_7^{2-}] = K_c[CrO_4^{2-}]^2[H^+]^2 = (4.2 \times 10^{14})(0.010)^2(1.0 \times 10^{-12})^2$$
$$= 4.2 \times 10^{-14} \; M$$

Therefore,

$$\frac{[Cr_2O_7^{2-}]}{[CrO_4^{2-}]} = \frac{4.2 \times 10^{-14}}{0.010} = 4.2 \times 10^{-12}$$

Thus, in basic solution, the CrO_4^{2-} ion predominates.

28-24. **Refer to Section 28-6, Table 28-7 and Exercise 28-12 Solution.** • • • • • • •

Ionic oxides (usually metal oxides) are termed basic because they react with water to form basic solutions, whereas covalent oxides (usually nonmetal oxides) are termed acidic because they react with water to form acidic solutions. In Exercise 28-12 Solution, we stated that the oxides of transition metals become more acidic with increasing oxidation number. And so, in order of increasing acidity, we have

manganese(II) oxide <	manganese(III) oxide <	manganese(VII) oxide
MnO	Mn_2O_3	Mn_2O_7

In fact, MnO is basic, Mn_2O_3 is weakly basic and Mn_2O_7 is acidic.

28-26. **Refer to Section 21-21 and Appendix J.** • • • • • • • • • • • • • • •

		E^o
oxidation	$4(Fe^{2+} \rightarrow Fe^{3+} + e^-)$	-0.771 V
reduction	$O_2 + 4H^+ + 4e^- \rightarrow 2H_2O$	1.229 V
cell reaction	$4Fe^{2+} + 4H^+ + O_2 \rightarrow 4Fe^{3+} + 2H_2O$	0.458 V = E^o_{cell}

$$\Delta G^o_{rxn} = -nFE^o_{cell} = -(4 \text{ mol } e^-)(96{,}500 \text{ J/V}\cdot\text{mol } e^-)(0.458 \text{ V}) = \mathbf{-1.77 \times 10^5 \text{ J}}$$

or -177 kJ

28-28. **Refer to Section 10-17.** •

Balanced equation: $2Sc(s) + 6HCl(aq) \rightarrow 2ScCl_3(aq) + 3H_2(g)$

$$? \text{ L}_{STP} \text{ H}_2 = 6.00 \text{ g Sc} \times \frac{1.00 \text{ mol Sc}}{45.0 \text{ g Sc}} \times \frac{3 \text{ mol H}_2}{2 \text{ mol Sc}} \times \frac{22.4 \text{ L}_{STP} \text{ H}_2}{1 \text{ mol H}_2} = \mathbf{4.48 \text{ L}_{STP} \text{ H}_2}$$

29 Coordination Compounds

29-2. **Refer to Section 29-1 and Table 29-1.** . . . .

$NiSO_4 \cdot 6H_2O \equiv [Ni(OH_2)_6]SO_4$

$Cu(NO_3)_2 \cdot 4NH_3 \equiv [Cu(NH_3)_4](NO_3)_2$

$Ni(NO_3)_2 \cdot 6NH_3 \equiv [Ni(NH_3)_6](NO_3)_2$

29-4. **Refer to Section 29-1 and Table 29-1.** . . . .

Formula	Coordination Sphere	Name
$[Pt(NH_3)_2Cl_4]$	$[Pt(NH_3)_2Cl_4]^0$	diamminetetrachloroplatinum(IV)
$[Pt(NH_3)_3Cl_3]Cl$	$[Pt(NH_3)_3Cl_3]^+$	triamminetrichloroplatinum(IV) chloride
$[Pt(NH_3)_4Cl_2]Cl_2$	$[Pt(NH_3)_4Cl_2]^{2+}$	tetraamminedichloroplatinum(IV) chloride
$[Pt(NH_3)_5Cl]Cl_3$	$[Pt(NH_3)_5Cl]^{3+}$	pentaamminechloroplatinum(IV) chloride
$[Pt(NH_3)_6]Cl_4$	$[Pt(NH_3)_6]^{4+}$	hexaammineplatinum(IV) chloride

29-6. **Refer to Sections 29-3 and 29-4.** . . . .

(a) $[Ni(CO)_4]$ tetracarbonylnickel(0)

(b) $Na_2[Co(OH_2)_2(OH)_4]$ sodium diaquatetrahydroxocobaltate(II)

(c) $[Ag(NH_3)_2]Br$ diamminesilver(I) bromide

(d) $[Cr(en)_3](NO_3)_3$ tris(ethylenediamine)chromium(III) nitrate

(e) $[Pt(NH_3)_4(NO_2)_2](NO_3)_2$ tetraamminedinitroplatinum(IV) nitrate

(f) $K_2[Cu(CN)_4]$ potassium tetracyanocuprate(II)

29-8. **Refer to Sections 29-3 and 29-4.** . . . .

(a) sodium tetracyanocadmate $Na_2[Cd(CN)_4]$

(b) hexaamminecobalt(III) chloride $[Co(NH_3)_6]Cl_3$

(c) diaquadicyanocopper(II) $[Cu(OH_2)_2(CN)_2]$

(d) potassium hexachloropalladate(IV) $K_2[PdCl_6]$

(e) *cis*-diaquabis(ethylenediamine)cobalt(III) perchlorate *cis*-$[Co(OH_2)_2(en)_2](ClO_4)_3$

(f) ammonium *trans*-dibromobis(oxalato)chromate(III) $(NH_4)_3$-*trans*-$[CrBr_2(ox)_2]$

29-10. **Refer to Section 29-2.** . . . .

The metal hydroxides that dissolve in an excess of aqueous NH_3 to form ammine complexes are derived from the twelve metals of the cobalt, nickel, copper and zinc families. The only cation of these metals that behaves differently is Hg_2^{2+}. Therefore, when excess NH_3 is added:

(a) $Zn(OH)_2(s)$ will dissolve

(b) $Cr(OH)_3(s)$ will not dissolve

(c) $Fe(OH)_2(s)$ will not dissolve

(d) $Ni(OH)_2(s)$ will dissolve

(e) $Cd(OH)_2(s)$ will dissolve

29-12. **Refer to Section 29-2.** . . . .

In general terms, we may represent the reaction in which a metal cation reacts in aqueous NH_3 to form an insoluble metal hydroxide by the following reaction:

$$M^{n+} + nNH_3 + nH_2O \rightarrow M(OH)_n(s) + nNH_4^+.$$

(a) $Cu^{2+} + 2NH_3 + 2H_2O \rightarrow Cu(OH)_2(s) + 2NH_4^+$

(b) $Zn^{2+} + 2NH_3 + 2H_2O \rightarrow Zn(OH)_2(s) + 2NH_4^+$

(c) $Fe^{3+} + 3NH_3 + 3H_2O \rightarrow Fe(OH)_3(s) + 3NH_4^+$

(d) $Hg^{2+} + 2NH_3 + 2H_2O \rightarrow Hg(OH)_2(s) + 2NH_4^+$

(e) $Mn^{3+} + 3NH_3 + 3H_2O \rightarrow Mn(OH)_3(s) + 3NH_4^+$

29-14. **Refer to Sections 29-3 and 29-4.** . . . .

(a)	$[Ag(NH_3)_2]^+$	diamminesilver(I) ion
	$[Pt(NH_3)_4]^{2+}$	tetraammineplatinum(II) ion
	$[Cr(OH_2)_6]^{3+}$	hexaaquachromium(III) ion
(b)	$[Ni(en)_3]^{2+}$	tris(ethylenediamine)nickel(II) ion
	$[Co(en)_3]^{3+}$	tris(ethylenediamine)cobalt(III) ion
	$[Cr(en)_3]^{3+}$	tris(ethylenediamine)chromium(III) ion
(c)	$[Co(en)_2(NO_2)_2]^+$	bis(ethylenediamine)dinitrocobalt(III) ion
	$[CoBr_2(en)_2]^+$	dibromobis(ethylenediamine)cobalt(III) ion
	$[Ni(en)_2(NO)_2]^{2+}$	bis(ethylenediamine)dinitrosylnickel(II) ion
(d)	$[FeCl(dien)(en)]^{2+}$	chlorodiethylenetriamineethylenediamineiron(III) ion
	$[Cr(OH_2)(dien)(ox)]^+$	aquadiethylenetriamineoxalatochromium(III) ion
	$[RuCl(dien)(en)]^{2+}$	chlorodiethylenetriamineethylenediamineruthenium(III) ion
(e)	$[Co(OH_2)_3(dien)]^{3+}$	triaquadiethylenetriaminecobalt(III) ion
	$[Cr(NH_3)_3(dien)]^{3+}$	triamminediethylenetriaminechromium(III) ion
	$[Fe(NH_3)_3(dien)]^{2+}$	triamminediethylenetriamineiron(II) ion

29-16. **Refer to the Introduction to Section 29-6 and the Key Terms for Chapter 29. . .**

Isomers are substances that have the same number and kinds of atoms, but arranged differently. Structural isomers, as applied to coordination compounds, are isomers whose differences involve having more than a single coordination sphere or different donor atoms on the same ligand. They contain different atom-to-atom bonding sequences. Stereoisomers, on the other hand, are isomers that differ only in the way that atoms are oriented in space, and therefore involve only one coordination sphere and the same ligands and donor atoms.

29-18. **Refer to Sections 29-3, 29-4 and 29-6. . . . .**

An ionization isomer results from the exchange of ions inside and outside the coordination sphere.

(a) $[Cr(NH_3)_4BrI]I$ tetraamminebromoiodochromium(III) iodide

(b) $[NiCl_2(en)_2](NO_2)_2$ dichlorobis(ethylenediamine)nickel(IV) nitrite

(c) $[Fe(NH_3)_5SO_4]CN$ pentaamminesulfatoiron(III) cyanide

29-20. **Refer to Sections 29-3, 29-4 and 29-6. . . . .**

A coordination isomer involves the exchange of ligands between a complex cation and a complex anion of the same compound, forming another complex cation and complex anion.

(a) There are 5 possible coordination isomers of $[Co(NH_3)_6][Cr(CN)_6]$:

$[Co(NH_3)_5(CN)][Cr(NH_3)(CN)_5]$	pentaamminecyanocobalt(III) amminepentacyanochromate(III)
$[Co(NH_3)_4(CN)_2][Cr(NH_3)_2(CN)_4]$	tetraamminedicyanocobalt(III) diamminetetracyanochromate(III)
$[Cr(NH_3)_4(CN)_2][Co(NH_3)_2(CN)_4]$	tetraamminedicyanochromium(III) diamminetetracyanocobaltate(III)
$[Cr(NH_3)_5(CN)][Co(NH_3)(CN)_5]$	pentaamminecyanochromium(III) amminepentacyanocobaltate(III)
$[Cr(NH_3)_6][Co(CN)_6]$	hexaamminechromium(III) hexacyanocobaltate(III)

(b) $[Cu(en)_3][Ni(CN)_4]$, tris(ethylenediamine)copper(II) tetracyanonickelate(II)

29-22. **Refer to Sections 29-3, 29-4 and 29-7. . . . .**

Geometrical isomers, also called position isomers, are stereoisomers that are not mirror images of each other.

(a) $[Pt(NH_3)_2Cl_2]$ (square planar) has 2 geometrical isomers:

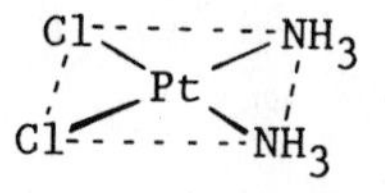

cis-diamminedichloroplatinum(II)

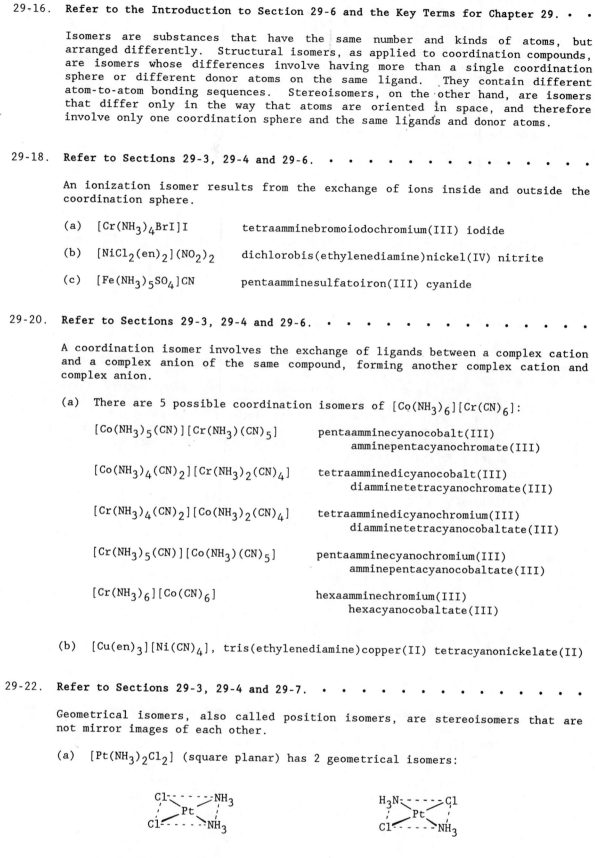

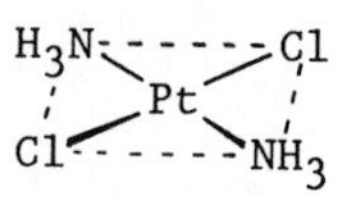

trans-diamminedichloroplatinum(II)

(b) The amminetrichloroplatinate(II) ion, $[Pt(NH_3)Cl_3]^-$, cannot have geometrical isomers. No matter where the NH_3 ligand is placed in the square planar complex, it will always have the same position relative to the other Cl atoms.

(c) Diamminedichlorozinc(II), $[Zn(NH_3)_2Cl_2]$, has no geometrical isomers because it is tetrahedral. All 4 ligands are equidistant from each other.

29-24. **Refer to Sections 29-3, 29-4 and 29-7.** . . . .

(a) $[Co(NH_3)_2Cl_4]^-$

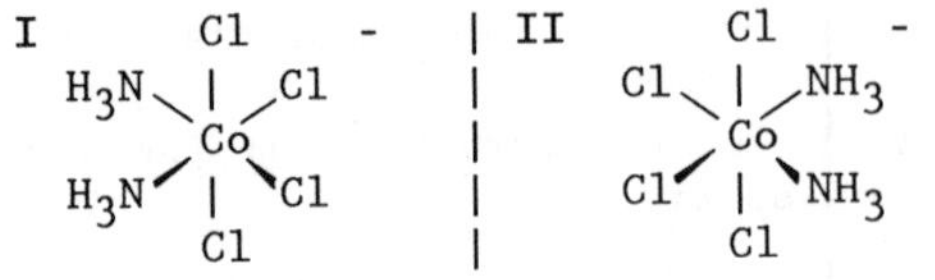

cis-diamminetetrachlorocobaltate(III) ion

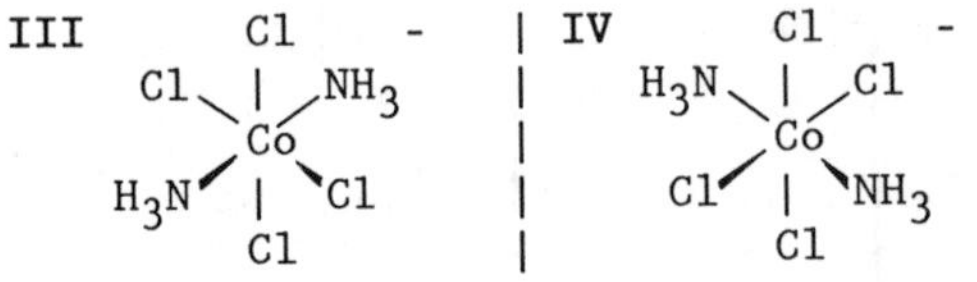

trans-diamminetetrachlorocobaltate(III) ion

Structures I and II are mirror images that are superimposable if the Cl-Co-Cl axis is rotated 180°. Structures III and IV are superimposable if the Cl-Co-Cl axis is rotated 90°. Therefore, structures I and II are simply different representations of the same compound, as are structures III and IV. Hence, both the *cis* and the *trans* geometric isomers of $[Co(NH_3)_2Cl_4]^-$ have no optical isomer.

(b) $[Co(NH_3)_3Cl_3]$

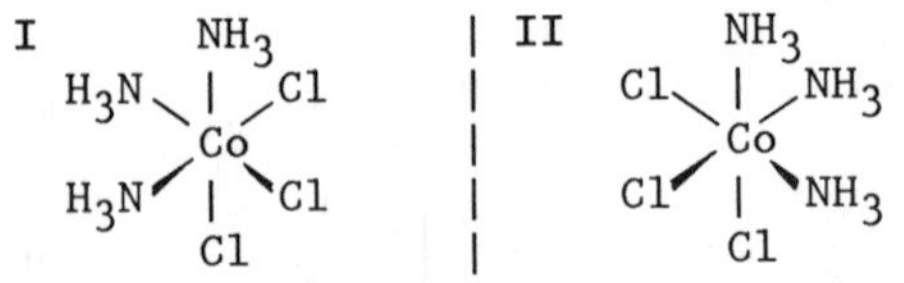

cis-triamminetrichlorocobalt(III)

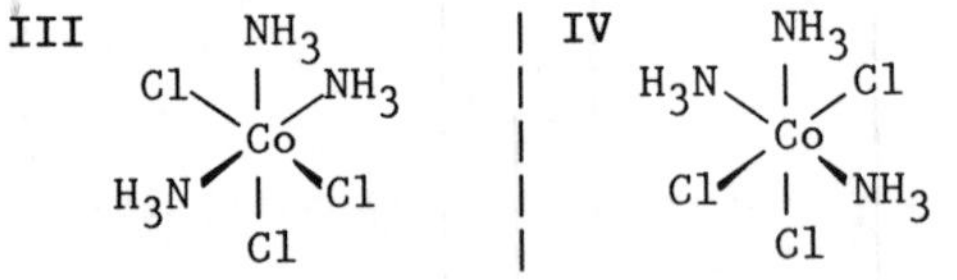

trans-triamminetrichlorocobalt(III)

Structures I and II are identical. Hence, the *cis* geometric isomer has no optical isomer. Structures III and IV are also identical. So, the *trans* geometric isomer also has no optical isomer.

(c) $[Co(NH_3)(en)Cl_3]$

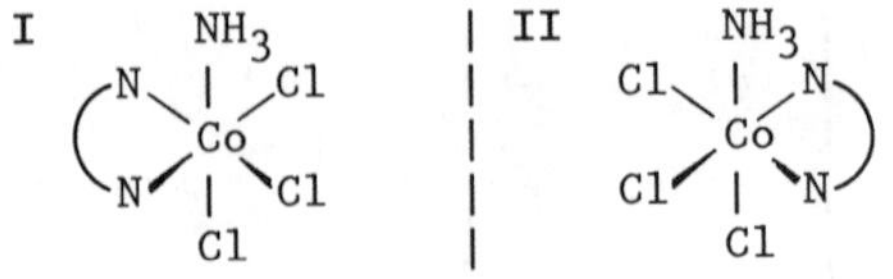

cis-amminetrichloroethylenediaminecobalt(III)

III
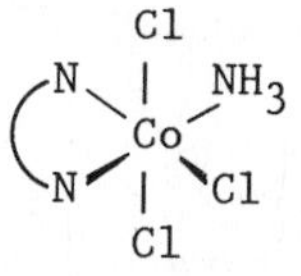

IV
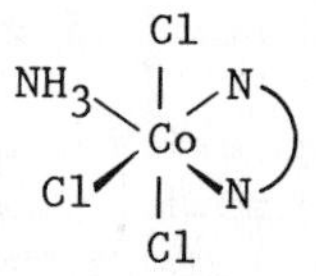

trans-amminetrichloroethylenediaminecobalt(III)

Structures I and II are identical. Hence, the *cis* geometric isomer has no optical isomer. Likewise, structures III and IV are superimposable mirror images of each other, if one of them is turned upside down. Therefore, the *trans* geometric isomer also has no optical isomer.

29-26. **Refer to Section 29-5 and Exercises 29-6 and 29-8.** . . . .

	Molecular or Ionic Geometry	Hybridization of Central Metal Atom
For Exercise 29-6:		
(a) $[Ni(CO)_4]$	tetrahedral	sp^3
(b) $[Co(OH_2)_2(OH)_4]^{2-}$	octahedral	d^2sp^3 or sp^3d^2
(c) $[Ag(NH_3)_2]^+$	linear	sp
(d) $[Cr(en)_3]^{3+}$	octahedral	d^2sp^3 or sp^3d^2
(e) $[Pt(NH_3)_4(NO_2)_2]^{2+}$	octahedral	d^2sp^3 or sp^3d^2
(f) $[Cu(CN)_4]^{2-}$	square planar	dsp^2
For Exercise 29-8:		
(a) $[Cd(CN)_4]^{2-}$	tetrahedral	sp^3
(b) $[Co(NH_3)_6]^{3+}$	octahedral	d^2sp^3 or sp^3d^2
(c) $[Cu(OH_2)_2(CN)_2]$	square planar	dsp^2
(d) $[PdCl_6]^{2-}$	octahedral	d^2sp^3 or sp^3d^2
(e) $[Co(OH_2)_2(en)_2]^{3+}$	octahedral	d^2sp^3 or sp^3d^2
(f) $[CrBr_2(ox)_2]^{3-}$	octahedral	d^2sp^3 or sp^3d^2

The general geometric shapes corresponding to square planar, tetrahedral and octahedral complexes are given below:

M = metal

L = ligand

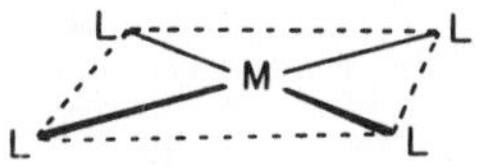

square planar

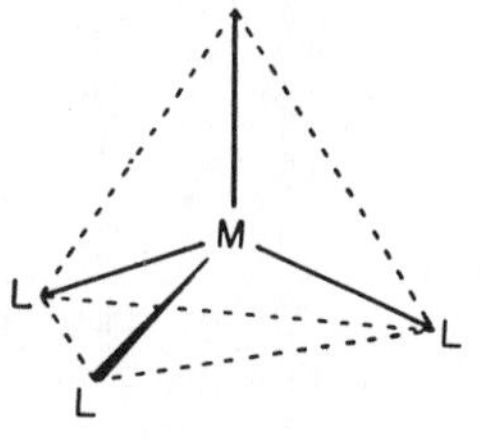

tetrahedral

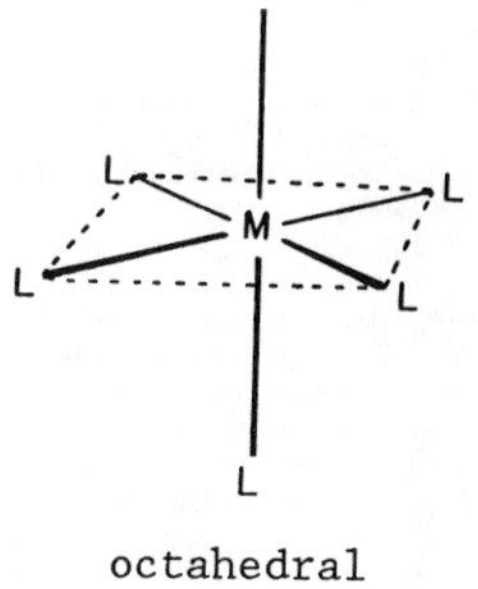

octahedral

29-28. Refer to Section 29-8 and Figure 4-22. . . . .

The spatial orientation of an atom's *d* orbitals have the lobes of the $d_{x^2-y^2}$ and d_{z^2} orbitals directed along the x, y and z axes, while those of the other three *d* orbitals lie between the axes. Valence Bond Theory postulates that in octahedral complexes, the six donor atoms must donate electrons into a set of 6 hybridized d^2sp^3 or sp^3d^2 metal orbitals along each of the axes. Therefore, the $d_{x^2-y^2}$ and d_{z^2} orbitals are the only ones which can be used to combine with *s*, p_x, p_y and p_z orbitals to provide six hybridized orbitals along the three axes.

29-30. Refer to Sections 29-8, 29-9 and 29-10, and Tables 29-6 and 29-7. . . . . .

According to Crystal Field Theory, different ligands repel the metal electrons to different extents, thus removing the degeneracy of the *d* orbitals on the metal. The energy separation between the *d* orbitals, Δ_{oct}, depends on the crystal field strength of the ligands. Weak field ligands form high spin complexes (outer orbital complexes in Valence Bond Theory) and strong field ligands form low spin complexes (inner orbital complexes in Valence Bond Theory). The spectrochemical series is the arrangement of ligands in order of increasing ligand field strength:

$$I^- < Br^- < Cl^- < F^- < OH^- < H_2O < (COO)_2^{2-} < NH_3 < en < NO_2^- < CN^-$$

weak field ligands — strong field ligands

As shown in Table 29-7, low spin configurations exist only for octahedral complexes having metal ions with d^4 - d^7 configurations. For d^1 - d^3 and d^8 - d^{10} ions, only one possibility exists, which is designated as high spin.

Using the spectrochemical series, we can propose the following electron configurations and hybridizations:

(a) $[Cu(OH_2)_6]^{2+}$ — outer orbital (weak field) complex ion of Cu^{2+} (d^9)
paramagnetic (1 unpaired e^-)

outer electronic configuration: 3d: ↑↓ ↑↓ ↑↓ ↑↓ ↑_ ; sp^3d^2: xx xx xx xx xx xx ; 4d: __ __ __

(b) $[MnF_6]^{3-}$ — outer orbital (weak field) complex ion of Mn^{3+} (d^4)
paramagnetic (4 unpaired e^-)

outer electronic configuration: 3d: ↑_ ↑_ ↑_ ↑_ __ ; sp^3d^2: xx xx xx xx xx xx ; 4d: __ __ __

(c) $[Co(CN)_6]^{3-}$ — inner orbital (strong field) complex ion of Co^{3+} (d^6)
diamagnetic (0 unpaired e^-)

outer electronic configuration: 3d: ↑↓ ↑↓ ↑↓ ; d^2sp^3: xx xx xx xx xx xx ; 4d: __ __ __ __ __

(d) $[Cr(NH_3)_6]^{3+}$ — inner orbital (strong field) complex ion of Cr^{3+} (d^3)
paramagnetic (3 unpaired e^-)

outer electronic configuration: 3d: ↑_ ↑_ ↑_ ; d^2sp^3: xx xx xx xx xx xx ; 4d: __ __ __ __ __

(e) $[CrBr_2Cl_4]^{3-}$ — inner orbital complex ion of Cr^{3+} (d^3)
paramagnetic (3 unpaired e^-)

outer electronic configuration: 3d: ↑_ ↑_ ↑_ ; d^2sp^3: xx xx xx xx xx xx ; 4d: __ __ __ __ __

(f) $[Co(en)_3]^{3+}$ inner orbital (strong field) complex ion of Co^{3+} (d^6) diamagnetic (0 unpaired e^-)

outer electronic configuration: ↑↓ ↑↓ ↑↓ (3d) xx xx xx xx xx xx (d^2sp^3) __ __ __ __ __ (4d)

(g) $[Fe(OH)_6]^{3-}$ outer orbital (weak field) complex ion of Fe^{3+} (d^5) paramagnetic (5 unpaired e^-)

outer electronic configuration: ↑ ↑ ↑ ↑ ↑ (3d) xx xx xx xx xx xx (sp^3d^2) __ __ __ (4d)

(h) $[Fe(NO_2)_6]^{3-}$ inner orbital (strong field) complex ion of Fe^{3+} (d^5) paramagnetic (1 unpaired e^-)

outer electronic configuration: ↑↓ ↑↓ ↑ (3d) xx xx xx xx xx xx (d^2sp^3) __ __ __ __ __ (4d)

29-32. **Refer to Sections 29-9 and 29-10 and the Key Terms for Chapter 29. . . . . .**

Crystal Field Theory is a theory of bonding in transition metal complexes in which the bonds between metal ions and ligands are strictly electrostatic interactions. During bonding, the repulsions between ligand electrons and metal electrons in *d* orbitals cause an electric field, i.e., the crystal field, to be set up by the approach of the ligands to split the *d* orbitals into two sets, the t_{2g} set at lower energy and the e_g set at higher energy. The energy separation between the two sets is named Δ_{oct} and is proportional to the crystal field strength of the ligands, that is, how strongly the ligand electrons repel the metal electrons.

When electrons undergo transitions from a lower energy t_{2g} *d* orbital to a higher energy e_g *d* orbital, an amount of energy equivalent to the wavelengths of visible light are absorbed, resulting in transition metal complexes with the complementary color of the light absorbed. Δ_{oct} can be determined experimentally from the wavelength of the light absorbed:

$$\Delta_{oct} = EN_A = \frac{hcN_A}{\lambda}$$

where E = energy of the absorbed photon
N_A = Avogadro's Number
h = Planck's constant
c = speed of light
λ = wavelength

It is possible to arrange the common ligands in the order of increasing crystal field strengths, by interpreting the visible spectra of many complexes. This is the spectrochemical series, shown in Exercise 29-30. From the above equation, we can deduce that transition metal complexes that are colored, e.g. red or orange, are absorbing the complementary colors, blue or purple. These absorbed wavelengths are at the shorter end of the visible light range and correspond to larger Δ_{oct} values. The ligands in these complexes have larger crystal field strengths and are located at the high end of the spectrochemical series.

29-34. **Refer to Sections 29-9 and 29-10, and Exercises 29-29 and 29-30 Solutions. . .**

A high spin complex is the Crystal Field designation for an outer orbital complex where all t_{2g} and e_g orbitals are singly occupied before pairing begins. A low spin complex, on the other hand, is the Crystal Field designation for an inner orbital complex. It contains electrons paired in t_{2g} orbitals before e_g orbitals are occupied. However, the low spin configuration exists only for octahedral complexes having metal ions with d^4 - d^7 configurations. For d^1 - d^3

and d^8 - d^{10} ions, only one possibility exists which is designated as high spin. In the case of d^4 - d^7 configurations: (1) weak ligand field strength is associated with high spin (outer orbital) complexes, whereas strong ligand field strength is associated with low spin (inner orbital) complexes, and (2) the spectrochemical series ranks ligands in order of increasing ligand field strength. The following are the predictions:

	Complex Ion	Metal Ion Configuration	Complex Configuration
(a)	$[Cu(OH_2)_6]^{2+}$	d^9	high spin
(b)	$[MnF_6]^{3-}$	d^4	high spin (weak field strength ligands)
(c)	$[Co(CN)_6]^{3-}$	d^6	low spin (strong field strength ligands)
(d)	$[Cr(NH_3)_6]^{3+}$	d^3	high spin (by convention)
(e)	$[CrBr_2Cl_4]^{3-}$	d^3	high spin
(f)	$[Co(en)_3]^{3+}$	d^6	low spin (strong field strength ligands)
(g)	$[Fe(OH)_6]^{3-}$	d^5	high spin (weak field strength ligands)
(h)	$[Fe(NO_2)_6]^{3-}$	d^5	low spin (strong field strength ligands)

29-36. **Refer to Sections 29-4, 29-9 and 29-10, and Exercise 29-35.** . . . .

Metal Ion	Ligand field Strength	Example	
V^{2+}	weak	$[VF_6]^{4-}$	hexafluorovanadate(II) ion
		$[V(OH_2)_6]^{2+}$	hexaaquavanadium(II) ion
Mn^{2+}	strong	$[Mn(en)_3]^{2+}$	tris(ethylenediamine)manganese(II) ion
		$[Mn(NH_3)_6]^{2+}$	hexaamminemanganese(II) ion
Mn^{2+}	weak	$[MnF_6]^{4-}$	hexafluoromanganate(II) ion
		$[MnBr_6]^{4-}$	hexabromomanganate(II) ion
Ni^{2+}	weak	$[Ni(OH_2)_6]^{2+}$	hexaaquanickel(II) ion
		$[NiF_6]^{4-}$	hexafluoronickelate(II) ion
Cu^{2+}	weak	$[Cu(OH_2)_6]^{2+}$	hexaaquacopper(II) ion
		$[CuF_6]^{4-}$	hexafluorocuprate(II) ion
Fe^{3+}	strong	$[Fe(CN)_6]^{3-}$	hexacyanoferrate(III) ion
		$[Fe(NH_3)_6]^{3+}$	hexaammineiron(III) ion
Cu^{+}	weak	$[CuCl_6]^{5-}$	hexachlorocuprate(I) ion
		$[Cu(OH_2)_6]^{+}$	hexaaquacopper(I) ion
Ru^{3+}	strong	$[Ru(NH_3)_6]^{3+}$	hexaammineruthenium(III) ion
		$[Ru(en)_3]^{3+}$	tris(ethylenediamine)ruthenium(III) ion

29-38. **Refer to Section 29-11, Table 29-9 and the Key Terms for Chapter 29. . . . .**

Crystal field stabilization energy (CFSE) is a measure of the net energy of stabilization gained by a metal ion's nonbonding *d* electrons as a result of complex formation.

(a) $[Co(NH_3)_6]^{3+}$ is a low spin d^6 complex ion with a $t_{2g}{}^6e_g{}^0$ configuration.

$$\begin{aligned} \text{CFSE} &= (6)(-2/5)(\Delta_{oct}) + (0)(+3/5)(\Delta_{oct}) \\ &= (6)(-2/5)(22{,}900\ \text{cm}^{-1} \times 0.01196\ \text{kJ/mol}\cdot\text{cm}^{-1}) \\ &= \mathbf{-657\ kJ/mol} \end{aligned}$$

(b) $[Ti(OH_2)_6]^{2+}$ is a high spin d^2 complex ion with a $t_{2g}{}^2e_g{}^0$ configuration.

$$\begin{aligned} \text{CFSE} &= (2)(-2/5)(\Delta_{oct}) + (0)(+3/5)(\Delta_{oct}) \\ &= (2)(-2/5)(20{,}300\ \text{cm}^{-1} \times 0.01196\ \text{kJ/mol}\cdot\text{cm}^{-1}) \\ &= \mathbf{-194\ kJ/mol} \end{aligned}$$

(c) $[Cr(OH_2)]_6{}^{3+}$ is a high spin d^3 complex ion with a $t_{2g}{}^3e_g{}^0$ configuration.

$$\begin{aligned} \text{CFSE} &= (3)(-2/5)(\Delta_{oct}) + (0)(+3/5)(\Delta_{oct}) \\ &= (3)(-2/5)(17{,}600\ \text{cm}^{-1} \times 0.01196\ \text{kJ/mol}\cdot\text{cm}^{-1}) \\ &= \mathbf{-253\ kJ/mol} \end{aligned}$$

(d) $[Co(CN)_6]^{3-}$ is a low spin d^6 complex ion with a $t_{2g}{}^6e_g{}^0$ configuration.

$$\begin{aligned} \text{CFSE} &= (6)(-2/5)(\Delta_{oct}) + (0)(+3/5)(\Delta_{oct}) \\ &= (6)(-2/5)(33{,}500\ \text{cm}^{-1} \times 0.01196\ \text{kJ/mol}\cdot\text{cm}^{-1}) \\ &= \mathbf{-962\ kJ/mol} \end{aligned}$$

(e) $[Co(OH_2)_6]^{2+}$ is a high spin d^7 complex ion with a $t_{2g}{}^5e_g{}^2$ configuration.

$$\begin{aligned} \text{CFSE} &= (5)(-2/5)(\Delta_{oct}) + (2)(+3/5)(\Delta_{oct}) \\ &= (5)(-2/5)(10{,}000\ \text{cm}^{-1} \times 0.01196\ \text{kJ/mol}\cdot\text{cm}^{-1}) \\ &\qquad + (2)(+3/5)(10{,}000\ \text{cm}^{-1} \times 0.01196\ \text{kJ/mol}\cdot\text{cm}^{-1}) \\ &= \mathbf{-95.7\ kJ/mol} \end{aligned}$$

(f) $[Cr(en)_3]^{3+}$ is a high spin d^3 complex ion with a $t_{2g}{}^3e_g{}^0$ configuration.

$$\begin{aligned} \text{CFSE} &= (3)(-2/5)(\Delta_{oct}) + (0)(+3/5)(\Delta_{oct}) \\ &= (3)(-2/5)(21{,}900\ \text{cm}^{-1} \times 0.01196\ \text{kJ/mol}\cdot\text{cm}^{-1}) \\ &= \mathbf{-314\ kJ/mol} \end{aligned}$$

(g) $[Cu(OH_2)_6]^{2+}$ is a high spin d^9 complex ion with a $t_{2g}{}^6e_g{}^3$ configuration.

$$\begin{aligned} \text{CFSE} &= (6)(-2/5)(\Delta_{oct}) + (3)(+3/5)(\Delta_{oct}) \\ &= (6)(-2/5)(13{,}000\ \text{cm}^{-1} \times 0.01196\ \text{kJ/mol}\cdot\text{cm}^{-1}) \\ &\qquad + (3)(+3/5)(13{,}000\ \text{cm}^{-1} \times 0.01196\ \text{kJ/mol}\cdot\text{cm}^{-1}) \\ &= \mathbf{-93.3\ kJ/mol} \end{aligned}$$

(h) $[V(OH_2)_6]^{3+}$ is a high spin d^2 complex ion with a $t_{2g}^{\,2}e_g^{\,0}$ configuration.

$$\text{CFSE} = (2)(-2/5)(\Delta_{oct}) + (0)(+3/5)(\Delta_{oct})$$
$$= (2)\ (-2/5)(18{,}000\ \text{cm}^{-1} \times 0.01196\ \text{kJ/mol}\cdot\text{cm}^{-1})$$
$$= \mathbf{-172\ kJ/mol}$$

29-40. **Refer to Section 18-4 and Appendices G and I.** • • • • • • • • • • • • •

Balanced equations:

$$[Cu(NH_3)_4]Cl_2 \rightarrow [Cu(NH_3)_4]^{2+} + 2Cl^- \qquad \text{(to completion)}$$
$$[Cu(NH_3)_4]^{2+} \rightleftharpoons Cu^{2+} + 4NH_3 \qquad K_d = 8.5 \times 10^{-13}$$
$$NH_3 + H_2O \rightleftharpoons NH_4^+ + OH^- \qquad K_b = 1.8 \times 10^{-5}$$

Plan: (1) Calculate the concentration of NH_3 from the equilibrium expression for the dissociation of the complex ion. Note: this is possible only if we assume that the ionization of NH_3 does not appreciably alter the concentration of NH_3.

(2) Calculate the $[OH^-]$ and pH of the solution using the equilibrium expression for the ionization of NH_3. Note: we are ignoring the effect of the hydrolysis of Cu^{2+} ion on pH.

(1) Let x = $[Cu(NH_3)_4]^{2+}$ that dissociates. Then

	$[Cu(NH_3)_4]^{2+}$	$\rightleftharpoons$	Cu^{2+}	+	4 NH_3
initial	0.080 *M*		0 *M*		0 *M*
change	- x *M*		+ x *M*		+ 4x *M*
at equilibrium	(0.080 - x) *M*		x *M*		4x *M*

$$K_d = \frac{[Cu^{2+}][NH_3]^4}{[Cu(NH_3)_4]^{2+}} = \frac{(x)(4x)^4}{(0.080 - x)} = 8.5 \times 10^{-13} \approx \frac{(x)(4x)^4}{0.080} = \frac{256x^5}{0.080}$$

Solving, $x = (2.7 \times 10^{-16})^{1/5} = 7.7 \times 10^{-4}$

Therefore, $[NH_3] = 4x = 3.1 \times 10^{-3}\ M$

(2) Let $y = [NH_3]_{ionized}$. Then, $[NH_3] = 3.1 \times 10^{-3} - y$ and $[NH_4^+] = [OH^-] = y$

$$K_b = \frac{[NH_4^+][OH^-]}{[NH_3]} = \frac{y^2}{(3.1 \times 10^{-3} - y)} = 1.8 \times 10^{-5}$$

Solving the quadratic equation: $y^2 + (1.8 \times 10^{-5})y - 5.6 \times 10^{-8} = 0$

$y = 2.3 \times 10^{-4}$ or -2.4×10^{-4} (discard)

Therefore, $[OH^-] = 2.3 \times 10^{-4}\ M$; pOH = 3.64; pH = **10.36**

30 Nuclear Chemistry

30-2. **Refer to Sections 1-1 and 30-3, and Exercise 1-4 Solution.** . . . .

Einstein's equation relates matter and energy:

$E = mc^2$ where E = amount of energy released
m = mass of matter transformed into energy
c = speed of light in a vacuum, 3.00×10^8 m/s

If m is expressed in kg and c in m/s, the obtained E will be in units of J.

30-4. **Refer to Section 30-3, Table 30-1, and Examples 30-1 and 30-2.** . . . . . .

(a) One neutral atom of ^{64}Zn contains 30 e^-, 30 p^+ and 34 n^0.

electrons:	30 × 0.00054858 amu	= 0.016 amu
protons:	30 × 1.0073 amu	= 30.219 amu
neutrons:	34 × 1.0087 amu	= 34.296 amu
	sum	= 64.531 amu

Δm = (sum of masses of e^-, p^+ and n^0) - (actual mass of a ^{64}Zn atom)
= 64.531 amu - 63.929 amu
= 0.602 amu

Therefore, the mass deficiency for ^{64}Zn is **0.602 amu/atom** or **0.602 g/mol**.

(b) Note: 1 joule = 1 kg × $(1\ \text{m/s})^2$

The nuclear binding energy, $B.E.$ = $(\Delta m)c^2$
$= (0.602 \times 10^{-3}\ \text{kg/mol})(3.00 \times 10^8\ \text{m/s})^2$
$= 5.42 \times 10^{13}\ \text{kg}\cdot\text{m}^2/\text{mol}\cdot\text{s}^2$
$= 5.42 \times 10^{13}$ J/mol
or **5.42×10^{10} kJ/mol of ^{64}Zn atoms**

30-6. **Refer to Section 30-3, Table 30-1, Examples 30-1 and 30-2, and Appendix C.** . .

(a) A neutral atom of ^{63}Cu contains 29 e^-, 29 p^+ and 34 n^0.

electrons:	29 × 0.00054858 amu	= 0.016 amu
protons:	29 × 1.0073 amu	= 29.212 amu
neutrons:	34 × 1.0087 amu	= 34.296 amu
	sum	= 63.524 amu

Δm = (sum of masses of e^-, p^+ and n^0) - (actual mass of a ^{63}Cu atom)
= 63.524 amu - 62.9298 amu
= **0.594 amu/atom**

(b) Δm = **0.594 g/mol**

(c) Note: 1 erg = 1 g × $(1\ \text{cm/s})^2$

$$B.E. = (\Delta m)c^2 = \left(0.594\ \text{g/mol} \times \frac{1\ \text{mol}}{6.02 \times 10^{23}\ \text{atoms}}\right)(3.00 \times 10^{10}\ \text{cm/s})^2$$
$$= 8.88 \times 10^{-4}\ \text{g}\cdot\text{cm}^2/\text{atom}\cdot\text{s}^2$$
$$= \mathbf{8.88 \times 10^{-4}\ erg/atom}$$

(d) Note: 1 erg = 10^{-7} joules

$B.E. = 8.88 \times 10^{-4}$ erg/atom $\times 10^{-7}$ J/erg = $\mathbf{8.88 \times 10^{-11}}$ **J/atom**

(e) $B.E. = (8.88 \times 10^{-11}$ J/atom)$(6.02 \times 10^{23}$ atoms/mol)$(10^{-3}$ kJ/J)
$= \mathbf{5.35 \times 10^{10}}$ **kJ/mol**

30-8. **Refer to the Introduction to Chapter 30 and Section 30-13.** . . . .

Nuclear fission is the process in which a heavy nucleus splits into nuclei of intermediate masses, whereas nuclear fusion involves the combination of light nuclei to produce a heavier nucleus. One or more neutrons are also emitted during the fission process. The similarity is that both processes are accompanied by the releases of huge amounts of energy which are the result of forming products with higher binding energies per nucleon. However, fusion produces much larger amounts of energy per unit mass of reacting atoms than fission.

30-10. **Refer to Section 30-2 and Figure 30-1.** . . . .

A plot of the number of neutrons versus the atomic number for the stable nuclides shows:

(1) at low atomic numbers, the most stable nuclides have the same number of neutrons and protons,

(2) at atomic numbers > 20, the most stable nuclides have more neutrons than protons, and

(3) a step-wise shape is associated with the plot due to the high stability of those nuclides with even numbers of neutrons and protons. Nuclides with odd numbers of both nucleons are least common (there are only 5), and those with odd-even combinations are intermediate in abundance.

30-12. **Refer to Section 30-12.** . . . .

In equations for nuclear reactions, the sums of the mass numbers and atomic numbers of the reactants must equal the sums for the products. Therefore,

(a) ${}^{23}_{11}\text{Na} + {}^{\mathbf{1}}_{\mathbf{1}}\mathbf{H} \rightarrow {}^{23}_{12}\text{Mg} + {}^{1}_{0}n$

(b) ${}^{96}_{42}\text{Mo} + {}^{4}_{2}\text{He} \rightarrow {}^{100}_{43}\text{Tc} + {}^{\mathbf{0}}_{\mathbf{+1}}\mathbf{e}$

(c) ${}^{232}_{90}\text{Th} + {}^{\mathbf{12}}_{\mathbf{6}}\mathbf{C} \rightarrow {}^{240}_{96}\text{Cm} + 4\,{}^{1}_{0}n$

(d) ${}^{\mathbf{28}}_{\mathbf{13}}\mathbf{Al} + {}^{1}_{1}\text{H} \rightarrow {}^{29}_{14}\text{Si} + {}^{0}_{0}\gamma$

(e) ${}^{209}_{83}\text{Bi} + {}^{\mathbf{2}}_{\mathbf{1}}\mathbf{H} \rightarrow {}^{210}_{84}\text{Po} + {}^{1}_{0}n$

(f) ${}^{238}_{92}\text{U} + {}^{16}_{8}\text{O} \rightarrow {}^{\mathbf{249}}_{\mathbf{100}}\mathbf{Fm} + 5\,{}^{1}_{0}n$

30-14. **Refer to Section 30-12 and Exercise 30-12 Solution.** . . . .

The equation for a nuclear reaction can be given in the following abbreviated form:

parent nuclei(bombarding particle,emitted particle)daughter nuclei

(a) ${}^{60}_{28}\text{Ni}(n,p){}^{\mathbf{60}}_{\mathbf{27}}\mathbf{Co}$

(b) ${}^{98}_{42}\text{Mo}(n,\beta){}^{\mathbf{99}}_{\mathbf{43}}\mathbf{Tc}$

(c) ${}^{35}_{17}\text{Cl}(p,\alpha){}^{\mathbf{32}}_{\mathbf{16}}\mathbf{S}$

(d) ${}^{20}_{10}\text{Ne}(\alpha,\gamma){}^{\mathbf{24}}_{\mathbf{12}}\mathbf{Mg}$

(e) ${}^{15}_{7}\text{N}(p,\alpha){}^{\mathbf{12}}_{\mathbf{6}}\mathbf{C}$

(f) ${}^{10}_{5}\text{B}(n,\alpha){}^{\mathbf{7}}_{\mathbf{3}}\mathbf{Li}$

30-16. **Refer to Section 30-8.** . . . .

(1) Photographic Detection: Radioactive substances affect photographic plates. Although the intensity of the affected spot is related to the amount of radiation, precise measurement by this method is tedious.

(2) Detection by Fluorescence: Fluorescent substances can absorb radiation and subsequently emit visible light. This is the basis for scintillation counting and can be used for quantitative detection.

(3) Cloud Chambers: A chamber containing air saturated with vapor is used. Radioactive particles ionize air molecules in the chamber. Cooling the chamber causes droplets of liquid to condense on these ions, giving observable fog-like tracks.

(4) Gas Ionization Counters: A common gas ionization counter is the Geiger-Muller counter where the electronic pulses derived from the ionization process are registered as counts.

30-18. **Refer to Section 30-11.** . . . .

The radioisotope carbon-14 is produced continuously in the atmosphere as nitrogen atoms capture cosmic-ray neutrons:

$$^{14}_{7}N + ^{1}_{0}n \rightarrow ^{14}_{6}C + ^{1}_{1}H.$$

The carbon-14 atoms react with O_2 to form $^{14}CO_2$. Like ordinary $^{12}CO_2$, it is removed from the atmosphere by living plants through the process of photosynthesis. As long as the cosmic-ray intensity remains constant, the amount of $^{14}CO_2$ in the atmosphere remains constant, and a certain fraction of carbon atoms in all living substances becomes carbon-14, a beta particle emitter with a half-life of 5730 years:

$$^{14}_{6}C \rightarrow ^{14}_{7}N + ^{0}_{-1}e$$

As a result, a steady state ratio of $^{14}C/^{12}C$ is attained in living plants and organisms. After death the plant no longer carries out photosynthesis, so it no longer takes up $^{14}CO_2$. The radioactive emissions from the carbon-14 in dead tissue then decrease with the passage of time. The activity per gram of carbon is a measure of the length of time elapsed since death. This is the basis of radiocarbon dating.

30-20. **Refer to Section 30-11.** . . . .

(a) Radionuclides are useful as tracers in chemical research to study pathways of chemical reactions. For example, if a compound containing several C atoms at non-equivalent positions undergoes decomposition to give several C-containing products, the use of C-14 to label one of the positions will reveal which product is derived from the C at that position.

(b) Radionuclides can be used to replace toxic insecticides for insect control by sterilizing male insects so that no offspring can be produced. Radionuclides can also be used to study nutrient uptake by plants. Gamma irradiation of some foods, including milk, allows them to be stored for long periods of time.

(c) Radionuclides have many applications in industry and engineering. The penetrating power of radioactive emissions are used to determine the thickness of metals with great precision. The flow of a liquid or gas through a pipeline can be monitored by injecting a sample containing a radioactive substance. Leaks in pipelines can also be similarly detected.

30-22. **Refer to Section 30-13, Exercise 30-8 Solution, Figure 30-10 and the Key Terms for Chapter 30.** . . . .

The energy that is released in fission and fusion processes is directly related to the binding energies of reactant and product nuclei in the nuclear reactions. Recall that nuclear binding energy is the amount of energy necessary to break up a nucleus into its individual protons and neutrons. Figure 30-10 is a plot of nuclear binding energies per nucleon against mass number, the highest binding energy belonging to the very stable isotope, Fe-56, an element of intermediate mass. Elements with higher binding energies are more stable, since it takes more energy to break them apart.

A nuclear fission process involves splitting a heavy nucleus, which has a lower nuclear binding energy, into more stable nuclei of intermediate masses, which have higher binding energies per nucleon. Because the products are more stable than the reactants, as seen by their nuclear binding energies, energy is released in nuclear fission.

Likewise, a nuclear fusion process involves the combination of light nuclei, which have lower nuclear binding energies, to produce a more stable heavier nucleus, which has a higher nuclear binding energy. Since the products are more stable than the reactants, as shown by their nuclear binding energies, energy is released in nuclear fusion.

30-24. **Refer to Section 30-15.** . . . .

The primary advantage of nuclear energy is that enormous amounts of energy are liberated per unit mass of fuel. Also, the air pollution (oxides of S, N, C and particulate matter) caused by fossil fuel electric power plants is not a problem with nuclear energy plants. In European countries, where fossil fuel reserves are scarce, most of the electricity is generated by nuclear power plants for these reasons.

There are, however, some disadvantages associated with nuclear power from controlled fission reactions. The radionuclides must be properly shielded to protect the workers and the environment from radiation and contamination. Spent fuel, containing long-lived radioisotopes, must be disposed of carefully using special containers placed underground in geologically inactive areas. This is because the radiation from the fuel is biologically dangerous and must be contained until the fuel has decayed to the point when it is no longer dangerous. The problem is that the time involved could be several hundred thousand years. If there is inadequate cooling in the reactor, there is the possibility of overheating the fuel and causing a "meltdown." This cooling water can cause biological damage to aquatic life if it is returned to the natural water system while it is still too warm. Finally, it is possible that Pu-239 could be stolen and used for bomb production.

In the future, when nuclear fusion power plants are in operation, most of these disadvantages will not be a concern. For example, fusion reactions produce only short-lived isotopes and so there would be no long-term storage problems. An added advantage is that there is a virtually inexhaustible supply of deuterium fuel in the world's oceans.

30-26. **Refer to Section 30-15 and the Introduction to the Rare Earths in Chapter 28.** .

Uranium ores contain only about 0.7% U-235 which is fissionable. Most of the rest is nonfissionable U-238. To enrich U-235 for use in nuclear power plants, the oxide is converted to UF_4 with HF and then oxidized to UF_6 by fluorine. The vapor of $^{235}UF_6$ and $^{238}UF_6$ is then subjected to repeated diffusion through porous barriers to concentrate $^{235}UF_6$ (Graham's Law). Gas centrifuges are now used for the concentration process which is also based upon the difference in masses of the two U isotopes.

30-28. Refer to Section 30-16. . . . .

The major advantages of fusion as a potential energy source are three-fold:

(1) Fusion reactions are accompanied by much greater energy production per unit mass of reacting atoms than fission reactions.

(2) The deuterium fuel for fusion reactions is present in a virtually inexhaustible supply in the world oceans.

(3) Fusion reactions produce only short-lived radionuclides; there would be no long-term waste-disposal problem.

The only disadvantage of fusion is that extremely high temperatures are required to initiate the fusion process. A structural material that can withstand the high temperatures (4×10^7 K or more) and contain the fusion reaction, does not as yet exist.

30-30. Refer to Sections 30-9 and Section 16-8, and Example 30-3. . . . .

For first-order kinetics, $\log\left(\frac{A_o}{A}\right) = \frac{kt}{2.303}$ and $t_{1/2} = \frac{0.693}{k}$

A_o = initial amount of isotope
A = amount remaining after time, t
k = rate constant (units of time^{-1})
t = time
$t_{1/2}$ = half-life

Plan: (1) Calculate the rate constant, k, from the half-life for Fr-223.
(2) Calculate the amount of Fr-223 remaining.

(1) $k = \frac{0.693}{t_{1/2}} = 0.693/22 \text{ min} = 0.032 \text{ min}^{-1}$

(2) Substituting into the first-order rate equation,

$$\log\left(\frac{26\ \mu g}{A}\right) = \frac{(0.032 \text{ min}^{-1})(1.5 \text{ hr} \times 60 \text{ min/hr})}{2.303} = 1.3$$

$$\frac{26\ \mu g}{A} = 20$$

$$\mathbf{A = 1.3\ \mu g \text{ Fr-223 remaining}}$$

30-32. Refer to Sections 30-9 and 16-8, Example 30-4 and Exercise 30-30 Solution. . .

Plan: (1) Calculate the rate constant, k, from the half-life for carbon-14.
(2) Calculate the age of the object, t.

(1) $k = \frac{0.693}{t_{1/2}} = 0.693/5730 \text{ yr} = 1.21 \times 10^{-4} \text{ yr}^{-1}$

(2) The first-order rate equation: $\log\left(\frac{A_o}{A}\right) = \frac{kt}{2.303}$

Substituting, $\log\left(\frac{7.50\ \mu g}{0.67\ \mu g}\right) = \frac{(1.21 \times 10^{-4} \text{ yr}^{-1})(t)}{2.303}$

$$\mathbf{t = 2.00 \times 10^4 \text{ yr}}$$

30-34. **Refer to Sections 30-9 and 16-8, and Exercise 30-30 Solution.** • • • • • •

Plan: (1) Calculate the rate constant, k, using the first-order rate equation.
(2) Calculate the half-life, $t_{1/2}$, for potassium-42 from k.

(1) The first-order rate equation: $\log\left(\frac{A_o}{A}\right) = \frac{kt}{2.303}$

Substituting, we have

$$\log\left(\frac{4.000 \text{ g}}{0.232 \text{ g}}\right) = \frac{(k)(50.95 \text{ hr})}{2.303}$$ Solving, $k = 5.589 \times 10^{-2} \text{ hr}^{-1}$

(2) $t_{1/2} = 0.6933/k = 0.6933/(5.589 \times 10^{-2} \text{ hr}^{-1}) = \mathbf{12.40 \text{ hr}}$

Note: in Section 16-8, the half-life equation was derived, $t_{1/2} = \frac{\text{constant}}{k}$

where the constant = $2.303 \times \log 2 = 0.6933$

$= 0.693$ (3 significant figures)

31 Organic Chemistry I: Hydrocarbons

31-2. **Refer to the Introduction to Chapter 31.** . . . .

(a) Catenation means "chain-making" which describes the ability of an element to bond to itself.

(b) Carbon exhibits catenation to a much greater extent than any other element. Carbon atoms bond to each other to form long chains, branched chains and rings which may also contain chains attached to them. Millions of such compounds are known which constitutes the study of organic chemistry.

31-4. **Refer to the Introduction to Chapter 31.** . . . .

(a) Most synthetic organic materials are derived from petroleum, coal and natural gas.

(b) Most geochemists believe that petroleum, natural gas and coal are derived from plant matter, buried millions of years ago. Since the source of carbon for plants is CO_2, we can say that the ultimate source of many naturally occurring organic compounds which are based on carbon, is CO_2.

31-6. **Refer to Section 31-1.** . . . .

In alkanes such as (a) methane, CH_4, (b) ethane, C_2H_6, (c) propane, C_3H_8, and (d) *n*-butane, C_4H_{10} (Figures 31-1 to 31-4), the geometry about each C atom is tetrahedral. All the carbon atoms are connected to each other to form chains, and in the case of *n*-C_4H_{10}, a straight chain of 4 carbon atoms without branching is formed. Each of the C atoms undergoes sp^3 hybridization, and forms σ bonds with each other by using the sp^3 hybrid orbitals. The C atoms at the end of each chain are in the form of CH_3, each bonded to 3 H atoms by overlapping with their 1*s* orbitals to give σ bonds. The C atoms in the interior of each chain are in the form of CH_2, each bonded to 2 H atoms in the same fashion.

These four molecules are the first four members of the alkanes, a homologous series of saturated hydrocarbons with the general formula, C_nH_{2n+2}. The difference between them is in the number of C atoms in the compound; the formula of each alkane differs from the next by one CH_2 group.

31-8. **Refer to Section 31-1.** . . . .

(a) A homologous series is a series of compounds in which each member differs from the next by a specific number and kind of atoms.

(b) The alkane series contains saturated hydrocarbons such as CH_4, C_2H_6, C_3H_8 and C_4H_{10}. Each member differs from the next by CH_2 and are therefore examples of compounds that are members of a homologous series. Refer to Table 31-1 for the names and formulas of more members of this homologous series.

(c) A methylene group is a CH_2 group.

(d) The structures of homologous series members such as the alkanes, C_nH_{2n}, the alkenes, C_nH_{2n}, and the alkynes, C_nH_{2n-2}, all differ by a CH_2 unit from one member to the next. The properties of the members of a homologous series are closely related. For example, the boiling point of a compound in the homologous series given in (b) is higher than the compounds before it in the series, but less than the compounds after it due to increasing London dispersion forces.

31-10. **Refer to Section 31-1, Figure 31-6, Tables 31-2 and 31-3, and the Key Terms for Chapter 31.** •

(a) Structural isomers are compounds that contain the same number of the same kinds of atoms in different geometric arrangements.

(b) C_4H_{10}

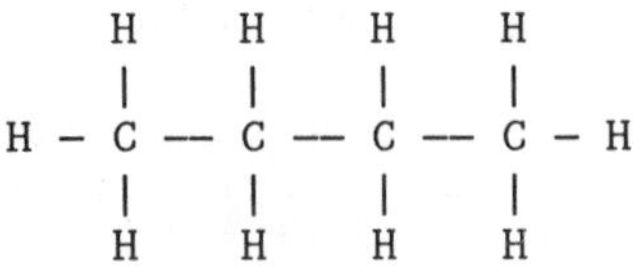

n-**butane**

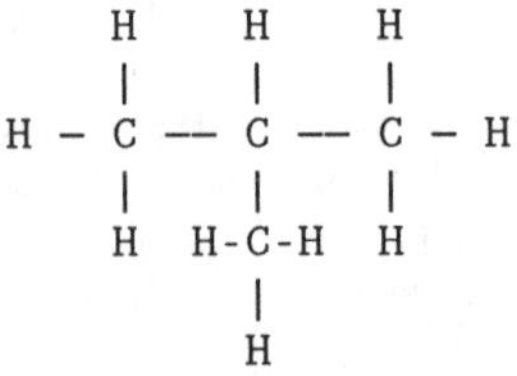

2-methylpropane

C_5H_{12}

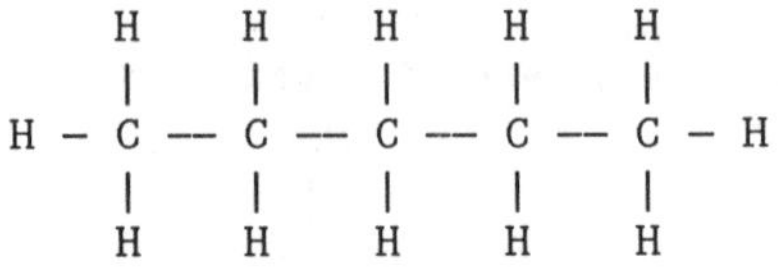

n-**pentane**

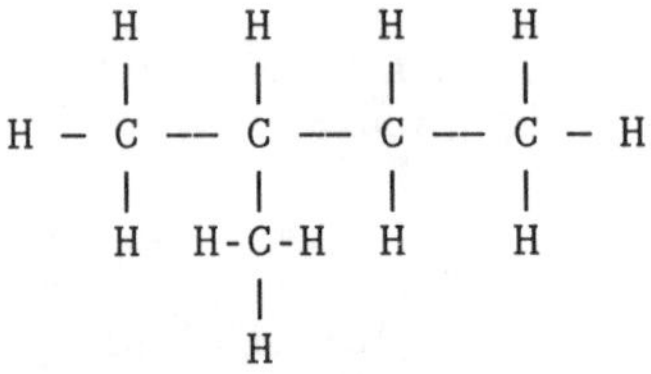

2-methylbutane

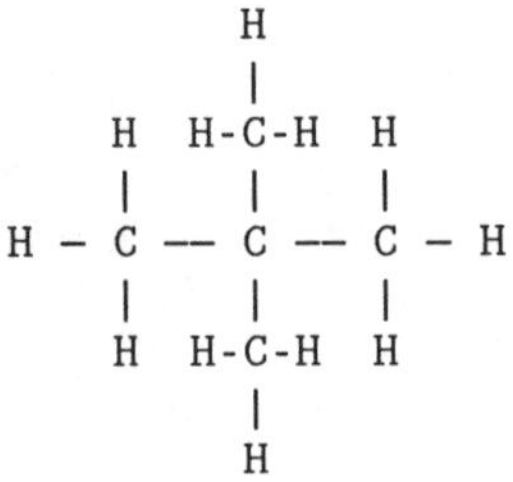

2,2-dimethylpropane

C_6H_{14}

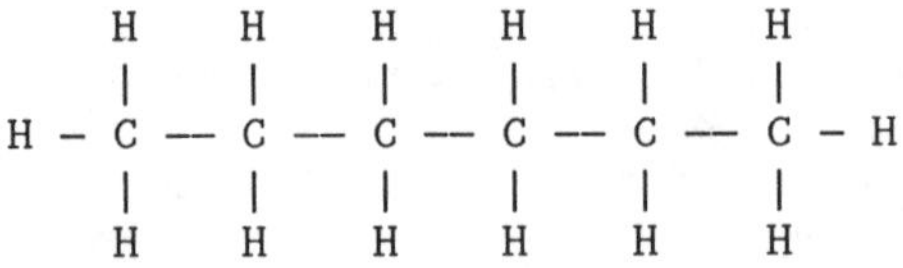

n-**hexane**

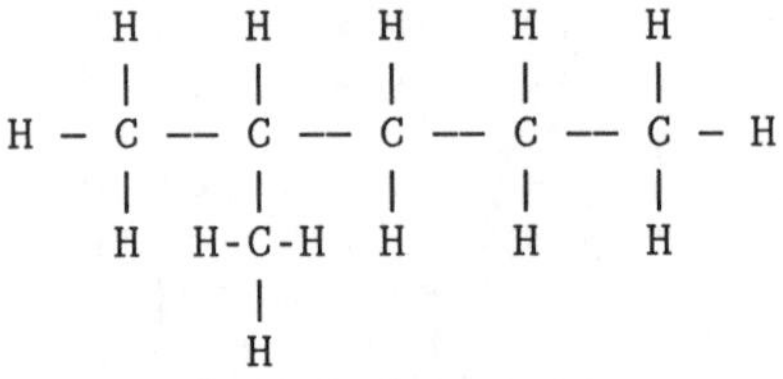

2-methylpentane

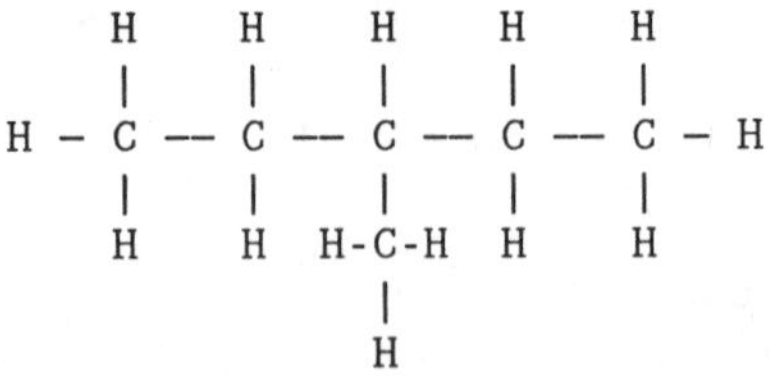

3-methylpentane

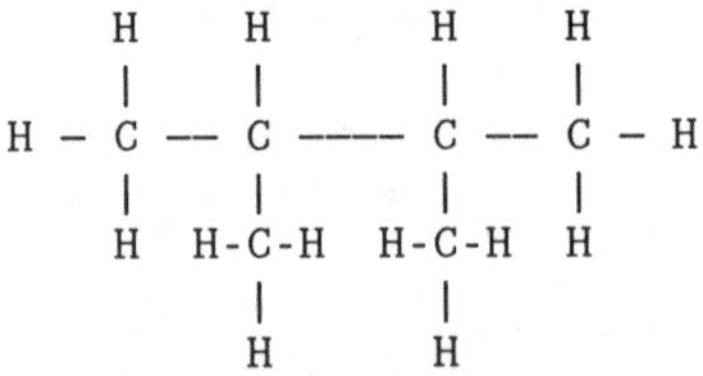

2,3-dimethylbutane

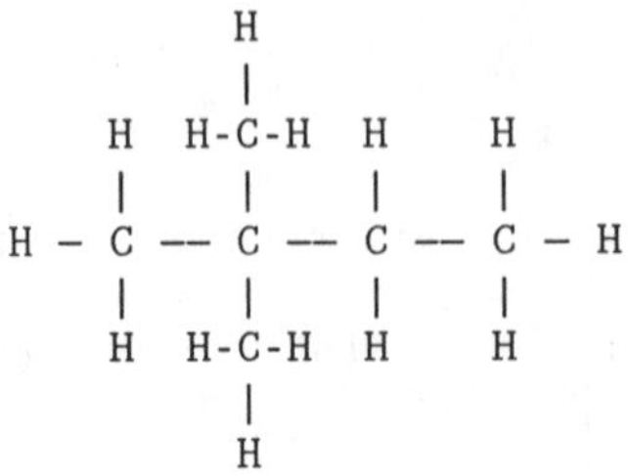

2,2-dimethylbutane

(c) Propane, C_3H_8, has no structural isomers because there is only one way to arrange the carbon and hydrogen atoms.

31-12. **Refer to Section 31-2 and the Key Terms for Chapter 31.** • • • • • • • • •

(a) An alkyl group (C_nH_{2n+1}) is derived from an alkane (C_nH_{2n+2}) by the removal of one hydrogen atom.

(b) methyl, CH_3-

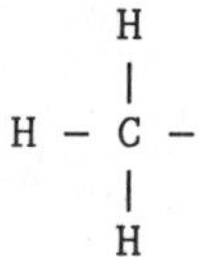

```
        H
        |
  H  –  C  –
        |
        H
```

ethyl, C_2H_5-

```
     H   H
     |   |
 H – C – C –
     |   |
     H   H
```

n-propyl, C_3H_7-

```
     H   H   H
     |   |   |
 H – C – C – C –
     |   |   |
     H   H   H
```

n-butyl, C_4H_9-

```
     H   H   H   H
     |   |   |   |
 H – C – C – C – C –
     |   |   |   |
     H   H   H   H
```

n-pentyl, C_5H_{11}-

```
     H   H   H   H   H
     |   |   |   |   |
 H – C – C – C – C – C –
     |   |   |   |   |
     H   H   H   H   H
```

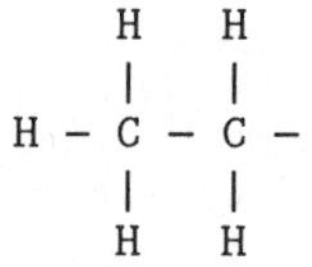

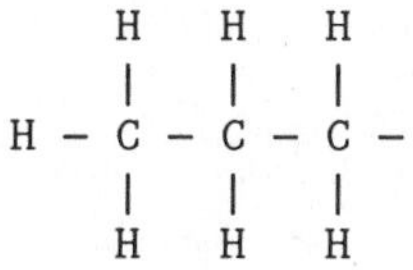

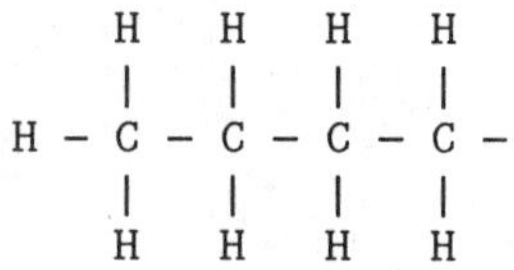

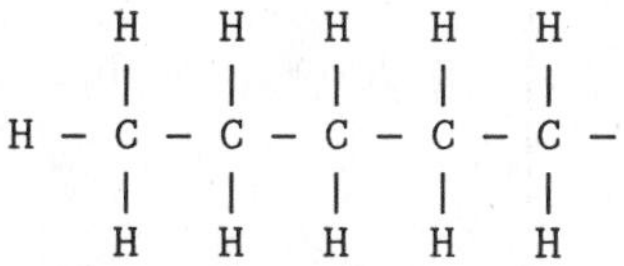

(c) The alkyl group names are derived from their parent hydrocarbon by changing the name ending from -ane to -yl.

31-14. **Refer to Section 31-3 and the Key Terms for Chapter 31.** • • • • • • • •

A conformation is one particular geometry of a molecule. Conformations of a molecule differ from one another by the extent of rotation about a single bond. This is illustrated in Figures 31-7 and 31-8.

31-16. **Refer to Section 31-5.** • • • • • • • • • • • • • • • • • • •

(a) A substitution reaction is a reaction in which an atom (or group of atoms) replaces another atom (or group of atoms) in an organic compound.

(b) A halogenation reaction is a reaction in which one or more hydrogen atoms of a hydrocarbon is replaced by the corresponding number of halogen atoms.

31-18. **Refer to Section 31-5 and Exercise 31-20 Solution.** • • • • • • • • • •

(a) The chlorination of ethane in ultraviolet light is a free radical chain reaction. It begins when the chlorine molecule is split into two very reactive Cl atoms, which can attack ethane, removing one of its H atoms to form HCl and a monosubstituted chloroethane. When a second hydrogen atom is replaced, a mixture of two disubstituted ethanes are produced as shown below.

(b),(c)

```
                  H   H                              H   H
                  |   |                              |   |
Cl – Cl   +   H – C – C – H   ─────────→   H – C – C – Cl   +   HCl
                  |   |           uv               |   |
                  H   H                            H   H

  chlorine        ethane                       chloroethane       hydrogen
                                                                  chloride
```

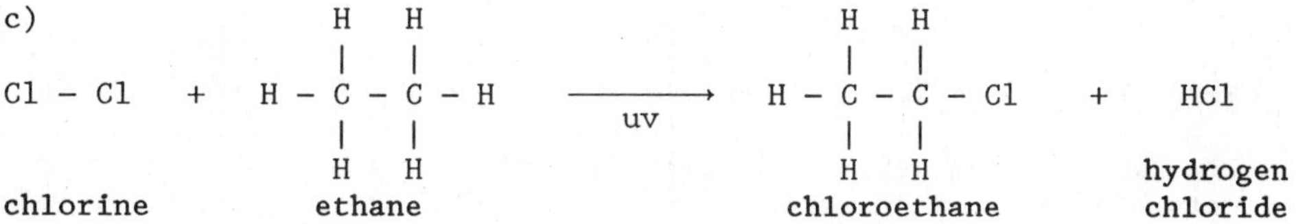

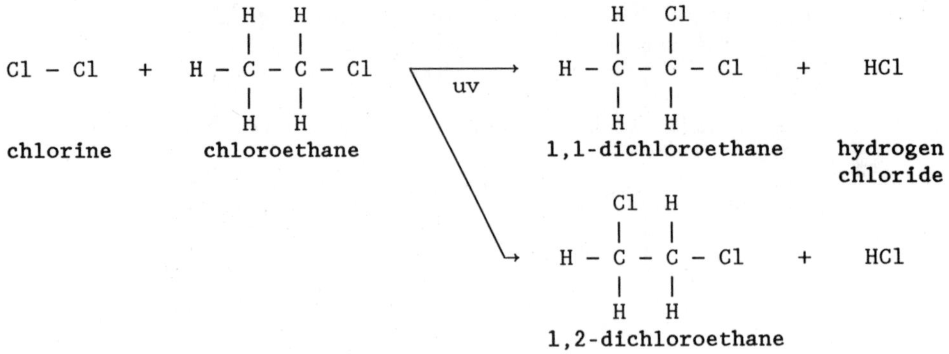

The substitution reaction will continue in the presence of excess Cl_2 to give the following chlorinated compounds:

H Cl	Cl Cl	Cl Cl
| |	| |	| |
H – C – C – Cl	H – C – C – Cl	H – C – C – Cl
| |	| |	| |
H Cl	H H	H Cl
1,1,1-trichloroethane	1,1,2-trichloroethane	1,1,1,2-tetrachloroethane

Cl Cl	Cl Cl	Cl Cl
| |	| |	| |
H – C – C – H	H – C – C – Cl	Cl – C – C – Cl
| |	| |	| |
Cl Cl	Cl Cl	Cl Cl
1,1,2,2-tetrachloroethane	pentachloroethane	hexachloroethane

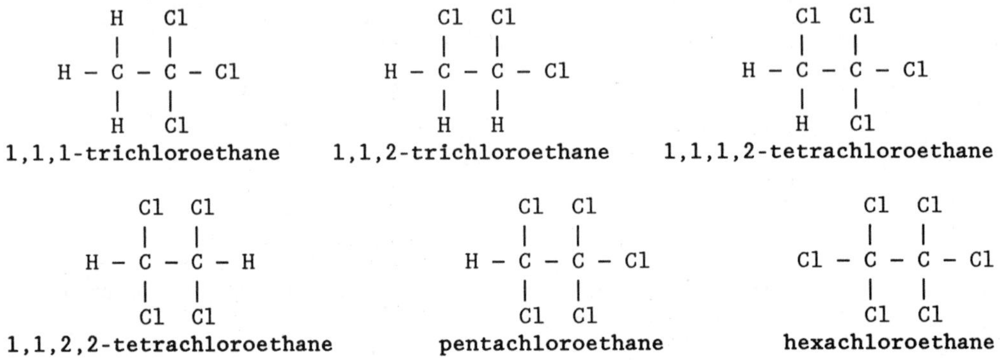

31-20. **Refer to Sections 31-5 and 24-10.** $\cdot\ \cdot\ \cdot\ \cdot\ \cdot\ \cdot\ \cdot\ \cdot\ \cdot\ \cdot\ \cdot\ \cdot\ \cdot$

A free radical chain reaction is one with a free radical or atom as a chain carrier, which perpetuates the reaction. It normally consists of three steps. For example, for the overall reaction: $A_2 + B_2 \rightarrow 2AB$

   (1) Chain Initiation: $A_2 \rightarrow 2A$

   (2) Propagation: $A + B_2 \rightarrow AB + B$
                           $B + A_2 \rightarrow AB + A$

   (3) Termination: $2A \rightarrow A_2$
                           $A + B \rightarrow AB$
                           $2B \rightarrow B_2$

31-22. **Refer to Section 31-5 and Exercise 31-18.** $\cdot\ \cdot\ \cdot\ \cdot\ \cdot\ \cdot\ \cdot\ \cdot\ \cdot\ \cdot\ \cdot\ \cdot$

Halogenation of cycloalkanes results in a single monosubstituted halogenation product because all the carbon atoms in cycloalkanes are equivalent and have the same number of hydrogen atoms bonded to each of them, i.e., they are all CH_2 groups. This is not the case with normal alkanes which contain not only CH_3, but also possibly CH_2 groups. Therefore, the halogenation of normal alkanes results in a mixture of different products.

31-24. **Refer to Section 31-6 and Table 31-4.** $\cdot\ \cdot\ \cdot\ \cdot\ \cdot\ \cdot\ \cdot\ \cdot\ \cdot\ \cdot\ \cdot\ \cdot$

The heat of combustion of ethyl alcohol (ethanol) is lower than that of the saturated alkanes on a per gram basis. On a per mole basis, ethanol's heat of combustion is lower than all the saturated alkanes except methane.

31-26. Refer to Section 31-7. . . . .

(a) Cracking refers to the process in which higher molecular-weight hydrocarbons (C_{12} and higher) are heated in the absence of air with catalysts to produce a mixture of more highly branched lower molecular-weight hydrocarbons.

(b) Cracking processes are necessary to increase the supply of gasoline from petroleum. Petroleum is a mixture of many hydrocarbons and only a fraction of them (C_5 to C_{12}) is initially suitable for use as gasoline.

31-28. Refer to Section 31-7. . . . .

Normal or "straight-chain" hydrocarbons have low octane ratings, while branched-chain hydrocarbons have higher octane ratings. The latter produces less knock in standard test engines.

31-30. Refer to Sections 31-1, 31-5 and 31-8. . . . .

(a) Alkenes contain C=C double bonds formed at the expense of two hydrogen atoms. Therefore, the general formula for alkenes is C_nH_{2n}, while that for alkanes is C_nH_{2n+2}.

(b) When an alkane loses two H atoms, the resulting species could undergo either ring-enclosure to give a cycloalkane, or H-shifting to give an alkene. Therefore, cycloalkanes and alkenes are isomers, both having the general formula, C_nH_{2n}.

31-32. Refer to Section 31-8. . . . .

(a) Dienes are alkenes with 2 double bonds per molecule.

(b) CH_2=CH-CH=CH_2 1,3-butadiene CH_3-CH=C=CH_2 1,2-butadiene

31-34. Refer to Sections 7-11 and 31-8, and Figures 7-2 and 7-3. . . . .

(a) ethene

CH_2=CH_2

Both carbon atoms in ethene undergo sp^2 hybridization. The C-H bonds involve overlap of sp^2 carbon orbitals with 1*s* orbitals of the H atoms. The carbon-carbon double bond involves the overlap of sp^2 orbitals from each carbon to give the σ bond and the side-on overlap of a *p* orbital from each carbon atom to give the π bond.

(b) propene

$\overset{1}{C}H_3$-$\overset{2}{C}H$=$\overset{3}{C}H_2$

Carbon atom (1) uses sp^3 hybrid orbitals to form four sigma bonds, three by overlap with the hydrogen 1*s* orbitals and one by overlap with a sp^2 orbital from the central carbon (2). The two carbon atoms involved in the double bond undergo sp^2 hybridization. They form C-H bonds by overlapping with 1*s* orbitals of the H atoms. The C=C double bond is formed similarly to that described in (a). The C(1)-C(2) single bond involves sp^2-sp^3 overlap.

(c) 1-butene

$\overset{1}{C}H_2$=$\overset{2}{C}H$-$\overset{3}{C}H_2$-$\overset{4}{C}H_3$

Carbon atoms (1) and (2) undergo sp^2 hybridization, while carbon atoms (3) and (4) undergo sp^3 hybridization. The overlap of the hybrid orbitals with the 1*s* orbitals of H atoms gives the C-H bonds. The C=C double bond is formed similarly to that described in (a). The C-C single bonds involve sp^2-sp^3 overlap for the C(2)-C(3) single bond and sp^3-sp^3 overlap for the C(3)-C(4) single bond.

(d) 2-butene

$\overset{1}{CH_3}-\overset{2}{CH}=\overset{3}{CH}-\overset{4}{CH_3}$

The orbital overlaps are similar to those in 1-butene except carbon atoms (2) and (3) are involved in the double bond and carbon atoms (2) and (3) are also bonded to another carbon atom by sp^2-sp^3 overlaps. A pair of geometric isomers (*trans-* and *cis-*) exists for this compound; exactly which isomer depends on which sp^2 orbitals on the middle C atoms are used to overlap with orbitals of their hydrogen atoms.

31-36. **Refer to Sections 31-1, 31-3 and 31-8.** • • • • • • • • • • • • • • •

Conformations are different orientations in space of a single molecule that result from rotation around single (sigma) bonds. Isomers are different compounds with different bonding arrangements that have the same molecular formula.

31-38. **Refer to Sections 31-5 and 31-9.** • • • • • • • • • • • • • • • • •

The characteristic reaction of the relatively unreactive alkanes is the substitution reaction which involves the replacement of one σ bonded atom for another and requires heat or light. The more reactive alkenes are characterized by addition reactions to the double bond, many of which occur easily at room temperature. The carbon-carbon double bond is a reaction site and is classified as a functional group. The π portion of the double bond can be utilized to accommodate two incoming atoms, converting the double bond into one single σ bond.

31-40. **Refer to Section 31-9.** • • • • • • • • • • • • • • • • • • •

(1) $Cl-Cl \quad + \quad CH_2=CH-CH_3 \quad \rightarrow \quad \underset{Cl}{CH_2}-\underset{Cl}{CH}-CH_3$

chlorine   **propene**   **1,2-dichloropropane**

(2) $Br-Br \quad + \quad CH_3-CH=CH-CH_3 \quad \rightarrow \quad CH_3-\underset{Br}{CH}-\underset{Br}{CH}-CH_3$

bromine   **2-butene**   **2,3-dibromobutane**

31-42. **Refer to Section 31-9 and the Key Terms for Chapter 31.** • • • • • • • • •

(a) Hydrogenation refers to the reaction in which molecular hydrogen, H_2, adds across a double or triple bond. This reaction requires elevated temperatures, high pressure and the presence of an appropriate heterogeneous catalyst (finely divided Pt, Pd or Ni).

(b) Hydrogenation is an important industrial process in many areas. For example, unsaturated hydrocarbons can be converted to saturated hydrocarbons by hydrogenation to manufacture high octane gasoline and aviation fuels. It is also employed to convert unsaturated vegetable oils to solid cooking fats.

(c),(d) $\underset{\textbf{ethene}}{CH_2=CH_2} \quad + \quad H_2 \quad \xrightarrow[\Delta]{\text{catalyst}} \quad \underset{\textbf{ethane}}{CH_3-CH_3}$

$\underset{\textbf{2-butene}}{CH_3-CH=CH-CH_3} \quad + \quad H_2 \quad \xrightarrow[\Delta]{\text{catalyst}} \quad \underset{\textbf{butane}}{CH_3-CH_2-CH_2-CH_3}$

31-44. **Refer to Section 31-9.** . . . .

(a) Polymerization is the combination of many monomers, usually small molecules, to form large molecules called polymers which contain repetitive units of the monomers.

(b) Examples of polymerization reactions:

(1) $n\text{CH}_2\text{=CH}_2 \xrightarrow{\text{catalyst}} \left[\text{CH}_2\text{-CH}_2\right]_n$

ethylene → **polyethylene**

(2) $n\text{CF}_2\text{=CF}_2 \xrightarrow[\Delta]{\text{catalyst}} \left[\text{CF}_2\text{-CF}_2\right]_n$

tetrafluoroethene → **"Teflon"**

(3) $n\text{CH}_2\text{=CH-C(Cl)=CH}_2 \longrightarrow \left[\text{CH}_2\text{-CH=C(Cl)-CH}_2\right]_n$

chloroprene → **neoprene**

31-46. **Refer to Section 31-9.** . . . .

(a) An elastomer is an elastic polymer with properties similar to natural rubber.

(b) Neoprene is an elastomer. It is used to make hoses for oil and gasoline, electrical insulation, and automobile and refrigerator parts.

(c) Neoprene is less affected by gasoline and oil and more elastic than natural rubber.

31-48. **Refer to Sections 31-11 and 31-8.** . . . .

(a) Alkynes are hydrocarbons containing carbon-carbon triple bonds.

(b) Since the first member of the series, C_2H_2, is commonly called acetylene, alkynes are also called acetylenic hydrocarbons.

(c) Alkynes have the general formula C_nH_{2n-2}.

(d) The general formula for alkynes is the same as that for cycloalkenes. If we compare the general formulas of both compounds to that of an alkene, C_nH_{2n}, we can see that there are 2 less H atoms in the alkyne and cycloalkene, due to the formation of another carbon-carbon bond. In the case of alkynes, the double bond becomes a triple bond, whereas for the cycloalkene, the extra carbon bond caused the cyclization of the compound.

31-50. **Refer to Section 31-11 and Exercise 31-49.** . . . .

(1) ethyne (acetylene): $\underset{sp}{\text{HC}}\equiv\underset{sp}{\text{CH}}$

(2) propyne: $\underset{sp^3}{\text{CH}_3}\text{-}\underset{sp}{\text{C}}\equiv\underset{sp}{\text{CH}}$

(3) 1-butyne: $\underset{sp^3}{\text{CH}_3}\text{-}\underset{sp^3}{\text{CH}_2}\text{-}\underset{sp}{\text{C}}\equiv\underset{sp}{\text{CH}}$

(4) 1-pentyne: $\underset{sp^3}{\text{CH}_3}\text{-}\underset{sp^3}{\text{CH}_2}\text{-}\underset{sp^3}{\text{CH}_2}\text{-}\underset{sp}{\text{C}}\equiv\underset{sp}{\text{CH}}$

The hybridization of the various C atoms in the four alkynes is shown above. For the C-H bonds, the C atom uses its hybridized orbitals to overlap with the 1*s* atomic orbital of H. All C-C single bonds are σ bonds, while the C≡C triple bond consists of one σ and two π bonds. The σ bonds are formed by overlapping the hybridized orbitals of the corresponding C atoms. The π bonds are formed by overlapping the remaining *p* orbitals. Ethyne has a linear molecular geometry.

For the other alkynes, the sp^3 hybridized carbon atoms have a basic geometry of a distorted tetrahedron centered on the carbon atom, while the *sp* hybridized carbon atoms lie on a straight line with *sp* carbon atoms in the center.

31-52. **Refer to Section 31-11.** . . . .

(a) The unsaturated π bonds are very susceptible to addition reactions because they are sources of electrons. Alkynes contain two π bonds, while alkenes contain only one π bond. Therefore, alkynes are more reactive.

(b) The most common kind of reaction that alkynes undergo is addition of atoms or groups across the triple bond.

(c),(d)

1. $$\underset{\textbf{1-butyne}}{CH_3\text{-}CH_2\text{-}C{\equiv}CH_3} \xrightarrow[Pt]{H_2} \underset{\textbf{1-butene}}{CH_3\text{-}CH_2\text{-}CH{=}CH_2} \xrightarrow[Pt]{H_2} \underset{\textbf{butane}}{CH_3\text{-}CH_2\text{-}CH_2\text{-}CH_3}$$

2. $$\underset{\textbf{propyne}}{CH_3\text{-}C{\equiv}CH} \xrightarrow{Br_2} \underset{\textbf{1,2-dibromopropene}}{CH_3\text{-}CBr{=}CHBr} \xrightarrow{Br_2} \underset{\textbf{1,1,2,2-tetrabromopropane}}{CH_3\text{-}CBr_2\text{-}CHBr_2}$$

3. $$\underset{\textbf{ethyne}}{HC{\equiv}CH} \xrightarrow{Cl_2} \underset{\textbf{1,2-dichloroethene}}{ClHC{=}CHCl} \xrightarrow{Cl_2} \underset{\textbf{1,1,2,2-tetrachloroethane}}{Cl_2HC\text{-}CHCl_2}$$

4. $$\underset{\textbf{propyne}}{CH_3\text{-}C{\equiv}CH} \xrightarrow{Cl_2} \underset{\textbf{1,2-dichloropropene}}{CH_3\text{-}CCl{=}CHCl} \xrightarrow{Cl_2} \underset{\textbf{1,1,2,2-tetrachloropropane}}{CH_3\text{-}CCl_2\text{-}CHCl_2}$$

31-54. **Refer to Section 31-12.** . . . .

(a) Benzene, C_6H_6, is the simplest, most common aromatic hydrocarbon.

(b) Benzene was discovered in 1825 by Michael Faraday when he fractionally distilled a by-product oil obtained in the manufacture of illuminating gas from whale oil.

31-56. **Refer to Section 31-12 and Figure 31-12a,b.** . . . .

(a) The structure of benzene, C_6H_6, is the combination of the following two resonance structures (left) which can be represented by the notation as shown on the right side.

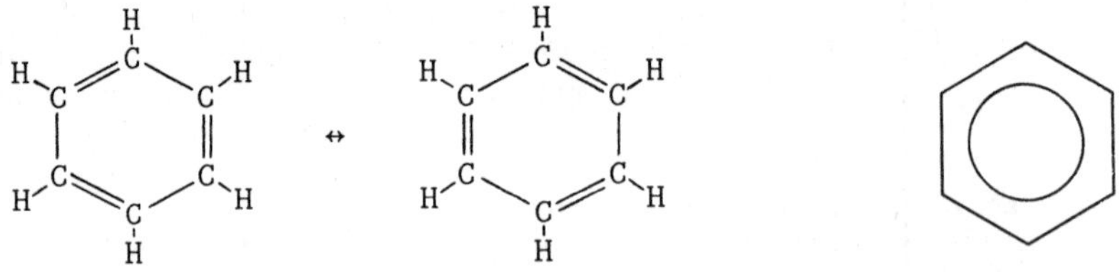

In a benzene molecule, all 12 atoms lie in a plane. This suggests sp^2 hybridization of each C. The six sp^2 hybridized C atoms lie in a plane, and the unhybridized *p* orbitals extend above and below the plane. Side-on overlap of the *p* orbitals form pi orbitals. The electrons associated with the pi bonds are delocalized over the entire benzene ring.

(b) The facts that only one monosubstitution product is obtained in numerous reactions and that no addition products can be prepared indicate conclusively that benzene has a symmetrical ring structure.

(a)

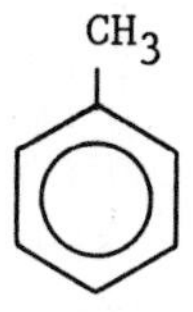

toluene   ethylbenzene   *ortho*-xylene   *meta*-xylene

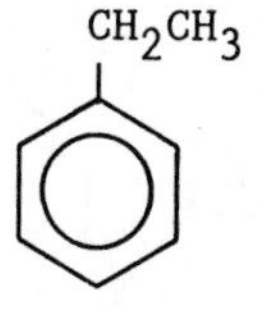

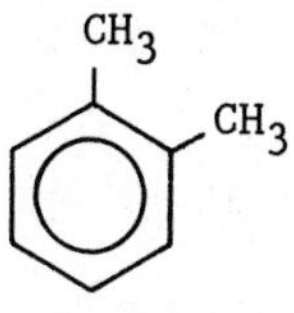

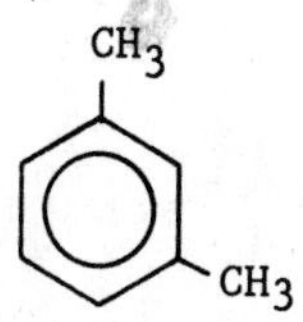

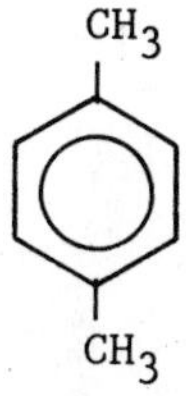

para-xylene   naphthalene   anthracene

(b) The aromatic hydrocarbons shown above are either substituted benzenes or systems with fused benzene rings.

31-60. Refer to Section 31-13. •

The prefixes, *ortho-*, *meta-* and *para-* refer to relative positions of the substitutents on a benzene ring. The *ortho-* (*o*-) prefix refers to two substituents located on adjacent carbons. The *meta-* (*m*-) prefix refers to substituents on carbon atoms (1) and (3). The *para-* (*p*-) prefix refers to substituents on carbon atoms (1) and (4). For example, the following names are interchangeable:

1,2-dichlorobenzene ≡ *o*-dichlorobenzene
1,3-dimethylbenzene ≡ *m*-xylene
4-nitrotoluene ≡ *p*-nitrotoluene

31-62. Refer to Section 31-14. •

(a) aliphatic substitution:

$$\text{toluene} + Cl_2 \xrightarrow[\text{(no catalyst)}]{\text{uv}} \text{benzyl chloride} + HCl$$

aromatic substitution:

$$\text{toluene} + Cl_2 \xrightarrow[\text{Fe catalyst}]{\text{(dark)}} \textit{p}\text{-chlorotoluene} + HCl$$

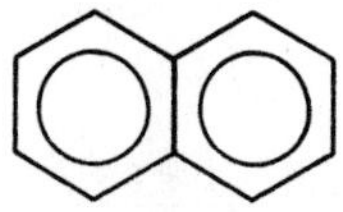

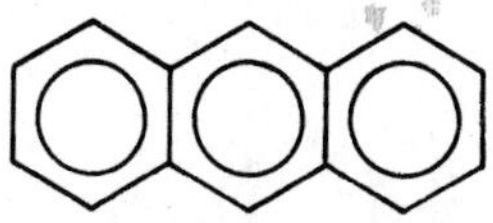

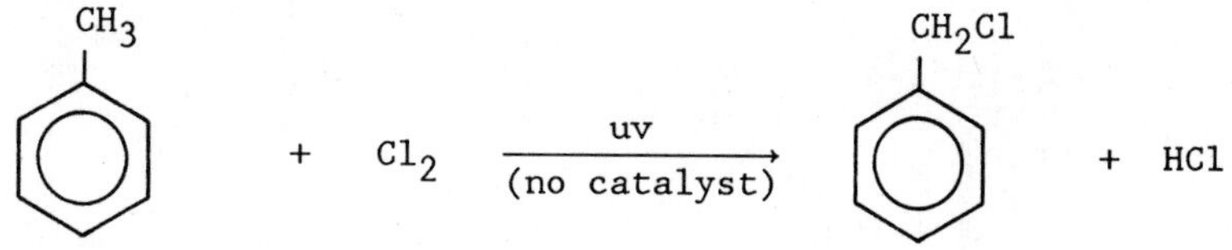

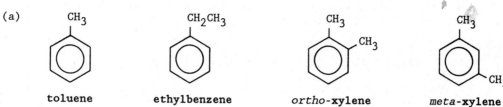

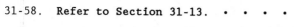

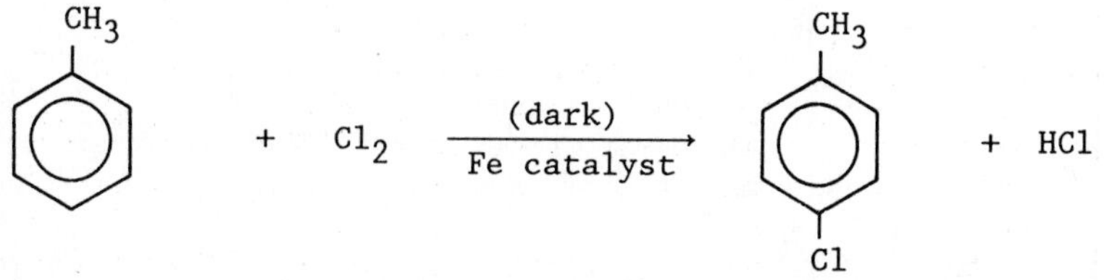

(b) aliphatic substitution:

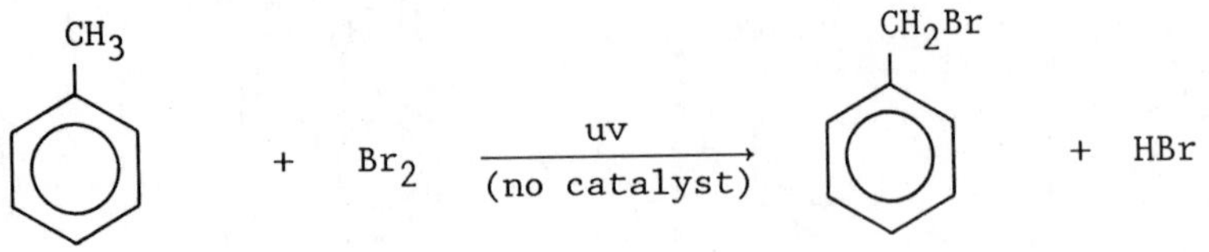

benzyl bromide

aromatic substitution:

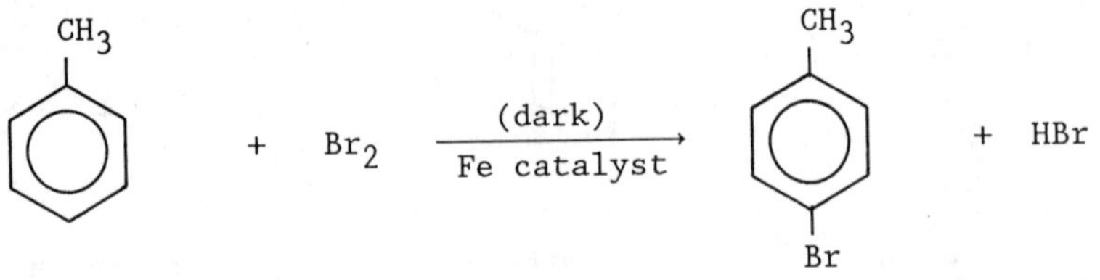

p-**bromotoluene**

31-64. **Refer to Section 31-14.** . . . .

(a) Elemental carbon particles (soot) are produced in burning if there is incomplete combustion. Aromatic hydrocarbons, such as benzene, are very stable due to resonance and therefore when combusted, they release less energy to the combustion process than expected. This in turn causes the carbon atoms to be less efficiently oxidized. Carbon atoms then are oxidized to oxidation state 0, producing soot, rather than to oxidation state +4, the oxidation state of carbon in CO_2.

(b) The flames would be expected to be yellow (a reducing flame), a sign of incomplete combustion, rather than blue (an oxidizing flame), a sign of complete combustion.

32 Organic Chemistry II: Functional Groups

32-2. Refer to the Introduction to Alcohols and Phenols in Chapter 32. • • • • • •

(a) Alcohols and phenols are hydrocarbon derivatives which contain the hydroxyl group (-OH) as their functional group.

(b) Alcohols are derived from hydrocarbons by replacing at least one hydrogen atom with a hydroxyl (-OH) group. On the other hand, in phenols, the -OH group must attach directly to an aromatic ring. Phenols are weak acids, while alcohols are neutral.

(c) Alcohols and phenols can be viewed as derivatives of hydrocarbons in which a hydrogen atom is replaced by an -OH group. On the other hand, they can also be viewed as derivatives of water in which a hydrogen atom is replaced by an organic group.

32-4. Refer to Section 32-1. •

Alcohols: CH_3OH CH_3CH_2OH $CH_3CH_2CH_2OH$ $CH_3CH_2(OH)CH_3$
             methanol   ethanol   1-propanol   2-propanol

             $CH_3CH_2CH_2CH_2OH$ $CH_3CH_2CH(OH)CH_3$
             1-butanol   2-butanol

Phenols: OH CH_3 OH

                                      OH

                                                         OH
             phenol   o-cresol   hydroquinone

32-6. Refer to Section 32-1. • • • • • • • • • • • • • • • • • • •

(a) The four saturated alcohols that contain four carbon atoms and one -OH group per molecule are:

$CH_3CH_2CH_2CH_2OH$ CH_3CHCH_2OH $CH_3CH_2CHCH_3$ CH_3
                              | | CH_3COH
                             CH_3 OH |
                                                                       CH_3

   1-butanol   2-methyl-1-propanol   2-butanol   2-methyl-2-propanol
      1°   1°   2°   3°

(b) The eight saturated alcohols that contain five carbon atoms and one -OH group per molecule are:

$CH_3CH_2CH_2CH_2CH_2OH$ $CH_3CH_2CH_2CHCH_3$ $CH_3CH_2CHCH_2CH_3$ $CH_3CH_2CHCH_2OH$
                                     | | |
                                    OH   OH CH_3

   1-pentanol   2-pentanol   3-pentanol   2-methyl-1-butanol
      1°   2°   2°   1°

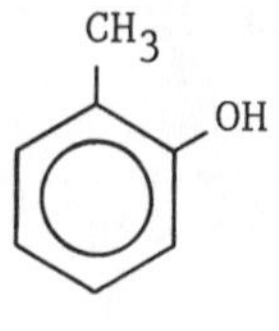

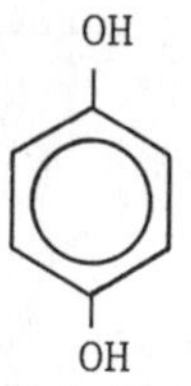

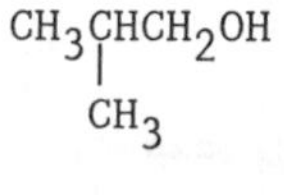

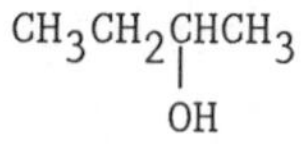

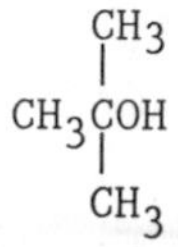

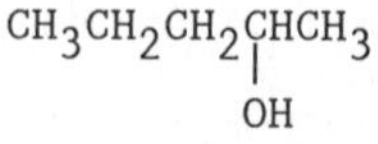

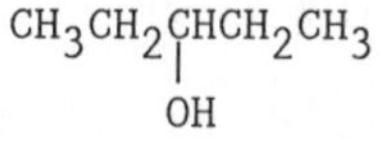

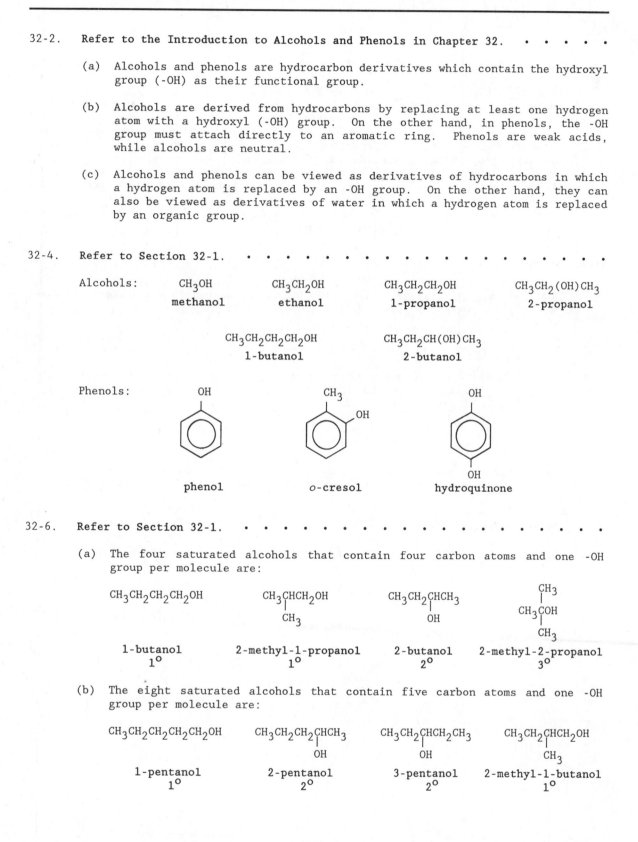

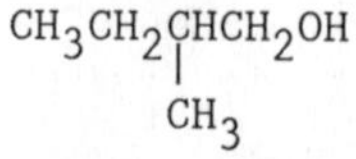

$$CH_3CHCH_2CH_2OH$$
$$CH_3$$

$$CH_3CH-CHCH_3$$
$$CH_3 \; OH$$

$$OH$$
$$CH_3CH_2CCH_3$$
$$CH_3$$

$$CH_3$$
$$CH_3CCH_2OH$$
$$CH_3$$

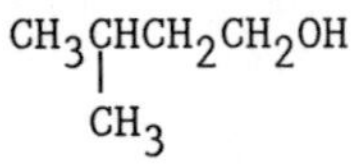

3-methyl-1-butanol
1°

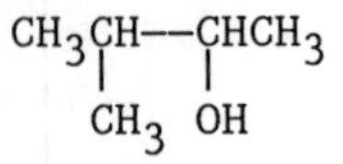

3-methyl-2-butanol
2°

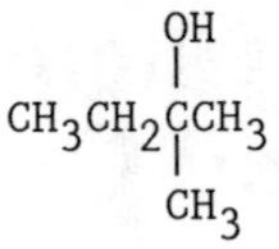

2-methyl-2-butanol
3°

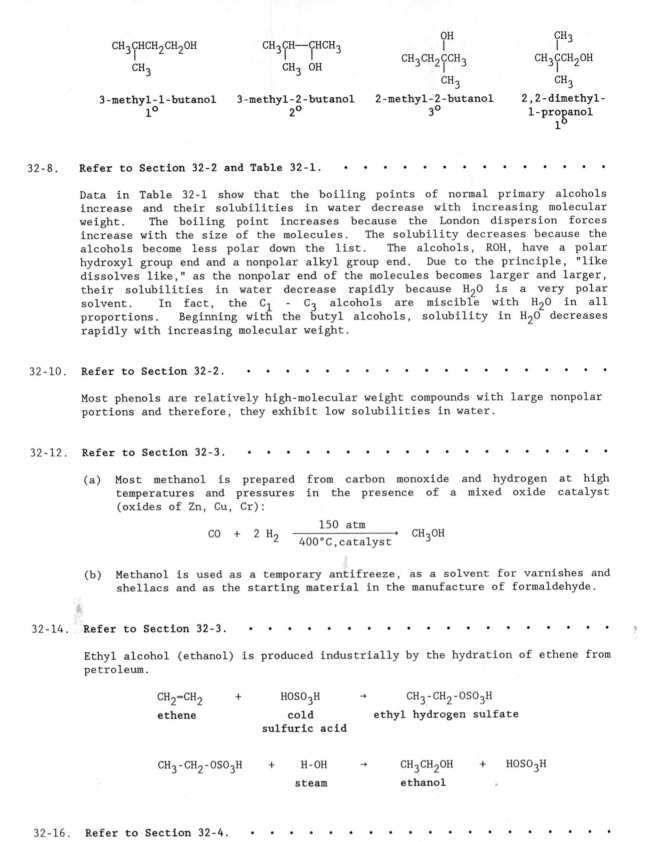

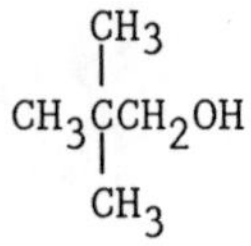

2,2-dimethyl-
1-propanol
1°

32-8. **Refer to Section 32-2 and Table 32-1.** • • • • • • • • • • • • • • • • •

Data in Table 32-1 show that the boiling points of normal primary alcohols increase and their solubilities in water decrease with increasing molecular weight. The boiling point increases because the London dispersion forces increase with the size of the molecules. The solubility decreases because the alcohols become less polar down the list. The alcohols, ROH, have a polar hydroxyl group end and a nonpolar alkyl group end. Due to the principle, "like dissolves like," as the nonpolar end of the molecules becomes larger and larger, their solubilities in water decrease rapidly because H_2O is a very polar solvent. In fact, the C_1 - C_3 alcohols are miscible with H_2O in all proportions. Beginning with the butyl alcohols, solubility in H_2O decreases rapidly with increasing molecular weight.

32-10. **Refer to Section 32-2.** •

Most phenols are relatively high-molecular weight compounds with large nonpolar portions and therefore, they exhibit low solubilities in water.

32-12. **Refer to Section 32-3.** •

(a) Most methanol is prepared from carbon monoxide and hydrogen at high temperatures and pressures in the presence of a mixed oxide catalyst (oxides of Zn, Cu, Cr):

$$CO + 2 H_2 \xrightarrow[\text{400°C,catalyst}]{\text{150 atm}} CH_3OH$$

(b) Methanol is used as a temporary antifreeze, as a solvent for varnishes and shellacs and as the starting material in the manufacture of formaldehyde.

32-14. **Refer to Section 32-3.** •

Ethyl alcohol (ethanol) is produced industrially by the hydration of ethene from petroleum.

$$CH_2=CH_2 \quad + \quad HOSO_3H \quad \rightarrow \quad CH_3-CH_2-OSO_3H$$
ethene   cold   ethyl hydrogen sulfate
sulfuric acid

$$CH_3-CH_2-OSO_3H \quad + \quad H-OH \quad \rightarrow \quad CH_3CH_2OH \quad + \quad HOSO_3H$$
steam   ethanol

32-16. **Refer to Section 32-4.** •

(a),(b) (1) $2CH_3OH \quad + \quad 2Na \quad \rightarrow \quad 2[Na^+ + CH_3O^-] \quad + \quad H_2$
methanol   sodium methoxide

(2) $2CH_3CH_2OH$ + $2Na$ → $2[Na^+ + CH_3CH_2O^-]$ + H_2
ethanol **sodium ethoxide**

(3) $2CH_3CH_2CH_2OH$ + $2Na$ → $2[Na^+ + CH_3CH_2CH_2O^-]$ + H_2
1-propanol **sodium propoxide**

(c) The reactions of alcohols with sodium are similar to the reaction of metallic sodium with water. Both types of reactions are oxidation-reduction reactions involving the displacement of hydrogen by sodium.

32-18. **Refer to Section 32-4.** . . . .

(a),(b) (1) CH_3OH + $HONO_2$ → CH_3ONO_2 + H_2O
methyl nitrate

(2) CH_3CH_2OH + $HONO_2$ → $CH_3CH_2ONO_2$ + H_2O
ethyl nitrate

(3) $CH_3CH_2CH_2OH$ + $HONO_2$ → $CH_3CH_2CH_2ONO_2$ + H_2O
***n*-propyl nitrate**

(4) $CH_3CH_2CH_2CH_2OH$ + $HONO_2$ → $CH_3CH_2CH_2CH_2ONO_2$ + H_2O
***n*-butyl nitrate**

(c) An inorganic ester may be thought of as a compound that contains one or more alkyl groups covalently bonded to the anion of a ternary inorganic acid.

32-20. **Refer to Sections 12-19 and 32-4.** . . . .

(a) An alkyl hydrogen sulfate is an inorganic ester which is composed of an alkyl group covalently bonded to the hydrogen sulfate anion.

(b) Alkyl hydrogen sulfates are prepared by the reaction of alcohols with cold concentrated sulfuric acid.

(c) Alkyl hydrogen sulfates can be classified as acids because the hydrogen atom in the hydrogen sulfate ion is acidic. This is supported by the fact that reactions of alkyl hydrogen sulfates with sodium hydroxide produce sodium salts of the alkyl hydrogen sulfates.

(d) Detergents are soap-like emulsifiers that contain sulfonate, $-SO_3^-$, sulfate, $-OSO_3^-$, or phosphate groups, $-OPO_3^{2-}$. They do not form precipitates with the ions of hard water, Ca^{2+}, Mg^{2+} and Fe^{3+}. Sodium salts of alkyl hydrogen sulfates that contain approximately 12 carbon atoms are excellent detergents.

(e) Sodium lauryl sulfate is $CH_3(CH_2)_{10}CH_2\text{-}OSO_3^-Na^+$.

(f) Sodium lauryl sulfate is a commonly used detergent.

32-22. **Refer to Section 32-5 and 32-6, and Table 32-2.** . . . .

(a) An acyl chloride, sometimes called an acid chloride, is a compound that is a derivative of a carboxylic acid by replacing the -OH by a Cl atom. It has the general formula:

$$R-\overset{\overset{\displaystyle O}{\|}}{C}-Cl$$

(b) Acyl chlorides are usually prepared by treating carboxylic acids with PCl_3, PCl_5 or $SOCl_2$ (thionyl chloride).

(c)

$$\text{(1)} \quad \underset{\text{acetic acid}}{CH_3\overset{\displaystyle O}{\overset{\|}{C}}OH} \quad + \quad \underset{\substack{\text{phosphorus} \\ \text{pentachloride}}}{PCl_5} \quad \rightarrow \quad \underset{\text{acetyl chloride}}{CH_3\overset{\displaystyle O}{\overset{\|}{C}}Cl} \quad + \quad \underset{\substack{\text{phosphorus} \\ \text{oxychloride}}}{POCl_3} \quad + \quad HCl$$

$$\text{(2)} \quad \underset{\text{acetic acid}}{CH_3\overset{\displaystyle O}{\overset{\|}{C}}OH} \quad + \quad \underset{\substack{\text{thionyl} \\ \text{chloride}}}{SOCl_2} \quad \rightarrow \quad \underset{\text{acetyl chloride}}{CH_3\overset{\displaystyle O}{\overset{\|}{C}}Cl} \quad + \quad \underset{\substack{\text{sulfur} \\ \text{dioxide}}}{SO_2} \quad + \quad HCl$$

$$\text{(3)} \quad \underset{\substack{\text{propanoic acid} \\ \text{(propionic acid)}}}{CH_3CH_2\overset{\displaystyle O}{\overset{\|}{C}}OH} \quad + \quad PCl_5 \quad \rightarrow \quad \underset{\text{propanoyl chloride}}{CH_3CH_2\overset{\displaystyle O}{\overset{\|}{C}}Cl} \quad + \quad POCl_3 \quad + \quad HCl$$

$$\text{(4)} \quad \underset{\text{benzoic acid}}{C_6H_5\overset{\displaystyle O}{\overset{\|}{C}}OH} \quad + \quad PCl_5 \quad \rightarrow \quad \underset{\text{benzoyl chloride}}{C_6H_5\overset{\displaystyle O}{\overset{\|}{C}}Cl} \quad + \quad POCl_3 \quad + \quad HCl$$

32-24. Refer to Section 32-7. • • • • • • • • • • • • • • •

$$\text{(1)} \quad \underset{\text{acetic acid}}{CH_3\text{-}\overset{\displaystyle O}{\overset{\|}{C}}\text{-}OH} \quad + \quad \underset{\text{ethanol}}{CH_3\text{-}CH_2\text{-}OH} \quad \rightarrow \quad \underset{\text{ethyl acetate}}{CH_3\text{-}\overset{\displaystyle O}{\overset{\|}{C}}\text{-}O\text{-}CH_2\text{-}CH_3} \quad + \quad H_2O$$

$$\text{(2)} \quad \underset{\substack{\text{propanoic acid} \\ \text{(propionic acid)}}}{CH_3\text{-}CH_2\text{-}\overset{\displaystyle O}{\overset{\|}{C}}\text{-}OH} \quad + \quad \underset{\text{methanol}}{CH_3\text{-}OH} \quad \rightarrow \quad \underset{\substack{\text{methyl propanoate} \\ \text{(methyl propionate)}}}{CH_3\text{-}CH_2\text{-}\overset{\displaystyle O}{\overset{\|}{C}}\text{-}O\text{-}CH_3} \quad + \quad H_2O$$

$$\text{(3)} \quad \underset{\text{benzoic acid}}{C_6H_5\text{-}\overset{\displaystyle O}{\overset{\|}{C}}\text{-}OH} \quad + \quad \underset{\text{ethanol}}{CH_3\text{-}CH_2\text{-}OH} \quad \rightarrow \quad \underset{\text{ethyl benzoate}}{C_6H_5\text{-}\overset{\displaystyle O}{\overset{\|}{C}}\text{-}O\text{-}CH_2\text{-}CH_3} \quad + \quad H_2O$$

32-26. Refer to Section 32-7 and Table 32-5. • • • • • • • • • • •

Ester	Formula	Odor or Origin
n-butyl acetate	$CH_3COOC_4H_9$	bananas
ethyl butyrate	$C_3H_7COOC_2H_5$	pineapples
n-amyl butyrate	$C_3H_7COOC_5H_{11}$	apricots
n-octyl acetate	$CH_3COOC_8H_{17}$	oranges
isoamyl isovalerate	$C_4H_9COOC_5H_{11}$	apples
methyl anthranilate	$C_6H_4(NH_2)(COOCH_3)$	grapes

32-28. Refer to Section 32-7. •

(a) $CH_3-\overset{\overset{\textstyle O}{\|}}{C}-O-CH_3$ + Na^+OH^- $\overset{\Delta}{\rightarrow}$ $CH_3-\overset{\overset{\textstyle O}{\|}}{C}-O^-Na^+$ + CH_3-OH
     methyl acetate   sodium acetate   methanol

(b) $H-\overset{\overset{\textstyle O}{\|}}{C}-O-CH_2-CH_3$ + Na^+OH^- $\overset{\Delta}{\rightarrow}$ $H-\overset{\overset{\textstyle O}{\|}}{C}-O^-Na^+$ + CH_3-CH_2-OH
     ethyl formate   sodium formate   ethanol

(c) $CH_3-\overset{\overset{\textstyle O}{\|}}{C}-O-CH_2-CH_2-CH_2-CH_3$ + Na^+OH^- $\overset{\Delta}{\rightarrow}$ $CH_3-\overset{\overset{\textstyle O}{\|}}{C}-O^-Na^+$ + $CH_3-CH_2-CH_2-CH_2-OH$
     n-butyl acetate   sodium acetate   1-butanol

(d) $CH_3-\overset{\overset{\textstyle O}{\|}}{C}-O-(CH_2)_7CH_3$ + Na^+OH^- $\overset{\Delta}{\rightarrow}$ $CH_3-\overset{\overset{\textstyle O}{\|}}{C}-O^-Na^+$ + $CH_3-(CH_2)_7-OH$
     n-octyl acetate   sodium acetate   *n*-octanol

32-30. Refer to Section 32-7. •

As shown below, glycerides are the triesters of glycerol. Simple glycerides are esters in which all three R groups are identical whereas mixed glycerides contain a mixture of various R groups.

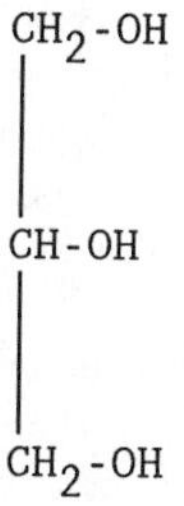

$$CH_2-OH$$
$$|$$
$$CH-OH$$
$$|$$
$$CH_2-OH$$
glycerol

$$CH_2-O-\overset{\overset{\textstyle O}{\|}}{C}-R$$
$$|$$
$$CH-O-\overset{\overset{\textstyle O}{\|}}{C}-R$$
$$|$$
$$CH_2-O-\overset{\overset{\textstyle O}{\|}}{C}-R$$
glycerides

32-32. Refer to Section 32-7. •

$$CH_2O\overset{\overset{\textstyle O}{\|}}{C}(CH_2)_{14}CH_3$$
$$|$$
$$CHO\overset{\overset{\textstyle O}{\|}}{C}(CH_2)_{14}CH_3$$
$$|$$
$$CH_2O\overset{\overset{\textstyle O}{\|}}{C}(CH_2)_{14}CH_3$$
glyceryl tripalmitate

$$CH_2O\overset{\overset{\textstyle O}{\|}}{C}(CH_2)_{16}CH_3$$
$$|$$
$$CHO\overset{\overset{\textstyle O}{\|}}{C}(CH_2)_{16}CH_3$$
$$|$$
$$CH_2O\overset{\overset{\textstyle O}{\|}}{C}(CH_2)_{16}CH_3$$
glyceryl tristearate

$$CH_2O\overset{\overset{\textstyle O}{\|}}{C}(CH_2)_{10}CH_3$$
$$|$$
$$CHO\overset{\overset{\textstyle O}{\|}}{C}(CH_2)_{10}CH_3$$
$$|$$
$$CH_2O\overset{\overset{\textstyle O}{\|}}{C}(CH_2)_{10}CH_3$$
glyceryl trilaurate

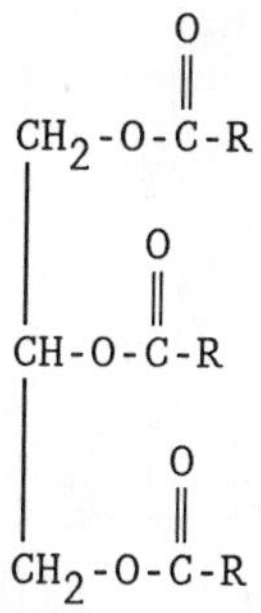

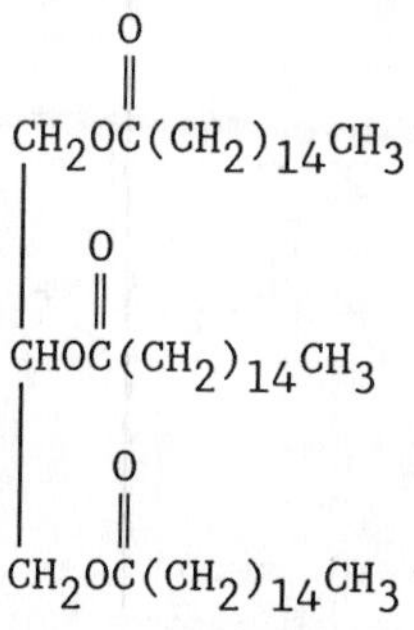

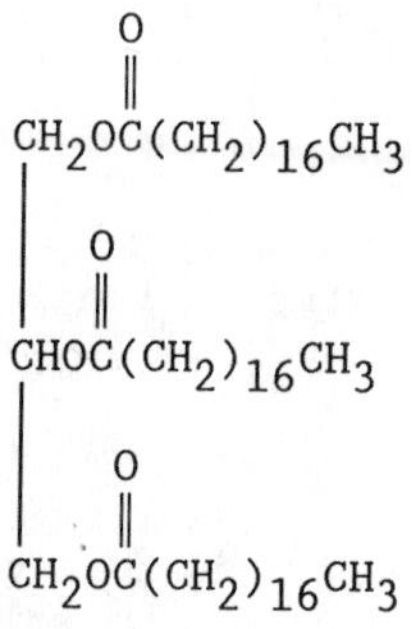

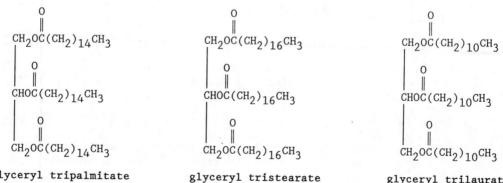

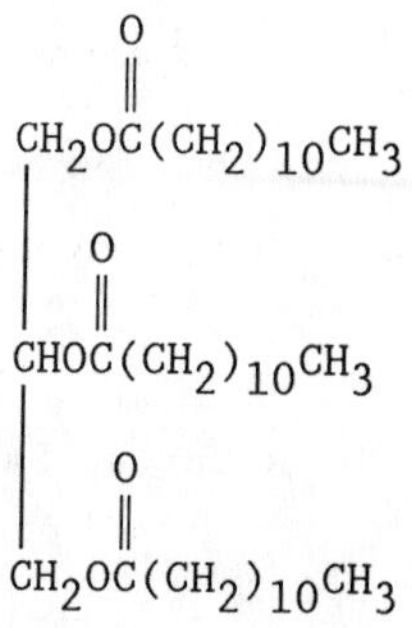

32-34. **Refer to Section 12-19 and 32-7.** . . . .

(a) Soaps are salts of long chain fatty acids, produced by the hydrolysis of fats in strongly basic solution.

(b) Detergents are salts of sulfonate, sulfate or phosphate groups, whereas soaps are salts of carboxylic acids.

32-36. **Refer to Section 32-8.** . . . .

(a) Polyesters are polymeric esters which are the reaction products between dihydric alcohols and dicarboxylic acids.

(b) Dacron is polyethylene terephthalate. It is not an absorbent and its properties are very much the same when wet or dry. It is very elastic, so it is used to make "permanent press" fabrics.

(c) Dacron is a polyester prepared from the reaction of ethylene glycol and terephthalic acid.

(d) Mylar is the name for polyethylene terephthalate (Dacron) made into a film, which has great strength.

32-38. **Refer to the Introduction to Amines in Chapter 32.** . . . .

(a) The amines are derivatives of ammonia, NH_3, in which one or more H atoms have been replaced by organic groups. They have the general formula: RNH_2, R_2NH or R_3N, where R is any alkyl or aryl group. Amines are basic; their basicity is derived from the lone pair of electrons on the N atoms.

(b) Amines are considered to be derivatives of ammonia. The structures of NH_3, primary, secondary and tertiary amines are shown below. From the comparison, it is obvious that amines can be treated as if one, two or three hydrogen atoms of ammonia have been replaced by organic groups.

ammonia | **primary amine** | **secondary amine** | **tertiary amine**

32-40. **Refer to the Introduction to Amines and Section 18-4.** . . . .

Amines react (hydrolyze) with water in a similar way as ammonia to form a basic solution. For example,

$$CH_3NH_2 + H_2O \rightleftharpoons CH_3NH_3^+ + OH^-$$

methylamine — **methylammonium ion**

The amines are basic because the lone pair of electrons on the N atom can be used to accommodate an incoming H^+ ion, thereby acting as a proton accepter.

32-42. **Refer to Section 32-11.** . . . .

(a) Amides are thought of as derivatives of primary or secondary amines and organic acids. They contain the following grouping of atoms:

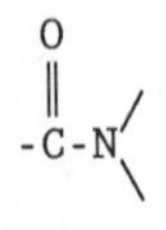

32-44. Refer to Section 32-11. . . . .

Polyamides are polymeric amides. Nylon, a very important fiber product, is the best known polymeric amide.

32-46. Refer to Section 32-11. . . . .

(a) Common Nylon is called Nylon 6-6 because the parent diamine and dicarboxylic acid of Nylon 6-6 each contain six carbon atoms.

(b) The parent diamine and dicarboxylic acid of Nylon x-y would contain x and y carbon atoms, respectively.

$$\text{Nylon 4-5:} \quad -NH \left[\begin{matrix} O \\ \| \\ C \end{matrix} - (CH_2)_3 - \begin{matrix} O \\ \| \\ C \end{matrix} - NH - (CH_2)_4 - NH \right] \begin{matrix} O \\ \| \\ C \end{matrix} -$$

$$\text{Nylon 6-4:} \quad -NH \left[\begin{matrix} O \\ \| \\ C \end{matrix} - (CH_2)_2 - \begin{matrix} O \\ \| \\ C \end{matrix} - NH - (CH_2)_6 - NH \right] \begin{matrix} O \\ \| \\ C \end{matrix} -$$

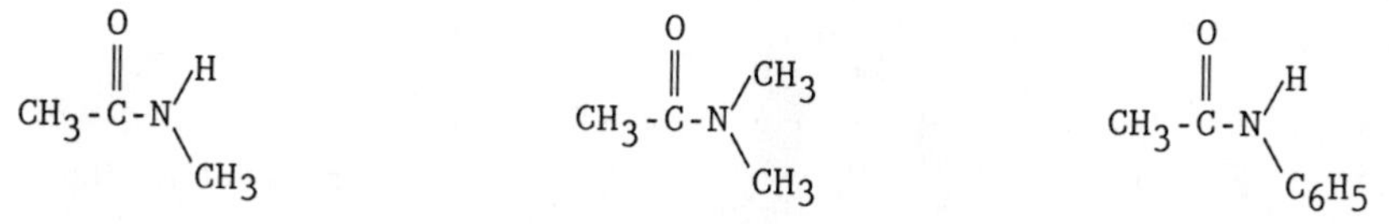

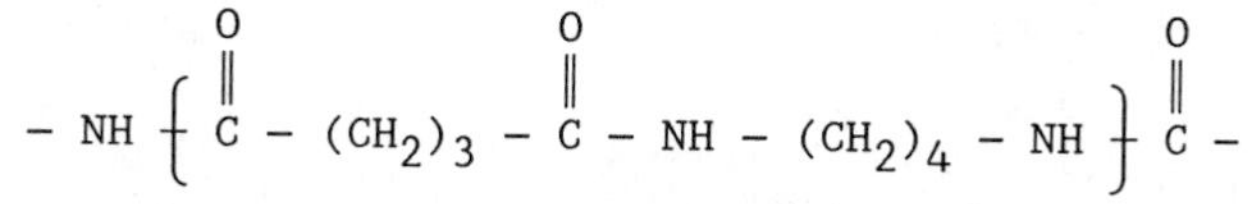

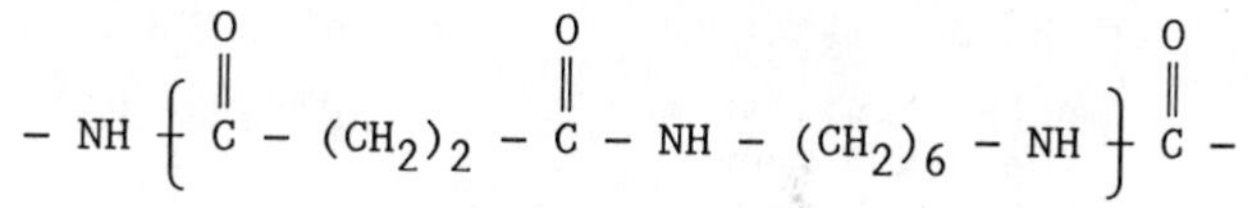

32-48. Refer to Section 32-12. . . . .

(a),(b)

Classification	Examples	Sources
Aldehyde	benzaldehyde	almonds
	cinnamaldehyde	cinnamon
	vanillin	vanilla bean
Ketone	muscone	musk deer
	testosterone	male sex hormone
	camphor	camphor tree

(c) The aldehydes listed in (a) have many uses. The three aldehydes can be used to add flavor to food. Muscone is the compound that gives the scent to Musk perfumes, deodorants, cologne and aftershave lotions. Testosterone is used to regulate male sexual and reproductive functions. Camphor is used in medicine as a diaphoretic, stimulant and sedative.

32-50. Refer to Section 32-13. . . . .

$$\begin{matrix} OH \\ | \\ \text{(1)} \quad CH_3\text{-}CH\text{-}CH_3 \end{matrix} \xrightarrow[OH^-]{KMnO_4} \begin{matrix} O \\ \| \\ CH_3\text{-}C\text{-}CH_3 \end{matrix}$$

2-propanol $\qquad\qquad\qquad$ propanone (dimethyl ketone or acetone)

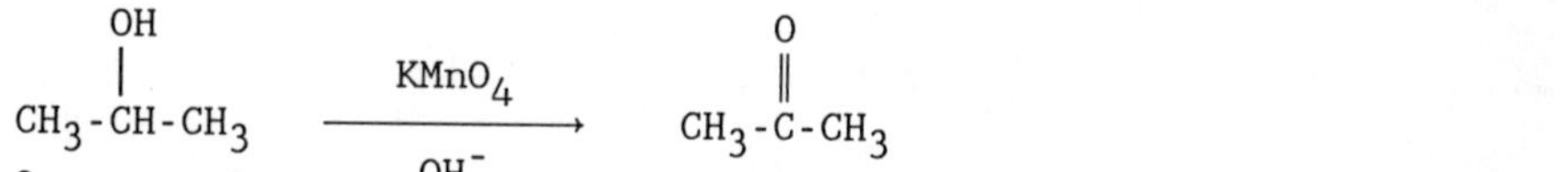

OH
                    |
(2) $CH_3-CH_2-CH-CH_3$ $\xrightarrow[\text{OH}^-]{\text{KMnO}_4}$ $CH_3-CH-\overset{\displaystyle O}{\overset{\|}{C}}-CH_3$
              2-butanol   2-butanone (methyl ethyl ketone)

(3)   cyclohexanol $\xrightarrow[\text{dil. H}_2\text{SO}_4]{\text{K}_2\text{Cr}_2\text{O}_7}$ cyclohexanone

32-52.   **Refer to the Introduction to Ethers.** · · · · · · · · · · · · · · · ·

(a)   Ethers are compounds in which an oxygen atom is bonded to two organic groups, as shown below.

                              R-O-R'

(b) CH_3-O-CH_3 $CH_3-CH_2-O-CH_2-CH_3$
           dimethyl ether   **diethyl ether**
          (methoxymethane)   **(ethoxyethane)**

           $CH_3-O-CH_2-CH_3$ $C_6H_5-O-CH_3$
          methyl ethyl ether   **methyl phenyl ether**
           (methoxyethane)   **(methoxybenzene)**

32-54.   **Refer to the Introduction to Ethers.** · · · · · · · · · · · · · · · ·

Diethyl ether is used both as an anesthetic and as a solvent for extracting organic compounds from plants and other sources. Other ethers are used as artificial flavors and refrigerants.

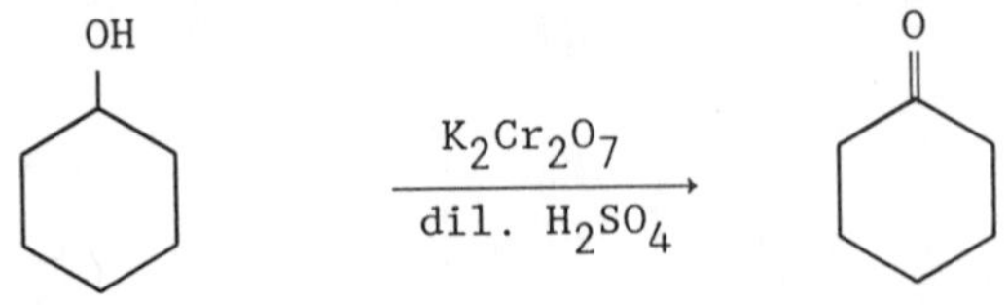